Chemistry

Fifth Edition of *Chemis*
and Schools

D. N. Underwood

Headmaster, Wolfreton School, N. Humberside

D. E. Webster

Senior Lecturer in Chemistry, The University, Hull

Edward Arnold

First published 1959
by Edward Arnold (Publishers) Ltd
41 Bedford Square, London WC1B 3DQ

Edward Arnold (Australia) Ltd
80 Waverley Road, Caulfield East
Victoria 3145, Australia

Second Edition 1967
Reprinted 1967
Third Edition 1970
Reprinted 1972, 1974
Fourth Edition 1975
Reprinted 1976, 1977, 1978, 1979, 1980

Fifth Edition 1981
Reprinted 1982, 1984, 1985

ISBN 0 7131 0516 X

Set in 10/11 Times N.R. by Macmillan India Ltd., Bangalore

Printed by Singapore National Printers Pte Ltd.

Contents

Introduction to the Fourth Edition

This book is intended for use during the last two or three years of an Ordinary Level Chemistry course and covers the syllabus requirements of the important examining bodies. The theoretical principles are discussed in the earlier chapters followed by a detailed description of the elements and compounds required for the O-Level examination. A final chapter "Chemistry and Society" has been added to give readers a glimpse of the relationship between the chemistry dealt with in the earlier chapters and other areas of human activity. A considerable number of experiments is given throughout, but this book is not intended to provide a systematic practical course.

The first edition of this book was written by one of us (D.N.U.) in 1959, and with modifications was published as a second and a third edition. Continuing changes in the chemistry syllabus have necessitated a further edition, and we decided that the time had come for an almost complete rewriting of the book. Those familiar with the old editions will, therefore, find many changes. These we hope are all improvements.

We have written this book fully aware of the ASE booklet "Chemical Nomenclature, Symbols and Terminology for use in School Science", and related publications by the examining boards. It is evident that although there are many topics therein where there is agreement as to the "best" practice, there are several areas where agreement has not been reached and where no "best" system is to be found. It must also be recognized that the authors of a text-book face a further dilemma. This is that the requirements of a student during the months immediately prior to his or her examination at 16, may be too complex and quite unnecessary when the book is first used some 3 years earlier.

Some of the features of the book are worthy of comment here.

(*a*) We have treated the periodic table more fully than in most O-level text books (and in previous editions of this book). Considerable emphasis is placed on comparative chemistry throughout the text. We hope that readers will find enjoyable the exercises that we have designed to familiarize them with this most important method of chemical classification.

(*b*) The mole is the basis of the quantitative features of this book.

(*c*) For simplicity we have throughout used H^+ for the proton, even in aqueous solution.

(*d*) For the five solid lattices sodium chloride, caesium chloride, ice, diamond and graphite we have used stereoscopic pairs of computer drawn pictures. We hope that readers will find these both interesting and instructional.

(*e*) A large selection of questions is included at the end of each chapter and numerical answers are given at the end of the book.

(*f*) The final chapter "Chemistry and Society" includes material that is not in most O-level syllabuses. It has not been included to cover the London O-level syllabus of the same name, although readers will find that this book does cover a large part of that syllabus.

(*g*) We have tried to face up to the problems of nomenclature, symbols and terminology and throughout the reader will find that we have acted as indicated in (1), (2), and (3) below.

1. I.U.P.A.C. (International Union of Pure and Applied Chemistry) nomenclature has been used but other common names for organic compounds are often given in brackets. However, we have (*a*) not given the oxidation state of inorganic ions where only one such oxidation state exists, e.g. we use sodium and calcium and not sodium(I) and calcium(II), (*b*) not given the oxidation state of non-metals in common anions and (*c*) retained the names nitrite and sulphite.

2. We have given the chemical equations in pairs where this is appropriate. This is usually when considering reactions in water. Of these pairs of equations the top one is the simplest form and suitable for use in the early years with the book. The bottom equation is in its ionic form, omits spectator ions (ions that do not participate in the reaction) and includes the state symbols (s), (l), (g) or (aq) for solid, liquid, gas or aqueous solution. These should be useful in later years of the course. It is also educationally valuable to have the two equations together, as the reader can compare them and see whether or not the second conveys more information than the first.

3. In the formulae of ions we never put superscripts for charges immediately above subscripts for the number of atoms. Nor have we, as a general rule, written the formulae of inorganic solids in an ionic form, as these imply, sometimes incorrectly, the nature of the particles (atoms or ions) present in a solid.

We realize that in this area of nomenclature and terminology whatever choice is made will find its critics. But it is our hope that the approach that we have adopted is both rational and *useful* to the 13 to 16 year old pupil and his teacher.

A further noticeable change, in this from previous editions, is the inclusion of a large number of relevant photographs. We wish to express our thanks to the many firms and institutions who have so willingly supplied pictures and given permission for us to reproduce them. Individual acknowledgements occur throughout the book. We are particularly grateful to Aubrey Rendell for taking a number of excellent photographs specially for us including the flame pictures on the back cover. Our thanks are also due to David Binnington for help in testing the experiments; to Tony Dixon for the nitrogen and carbon cycle drawings; to Jenny Berry for typing most of the manuscript and Roger Hateley for reading it all.

Thanks are due to the following examining boards for permission to reprint questions from past examination papers: Associated Examining Board; University of Cambridge Local Examinations Syndicate; University of Durham Schools Examination Board; University of London; Joint Matriculation Board; Oxford Local Examinations Board; Oxford and Cambridge Schools Examinations Board; Southern Universities Joint Board for School Examinations; Welsh Joint Education Committee. (At the request of the A.E.B. the terminology of their questions has been brought up-to-date.)

Thanks are also due for permission to base diagrams on others published elsewhere to: John Murray (figures 6.25, 6.26 and 6.27 (from "Success in Chemistry" by J. Bandtock and P. Hanson)); Pilkington Brothers Ltd. (figure

18.11); British Steel Corporation (cover and figures 24.4, 24.5 and 24.8) and B.P. Company Ltd. (figures 1.13(b) and 25.10).

Finally we thank our wives for their help and encouragement during the past months.

D.N.U.
D.E.W.

Introduction to the Fifth Edition

The fourth edition of this book 6 years ago was almost a complete rewrite of the third. Its content and form have been widely accepted and are retained for this edition. There are two major differences between the fourth edition and this one.

The last chapter "Chemistry and Society" has been completely rewritten, expanded and questions have been added. The format of this chapter has been retained but we have considerably expanded the sections on air pollution, water pollution, mineral resources, energy resources, and food resources and have added a section on the chemical industry. The need for these changes is a reflection of the new O-level syllabuses, the J.M.B. and Hong Kong for example, which place considerable emphasis on the topics in this chapter.

A chapter entitled "The Identification of Common Gases and Ions" has been added. This is a practical chapter. Various tests, that are spread throughout the book, with others added, have been brought together in a form that can be used in the school chemistry laboratory.

Other changes include substantial alterations to the chapter on fuels, changes and simplification to the introduction to organic chemistry and to energy and chemistry, and the addition of sections on oxidation numbers, the diaphragm cell, the fabrication of plastics, physical properties of the elements and Brownian motion. There are also very many minor changes scattered throughout the book.

Although any particular examination board removes some topics from its syllabus when new ones are added, the large number of examination boards, each with a different syllabus, has meant that as we have added topics it has not been possible to remove others; nothing that was in the fourth edition has been removed.

Thanks are due to those who have told us of mistakes, omissions, and obscurities, particularly John Peace and Stuart Warburton. We hope that they will all recognise where their helpful advice has been heeded.

It has been pointed out to us that we have not continually stressed the need to take care when carrying out chemical experiments. This is so. We believe that it is the task of the teacher to develop a sense of experimental responsibility (e.g. the wearing of safety spectacles) in his or her embryo chemists–to be safe and yet not so safe that they never discover, for example, what the common gases smell like. Chemistry is exciting–chemistry at the level for which this book is intended should be very exciting–it is the teacher's task not to make it appear so dangerous (*which it is not*) that the excitement is removed.

D. N. U.
D. E. W.
1981

1

The Nature of Matter

1.1 What is chemistry?

Chemistry is the study of the properties of the matter that makes up the earth and the rest of the universe. The chemist is particularly concerned with the ways in which different substances react together, and with the ways in which some substances are made up from other simpler ones. When beginning a study of chemistry we need to consider the different types of matter.

Matter is anything that occupies space. The amount of matter is its *mass*. Matter is made up of different *substances*. All parts of a substance are chemically the same. Examples of pure substances are diamonds, common salt and water.

1.2 The states of matter

A substance can be either a solid, a liquid or a gas. These are known as *the three states of matter*.

A *solid* has a definite shape with distinct boundaries. Its volume is hardly affected by changes in temperature and pressure.

A *liquid* takes up the shape of any vessel in which it is placed, but like a solid it has a definite volume and is hardly affected by changes in temperature and pressure.

A *gas* has neither a definite shape nor volume. It assumes the shape of any vessel in which it is placed and, moreover, spreads itself evenly throughout any containing vessel. Thus if a gas is put into a large flask the gas will, in a very short time, have spread itself evenly throughout the flask. The volumes of gases are considerably affected by changes in temperature and pressure.

The differing properties of solids, liquids and gases are explained by assuming that all matter is composed of minute particles. In *solids* these particles are in definite positions. They vibrate in these positions but do not move from one part of the solid to another. The state that a substance is in depends on the energy that the substance contains. In the solid state substances have the minimum energy content. As a solid is heated, the energy of the solid is increased and the particles vibrate more vigorously, until at a certain temperature, which is always the same for the same substance, the solid changes to a liquid, i.e., it melts. For the solid this temperature is called the *melting point*. If the source of heat is removed the liquid will cool and at the same temperature as that at which the solid melted, the liquid will solidify. For the liquid this temperature is called the *freezing point*.

In *liquids* the particles move about freely and do not merely vibrate about a fixed position. The particles do, however, attract one another, and these attractive forces give liquids a definite volume. All the particles in a liquid do not move at the same speed; some move faster than others. Moreover, the particles are continually colliding and exchanging energy and hence their speeds are continually changing. Near the surface of the liquid some particles that happen to be moving rapidly will fly out of the liquid and escape. In this way the liquid *evaporates*. If the temperature of a liquid is increased the particles move more rapidly, and so the rate of evaporation increases. For any substance at a certain temperature, and at a certain pressure, the addition of heat produces no further rise in temperature. At this temperature all of the energy being put into the substance is converting the liquid to a gas, i.e., the liquid is said to vapourize or boil. This temperature, which is markedly affected by pressure, is called the *boiling point* of the liquid. If the heat is removed the gas will cool, and at the same temperature as that at which the liquid boils, the gas will revert to a liquid. For a gas this process is called *condensation*. Gases that are at so low a temperature that they can be condensed by pressure alone are called *vapours*.

In *gases* the particles are much farther apart from one another than in liquids or solids, and consequently the attractive forces between particles are much less important. Each gas particle moves more or less independently of all other gas particles around it. This is why gases have no definite shape or volume, and it also accounts for the remarkable property of *diffusion* of gases. Thus if some hydrogen sulphide is prepared in one part of a laboratory, its smell is very rapidly detected all over the room. The properties of gases are discussed further in chapter 7.

Occasionally when a solid is heated it becomes a gas without passing through the liquid state. This process is called *sublimation*. It occurs when solid carbon dioxide or iodine are heated (see sections 18.3 and 20.3). Carbon dioxide or iodine vapour on cooling condense directly to the solid state. Water vapour goes directly to the solid state when snow flakes (figure 1.1) and hoar frost are formed, in the upper atmosphere and on the earth's surface respectively. The most

Figure 1.1 Snowflakes. (Photo: Radio Times Hulton Picture Library).

common form of sulphur, a fine yellow powder known as "Flowers of sulphur" is also produced by the condensation of the vapour direct to the solid on a cold surface. The relationships between the states of matter are summarized in figure 1.2.

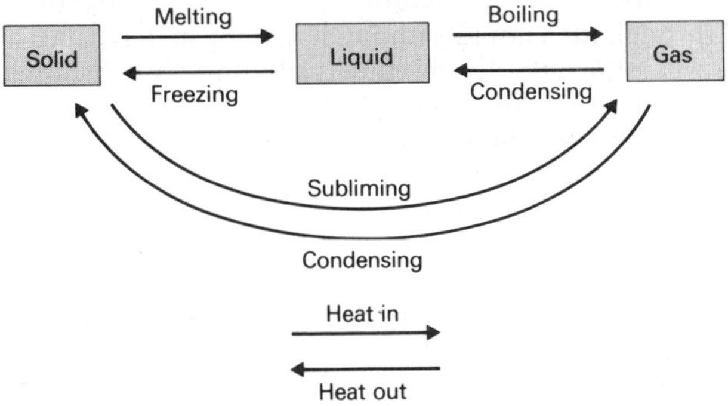

Figure 1.2 The three states of matter.

1.3 Solutions

Some substances, when they are mixed with water, seem to disappear. For example, if a little common salt is mixed with water the most careful microscopic examination of the resulting water-like liquid fails to reveal any trace of the salt. Yet tasting the liquid gives immediate proof of its presence. It is apparent that the salt and water have formed a mixture which is the same throughout, i.e., *homogeneous*, and the mixture will remain homogeneous even if it is left for ever. Such mixtures are called *solutions*, and the salt is said to have dissolved in the water. The dissolved substance, here the salt, is called the *solute*, and the substance that has dissolved it, here the water, is called the *solvent*.

Water is the most common solvent, mainly because it is so widespread on the earth, but there are many others. If you get a stain on your clothes you might try to remove it with a "stain-remover", or if the clothes are very dirty you may take them to be "dry-cleaned". The stain remover and the liquid used by the dry cleaner will be trichloroethene or some similar organic solvent.

There are many other solvents. Polystyrene, used for making toy model kits, is soluble in propanone (acetone). Cellulose ethanoate (acetate), a substance made from wood pulp (see section 26.3.1), is also soluble in propanone or a mixture of propanone and pentyl ethanoate (acetate). The artificial fibre, rayon, is made by squirting a solution of cellulose ethanoate in propanone through a small hole into a warm atmosphere, when the propanone evaporates and leaves a long fibre behind. Nail varnish is a solution of cellulose ethanoate in propanone, as are "balsa cement" and some paints.

Solutions of iodine are often in an alcohol solvent. Perfumes also contain pure alcohol as a solvent. Turpentine, or its substitute, white spirit, is a useful solvent for paint; it is often used for diluting paint, and for removing fresh paint stains.

There are many more solvents—you can probably think of others in addition to those given here.

1.4 Suspensions

Many substances do not dissolve in a particular solvent, and so they are said to be *insoluble* in the solvent. Water, for example, has no effect on glass, sand, wood and many other substances. However, if some fine chalk powder, which is insoluble in water, is shaken with it a milky mixture is formed. Such a mixture, where particles are spread throughout a liquid without dissolving in it, is called a *suspension*. These particles will settle out if the mixture is left long enough. Some medicines are suspensions; hence the instruction to "shake the bottle". Other common examples of suspensions are muddy water and most paints.

1.5 Filtration

If a mixture of water and chalk is poured onto a filter paper, the water passes through the filter paper and the chalk particles are left behind. Filter paper is unglazed paper, similar to blotting paper. It contains many small holes, which, although large enough to allow the water to pass through, retain the much larger particles of chalk. This process is called *filtration*; the liquid that has passed through the paper is called a *filtrate* and the solid left behind on the filter paper a *residue* (figure 1.3).

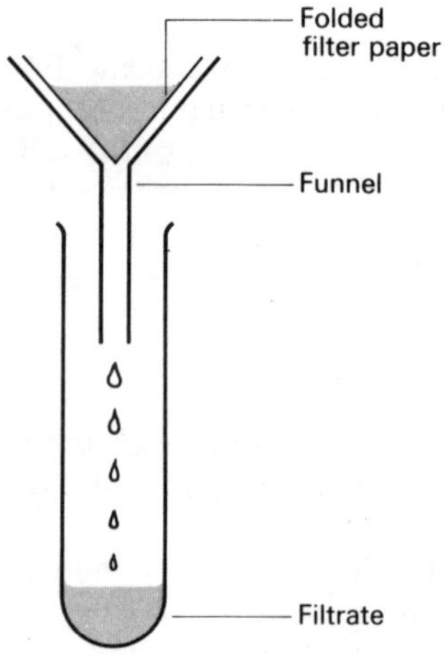

Figure 1.3 Filtration.

Filtration is a much used process, and there are various substitutes for the filter paper depending on the size of the particles to be removed. The most widespread application is in the treatment of our water supplies, where layers of sand and gravel are used to remove suspended particles. In the engine of a car the oil passes through a filter to remove solid particles that if left in the oil would cause damage, and for the same reason dirt particles in the air and petrol are removed by filters before they pass into the engine (figure 1.4).

Dust particles are filtered from the waste gases of power stations and other industrial plants to stop this dirt being expelled into the air where it would be a major source of atmospheric pollution. For these industrial uses, where the volume of air to be filtered is very large, a filter paper method would be much too slow. A special type of filter is used called an electrostatic precipitator (figure 1.5).

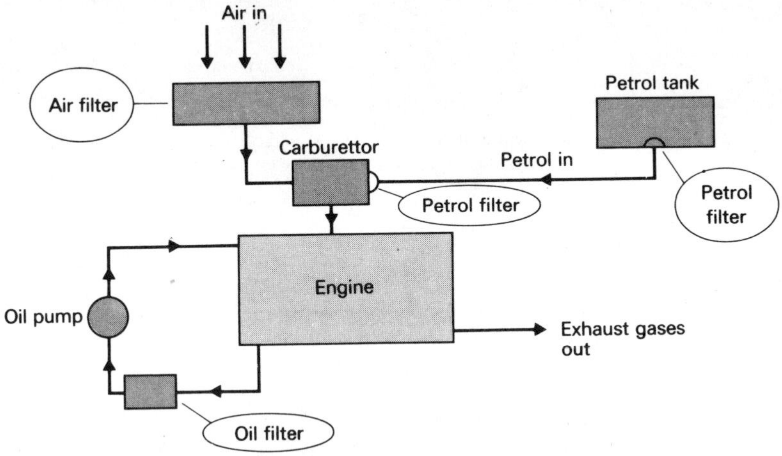

Figure 1.4 Schematic diagram of a car engine showing the air, oil and petrol filters.

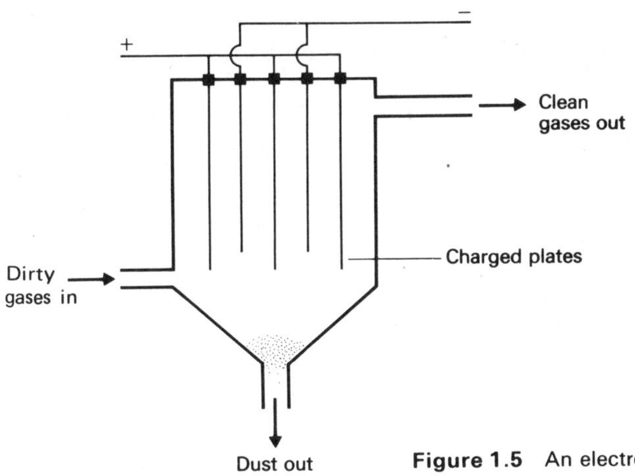

Figure 1.5 An electrostatic precipitator.

In this the dust particles are electrically charged and they then become attached to metal plates of opposite charge. After a time the electric current through the plates is switched off, and the dust then falls to the bottom of the precipitator when it can be removed.

1.6 Evaporation

In section 1.3 we saw that salt dissolves in water to give a solution. The salt cannot be separated from the water by filtration; they both pass straight through the filter paper. To obtain the salt the solution must be heated, when the water escapes as steam leaving the involatile salt. A porcelain dish called an evaporating dish is often used (figure 1.6). If the solution is heated until all the water has been vapourized the process is called *evaporation to dryness*. When doing this care must be taken towards the end; the solution becomes very

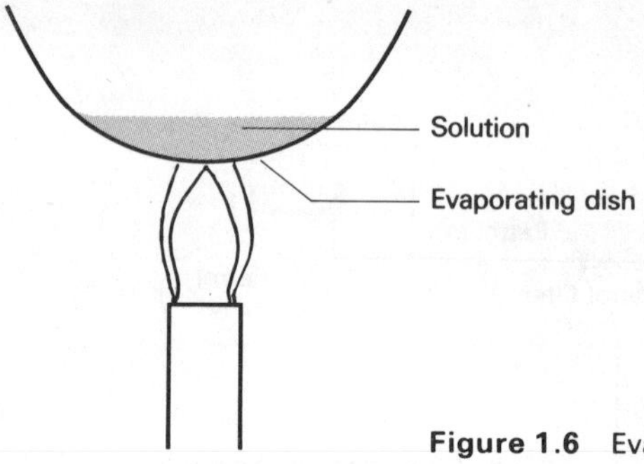

Figure 1.6 Evaporation over a bunsen burner.

concentrated and will spit out of the dish as steam is formed beneath semi-solid parts of the mixture. To minimize spitting gentle heating must be used.

As the process of evaporation removes the water, it concentrates the salt. If a solution that has been concentrated is allowed to cool, crystals of the salt may form. This process is called *crystallization* and is discussed further in section 1.14.

When an inflammable solvent, such as ether or alcohol, is being evaporated, the solution must not be heated with a naked flame, but instead a hot water-bath or a steam-bath is used (figure 1.7).

A beaker containing water at a high enough temperature to cause evaporation can be used as the steam-bath. Even with a steam-bath great care must be taken when evaporating liquids whose vapours are inflammable and heavier than air, such as ether or carbon disulphide, as these vapours can fall around the water-bath and become ignited. For such liquids the only safe procedure is to completely remove the flame before putting the solution on the water-bath.

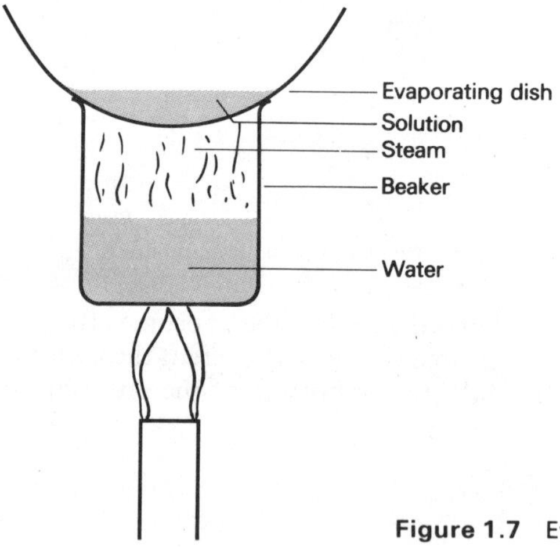

Figure 1.7 Evaporation on a steam bath.

1.7 Distillation

Evaporation gives only the involatile salt from the salt solution; the water has boiled away into the atmosphere. To obtain the water, it is necessary to condense the steam by cooling. A Liebig condenser, fitted to a distillation flask is normally used, and the apparatus is assembled as shown in figure 1.8. A small piece of broken pottery in the flask enables the solution to boil smoothly.

This process is called *distillation*, and the water obtained in this way is *distilled water*. Distilled water is very pure water and is usually obtained by the distillation of tap water. It is used when dissolved solids in the water are undesirable, such as in experiments in the chemistry laboratory and in car batteries.

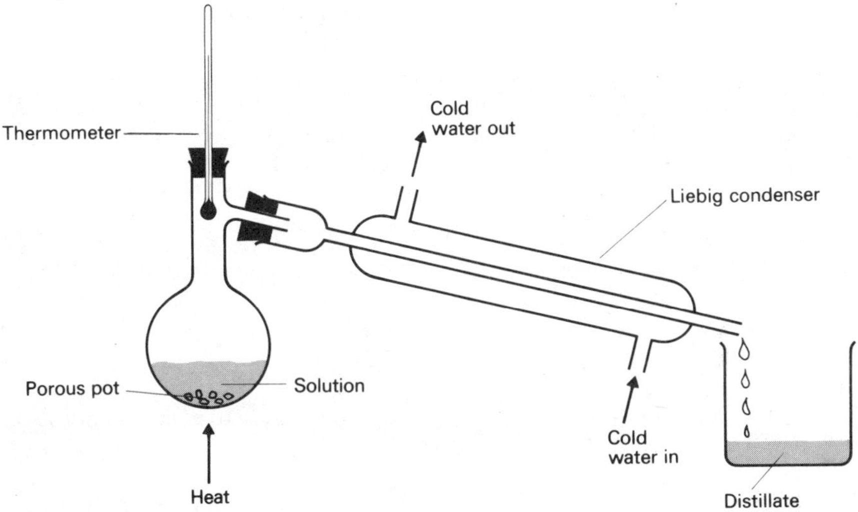

Figure 1.8 Distillation.

The process of distillation has been used for several thousands of years, mainly to concentrate the alcohol in wine by making spirits (see section 1.8). Also, over the past centuries distillation was an important process for the chemists' predecessors—the alchemists. The alchemist, in his search for methods of turning other metals into gold (a process that, we hope, the readers of this book will appreciate is rather difficult—see section 6.3.4) developed the very simple, all-glass piece of distillation apparatus called the retort (figure 1.9). The retort has become one of the symbols of the alchemist. Paintings of alchemists, such as that

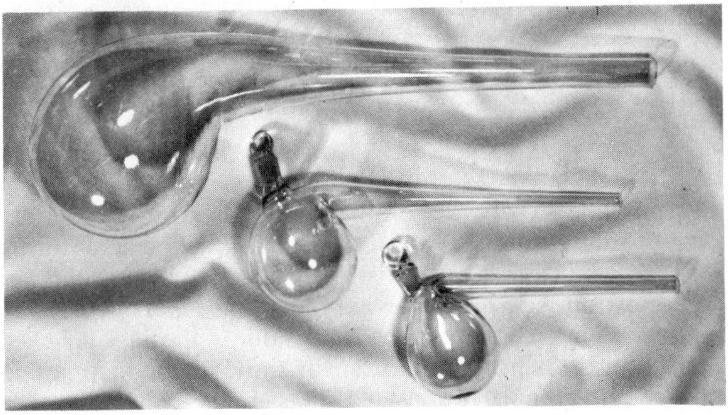

Figure 1.9 Retorts. The large one was made in Germany during the 19th century. The smaller ones, with ground glass stoppers, are modern. (Photo: A. Rendell).

by J. Stradanus in the Palazzo Vecchio in Florence (figure 1.10), abound with retorts and distillation equipment.

Retorts are not used much nowadays, but they are useful when distilling a liquid that attacks metal, rubber and other materials. For example, retorts are sometimes used in the school laboratory for preparing nitric acid and bromine, both of which are highly corrosive liquids. Modern all-glass apparatus with ground joints is, however, just as good and much more convenient to assemble and to dismantle and clean (see sections 16.6.2 and 20.2.3).

Figure 1.10 "The Alchemist" by J. Stradanus (from the Palazzo Vecchio, Florence) Photo: Mansell Collection.

1.8 Fractional distillation

By distillation it is also possible to separate a mixture of two or more liquids, provided that they have different boiling points. Ethanol boils at a lower temperature than water and when a mixture is distilled the vapour will contain a comparatively high proportion of the lower-boiling ethanol. This vapour condenses to give a distillate richer in ethanol, and the residue in the flask is richer in water. A complete separation is not effected, but if some of the distillate is redistilled a liquid still richer in ethanol is obtained. This is exactly what is done during the making of malt whisky.

The ethanol solution produced by the fermentation process is distilled twice, the final liquid distillate being whisky. The distillation flasks used are much larger than those used in the laboratory, are made of copper and are called stills. Figure 1.11 is a picture of two whisky stills. These stills each holds over 20 000 litres of solution. Instead of a Liebig condenser, a copper spiral inside a container of cold water, shown on the right hand side of figure 1.11, is used to condense the vapour.

Figure 1.11 Malt whisky stills at the Royal Brackla Distillery at Nairn in Scotland. (Photo: The Distillers' Co. Ltd.).

Often, however, it is necessary to obtain each of the liquids that make up the mixture in as pure a form as possible. To achieve this a *fractionating column* is used. There are many different types; one of the simplest consists of a long tube packed with small glass beads. This tube is fitted to the neck of the flask containing the liquid to be distilled (figure 1.12). When a mixture of vapours passes up the column those of the less volatile liquid condense on the glass beads more readily than those of the more volatile liquid. Liquid that has condensed on the column is evaporated by the hot vapours, and the more volatile liquid evaporates most readily. As a result of the many evaporations and condensations the vapour moving up the column becomes progressively richer in the more volatile liquid. At the top of the column the vapour will be mainly the more volatile liquid, and this is then collected in the receiver.

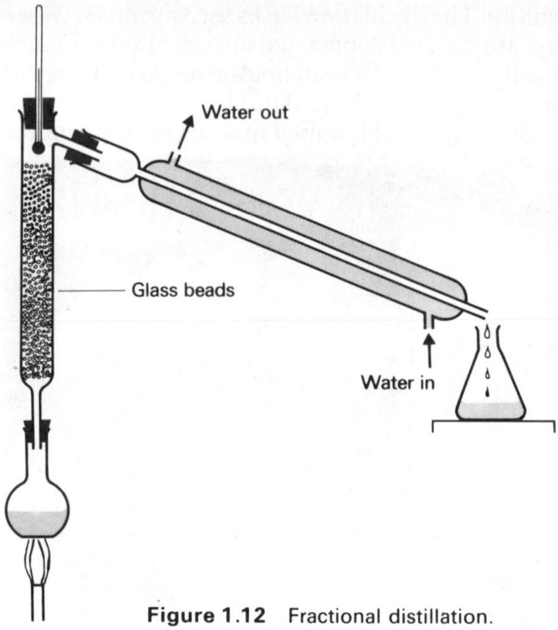

Figure 1.12 Fractional distillation.

Fractional distillation, as the process is called, is used widely in the chemical industry. Petroleum for cars, aviation spirit for aeroplanes, diesel oil for lorries, paraffin for heaters, oil for lubrication, asphalt for road-surfacing and many other organic compounds are all produced from the oil obtained out of the earth. Of the processes involved in separating the complex mixtures that make up this Crude Oil, the process of fractional distillation is the most important. For these industrial uses the fractionating columns are very large. Figure 1.13(*a*) shows such a *fractionating tower* and figure 1.13(*b*) is a diagram that illustrates the parts of the tower. The fractional distillation of crude oil is discussed further in section 25.3.2.

The liquefaction of air is discussed in section 13.4. If liquid air is fractionally distilled the lower boiling nitrogen is more volatile than the oxygen and collects at the top of the fractionating column; thus these two major components of air can be separated. With very efficient fractionating columns the other minor components of air (i.e. the noble gases) can also be obtained in a pure form from the mixture.

1.9 Sublimation

We saw in section 1.2 that sometimes, when a solid is heated, it becomes a vapour without passing through the liquid state. The vapour will also condense back to the solid without becoming a liquid. This process can be used to purify a mixture of compounds in which one sublimes, but the other does not. For example, if a mixture of ammonium chloride and sodium chloride is heated in an evaporating dish only the ammonium chloride sublimes. The vapour can be cooled by using an inverted filter funnel (figure 1.14). Solid ammonium chloride collects on the cool sides of the funnel, the sodium chloride remains in the dish.

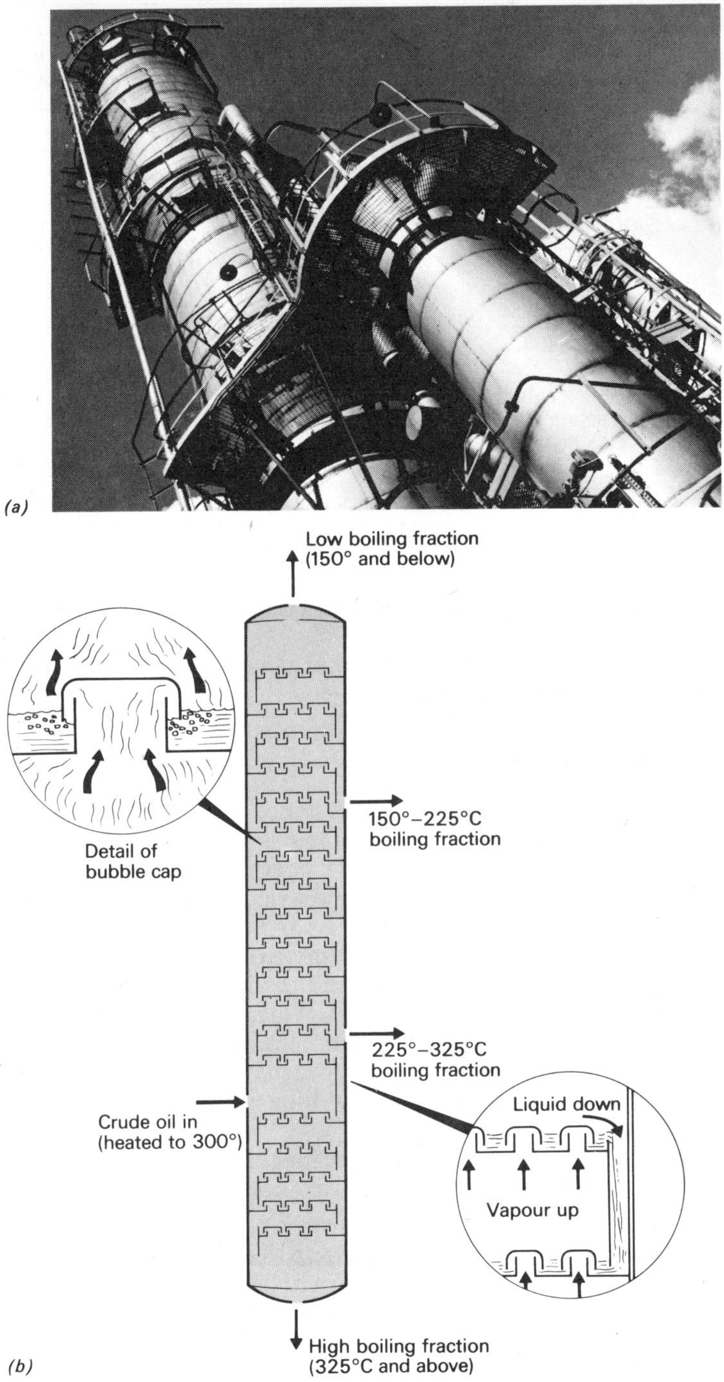

Figure 1.13(*a*) Fractionating towers at the Europoort Refinery at Rotterdam. (Photo: British Petroleum Co. Ltd.). (*b*) Diagram of a "bubble-cap" fractionating tower for distilling crude oil.

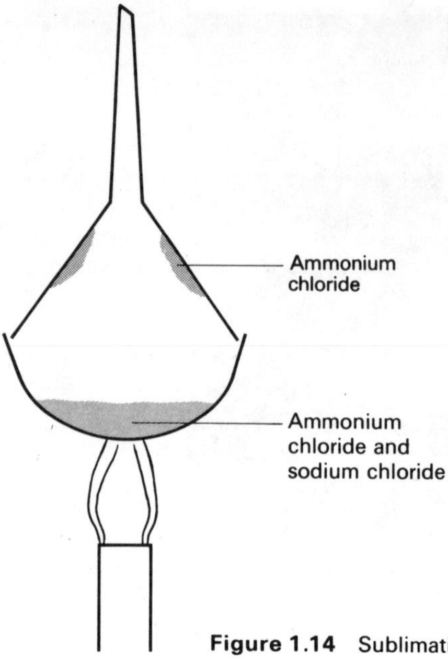

Figure 1.14 Sublimation of ammonium chloride.

Iodine (see section 20.3.1) and naphthalene (an organic compound) are two other examples of substances that sublime.

1.10 Paper chromatography

This technique is used to separate two, or more, substances that are both soluble in the same solvent. A mixture of dyes can be separated in this way; for example the dyes in a sample of ink (black ink is very good), or screened methyl orange, which is a mixture of two dyes (methyl orange and xylene cyanol (FF)). One method is to cut a filter paper as shown in figure 1.15, place a drop of the dye mixture near the centre of the paper and allow the spot to dry. The paper is then placed flat over the top of an evaporating dish with the wick (*A*) dipping into a solvent and covered with a clock glass (figure 1.15). Water alone can be used as the solvent but a mixture of water, ethanol, butanol and ammonia is better.

The solvent moves up the wick and then outwards on the filter paper. As it does so it carries the components of the dye with it to different distances from the centre of the paper.

The name *chromatography* comes from the Greek for colour (*chroma*) because the method was first used for mixtures of coloured compounds. Chromato-

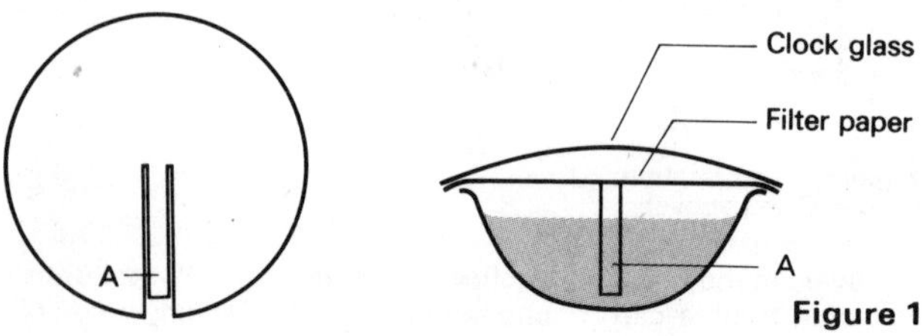

Figure 1.15 Paper chromatography.

graphy is now used very widely, and by choosing the right equipment and technique it is possible to separate mixtures of very closely related substances, such as the amino-acids produced when a protein is hydrolysed. Sometimes the substances being separated are colourless, and the chromatogram must then be "developed" to make the different "spots" on the paper visible. An example that requires development is given in the experiment at the end of this section.

The technique of paper chromatography works because the different substances in the mixture show different relative solubilities in two solvents. Within the filter paper is a lot of water; much of this is combined with the cellulose of the paper (see section 26.3.1). As the solvent containing the mixture of solutes passes along the filter paper, the different solutes are soluble to different extents in the water of the paper; they are "partitioned" between the two solvents. The dyes that are most soluble in the water are held back more than the dyes that are less soluble and hence a separation is achieved. It is found that the distance travelled by each solute divided by the distance travelled by the solvent is a constant for a particular solvent and this ratio is called the R_f value. The R_f value will enable a chemist to say exactly what the substance is.

A chromatography experiment—The separation of iron, cobalt, nickel and copper ions.

In this experiment a more useful technique than the radial method described above, is used. The solvent climbs up the sheet of filter paper from a small volume in the bottom of a beaker (figure 1.16(*a*)). The beaker is kept covered so that there is solvent vapour surrounding the chromatogram. Also, the final chromatogram must be developed so that the different spots are clearly visible.

Put into the bottom of a 600 cm^3 beaker a depth of 0·5 cm of a solvent consisting of 90 parts acetone, 5 parts concentrated hydrochloric acid and 5 parts water (all by volume). On a piece of filter paper, or chromatography paper (21 cm × 11 cm), put a line about 1 cm from the bottom and along this line place small spots of solutions of iron(II) ammonium sulphate (30 %), cobalt(II) sulphate (10 %), nickel(II) sulphate (10 %) and copper(II) sulphate (10 %). To obtain small spots use a melting point tube that has been drawn in a Bunsen flame to a fine capillary. Dip the capillary into the solution and then apply the tip to the paper. Put the spots at least 2 cm apart. Allow the spots to dry, roll the paper into a cylinder, fix it together with paper clips and put the paper in the beaker containing the solvent (figure 1.16(*a*)). Put a clock glass on the beaker as a cover.

Allow the solvent to climb up the paper until the front edge of the solvent is near the top of the paper; this should take about 15 minutes. Remove the paper from the beaker, remove the paper clips, carefully mark on the paper the position of the front edge of the solvent, and allow the paper to dry.

Place the paper in a large covered beaker in which is a small beaker containing concentrated ammonia for about 2 minutes (figure 1.16(*b*)). Remove it from the ammonia and in a fume cupboard spray lightly with an atomiser spray containing a solution of 0·5 g of dithiooxamide (rubeanic acid) in 100 cm^3 ethanol or industrial spirit, when the spots as shown in figure 1.16(*c*) should appear. You may find that your samples of iron, cobalt, nickel

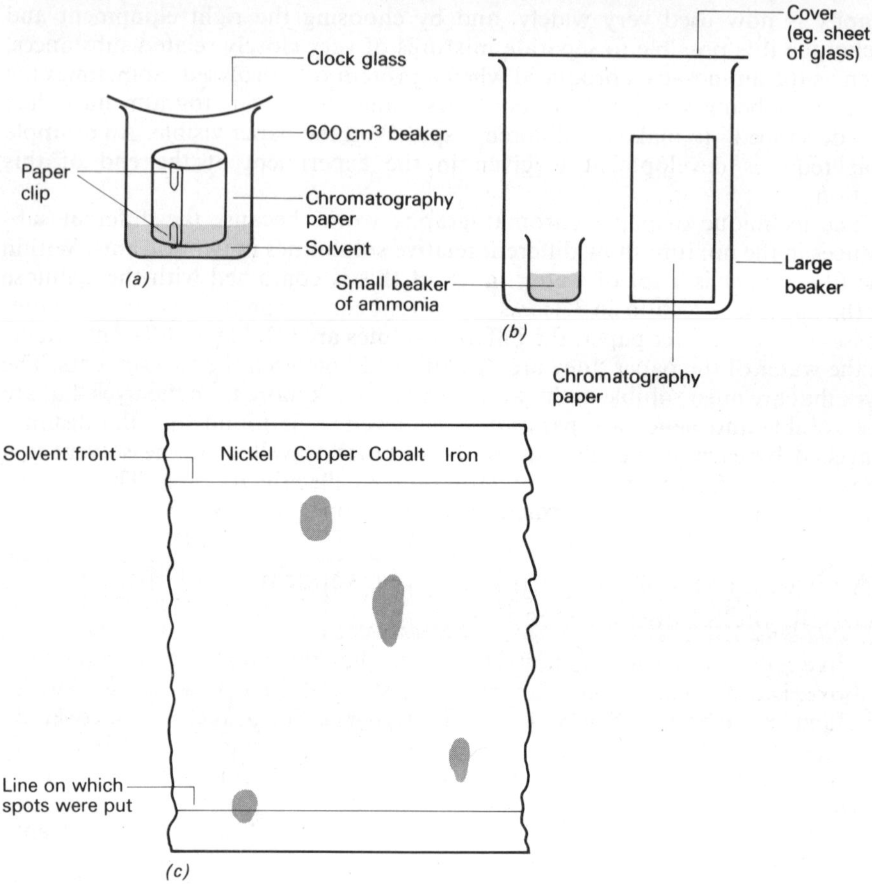

Figure 1.16 Paper chromatography.

and copper salts are impure. Look at your chromatograms carefully. Does the nickel(II) sulphate in your laboratory contain any cobalt(II) sulphate, and vice versa? They often do, because cobalt and nickel occur together in nature and it is very difficult to get them pure. Also, if you used metal paper clips there will be stains round them. What metal are the paper clips made from?

1.11 Gas chromatography

The technique of chromatography can also be applied to mixtures of gases, and *gas chromatography* is very widely used in chemical research and industry. Substances that are gases at room temperature can be separated, or more often liquids can be vaporized and then separated. In *gas chromatography* the liquid in which the gases are to be "partitioned" is put onto an inert solid powder support and put into a tube. The gas mixture is blown down the tube by an inert carrier gas (usually nitrogen or helium) and the gases emerging from the end of the tube are analyzed by one of a number of different types of detector. A simple gas-

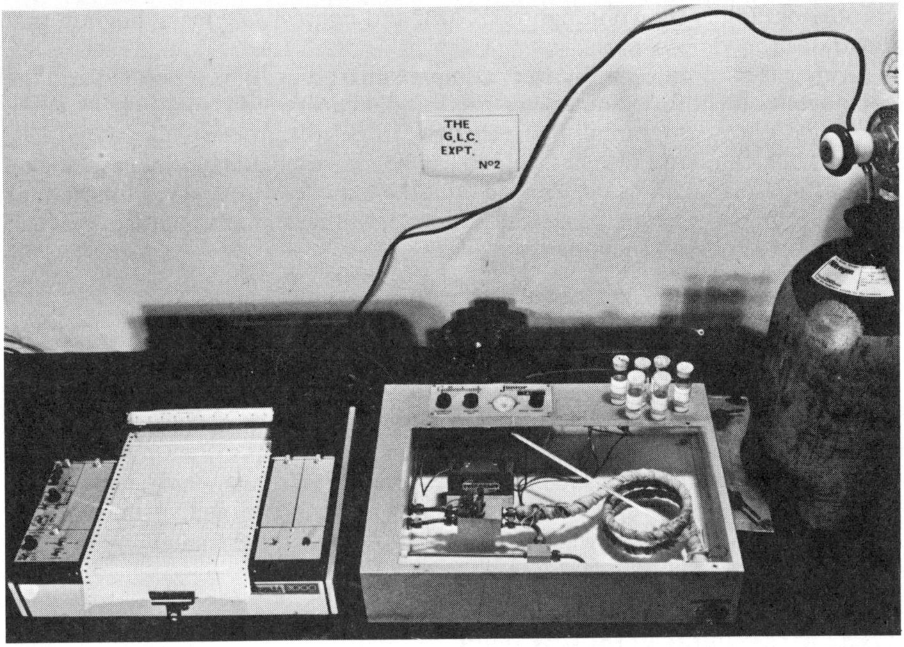

(a)

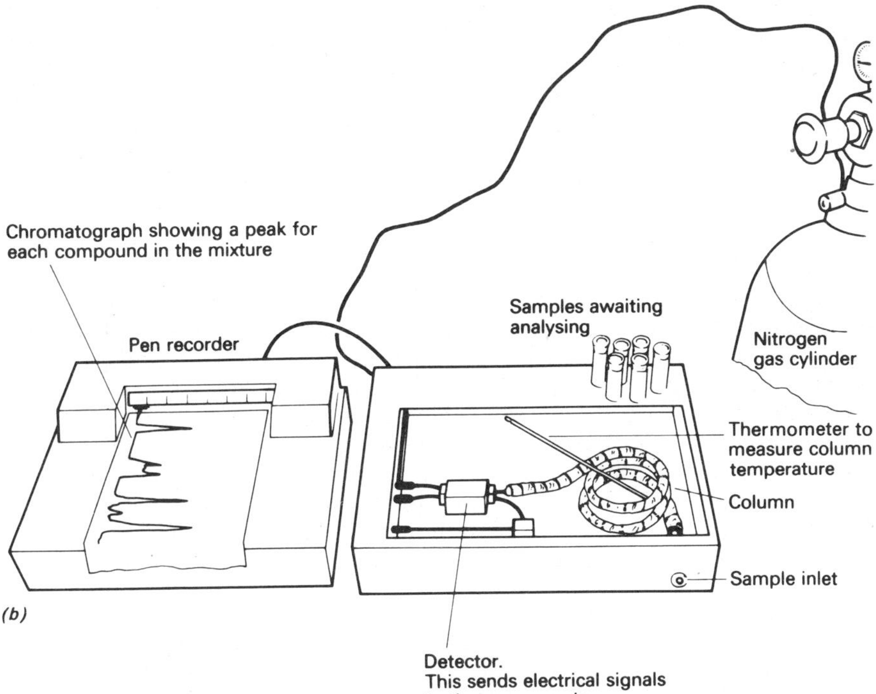

(b)

Figure 1.17 (*a*) and (*b*) Gas chromatography. (Photo: A. Rendell).

chromatograph is shown in figure 1.17(*a*), and figure 1.17(*b*) is a diagram that illustrates the various parts.

Using gas chromatography, very complex mixtures can be separated and the components identified. Sometimes the sample can be very small (10^{-6} cm^3). Using gas chromatography, hydrocarbons in petroleum can be analyzed, the amount of alcohol in a sample of blood can be measured, the type of pesticide on the leaf of a plant can be determined, and the many constituents in tobacco tar can be identified. These are just four examples of the complex mixtures that can be separated by this technique.

1.12 Saturated solutions

If a small amount of salt, or any other soluble solid, is added to water, the salt dissolves. The solution obtained is said to be unsaturated. If further quantities of salt are continually added, then a stage will be reached when no more salt will dissolve; the water is then said to be saturated with the salt. A *saturated solution* is defined as one that contains the maximum amount of solute which, with excess solute present, can be dissolved at the temperature concerned. If the water is heated, more of the solute will probably dissolve because the solubility of most solids increases with temperature.

1.13 Fractional crystallization

It is possible to measure exactly how much of a solid dissolves in a particular mass of a solvent. Usually the quantity measured is the number of grams of solute required to saturate 100 g of solvent at a particular temperature. These measurements are made by preparing saturated solutions at various temperatures and then analyzing them. The measured solubilities can be plotted against temperature on a graph. Such graphs are called *solubility curves*, and these are given for a few common substances (figure 1.18). The graphs show, for example, that in boiling water (at 100 °C) the least soluble salt, of those given, is potassium sulphate (25 g/100 g water). Copper(II) sulphate is more soluble (73 g/100 g water), and potassium nitrate is much more soluble than any of the other substances (well over 160 g/100 g water—it is off of the top of the graph).

Another useful piece of information that can be obtained from these curves concerns the behaviour of a solution on changing the temperature. For example, if you have a solution in 100 g of water at 100 °C containing 25 g of potassium chlorate, then the graph tells us that the solution will not be saturated, as 100 g of water will dissolve 53 g of potassium chlorate at 100 °C. Now, if the solution is cooled the graph tells us that it will become saturated when the temperature reaches 58 °C, and below 58 °C less than 25 g of potassium chlorate will be needed to give a saturated solution. So as the solution is cooled further, solid potassium chlorate will be deposited out of the solution until at room temperature, about 18 °C, only 7 g will remain in solution; 18 g will have been deposited in the container and could be filtered off.

Now suppose that in addition to 25 g of potassium chlorate, 25 g of sodium chloride was dissolved in the boiling water. The graph tells us that even at room temperature 37 g of sodium chloride are required to give a saturated solution, hence on cooling all of the sodium chloride would remain in solution. By dissolving in boiling water and cooling, therefore, pure potassium chlorate could

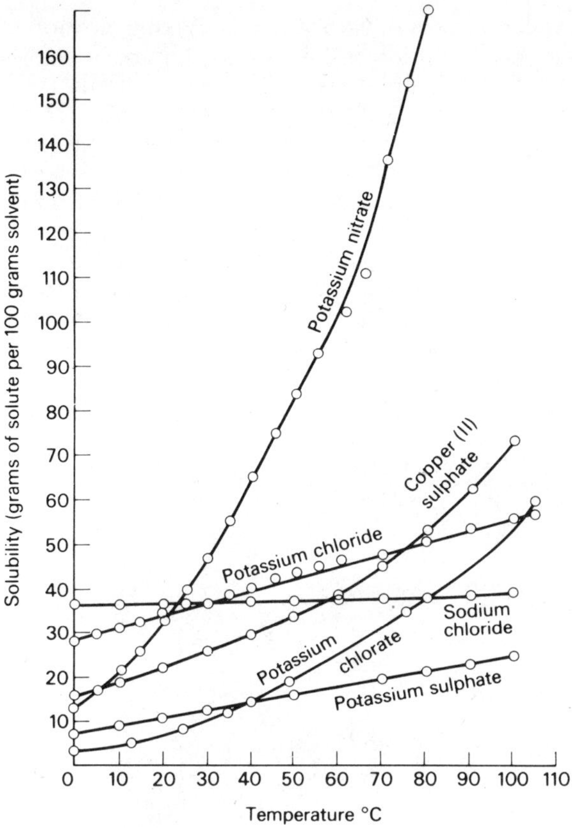

Figure 1.18 Solubilities of some substances in water.

be obtained from a mixture of potassium chlorate and sodium chloride.

Usually the solubilities of the substances in a mixture are such that a mixture of the substances is deposited on cooling, but the composition of the mixture will be different from that of the solution. This is the basis of a method of separation of a mixture of solids known as *fractional crystallization*. To get substances pure further recrystallization are often necessary. This makes fractional crystallization a very tedious process, but it is a valuable method used for obtaining pure substances from mixtures.

1.14 Crystals

We have just seen that a cold solvent is unable to hold as much solute as a hot solvent. Therefore, if a hot saturated solution is cooled the excess solute must be deposited from the solution, and the deposit is obtained as crystals. Rapid cooling will give small crystals, slow cooling will give large crystals.

Experiment—Growing crystals of alum

Add a small quantity of alum (aluminium(III) potassium sulphate-12-water) to about 50 cm^3 of distilled water until, even with vigorous shaking, no more will dissolve. To this cold saturated solution add a few more grams of alum and warm the solution sufficiently to dissolve these. When the warm solution cools the extra quantity of alum added will crystallize out as quite large crystals. The solution left is, of course, saturated and is called the "mother liquor".

A much larger crystal can be grown by choosing a well shaped crystal and suspending it with a hair in the mother liquor (figure 1.19). Leave the solution where the temperature is even and as it slowly evaporates the extra alum will be deposited on the crystal and it will grow. With skill and care really large crystals can be grown by this method. The book recommended in the footnote below gives a variation on this method that gives particularly good results (see figure 1.20), and you may be interested in trying your skill at crystal growing in your spare time.

Figure 1.19 Growing a large alum crystal. (Photo: "Chemistry experiments" by H. L. Heys, Harrap, 1949).

Crystals can also be grown from vapours during sublimation. If iodine or naphthalene are gently heated in an evaporating dish covered by an inverted filter funnel, beautiful crystals are formed on the funnel.

Crystals are also formed when a liquid cools. This is often so for metal crystals. If you look carefully at a galvanized article (figure 22.6) (e.g. a galvanized bucket), you will see that the thin coating of zinc is in the form of crystals. These were formed as the molten zinc cooled. Crystals of metallic substances are used to

Footnote. An excellent book for practical chemistry at home is "Chemistry Experiments at Home for Boys and Girls" by H. L. Heys (Harrap). This describes many safe experiments and indicates how you can make apparatus and buy chemicals. Such a book is a better investment than an expensive chemistry set.

Figure 1.20 Home grown alum crystals, with a thimble for comparison. (Photo: "Chemistry experiments" by H. L. Heys, Harrap, 1949).

make transistors for radios and other electronic equipment. These are formed by slowly pulling a rod of the substance out from a bath of the molten liquid. In this way one large crystal is formed, although from its external appearance it does not look like a crystal (figure 1.21).

When the rocks and minerals of the earth were formed the liquid and vapours cooled very slowly and some very large crystals were formed. In figure 1.22 are some large quartz (silicon(IV) oxide) crystals.

1.15 Supersaturation

If hot saturated solutions of some substances are cooled, the expected crystallization does not occur, particularly if the cooling is slow, the solution is left undisturbed, and all dust is excluded. Sodium thiosulphate is one of the few substances that readily shows this phenomenon. If some sodium thiosulphate crystals are gently heated with a very little water they dissolve. The ease of solution is partly due to the high percentage of water of crystallization (see section 11.5) in the crystal. There is so much, in fact, that it is possible for the sodium thiosulphate to dissolve in its own water of crystallization. On cooling no crystals form.

Figure 1.21 A single "crystal" of indium phosphide, 40 mm long and 20 mm diameter, for use in making electronic devices. (Photo: Allen Clark Research Centre of The Plessey Co. Ltd.).

Figure 1.22 Large crystals of quartz. (Photo: Institute of Geological Sciences. Crown Copyright Reserved).

Such a solution is said to be *supersaturated.* It is in an unstable state. The addition of a single crystal of sodium thiosulphate acts as a nucleus on which other crystals can form, and crystallization occurs. There is a noticeable evolution of heat when the crystals form. This is because, when a solvent dissolves a solute to form a solution, heat is usually taken in by the solution. Hence when crystals form the solution is changing back to solvent and solute and heat is evolved.

Supersaturation does not occur if there is excess solute present. A *supersaturated solution* can be defined, therefore, as one containing more solute at a particular temperature than it contains in the presence of crystals of the solute.

1.16 Solubility of gases in water

So far we have considered only the solubility of solids in liquids. Solutions of gases in water are also of importance. The various gases show great differences in their solubilities. For example, 1 cm^3 of water at 0°C will dissolve approximately 1300 cm^3 of ammonia, but only 0.02 cm^3 of hydrogen. The gases that are readily soluble, such as ammonia, hydrogen chloride and sulphur dioxide, all react with the water, but no reaction occurs with the sparingly soluble ones, such as hydrogen, oxygen and nitrogen.

Water will inevitably contain small quantities of gases dissolved in it, as a result of its contact with air. This can be illustrated by filling a large flask and delivery tube with tap water and setting up the apparatus shown in figure 13.6. When the flask is warmed a small amount of gas collects in the test tube. The gas will re-ignite a glowing splint—indicative of a high proportion of oxygen and, therefore, of the fact that oxygen is more soluble in water than is nitrogen.

This experiment also illustrates that, unlike nearly all solids, gases are more soluble in cold water than in hot. In fact, a gas that does not react with water can be removed by boiling.

The presence of dissolved oxygen in water is of immense biological importance, since it is this oxygen that fish extract from water as it passes through their gills, and it is their sole means of respiration.

1.17 Colloidal solutions

We have seen that salt forms a solution in water that is completely homogeneous, i.e., the individual salt particles cannot be seen and the solution passes through a filter paper. We have also seen that finely divided chalk forms a suspension with water. The chalk particles can be readily seen under a microscope and are removable by filtration.

It is possible to prepare a solution where the particles of the substance dispersed in the liquid are intermediate in size between those of solute particles and the particles of a suspension. Such a solution is called a *colloidal solution.*

If a solution of sulphur dioxide is added to a solution of hydrogen sulphide, until the smell of the latter is removed, sulphur is formed. The sulphur is in colloidal solution and the mixture obtained will pass through a filter paper.

Colloidal solutions prepared in this manner are not very stable and the colloidal particles soon coagulate (i.e., join together) to form larger particles that then settle out from the solution. Some colloidal solutions, carefully prepared and stored, last indefinitely.

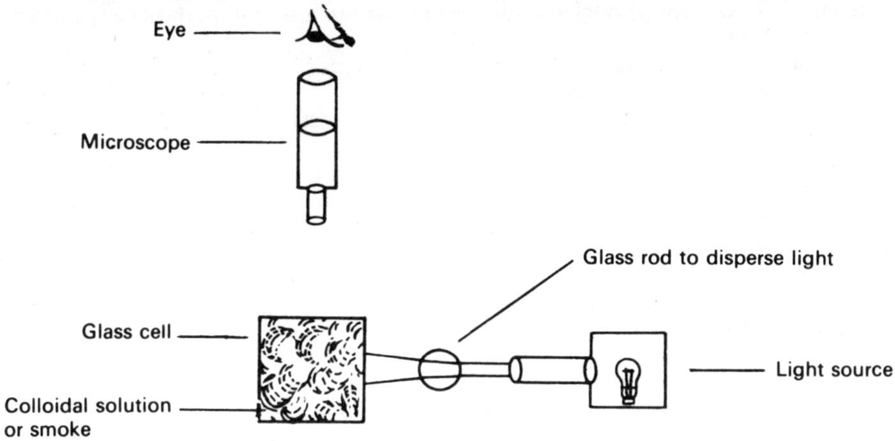

Figure 1.23 Brownian motion

Some colloidal solutions have beautiful colours. Colloidal gold is a rich deep red colour. This has been known for many centuries and made use of in the making of red glass for stained glass windows in churches.

When the substance dispersed in water is also a liquid the colloidal system obtained is called an *emulsion*. An emulsion is formed if a little paraffin is shaken with water, but it soon separates. If a little soap solution is also added the emulsion is much more stable. Soap is effective in removing oil and grease because it helps to take them into colloidal solution and so helps them to be washed away.

Milk is an important natural emulsion of fat in water. It is a fairly stable emulsion, but cow's milk in contact with gastric juices forms curds, making it less digestible. For this reason cow's milk is not suitable for young babies. Human mother's milk contains an albumin that prevents this coagulation.

1.18 Brownian motion

A colloidal solution (or smoke from a piece of burning string) is placed in the cell (figure 1.23) and illuminated by a strong light from the side. If it is then viewed from above with a microscope the particles are seen to be moving constantly first in one direction and then another. They appear as pin points of light because the particles scatter the light in the same way as dust and smoke particles scatter light from a projector in a cinema.

This effect was first noticed by the botanist Robert Brown in 1827 when he was examining with a microscope pollen grains suspended in water. The erratic motion, now called *Brownian motion*, occurs with any substance which is suspended in either a gas or liquid. It is regarded as good evidence that gas and liquid molecules are in constant motion, the colloidal or smoke particles being hit "from pillar to post" by the gas and liquid molecules.

Questions on chapter 1

1. What is a crystal? How would you make some crystals of copper(II) sulphate, starting

with powdered copper(II) sulphate, and then proceed to grow one of them? Explain why the crystal grows.

2. Describe, with diagrams, how you would obtain:
(*a*) pure water from tap water,
(*b*) ammonium chloride from a mixture of common salt and ammonium chloride,
(*c*) clean, dry sand from a mixture of sand and salt.

3. Gunpowder is a mixture of sulphur, potassium nitrate and carbon. Sulphur is soluble in the highly inflammable and poisonous liquid, carbon disulphide. Potassium nitrate is soluble in water. Describe experiments by which you would attempt to find the percentage composition by weight of gunpowder.

4. What factors influence the solubility of (*a*) a solid and (*b*) a gas in water?
Describe, with diagrams, how you would show that tap water contains (*a*) dissolved solid, (*b*) dissolved gas.

5. (*a*) Two compounds in a mixture were known to be compounds *A* or *B* or *C* or *D*. Drops of the four pure compounds *A*, *B*, *C*, and *D* and of the mixture were placed on a sheet of chromatography paper and the paper placed in a suitable solvent as in figure 1.16(*a*). When the solvent front was 15·0 cm up the paper from the starting line, the four compounds *A*, *B*, *C*, and *D* were 5·0 cm, 12·0 cm, 1·0 cm, and 7·0 cm respectively up the paper. The mixture had separated into 2 spots 7·0 and 12·0 cm respectively up the paper.
Of what two compounds is the mixture composed and what is the R_f value of compound *A*?
(*b*) A solution contains a red dye and a blue dye dissolved in water. The R_f value of the red dye is 0·75 and that of the blue dye is 0·4. A drop of the solution was placed in the centre of a filter paper that was used as in figure 1.15. Draw a diagram of the filter paper after a little water had been used. Show the position of the solvent front, and the red and blue dyes.

6. The following table gives some details about the properties of three compounds. Use this information to devise a scheme for obtaining pure dry samples of the compounds from a mixture of the three, and carefully describe the procedure.

Compound	*Heat*	*Cold water*	*Hot water*
Naphthalene	Sublimes	Insoluble	Insoluble
Calcium fluoride	No effect	Insoluble	Insoluble
Potassium chloride	No effect	Soluble	More soluble

Give one property of calcium fluoride in which it differs from calcium chloride.
How would you prepare a dry sample of calcium fluoride from potassium fluoride solution? (*J.M.B.*)

7. It is required to separate a mixture of *three* solid dyes; one red, one yellow and one blue. The following facts are known about the dyes. The blue and yellow dyes are soluble in cold water, while the red dye is insoluble. When an excess of aluminium oxide is added to a stirred, green aqueous solution of the mixed blue and yellow dyes and the aluminium oxide is filtered off and washed with water, it is found that the solid residue is yellow and the filtrate blue. When the yellow solid is stirred with ethanol and the mixture filtered, the solid residue is white and the filtrate yellow.
Describe how you would obtain dry samples of the three dyes. (*J.M.B.*)

8. Spinach leaves were dried and then extracted with a solvent. The solution was filtered and the solvent removed from the green filtrate. The dry residue was then dissolved in a small quantity of propanone, and a few drops of this solution were then placed at the centre of a filter paper and allowed to dry. Drops of propanone were then added to the centre of the filter paper. This produced two concentric circles, one yellow and one green.

(*a*) What process could be used to remove the solvent after the filtration?
(*b*) Describe a convenient way to add a continuous series of drops of propanone to the filter paper.
(*c*) For what purpose is the process named in (*a*) used?
(*d*) What can be said about the colouring matter of the spinach leaves? (*J.M.B.*)

2

Physical and Chemical Change

2.1 Physical change

Everyone is familiar with the change of water into ice. No new substance is formed, and the change is readily reversed by change in temperature.

Many other changes similar to this will occur to you.

(*a*) When salt is dissolved in water, all that happens is a separation by the water of the particles which make up the salt. Evaporation of the solution to dryness will give the salt in its original form.

(*b*) When an electric current is passed through the filament in an electric-light bulb the filament becomes hot and emits light. On cooling the filament is unchanged.

(*c*) Iodine, when warmed in a test-tube, gives a violet vapour which cools to give iodine crystals.

(*d*) Zinc oxide when strongly heated changes from a white to a yellow colour. The white colour returns on cooling.

(*e*) If an unmagnetised clock-spring is stroked several times with one end of a magnet, as shown in figure 2.1, it becomes magnetized. The magnetism can be removed by hitting the spring a number of times with a hammer.

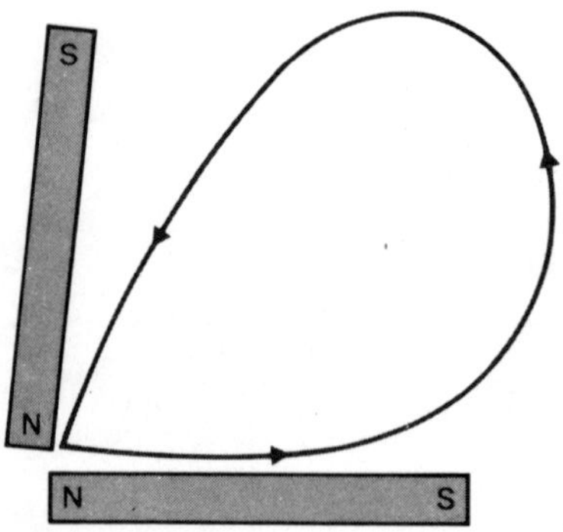

Figure 2.1 Magnetising a piece of iron with a magnet.

All such changes, that are readily reversed by changing the conditions, are said to be *physical changes*.

2.2 Chemical change

If a piece of copper foil is heated it becomes black on the surface. The black colour remains on cooling, and a new substance, copper(II) oxide, is formed.

This change is not reversed by any simple change of conditions. Other changes of this type are:

(*a*) When some iron, e.g. the blade of a penknife, is dipped into copper(II) sulphate solution it becomes coated with copper.

Copper(II) sulphate + Iron → Iron(II) sulphate + Copper

Some of the iron takes the place of the copper; this experiment is certainly not good for the penknife. A similar change occurs if some copper is placed in mercury(II) nitrate solution, when it becomes coated with mercury. Some years ago a London street-trader with a knowledge of this reaction did a brisk trade with a magic liquid for "silvering" metal articles. The silvery coating of mercury wears off after a few hours.

(*b*) A piece of magnesium ribbon, if held in a pair of tongs and heated in the flame of a Bunsen burner, burns with a brilliant flame. Clouds of white smoke are evolved and a white residue remains. The white substance consists mainly of magnesium oxide:

Magnesium + Oxygen → Magnesium oxide

(*c*) Mercury(II) oxide, when heated in a test-tube, changes to a black colour. Mercury metal distils and small droplets of it are formed at the top of the tube. If a glowing splint is held at the top of the tube it is rekindled. This shows that oxygen is being formed.

Mercury(II) oxide → Mercury + Oxygen

(*d*) If the hydrogen and chlorine in two gas jars are mixed and a lighted taper is applied to the mixture a violent explosion occurs. A flame shoots down the jar and a colourless gas, which fumes in damp air, is formed. This gas is hydrogen chloride.

Hydrogen + Chlorine → Hydrogen chloride

The last series of changes are said to be *chemical changes*. When a chemical change occurs a chemical reaction takes place. Most of the chemistry with which you will be concerned during the next few years will consist of changes of this nature.

2.3 Distinctions between physical and chemical changes

By far the most important distinction between these two changes is that in chemical changes new substances are formed. In all the changes listed above this fact is obvious. Thus, the magnesium oxide formed by burning magnesium in no way resembles magnesium. Ice may seem to be a different substance from water. It is, however, similar to water in its reactions and has exactly the same mass as the water from which it was produced. In nearly all chemical changes the mass of the product, or of one of the products, is different from the mass of any of the individual substances used at the beginning. Thus, magnesium oxide is heavier than the magnesium from which it is formed.

Physical changes are usually easily reversed, whereas chemical changes are not. This statement is always true for physical changes but for chemical changes it can be misleading. There are some chemical changes that are also easily reversed. For example, sulphur dioxide when passed into water combines with the water to form a solution of sulphurous acid. This acid is definitely a new substance. Its properties differ in important respects from those of sulphur dioxide. It follows that a chemical change has occurred. The solution, however, smells strongly of sulphur dioxide, and on heating decomposes completely to give sulphur dioxide and water, i.e. the change is readily reversible:

Sulphur dioxide + Water $\rightleftharpoons$ Sulphurous acid

A similar reaction occurs when ammonia combines with water:

Ammonia + Water $\rightleftharpoons$ Ammonium hydroxide

Further, when steam is passed over heated iron filings hydrogen and iron oxide are formed. When hydrogen is passed over iron oxide which is being heated, iron and steam are formed:

Iron + Steam $\rightleftharpoons$ Iron oxide + Hydrogen

On heating, calcium carbonate decomposes into calcium oxide and carbon dioxide. If these substances are cooled in contact with one another calcium carbonate is re-formed:

Calcium carbonate $\rightleftharpoons$ Calcium oxide + Carbon dioxide

Such reactions are said to be reversible, and are denoted by the double arrows, as shown. They are chemical changes and are more fully discussed later (section 10.6.2). In spite of these fairly common exceptions, we may say that in general physical changes are easily reversed and chemical changes are not.

When chemical reactions occur there is often a considerable energy change, involving the production or absorption of heat. In some cases there is also the production of light and sound. Little energy change occurs during most physical changes, but this, too, can be a misleading distinction. With many chemical reactions—for example most ionic reactions—little heat change occurs. On the other hand, those of you who study Physics will know of the heat evolved or absorbed with a change of state, e.g. a change of a liquid to a solid or of a liquid to a vapour. When 1 g of steam changes to water over 2000 joules of heat are evolved. There is also a cooling effect when most solids dissolve in water.

To sum up, we may say that if a change is reversible and involves little heat change it is likely to be physical. *It is, however, always necessary to consider whether new substances have been formed.*

2.4 Elements and compounds

The action of heat on mercury(II) oxide has already been mentioned. It decomposes to give two substances: mercury and oxygen. This means that mercury(II) oxide is made up of more than one simpler substance. Mercury and oxygen, however, although they undergo many reactions with other chemicals, cannot, by any known means, be converted into anything simpler. In all their reactions they gain in mass.

Copper(II) nitrate, when heated, gives a black solid, a brown gas and oxygen. The indications are that copper(II) nitrate is made up of at least three simpler substances. Two of these, the black solid and the brown gas, can be split up still further by other means. The former, copper(II) oxide, can be made to yield copper and oxygen. The latter, nitrogen dioxide, will yield nitrogen and oxygen. No further subdivision can occur.

Mercury, oxygen, copper and nitrogen are called *elements*. There are just over 100 such substances, most of which are present in the earth's crust, either in the free state or combined with other elements. A few of them have been discovered only in connection with atomic energy research, and do not occur naturally.

An element may be defined as a substance which cannot, by known chemical means, be split up into two or more simpler substances.

A list of the commoner elements is given in section 3.6.

Mercury(II) oxide and copper(II) nitrate are compounds. *A compound is a pure substance, containing two or more elements chemically combined together.* It is amazing that all the substances known—and there are probably millions of them—are made up from a mere 100 elements.

2.5 Compounds and mixtures

2.5.1 What are mixtures and compounds?

We meet with many substances which are, in fact, mixtures of various compounds or elements. Thus, air is a mixture of the elements oxygen and nitrogen, with much smaller quantities of other substances.

It is important to realize the exact difference between compounds and mixtures. Consideration of some examples should help to make this distinction clear.

A commonly quoted example, and one which is easy to investigate, deals with the differences between a mixture of iron and sulphur, and a compound of these two elements, iron(II) sulphide.

First, consider the mixture:

(*a*) A mixture of iron filings and sulphur can be made containing any proportions of the two elements. No heat change occurs during the mixing.

(*b*) The mixture is intermediate in colour between the colours of iron and sulphur. The individual particles of iron and sulphur can be distinguished with the aid of a magnifying glass.

(*c*) The mixture can be separated into its elements by various physical methods, e.g.

(i) By using a magnet, which attracts the iron but not the sulphur. This, incidentally, is an inefficient method for complete separation because particles of sulphur adhere mechanically to the iron filings.

(ii) By stirring the mixture with excess of carbon disulphide. The sulphur dissolves and iron is obtainable by filtration. If the solution is allowed to evaporate (NO FLAMES as carbon disulphide is very inflammable) in a fume cupboard (as carbon disulphide vapour is poisonous) the sulphur remains.

(*d*) The mixture possesses the properties of the individual elements contained in it. If dilute sulphuric acid is added to the mixture the iron combines with it and hydrogen is evolved. The sulphur remains unaltered. This is a typical reaction of iron, and it also gives, in this case, another method for isolating sulphur from the mixture.

Now consider the compound:

Iron(II) sulphide is made by heating together a well-powdered mixture of the elements (7 parts by mass of iron to 4 parts by mass of sulphur) in a test-tube. Much heat is evolved, and a bright red glow passes through the mixture.

Iron + Sulphur → Iron(II) sulphide

The experiments performed on the mixture can now be repeated using the compound with the following results:

(*a*) Iron(II) sulphide has a definite composition by mass, no matter how it is made. If in the preparation more than 7 parts by mass of iron are used for every 4 parts by mass of sulphur, then the excess iron will remain uncombined.

(*b*) The compound is greyish-black in colour and it is impossible to distinguish individual particles of iron and sulphur.

(*c*) It is impossible to obtain the elements from the compound by the methods used on the mixture:

(i) If the compound is finely powdered and a magnet is passed over it no separation into iron and sulphur occurs.

(ii) Carbon disulphide is without effect on the compound.

(*d*) Iron(II) sulphide has properties different from either of its constituents.

When dilute sulphuric acid is added to it the solid disappears with effervescence, and a poisonous gas with the odour of bad eggs—hydrogen sulphide—is evolved. Hydrogen is not formed.

This example illustrates many of the differences which usually exist between compounds and mixtures. They are:

(*a*) A compound is homogeneous, but mixtures usually are not.

(*b*) A mixture can be separated into its components by physical means; a compound can be separated into its elements only by chemical methods.

(*c*) A mixture gives the individual chemical reactions of each of its components. Certain physical properties, e.g. colour and density, are intermediate between those of these components. A compound, however, has properties peculiar to itself, and always quite different from those of its constituent elements.

(*d*) There is no heat change when a mixture is made. Formation of a compound is usually accompanied by considerable heat change.

(*e*) A compound has a fixed composition by mass. A mixture has not.

The last of these distinctions is by far the most important. It forms the basis of an important law that you will soon be studying. There are few exceptions to this law, and none that we need be concerned with at this stage.

It must be understood that, while the other rules listed above are usually true, more than one of them must be considered before it is possible to decide definitely in favour of a mixture or compound. Other examples will clarify this statement.

2.5.2 A solution of potassium nitrate in water

The solution is homogeneous. When potassium nitrate dissolves in water there is a cooling effect. To look at, the solution is indistinguishable from water, but it certainly bears no resemblance to potassium nitrate.

The evidence so far suggests that the solution is a compound. But within the limits imposed by the solubility of potassium nitrate in water it has a variable composition. This is the most important deciding factor, and enables us to say that the solution is in fact a mixture. Further, by evaporation to dryness the solution is separated into its components, potassium nitrate and water.

2.5.3 Brass

Brass is a homogeneous solid composed of copper and zinc. It cannot be separated into its components by physical means. But its properties are those of a mixture of copper and zinc, and it has a variable composition. Therefore brass is a mixture. A homogeneous mixture of this sort, usually of two or more metals, is called an *alloy*.

Later in this book we look at reasons for assuming air to be a mixture, and not a compound. Perhaps you can already think of a number of these.

Questions on chapter 2

1. Write a list of the differences between the following compounds and elements of which they are made.

Compound	*Elements*
Water	Hydrogen, oxygen
Common salt	Sodium, chlorine
Epsom salts	Magnesium, sulphur, oxygen, hydrogen
Sugar	Carbon, hydrogen, oxygen
Nylon	Carbon, hydrogen, oxygen, nitrogen
Vinegar	Carbon, hydrogen, oxygen
Copper(II) sulphate	Copper, sulphur, oxygen
Ammonia	Nitrogen, hydrogen

2. What do you understand by the term *element*? Compare in a table the properties of a *mixture* and those of a *compound*.

To which of these three classes of substance do the following belong: (*a*) chalk; (*b*) air; (*c*) red lead oxide; (*d*) iodine; (*e*) starch?

Choose one of these materials and indicate what experiments you would perform to show the class to which it belongs. (*D*.)

3. State *three* differences between physical and chemical changes.

State, with reasons, whether the following are physical or chemical changes: (*a*) the dissolving of phosphorus(V) oxide in water; (*b*) the lighting of an electric lamp; (*c*) the freezing of water; (*d*) the slaking of quicklime; (*e*) dissolving sugar in water; (*f*) the burning of coal gas. (*D*.)

4. You are supplied with a substance which is composed of iron and sulphur only. Describe *four* experiments you would carry out with this substance in order to find out whether it was a mixture or a compound of the elements, and in each case describe the results you would expect to obtain. (*O*.)

5. What is the essential difference between a chemical and a physical change?

Indicate clearly the chemical and physical changes involved in the following processes, giving full reasons in each case: (*a*) the addition of metallic sodium to water; (*b*) the solution of sodium chloride in water; (*c*) the heating of magnesium in air; (*d*) the heating of ammonium chloride; (*e*) the addition to water of concentrated sulphuric acid. (*L*.)

3

Atomic Theory and the Mole

3.1 What is matter?

People have been asking themselves this question since very early times, and even now the complete answer is not known. For the Greek philosophers the problem centred around the question of whether matter was composed of small indivisible particles or whether it was continuous; in other words, if suitable instruments were available, would a limit be reached beyond which further subdivision (cutting-up) would be impossible, or could the process be repeated indefinitely? There was no hope of reaching any sort of solution to this problem. Even if the Greeks had been prepared to resort to experiments, the lack of apparatus of a kind familiar to us would have rendered their task an impossible one.

The first real theoretical advance did not come until the beginning of the nineteenth century, when an English chemist, John Dalton, put forward his atomic theory. Before discussing this theory we must consider some important laws known to Dalton.

3.2 The law of conservation of matter (or mass)

This law states: *In a chemical reaction the total mass of the products equals the total mass of the reactants.* First put forward by the French scientist Antoine Lavoisier (see figure 13.10) in 1774 it seems to have been assumed by some scientists before that date. Lavoisier's great contributions to chemistry are due in large part to his careful quantitative experiments, i.e., he carefully weighed and measured things. Consequently he performed a number of experiments to verify the law. Although they were as accurate as Lavoisier could make them they would be considered very inadequate by present standards, and it was not until a century later that a series of really accurate experiments was performed by Landolt. His apparatus consisted of an H-tube, as shown in figure 3.1. The reagents were placed in different limbs of the tube and, after sealing each limb to ensure that nothing could enter or leave the apparatus, it was weighed. After tilting the tube to permit chemical reaction the apparatus was reweighed. Using a very accurate chemical balance, capable of accuracy to one part in ten million, Landolt found no change in mass.

Landolt was careful to choose reactions with which there was no evolution of gas and little evolution of heat. Such reactions occur on mixing solutions of silver nitrate and sodium chloride, or of sodium sulphate and barium chloride.

Silver nitrate + Sodium chloride → Silver chloride + Sodium nitrate

Barium chloride + Sodium sulphate → Barium sulphate + Sodium chloride

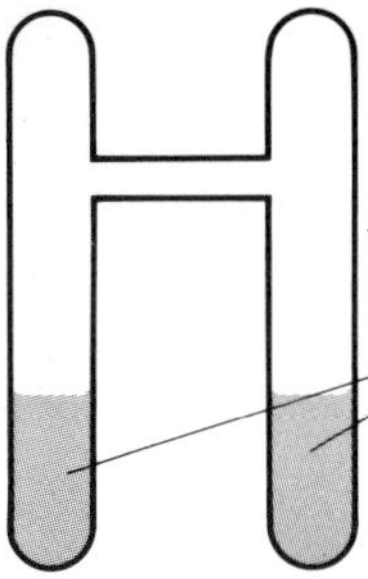

Figure 3.1 Landolt's H-tube.

The choice of such reactions was necessary, because a heavy apparatus would have been required to withstand the pressure of an evolved gas. Further, heat causes the evaporation of a thin film of moisture normally present on the surface of the glass. This means that if the apparatus is weighed before it has time to regain this moisture there is an apparent loss of mass. Because there is no change in mass in the comparatively small number of reactions investigated, it is assumed that the law holds true for all reactions.

The law can be verified in the school laboratory by a simple adaptation of Landolt's apparatus. Using a flask and test-tube as shown in figure 3.2 one of the reagents is placed in the bottom of the flask, the other in the small test-tube suspended by a thread. To mix the reagents the flask has to be tilted.

You may feel that there are several obvious objections to the law. Certain substances—for example, paper and coal— seem to disappear almost completely when they are burnt. The formation of invisible gaseous products is the cause of this apparent destruction. Thus, paper (a compound of carbon, hydrogen and oxygen) gives the gas carbon dioxide, and water vapour. If all the reactants, including the air which provides a proportion of the oxygen, and all the products were weighed the law would be found to be true in this case.

You might like to see if you can devise a suitable apparatus to verify the Law of Conservation of Mass, for a burning candle. You will find that there are many difficulties to overcome.

3.3 The law of constant composition

This law states: *No matter how a given chemical compound is prepared, it always contains the same elements in the same proportions by mass.* It was first stated by Proust in 1799, but was assumed by Lavoisier and other scientists before that date. In common with the Law of Conservation of Mass, it cannot possibly be proved in every case, but the truth of it is accepted on the basis of those compounds which have been investigated.

Figure 3.2 A laboratory version of Landolt's tube.

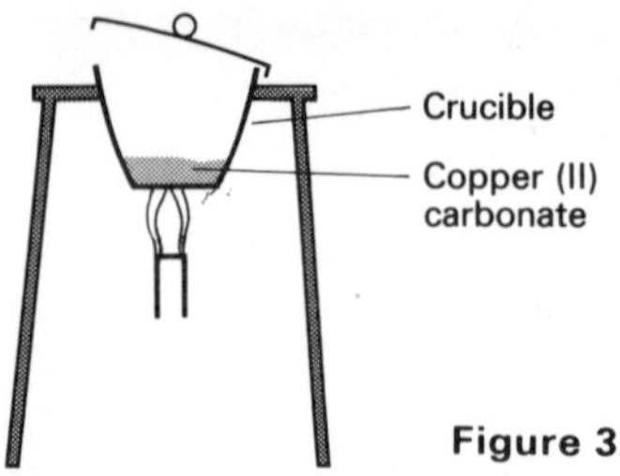

Figure 3.3 Preparation of copper(II) oxide.

Experiment—Verification of the law of constant composition

To verify the law it is necessary to obtain samples of a given compound from several different sources, and then to analyze each sample. Copper(II) oxide is suitable for use in the school laboratory, as it can be prepared readily by a variety of means. Use the following two methods; others you will learn later.

(*a*) *From copper(II) carbonate*. Place a few grams of the pale green copper (II) carbonate in a crucible and heat strongly (figure 3.3). The colour will change from green to black as the carbonate is converted into oxide.

Copper(II) carbonate → Copper(II) oxide + Carbon dioxide

Heat to constant mass to ensure complete decomposition.

As copper(II) oxide absorbs moisture from damp air, keep the product in a desiccator until ready for use.

(*b*) *From copper(II) nitrate*. Similarly, heat a few grams of copper(II) nitrate to constant mass. This time the solid will change from blue to black. Brown fumes of nitrogen dioxide are evolved during this chemical change.

Copper(II) nitrate → Copper(II) oxide + Nitrogen dioxide + Oxygen

N.B. It should be noted that copper(II) oxide cannot be prepared satisfactorily by heating copper in air or oxygen. Unless the copper is finely powdered and the heating prolonged, the copper(II) oxide forms only on the surface of the copper metal.

To analyze the samples of copper(II) oxide obtained, heat known masses in hydrogen or town gas, when the following reaction occurs:

Copper(II) oxide + Hydrogen → Copper + Water.

Then if the masses of copper formed are known it is a simple matter to calculate the composition of copper(II) oxide. Proceed as follows:

First weigh two porcelain boats empty and then containing the samples of copper(II) oxide. Place them in a hard glass tube (figure 3.4) and pass

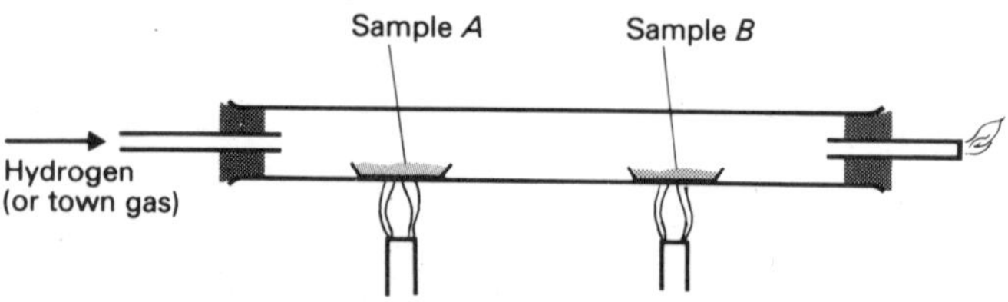

Figure 3.4 Verification of the Law of Constant Composition.

hydrogen through, until a test-tube full of the gas collected at the other end burns calmly. This shows that the hydrogen is not mixed with air—a necessary precaution, as mixtures of an inflammable gas with air are explosive.

Then light the hydrogen as shown and heat both samples. A glow will pass through the copper(II) oxide and clouds of steam replace the hydrogen, extinguishing the flame. When the reaction is complete no more steam will be formed. Then relight the hydrogen and allow the apparatus to cool with hydrogen still passing. Do not remove the copper whilst it is still hot, because it will combine to some extent with oxygen in the air, re-forming copper(II) oxide. List your data in tabular form and calculate as indicated:

	Sample A	Sample B
Mass of boat +copper(II) oxide	3·69 g	3·89 g
Mass of boat	2·01 g	1·85 g
Mass of copper(II) oxide	1·68 g	2·04 g
Mass of boat +copper	3·35 g	3·47 g
Mass of boat	2·01 g	1·85 g
Mass of copper	1·34 g	1·62 g
Percentage of copper in copper(II) oxide	$\frac{1.34 \times 100}{1.68}$ = 79·8	$\frac{1.62 \times 100}{2.04}$ = 79·4
Percentage of oxygen in copper(II) oxide	(100 − 79·8) = 20·2	(100 − 79·4) = 20·6

i.e. within the limits of experimental error, you should find your results to be in accordance with the Law of Constant Composition.

This method can be used with any other oxide capable of being reduced by hydrogen, e.g. lead(II) oxide.

3.4 Law of multiple proportions

This law, which was probably deduced by Dalton and only later subjected to experimental proof, states:

If two elements combine together to form more than one compound, then the masses of one of these elements combining with a fixed mass of the other are in a simple ratio to one another.

For example, carbon forms two common compounds with oxygen: carbon monoxide and carbon dioxide.

With carbon monoxide, 1·33 g of oxygen are combined with 1 g of carbon.

With carbon dioxide, 2·67 g of oxygen are combined with 1 g of carbon. That is, the masses of oxygen combining with a fixed mass of carbon are in the ratio 1 : 2. To verify the law in the school laboratory, examples that use solid substances are more suitable e.g. the oxides of lead.

3.4.1 Experiment—Verification of law of multiple proportions

Weigh three porcelain boats, first empty and then containing samples of lead(II) oxide, red lead oxide and lead(IV) oxide. Heat these oxides in a stream of hydrogen or town gas, taking the precautions outlined in the previous experiment. Each oxide is thereby converted into lead. Reweigh the boats containing the metal.

Here are some typical results:

	Lead(II) oxide	Red lead oxide	Lead(IV) oxide
Mass of boat	4·55 g	4·70 g	5·10 g
Mass of boat +oxide	11·50 g	11·32 g	12·14 g
Mass of boat +lead	11·00 g	10·71 g	11·19 g
Mass of lead (b)	6·45 g	6·01 g	6·09 g
Mass of oxide (a)	6·95 g	6·62 g	7·04 g
Mass of oxygen (a − b)	0·50 g	0·61 g	0·95 g

With lead(II) oxide:
0·50 g of oxygen is combined with 6·45 g of lead.

$$1{\cdot}00 \text{ g of oxygen is combined with } \frac{6{\cdot}45 \times 1{\cdot}00}{0{\cdot}50} \text{ g of lead}$$
$$= \underline{12{\cdot}90 \text{ g of lead}}$$

With red lead oxide:
0·61 g of oxygen is combined with 6·01 g of lead.

$$1{\cdot}00 \text{ g of oxygen is combined with } \frac{6{\cdot}01 \times 1{\cdot}00}{0{\cdot}61} \text{ g of lead}$$
$$= \underline{9{\cdot}85 \text{ g of lead}}$$

With lead(IV) oxide:
0·95 g of oxygen is combined with 6·09 g of lead.

$$1{\cdot}00 \text{ g of oxygen is combined with } \frac{6{\cdot}09 \times 1{\cdot}00}{0{\cdot}95} \text{ g of lead}$$
$$= \underline{6{\cdot}41 \text{ g of lead.}}$$

We see, therefore, that the masses of lead combining with a fixed mass of oxygen (1 g) are in the ratio of 12·90 : 9·85 : 6·41, i.e. 4 : 3 : 2 (within the limits of experimental error). The results obtained in a laboratory experiment are usually much poorer than this.

3.4.2 To see if you understand the law of multiple proportions, try the following problem:

A metal forms two chlorides containing respectively 65·6 and 55·9 per cent of chlorine. Show that these figures are in accordance with the Law of Multiple Proportions.

The compositions of the chlorides are:

	% chlorine	% metal
Chloride *A*	65·6	34·4
Chloride *B*	55·9	44·1

In chloride *A*,

65·6 g of chlorine combine with 34·4 g of metal.

$$1 \text{ g of chlorine combines with } \frac{34\cdot4}{65\cdot6} \text{ g of metal}$$
$$= 0\cdot525 \text{ g of metal}$$

In chloride *B*,
55·9 g of chlorine combine with 44·1 g of metal.

$$1 \text{ g of chlorine combines with } \frac{44\cdot1}{55\cdot9} \text{ g of metal.}$$
$$= 0\cdot789 \text{ g of metal}$$

i.e. the masses of metal combining with a fixed mass of chlorine are in the ratio 0·789 : 0·525 = 3 : 2. These figures are in accordance with the Law of Multiple Proportions.

It should be noted that it is immaterial which element, or what value for its fixed mass, is chosen. If with this same problem, 34·4 g of metal is taken as the fixed mass, we get:

In chloride *A*,
34·4 g of metal combine with 65·6 g of chlorine.

In chloride *B*,
44·1 g of metal combine with 55·9 g of chlorine.

$$34.4 \text{ of metal combine with } \frac{55\cdot9 \times 34\cdot4}{44\cdot1} \text{ g of chlorine}$$
$$= 43\cdot6 \text{ g of chlorine}$$

i.e. the masses of chlorine combining with a fixed mass of metal are in the ratio 65·6 : 43·6 = 3 : 2.

3.5 Dalton's atomic theory

We indicated earlier that there have been theories about atoms since very early times. Credit must go, however, to John Dalton (figure 3.5) for placing the atomic theory on a firm basis. His own account of his early history is interesting:

Figure 3.5 John Dalton, 1766–1844 (Photo: The Science Museum, London).

"The writer of this was born at the village of Eaglesfield, about two miles west of Cockermouth, Cumberland. Attended the village schools, there and in the neighbourhood, till eleven years of age, at which period he had gone through a course of mensuration, surveying, navigation, etc.; began about twelve years to teach the village school and continued it about two years; afterwards was occasionally employed in husbandry for a year or more; removed to Kendal at fifteen years of age as assistant in a boarding school; remained in that capacity for three or four years, then undertook the same school as principal and continued it for eight years; whilst at Kendal employed his leisure in studying Latin, Greek, French and the Mathematics, with natural philosophy; removed thence to Manchester in 1793 as tutor in mathematics and natural philosophy in the New College; was six years in that engagement and after was employed as private and public teacher of mathematics and chemistry in Manchester, but occasionally by invitation in London, Edinburgh, Glasgow, Birmingham and Leeds."

John Dalton was a Quaker, and, even later in life when he had achieved fame, remained completely unassuming, continuing, for example, to coach local boys and girls in their arithmetic.

His atomic theory states:

(*a*) *Matter is made up of small, indivisible particles called atoms.*

(*b*) *Atoms can neither be created nor destroyed.*

To use some of Dalton's own words: "We might as well attempt to introduce a new planet into the solar system, or to annihilate one already in existence, as to create or destroy one particle of hydrogen".

(*c*) *The atoms of a particular element are alike in every respect, including mass. They differ from the atom of any other element.*

(*d*) *When atoms combine they do so in small whole numbers to form "compound atoms". "Compound atoms" formed by such combination are alike in every respect. (These "compound atoms" are now called* MOLECULES.*)*

This theory leads us to a common definition of an atom:

An atom is the smallest particle of an element that can take part in a chemical change.

Both the words "law" and "theory" have now been used. It is essential to understand what is implied by these terms. A *law* is a general statement describing a way in which substances behave; it summarizes a number of isolated facts, which can be directly verified by experiment. A *theory* provides a generally accepted explanation of some way or other in which substances behave. A theory, once formulated, often leads to the discovery of fresh facts; in time, facts are found which are not in accordance with a particular theory. It then needs modifying, or may have to be abandoned altogether.

These points are well illustrated by reference to Dalton's theory. It provides satisfactory explanations for the laws mentioned earlier in the chapter.

(a) Law of conservation of mass

By Dalton's theory atoms can neither be created nor destroyed. It follows that a chemical reaction can consist only of a rearrangement of atoms, with no possible change in mass, i.e. the Law of Conservation of Mass will be obeyed.

(b) Law of constant composition

By Dalton's theory every molecule of a particular compound contains the same number of each type of atom present. Each atom has a fixed mass. It is therefore obvious that all the molecules of the compound should have the same percentage composition by mass, and that all samples of the compound should have the same percentage composition.

From the sort of result obtained from the experiment described in section 3.3, some idea of *relative atomic masses* may be obtained. Roughly 4 g of copper combine with 1 g of oxygen. If it is assumed (as Dalton did) that copper(II) oxide is composed of atoms with one atom of copper for every atom of oxygen, this means that a copper atom is four times as heavy as an oxygen atom. Atoms do not, however, always combine in the simple ratio of 1 : 1.

(c) Law of multiple proportions

If atoms combine together in *small whole numbers*, as suggested by Dalton, then the Law of Multiple Proportions must be true. For example, suppose two elements X and Y form two compounds.

Assume that in one compound, one atom of X combines with one atom of Y and that in the other compound, one atom of X combines with two atoms of Y.

As Dalton assumed all atoms to be alike in every respect, including mass, then it is obvious that the masses of Y combining with a fixed mass of X are in the ratio 1 : 2.

Whatever *small* number of atoms of X and Y are assumed to combine with one another, a simple ratio will exist for the masses of one of the elements combining with a fixed mass of the other.

3.6 Atoms, molecules and their symbols

The *atom* has already been defined as the smallest particle of an element that can take part in a chemical change, and a *molecule* as the species that is formed when atoms combine.

The substances made up of *atoms* are the elements.

The substances made up of *molecules* are elements if all the atoms are alike, or *compounds* if two or more different atoms are combined together.

It would obviously be convenient to have some simple shorthand system of representing atoms of the various elements. John Dalton himself invented a set of symbols consisting of letters and other characters enclosed in circles (figure 3.6). They were not generally adopted because of difficulties in remembering them and in writing them. J. J. Berzelius introduced a simpler system, which is the one now used. In general, the symbol for one atom of an element is the first letter of its Latin or English name. Where several elements have the same initial letter, the second letter or one of its other letters is added. It is worthwhile making an effort to memorize the symbols of the commoner elements; others you will easily learn as it becomes necessary.

Compounds are represented by using a combination of the appropriate symbols. Thus each molecule of hydrogen chloride is represented by the formula HCl, indicating that it contains one atom of hydrogen combined with one atom of chlorine. Each molecule of ethanol is represented by the formula C_2H_6O, indicating that it contains two atoms of carbon, six atoms of hydrogen and one atom of oxygen. Note that a formula represents a molecule of a substance, and its use as shorthand for an unlimited quantity of that substance is not strictly correct.

SYMBOLS OF COMMON ELEMENTS

Element	Symbol	Element	Symbol	Element	Symbol
Aluminium	Al	Iodine	I	Phosphorus	P
Barium	Ba	Iron	Fe	Potassium	K
Bromine	Br	Lead	Pb	Silver	Ag
Calcium	Ca	Magnesium	Mg	Sodium	Na
Carbon	C	Manganese	Mn	Sulphur	S
Chlorine	Cl	Mercury	Hg	Tin	Sn
Copper	Cu	Nitrogen	N	Zinc	Zn
Hydrogen	H	Oxygen	O		

(A more complete list of symbols is given at the end of this book.) (N.B. The first letter of every symbol is a capital letter; the second one is not.)

Figure 3.6 A list of some elements and their symbols prepared by John Dalton in 1806. Some of the members of the list, we now know, are not elements. See if you can decide which these are. (Photo: Crown copyright. Science Museum, London).

3.7 Relative atomic masses

Atoms are extremely small and the determination of their absolute masses was, in Dalton's time, impossible. We now know, for example, that the mass of a hydrogen atom is approximately $1{\cdot}67 \times 10^{-24}$ g.

It is therefore usual to refer to the *relative masses of atoms*, and by assuming that atoms usually combine together in the ratio 1 : 1 it was possible, as indicated in the previous section, to obtain these relative masses. Because this assumption is by no means always correct, incorrect atomic masses were often assigned to some elements in the early nineteenth century.

Until 1961 the mass of the hydrogen atom was taken as a standard and relative atomic mass defined as the number of times one atom of a particular element was heavier than one atom of hydrogen. In 1961, for reasons that need not concern us here, the definition was changed to: *The relative atomic mass of an element is the number of times one atom of that element is heavier than one twelfth of an atom of the main isotope of carbon (known as Carbon-12).* (See section 5.2.5 for a description of an isotope and a slightly modified definition of relative atomic mass.)

If this second definition is used the relative atomic mass of hydrogen changes from 1·000 to 1·008, so that in fact, whichever definition is used the relative atomic masses are almost the same.

Nowadays the relative atomic masses are obtained using a large expensive instrument known as a *mass-spectrometer* (figure 3.7). By shooting the atoms through some complicated electric and magnetic fields, this instrument is able to sort atoms according to their mass, and to measure very accurately the masses of the atoms. For this method it does not matter at all in what ratios atoms combine.

A full list of *relative atomic masses* is given at the back of this book.

(*Note*. Although *relative atomic mass* is the correct term, it is often abbreviated to *atomic mass*.)

Figure 3.7 A modern mass-spectrometer. (Photo: A. Rendell).

3.8 Relative molecular masses

In the water molecule, for example, two atoms of hydrogen combine with one atom of oxygen. The *molecular mass* of the water molecule is the sum of the atomic masses of the atoms, i.e., $2 \times$ (atomic mass of H) + (atomic mass of O) = $2 + 16 = 18$. Similarly, hydrogen chloride contains one hydrogen atom to each chlorine atom, hence its molecular mass = $(1 + 35{\cdot}5) = 36{\cdot}5$.

It follows from this and the definition of relative atomic mass that the *relative molecular mass is the number of times a molecule of an element or compound is heavier than one twelfth of one atom of the carbon-12 isotope*, (or one hydrogen atom, to use the old standard).

Molecular mass can also be used for compounds that do not exist as molecules (i.e., exist as ions). In these cases the *molecular mass* is the mass of a *"formula unit"*, a formula unit being the simplest statement of the combining ratio of the two ions. Thus the formula unit of sodium chloride is NaCl. This says there are equal numbers of sodium and chlorine units in sodium chloride. The *molecular mass* of sodium chloride is therefore $23 + 35{\cdot}5 = 58{\cdot}5$.

(*Note.* Although *relative molecular mass* is the correct term, it is often abbreviated to *molecular mass.*)

Further examples of the calculation of molecular masses are given in section 4.3.

3.9 Avogadro's constant

We have seen that the atomic mass of hydrogen is 1, and that of carbon is 12. This means, as we have just explained, that one atom of carbon is 12 times heavier than one atom of hydrogen. Clearly two atoms of carbon will be 12 times heavier than two atoms of hydrogen, and *n* atoms of carbon (where *n* is any number) will be 12 times heavier than *n* atoms of hydrogen. Therefore, *any mass of hydrogen will have in it the same number of atoms as 12 times that mass of carbon.*

Our standard unit of mass is the kilogram, hence we measure mass in kilograms and grams, i.e., *1 g of hydrogen will have in it the same number of atoms as 12 g of carbon.*

We could, of course, choose any other element and its atomic mass expressed in grams will contain the same number of atoms as 1 g of hydrogen and 12 g of carbon (e.g. 16 g of oxygen, or 23 g of sodium). Furthermore, we can choose any molecule and its molecular mass expressed in grams will contain the same number of molecules as there are atoms in 1 g of hydrogen (e.g. 18 g of water or 36·5 g of hydrogen chloride).

You might ask if it is known what this number is? The answer is that it is known very accurately. It is, as you might expect, a very big number, $6{\cdot}02 \times 10^{23}$. It is called *Avogadro's constant* after the Italian physicist Amedeo Avogadro, who has (you will see in section 7.7) contributed a great deal to this aspect of chemistry.

3.10 The mole

In the last section we saw that:

(*a*) there are $6{\cdot}02 \times 10^{23}$ atoms in the atomic mass expressed in grams of an element.

(*b*) there are $6{\cdot}02 \times 10^{23}$ molecules in the molecular mass expressed in grams of an element or compound.

Later (section 3.11.2) we shall see that there are $6{\cdot}02 \times 10^{23}$ ions in the moleoular mass expressed in grams of an ion.

"$6{\cdot}02 \times 10^{23}$ atoms or molecules or ions" is such an important quantity to the chemist that it is given a special name—*the mole* (abbreviated name: *mol*).

The mole is $6{\cdot}02 \times 10^{23}$ particles of anything. Usually the mole is used in referring to atoms or molecules or ions or electrons, e.g. 1 mole of copper atoms contains $6{\cdot}02 \times 10^{23}$ atoms, and 1 mole of water molecules contains $6{\cdot}02 \times 10^{23}$ molecules. The idea of a mole can be applied to anything although, of course, it is normally applied only to these small particles met with in science.

To give some idea of what an enormous number of particles are contained in a mole, try to imagine a mole of people i.e. $6{\cdot}02 \times 10^{23}$ people. If they stood packed tightly together over the whole surface of the earth (land and sea) there would be enough left over to cover another 250 000 000 earths (or if they all stood on one earth on everyone elses' shoulders the columns of people would stretch beyond the moon!). You might like to guess (or if you can, work out) the volume of a mole of footballs. (Answer: They have a volume 5 times that of the earth.)

In contrast we have a mole of hydrogen atoms ($6{\cdot}02 \times 10^{23}$ of them) in just 1 g, and a mole of water molecules ($6{\cdot}02 \times 10^{23}$ water molecules) in 18 g or a large tablespoonful (≡ 5 teaspoonsful).

Put another way—there are as many molecules of water in a teaspoonful as there would be footballs in the volume of the earth.

This gives you some idea of just how small are the building blocks of chemistry—atoms and molecules—that make up all matter.

3.11 Molar mass

3.11.1 Molar masses of atoms and molecules

We have defined a *mole* ($6{\cdot}02 \times 10^{23}$ particles), and we have seen that a mole of atoms or molecules is present in the atomic or molecular mass expressed in grams. This mass of any substance that contains a mole of particles is called the *molar mass*.

Thus a mole of sodium atoms (atomic mass = 23) has a mass of 23 g. To put it another way, the *molar mass* of sodium is $23\,g\,mol^{-1}$ ($g\ mol^{-1}$ means g per mole).

Similarly a mole of water molecules (molecular mass = 18) has a mass of 18 g. Or we can say that the *molar mass* of water is $18\ g\ mol^{-1}$.

(These *molar masses* were previously called *gram-atoms* and *gram-molecules*.)

3.11.2 Molar masses of ionic compounds

(This section may be clearer after you have read Chapter 6.)

Compounds made up of ions do not contain molecules and the 'particles' referred to in the definition of a mole of such a compound must be clearly specified. This is not usually difficult. We use, as explained in section 3.8, the "formula unit". For example, for sodium chloride, which has the formula NaCl (or Na^+Cl^-), the particle is taken to be everything represented in the usual formula. The *molar mass* of sodium chloride is, therefore, $(23 + 35{\cdot}5)\,g\,mol^{-1}$

$= 58{\cdot}5\,g\,mol^{-1}$. You will see that 1 mole of sodium chloride ($6{\cdot}02 \times 10^{23}$ formula units) consists of one mole, i.e. $6{\cdot}02 \times 10^{23}$, of sodium ions and one mole, i.e. $6{\cdot}02 \times 10^{23}$, of chloride ions.

Similarly the molar mass of sodium sulphate (Na_2SO_4 or $(Na^+)_2SO_4{}^{2-}$) is $(2 \times 23) + 32 + (4 \times 16)\,g\,mol^{-1} = 142\,g\,mol^{-1}$. Here 1 mole of sodium sulphate ($6{\cdot}02 \times 10^{23}$ formula units) consists of 2 moles, i.e. $12{\cdot}04 \times 10^{23}$, of sodium ions and one mole, i.e. $6{\cdot}02 \times 10^{23}$, of sulphate ions.

Summary

The *mole* is a *number* ($6{\cdot}02 \times 10^{23}$) of particles; it does not have any mass. The *mass* of a *mole* of particles is the *molar mass*. This has units of grams per mole ($g\,mol^{-1}$).

To emphasize the points that have been developed during the last few pages the following table lists them all for some atoms and molecules.

Atom	Symbol	Relative Atomic mass	Molar mass	Number of atoms in a mole
Hydrogen	H	1	$1\,g\,mol^{-1}$	$6{\cdot}02 \times 10^{23}$
Oxygen	O	16	$16\,g\,mol^{-1}$	$6{\cdot}02 \times 10^{23}$
Carbon	C	12	$12\,g\,mol^{-1}$	$6{\cdot}02 \times 10^{23}$
Sodium	Na	23	$23\,g\,mol^{-1}$	$6{\cdot}02 \times 10^{23}$
Sulphur	S	32	$32\,g\,mol^{-1}$	$6{\cdot}02 \times 10^{23}$

Molecule	Formula	Relative Molecular mass	Molar mass	Number of molecules in a mole
Hydrogen	H_2	2	$2\,g\,mol^{-1}$	$6{\cdot}02 \times 10^{23}$
Oxygen	O_2	32	$32\,g\,mol^{-1}$	$6{\cdot}02 \times 10^{23}$
Water	H_2O	18	$18\,g\,mol^{-1}$	$6{\cdot}02 \times 10^{23}$
Hydrogen chloride	HCl	36·5	$36{\cdot}5\,g\,mol^{-1}$	$6{\cdot}02 \times 10^{23}$
Carbon dioxide	CO_2	44	$44\,g\,mol^{-1}$	$6{\cdot}02 \times 10^{23}$
Sodium chloride	NaCl	58·5	$58{\cdot}5\,g\,mol^{-1}$	$6{\cdot}02 \times 10^{23}$
Sulphur dioxide	SO_2	64	$64\,g\,mol^{-1}$	$6{\cdot}02 \times 10^{23}$

Questions on chapter 3

Use the table of approximate relative atomic masses (at the back of this book) for the first six questions.

1. What are the *molar masses* of
(*a*) Magnesium (Mg), (*b*) Bromine (Br),
(*c*) Iron (Fe), (*d*) Sodium bromide (NaBr),
(*e*) Calcium bromide ($CaBr_2$), (*f*) Aluminium oxide (Al_2O_3)?
2. How many *atoms* are there in *1 mole* of
(*a*) Magnesium, (*b*) Bromine, (*c*) Chlorine?
3. How many *atoms* are there in
(*a*) 24 g of magnesium (*b*) 80 g of bromine (*c*) 35·5 g of chlorine?
4. How many *atoms* are there in
(*a*) 12 g of magnesium, (*b*) 48 g of magnesium,
(*c*) 0·24 g magnesium?
5. How many *moles* are there in
(*a*) 12 g of carbon, (*b*) 56 g of iron, (*c*) 160 g of bromine atoms,
(*d*) 120 g of calcium, (*e*) 8 g of sulphur, (*f*) 8 g of oxygen atoms?

6. Which contain the greater number of *moles*
(*a*) 24 g of magnesium or 80 g of calcium, (*b*) 1 g of magnesium or 1 g of calcium, (*c*) 5·6 g of iron or 6·5 g of zinc, (*d*) 14 g of nitrogen atoms or 14 g of aluminium, (*e*) 1 g of nitrogen atoms or 10 g of iodine atoms, (*f*) 2 g of nitrogen atoms or 10 g of iodine atoms?

7. State the Law of Multiple Proportions.

Weighed quantities of two oxides of the same metal were heated in a current of coal gas with the following results:

Mass of boat *A* empty	= 4·500 g
Mass of boat *A* with first oxide before heating	= 4·946 g
Mass of boat *A* with residue after final heating	= 4·914 g
Mass of boat *B* empty	= 5·000 g
Mass of boat *B* with second oxide before heating	= 5·462 g
Mass of boat *B* with residue after final heating	= 5·414 g

(*a*) Complete the calculation to show how the results are in accordance with the Law of Multiple Proportions.
(*b*) Sketch the apparatus which would be used. (*J.M.B.*)

8. State the Law of Multiple Proportions and describe with all necessary practical details, an experiment to verify it.

A metal forms two chlorides which contain respectively 44·16 and 34·58 per cent of metal. Show that these figures illustrate the Law of Multiple Proportions. (*C.W.B.*)

9. (*part*). State the Law of Multiple Proportions.

A metal forms two chlorides which contain 38·17 and 48·08 per cent of chlorine. Find out if these results illustrate the above law. (*C.W.B.*)

10. A metal forms two oxides, *A* and *B*. 1 g of *A* on reduction gave 0·881 g of the metal, while 1 g of *B* yielded 0·788 g of the metal. Show that these facts are in agreement with the Law of Multiple Proportions. (*O. and C.*)

11. (*part*). A metal *X* forms two chlorides, *A* and *B*. *A* contains 62·6 per cent by weight of metal, and *B* contains 45·6 per cent by weight. Show that these figures are in accordance with the Law of Multiple Proportions. (*O. and C.*)

12. (*part*). Analysis shows that two oxides *A* and *B* of the same metal contain 20 and 11·11 per cent respectively of oxygen. Show that these figures agree with the Law of Multiple Proportions. (*L.*)

13. State the Law of Multiple Proportions and show how it is explained by the Atomic Theory.

Two hydrocarbons (i.e. compounds of carbon and hydrogen only) contain 81·81 and 84·21 per cent of carbon respectively. Are these figures in accordance with the Law of Multiple Proportions?

14. Both water and hydrogen peroxide are compounds of hydrogen and oxygen. 17 g of hydrogen peroxide decompose to yield 8 g of oxygen and 9 g of water. Water contains 11·11 per cent of hydrogen. Show that these two compounds agree with the Law of Multiple Proportions.

4

Valency, Formulae and Equations. Calculations Involving Masses and Moles

4.1 Valency

The valency of an element is the number of atoms of hydrogen that will combine with or displace one atom of that element. In this chapter you will see how, with a knowledge of valency, formulae and then equations can be constructed. In the next chapter, an alternative definition of valency is developed.

The most usual valency, particularly of metals, is two. Valencies are rarely above four, and the highest known valency is eight. The valencies of the common elements are listed in the following table:

Valency				
1 (i.e. mono-or uni-valent elements)	2 (i.e. di-valent elements)	3 (i.e. tri-valent elements)	4 (i.e. tetra- valent elements)	5 (i.e. penta -valent elements
Hydrogen	Copper(II)	Aluminium	Tin(IV)	Phosphorus(V)
Sodium	Mercury(II)	Iron(III)	Carbon	
Potassium	Magnesium	Phosphorus(III)	Silicon	
Silver	Calcium	Nitrogen		
Mercury(I)	Oxygen			
Chlorine	Barium			
Bromine	Zinc			
Iodine	Tin(II)			
Copper(I)	Iron(II)			
	Sulphur			

It is also convenient to give valencies to certain groups of atoms which either exist only as ions (see section 6.2.2) or which have no independent existence.

(The charges given in the formulae of the ions in this table may be ignored at this stage if valency is treated initially as in this chapter. If valency is taught in

1	2	3
Ammonium (NH_4^+)	Sulphite (SO_3^{2-})	Phosphate (PO_4^{3-})
Hydroxide (OH^-)		
Nitrate (NO_3^-)	Sulphate (SO_4^{2-})	
Hydrogencarbonate (HCO_3^-)	Carbonate (CO_3^{2-})	

conjunction with the electronic theory of valency a knowledge of the charges is needed.)

You will need all of the valencies given; the sooner you know them by heart, the better.

4.2 Nomenclature

You will see in the table in the previous section that the name of the element followed by a roman number (I, II, III, or IV in the table) is given for certain elements. This is because these elements have *more than one common valency state*. You will see that there is copper(I) and copper(II), mercury(I) and mercury(II), tin(II) and tin(IV), iron(II) and iron(III) and phosphorus(III) and phosphorus(V). The roman number shows the valency state of the element. The other elements in the table have only one common valency state, and the number is usually omitted.

Thus throughout this book we write sodium chloride, magnesium chloride and aluminium chloride, as sodium, magnesium and aluminium each have only one valency state. But we write iron(II) chloride and iron(III) chloride as iron has two common valencies and can therefore form two different chlorides.

Other examples where roman numerals are used are sulphur(VI) oxide, manganese(IV) oxide, sodium chromate(VI), and potassium manganate(VII). Sulphur(VI) oxide means the oxide where sulphur has a valency of 6, and is used to distinguish this compound from the one where sulphur has a valency of 4 (this is sulphur(IV) oxide but it is usually called sulphur dioxide). Manganese(IV) oxide is the oxide where manganese has a valency of 4 (manganese can have several different valencies). Sodium chromate(VI) and potassium manganate(VII) are examples where the roman numerals come after the second word of the name (i.e. after the anion), and they show the valency (strictly the oxidation state) of the chromium or manganese in these oxo-anions (i.e. in these negative ions that contain the metal and oxygen).

4.3 Determination of formulae from valency

To determine the formula of a compound, both the symbol and valency of each element (or ion) in it must be known.

There is no difficulty with the formulae of compounds of hydrogen. These follow directly from the definition of valency, e.g.

Valency of chlorine = 1   Formula of hydrogen chloride is HCl
Valency of oxygen = 2   Formula of hydrogen oxide (water) is H_2O
Valency of nitrogen = 3   Formula of hydrogen nitride (ammonia) is H_3N
Valency of carbon = 4   Formula of hydrogen carbide (methane) is H_4C

Just as a symbol represents one atom of an element, so the formula represents one molecule of a compound.

Notice: (*a*) The ending *-ide* in the naming of the above compounds. A chloride is a compound containing chlorine and one other element or ion. Therefore hydrogen chloride contains hydrogen and chlorine combined together. Similarly, an oxide contains oxygen and one other element.

(*b*) In writing the formulae, the symbol for hydrogen is placed first. Usually, however, the formulae for ammonia and methane are written NH_3 and CH_4. This is merely a convention which it is best to follow. In general it is a convention to place the symbols of metals (or the formulae of positive ions) first in formulae, but apart from this there is no fixed rule.

When compounds do not contain hydrogen the construction of formulae is similar. Consider a few chlorides, remembering that chlorine has a valency of 1:

Valency of sodium	= 1	Formula of sodium chloride is NaCl
Valency of magnesium	= 2	Formula of magnesium chloride is $MgCl_2$
Valency of aluminium	= 3	Formula of aluminium chloride is $AlCl_3$
Valency of carbon	= 4	Formula of carbon chloride (now called tetrachloromethane) is CCl_4

All the compounds mentioned so far have formulae of the type AB_x (where x is the number of atoms of B) and in each case:

Valency of A = Valency of $B \times x$

i.e. the number of atoms is such that each element provides the same total number of 'valencies'.

This idea can be extended to formulae of the type A_xB_y. Then:

Valency of $A \times x$ = Valency of $B \times y$

Any compound containing two elements (or ions) must have a formula which obeys this relationship; e.g. if element A has a valency of 3 and element B of 2, the simplest formula is A_2B_3. The total number of 'valencies' supplied by $A(2 \times 3)$ then equals the total number of 'valencies' supplied by $B(3 \times 2)$.

Examples

Aluminium oxide:

Valency of aluminium = 3
Valency of oxygen = 2

If the formula of aluminium oxide is Al_xO_y, then $x \times 3 = y \times 2$. The lowest values for x and y are 2 and 3. Therefore the simplest formula for aluminium oxide is Al_2O_3. It will be seen later that the simplest formula is not necessarily the correct one but with most compounds this is the case.

Aluminium carbide

Valency of aluminium = 3
Valency of carbon = 4

If the formula of aluminium carbide is Al_xC_y, then $x \times 3 = y \times 4$. The lowest values for x and y are 4 and 3, and therefore the simplest possible formula is Al_4C_3.

Copper(II) nitrate

Valency of copper = 2
Valency of nitrate = 1

In copper(II) nitrate, therefore, there are two nitrate ions for each copper ion. This is represented by the formula $Cu(NO_3)_2$. The small 2 indicates that there are two of everything in the brackets immediately preceding it.

Magnesium sulphate

Valency of magnesium = 2
Valency of sulphate = 2

As the valencies are the same in both cases, the formula for magnesium sulphate is $MgSO_4$.

Iron(III) sulphate

Valency of iron = 3
Valency of sulphate = 2

If the formula of iron(III) sulphate is $Fe_x(SO_4)_y$, then $x \times 3 = y \times 2$. The lowest values of x and y are 2 and 3, giving iron(III) sulphate the formula $Fe_2(SO_4)_3$.

With a little practice it is found that there is no need to make an algebraic problem of these formulae, and that they can be written down almost automatically.

4.4 Calculation of molar masses

Molar mass was defined in section 3.11 as the mass in grams of a mole of particles. To calculate the molar mass of a compound, it is necessary to know the atomic mass of each of its constituent elements. The *molecular mass* is then the sum of the *atomic masses* of the individual atoms and the *molar mass* is this number of grams of the compound.

Note here that the use of formulae is very helpful. It tells us the number of each of the atoms present in the molecule and it defines the "particle" needed to specify a mole.

Carbon dioxide, CO_2:

Atomic mass of carbon = 12 (or, C = 12)
Atomic mass of oxygen = 16 (or, O = 16)

Each molecule of carbon dioxide contains one atom of carbon (atomic mass of carbon = 12) and two atoms of oxygen (atomic mass of oxygen = 16). The molecular mass is therefore, $12+16+16 = 44$, i.e. the molar mass of carbon dioxide is 44 g mol^{-1}.

It is convenient to set out such a calculation in this manner:

$$\text{molecular mass} = \overset{\text{C}\quad\;\; \text{O}_2}{\overline{12+(2\times 16)}} = 12+32 = 44$$

i.e., one mole of carbon dioxide has a mass of 44 g and contains $6{\cdot}02 \times 10^{23}$ molecules.

Aluminium sulphate, $Al_2(SO_4)_3$:

Al = 27; S = 32; O = 16.

With compounds like aluminium sulphate, which have a more complicated formula, the procedure is similar, but it is advisable to remove the brackets first.

$Al_2(SO_4)_3 \quad = \quad Al_2 \quad + \quad S_3 \quad + \quad O_{12}$

molecular mass $= (2 \times 27) + (3 \times 32) + (12 \times 16)$

$= \quad 54 \quad + \quad 96 \quad + \quad 192 \quad = 342$

molar mass $= 342 \text{ g mol}^{-1}$

Copper(II) sulphate crystals, $CuSO_4.5H_2O$:

Cu = 63·5; S = 32; O = 16; H = 1

This compound contains water of crystallization (see 11.5), and the formula indicates that five molecules of water are combined with each $CuSO_4$ unit.

$CuSO_4.5H_2O \quad = Cu \quad S \quad O_4 \quad 5H_2O$

molecular mass $= 63{\cdot}5 + 32 + (4 \times 16) + 5(2 + 16)$

$= 63{\cdot}5 + 32 + \quad 64 \quad + \quad 90 \quad = 249{\cdot}5$

molar mass $= 249.5 \text{ g mol}^{-1}$

4.5 Calculation of percentage composition by mass from formulae

If the preceding paragraphs have been understood, it is easy to see how the percentage composition of a compound may be calculated. Taking water as our example, we find that it has a molar mass of 18 g mol^{-1}. Of these 18 g, 16 g are accounted for by the oxygen content, and the remaining 2 g by the hydrogen.

i.e. $\frac{16}{18}$ of the total mass of water is oxygen,

i.e. $\frac{16}{18} \times 100$ per cent = 88·9 per cent.

And $\frac{2}{18}$ of the total mass of water is hydrogen,

i.e. $\frac{2}{18} \times 100$ per cent = 11·1 per cent.

Examples

(*a*) *Calculate the percentage composition by mass of sodium chloride Na = 23; Cl = 35·5*

NaCl = Na Cl

molar mass = 23 + 35·5 = 58·5 g mol^{-1}

Thus in every 58·5 g of sodium chloride there are 23 g of sodium and 35·5 g of chlorine.

$\therefore$ Fraction of sodium in sodium chloride $= \frac{23}{58{\cdot}5}$

and fraction of chlorine in sodium chloride $= \frac{35 \cdot 5}{58 \cdot 5}$

$\therefore$ *Percentage of sodium* $= \frac{23}{58 \cdot 5} \times 100 \qquad = 39 \cdot 3\,\%$

and *percentage of chlorine* $= \frac{35 \cdot 5}{58 \cdot 5} \times 100 \quad = 60 \cdot 7\%$

(*b*) *Calculate the percentage composition by mass of lead(II) nitrate. Pb = 207; N = 14; O = 16.*

$Pb(NO_3)_2 \quad = Pb \quad N_2 \quad O_6$

molar mass $= 207 + (2 \times 14) + (6 \times 16)$

$= 207 + 28 + 96 = 331\ g\,mol^{-1}$

Proceeding as in last example:

Percentage of lead $= \frac{207}{331} \times 100 = 62 \cdot 5\,\%$

Percentage of nitrogen $= \frac{28}{331} \times 100 = 8 \cdot 5\,\%$

Percentage of oxygen $= \frac{96}{331} \times 100 = 29 \cdot 0\,\%$

(*c*) *Calculate the percentage of water of crystallization in washing soda crystals, sodium carbonate-10-water,* $Na_2CO_3 . 10H_2O$

$Na_2CO_3 . 10H_2O = \quad Na_2 \quad C \quad O_3 \quad 10H_2O$

molar mass $= (2 \times 23) + 12 + (3 \times 16) + 10(2 + 16)$

$= \quad 46 \quad + 12 \quad 48 \quad + \quad 180$

$= 286\ g\ mol^{-1}$

i.e. in every 286 g of washing soda there are 180 g of water.

Percentage of water $= \frac{180}{286} \times 100 = 62 \cdot 9\,\%$

4.6 Determination of formulae by experiment

In the last sections the compositions by mass of compounds have been calculated from the formulae of compounds.

The reverse is obviously possible. If the masses of each element present in a certain mass of a substance is known (or if the percentage composition by mass of a substance is known), *the chemical formula can be calculated.* Chemists have a number of ways of determining experimentally the composition by mass of a compound and it is in this way that *chemical formulae are experimentally derived.*

For example, the percentage composition by mass of a certain oxide of sulphur is; oxygen, 50 per cent; sulphur, 50 per cent. Obviously the percentage of either of these elements depends on

(*a*) the atomic mass of that element,
(*b*) the number of atoms of that element in each molecule.

If the atomic mass of oxygen were equal to that of sulphur it is obvious that there would be one atom of oxygen present for each atom of sulphur. In actual fact, the atomic mass of oxygen is 16 and of sulphur 32, i.e., the atomic mass of oxygen is half the atomic mass of sulphur. For the oxide to contain 50 per cent by mass of each element there must, therefore, be two atoms of oxygen for each atom of sulphur. This means that the simplest formula for the oxide of sulphur is SO_2, but it is not possible, with the data given, to say that this is the actual formula. If the formula were S_2O_4, S_3O_6 (i.e., S_nO_{2n}) it would still give the same percentage composition by mass.

Another oxide of sulphur contains 60 per cent oxygen and 40 per cent sulphur by mass. Knowing the atomic masses of oxygen and sulphur to be 16 and 32 respectively, the formula can again be calculated. By similar reasoning there are $\frac{60}{16}$ oxygen atoms for every $\frac{40}{32}$ sulphur atoms in a molecule of this oxide:

$$\frac{60}{16}:\frac{40}{32} = \frac{60}{1}:\frac{40}{2} = 3:1$$

i.e. there are three oxygen atoms for each sulphur atom, giving the simplest formula as SO_3.

In general, we can find the relative numbers of each type of atom in a molecule of compound, and hence its simplest formula, by dividing the percentage by mass, or the actual masses, of the various elements by their atomic masses. Only rarely are the figures so convenient as in the examples quoted above. The following problems should, however, make the method clear.

4.6.1 Examples

(*a*) *A compound contains 29·1 per cent sodium, 40·5 per cent sulphur and 30·4 per cent oxygen. Calculate its formula (Na = 23; S = 32; O = 16.)*

The relative numbers of atoms of each element are obtained thus:

$$\text{Relative number of atoms} = \frac{\%}{\text{atomic mass}}$$

Percentage of sodium = 29·1. $\qquad \frac{29{\cdot}1}{23} = 1{\cdot}27$

Percentage of sulphur = 40·5. $\qquad \frac{40{\cdot}5}{32} = 1{\cdot}27$

Percentage of oxygen = 30·4. $\qquad \frac{30{\cdot}4}{16} = 1{\cdot}90$

Therefore the molecules of the compound must contain *whole* numbers of sodium, sulphur and oxygen atoms in the ratio 1·27:1·27:1·90. The simplest whole number ratio agreeing with these figures is probably obvious to you. If not it can be found by dividing all the figures by the lowest one,

$$\text{i.e. } 1{\cdot}27:1{\cdot}27:1{\cdot}90 = \frac{1{\cdot}27}{1{\cdot}27}:\frac{1{\cdot}27}{1{\cdot}27}:\frac{1{\cdot}90}{1{\cdot}27} = 1:1:1{\cdot}5$$

Therefore the simplest whole number ratio is 2:2:3. (N.B. The ratio is not absolutely exact because the figures given in the question are subject to experimental error.)

From these data we obtain $Na_2S_2O_3$ as the simplest formula for the compound. Such a formula is known as an *empirical formula.*

An empirical formula may be defined as the simplest formula which can be obtained for a compound using its composition by mass.

(*b*) *A compound containing water of crystallization has the following percentage composition: iron; 20·1%; sulphur, 11·5%; oxygen, 23·0%; water; 45·3%. Calculate the empirical formula of this compound. (Fe = 56, S = 32, O = 16, H = 1.)*

Here a molecule of water can be treated as a unit, in the same way as an atom. The molecular mass of water is 18.

	Relative number of atoms
Percentage of iron = 20·1	$\dfrac{20{\cdot}1}{56} = 0{\cdot}36$
Percentage of sulphur = 11·5	$\dfrac{11{\cdot}5}{32} = 0{\cdot}36$
Percentage of oxygen = 23·0	$\dfrac{23{\cdot}0}{16} = 1{\cdot}44$
Percentage of water = 45·3	$\dfrac{45{\cdot}3}{18} = 2{\cdot}52$

This indicates a formula $Fe_{0.36}S_{0.36}O_{1.44}.2{\cdot}52\,H_2O$. As before, the simplest ratio of atoms becomes obvious by dividing the relative numbers of atoms obtained by the smallest number.

$$0{\cdot}36:0{\cdot}36:1{\cdot}44:2{\cdot}52: = \frac{0{\cdot}36}{0{\cdot}36}:\frac{0{\cdot}36}{0{\cdot}36}:\frac{1{\cdot}44}{0{\cdot}36}:\frac{2{\cdot}52}{0{\cdot}36}$$

$$= 1:1:4:7$$

Therefore the simplest formula for this compound is $FeSO_4.7H_2O$

(*c*) *20·00 g of a compound containing carbon, hydrogen, and oxygen are found to contain 8·00 g of carbon and 1·33 g of hydrogen. Calculate the empirical formula of this compound. (C = 12, H = 1, O = 16.)*

Since the compound contains only carbon, hydrogen and oxygen, the mass of oxygen in 20·00 g of the compound

$= 20{\cdot}00 - (8{\cdot}00 + 1{\cdot}33)$ g

$= 10{\cdot}67$ g

Although we are not dealing with percentages in this compound, the method of solution is the same.

	Relative number of atoms
Mass of carbon = 8·00 g	$\dfrac{8{\cdot}00}{12} = 0{\cdot}67$
Mass of hydrogen = 1·33 g	$\dfrac{1{\cdot}33}{1} = 1{\cdot}33$
Mass of oxygen = 10·67 g	$\dfrac{10{\cdot}67}{16} = 0{\cdot}67$

$$0{\cdot}67:1{\cdot}33:0{\cdot}67 = \frac{0{\cdot}67}{0{\cdot}67}:\frac{1{\cdot}33}{0{\cdot}67}:\frac{0{\cdot}67}{0{\cdot}67} = 1:2:1$$

Therefore the empirical formula is CH_2O.

(*d*) *In sections* 3.3 *and* 3.4 *it was found by experiment that 1·68 g of a copper oxide contains 1·34 g of copper, that 0·50 g of oxygen combine with 6·45 g of lead to give lead oxide, that 0·61 g of oxygen combine with 6·01 g of lead to give a second lead oxide and that 0·95 g of oxygen combine with 6·09 g of lead to give a third lead oxide. Obtain the formulae of the copper oxide and of the three lead oxides. (Cu = 63·5, Pb = 207, O = 16.)*

Copper oxide

1·34 g of copper combine with (1·68 – 1·34) g of oxygen
1·34 g of copper combine with 0·34 g of oxygen

	Relative number of atoms
Mass of copper = 1·34 g	$\frac{1{\cdot}34}{63{\cdot}5} = 0{\cdot}0211$
Mass of oxygen = 0·34 g	$\frac{0{\cdot}34}{16} = 0{\cdot}0212$

0·0211 : 0·0212 = 1:1

i.e. the empirical formula is CuO.

First lead oxide

0·50 g of oxygen combine with 6·45 g of lead

	Relative number of atoms
Mass of lead = 6·45 g	$\frac{6{\cdot}45}{207} = 0{\cdot}0312$
Mass of oxygen = 0·50 g	$\frac{0{\cdot}50}{16} = 0{\cdot}0312$

0·0312 : 0·0312 = 1 : 1

i.e. the empirical formula is PbO.

We leave you to work out the formulae of the other two lead oxides.

4.7 Equations

An equation is a convenient method for representing a chemical change by using formulae and symbols. We have already represented a number of chemical reactions by using "word equations", and the process of building up equations will be understood by converting a number of these into normal chemical equations.

4.7.1 How to construct equations

(a) Mercury(II) oxide → Mercury + Oxygen.
To construct a chemical equation for this reaction, the formula for each substance involved must be known:

Mercury(II) oxide . . . . HgO
Mercury . . . . Hg
Oxygen . . . . O_2

(N.B. Most gaseous elements normally exist as aggregates of two atoms, see section 7.7.)

A provisional equation is then written:

$$HgO \rightarrow Hg + O_2$$

This equation does not "balance", i.e. the numbers of atoms and molecules involved must be so adjusted that the total numbers of each type of atom on either side of the equation are equal. To give two atoms of oxygen, two molecules of mercury(II) oxide are required. These will also give two atoms of mercury. Therefore the equation becomes:

$$2HgO \rightarrow 2Hg + O_2$$

i.e., two molecules of mercury(II) oxide give two atoms of mercury and one molecule of oxygen. Note that the figure 2 in $2HgO$ applies to the whole molecule, but that when a small figure is used as a subscript, e.g. in $PbCl_2$ and $Pb(NO_3)_2$, it applies only to the atom or ion preceding it. You might be tempted to alter a formula as an easy method of making an equation balance. *This must never be done. A particular compound has a definite and unalterable formula.*

(b) Potassium chlorate → Potassium chloride + Oxygen.
Inserting the formulae for these substances we get:

$$KClO_3 \rightarrow KCl + O_2$$

To balance this equation there must be an even number of oxygen atoms on the left-hand side of the equation (why?). Therefore it is possible that two molecules of potassium chlorate will be required.

Trying this, we see that two molecules of potassium chloride and three of oxygen will be obtained, and the equation then balances.

$$2KClO_3 \rightarrow 2KCl + 3O_2$$

(c) Iron + Copper(II) sulphate → Copper + Iron(II) sulphate.
Inserting the formulae we get:

$$Fe + CuSO_4 \rightarrow Cu + FeSO_4.$$

The equation balances without further adjustment.

(d) Hydrogen + Oxygen → Water.
Inserting the formulae we get:

$$H_2 + O_2 \rightarrow H_2O.$$

Two molecules of water will be formed from one molecule of oxygen. This means that two molecules of hydrogen will be required. Therefore the balanced equation is:

$$2H_2 + O_2 \rightarrow 2H_2O.$$

(e) Zinc + Hydrochloric acid → Zinc chloride + Hydrogen.
Inserting the formulae we get:

$$Zn + HCl \rightarrow ZnCl_2 + H_2$$

To give one molecule of zinc chloride, two atoms of chlorine, obtainable from two molecules of hydrochloric acid, are required. The equation then becomes:

$$Zn + 2HCl \rightarrow ZnCl_2 + H_2$$

This now balances.

The construction of equations becomes easy with practice, and to help with this your teacher will no doubt provide a number of examples for you to do. Do not try to learn correct equations in a parrot fashion. There are so many equations that have to be known that this method is never successful. A knowledge of the chemical reaction concerned, and of the way equations are built up, is all that is required.

4.7.2 State symbols

Substances can exist as solids, liquids or gases and all three states can take part in chemical reactions.

Thus in the reaction:

$$2H_2 + O_2 \rightarrow 2H_2O$$

two gases (hydrogen and oxygen) are reacting to form a liquid (water). In the reaction:

$$2HgO \rightarrow 2Hg + O_2$$

a solid (mercury(II) oxide) is reacting to give two gases (mercury vapour and oxygen).

It is often desirable to show the state (solid, liquid or gas) of the substances represented by a chemical equation. This is easily done by using the symbols s, for solid, l, for liquid, and g, for gas, and putting these symbols (known as *state symbols*) in brackets after the formula.

Thus the above two equations become:

$$2H_2(g) + O_2(g) \rightarrow 2H_2O(l)$$
and $2HgO(s) \rightarrow 2Hg(g) + O_2(g)$.

Very often we study reactions in solution in water (aqueous solutions). For example:

$$Zn + 2HCl \rightarrow ZnCl_2 + H_2$$

is a reaction between solid zinc and an aqueous solution of hydrogen chloride (i.e. hydrochloric acid) to give an aqueous solution of zinc chloride and hydrogen gas. The hydrogen chloride here is not a solid, a liquid or a gas but it is in aqueous solution, and a state symbol for aqueous solution is needed. The symbol used is (aq).

So the equation is:

$$Zn(s) + 2HCl(aq) \rightarrow ZnCl_2(aq) + H_2(g)$$

It must be remembered that many substances can exist as solids, liquids or gases depending on their temperature. Hence water may be formed as liquid water, but very often it is formed as steam (gaseous water). Hence the state symbol used may change if the reaction is carried out at a different temperature.

The reaction

$$2H_2(g) + O_2(g) \rightarrow 2H_2O(l)$$

is usually carried out by burning hydrogen in oxygen, and there is no doubt that in the hot flame the water formed is a gas. The above equation is better written:

$$2H_2(g) + O_2(g) \rightarrow 2H_2O(g)$$

Another commonly met situation where the substance formed is not in its usual (i.e. at room temperature) state is when metals are extracted from their ores. Here high temperatures are often used and the metal is produced as a liquid. In such a situation the appropriate equation would be, for example:

$$Fe_2O_3(s) + 3CO(g) \rightarrow 2Fe(l) + 3CO_2(g).$$

4.7.3 Calculations involving equations

(a) Moles

Chemical equations are of immense value to the chemist, as they tell him the combining ratios of different substances and the ratios of the products formed.

Consider the equation:

$$Fe + CuSO_4 \rightarrow Cu + FeSO_4$$

This equation says that *1 atom of iron* reacts with *1 molecule of copper(II) sulphate* to give *1 atom of copper* and *1 molecule of iron(II) sulphate.*

It also says that *1 mole of iron* reacts with *1 mole of copper(II) sulphate* to give *1 mole of copper* and *1 mole of iron(II) sulphate.*

There is 1 mole of each of the reactants and 1 mole of each of the products. *This 1 : 1 ratio does not always occur.*

For example, consider now the equation:

$$2H_2 + O_2 \rightarrow 2H_2O$$

This equation says that 2 molecules of hydrogen react with one molecule of oxygen to give 2 molecules of water, or that *2 moles of hydrogen* react with *one mole of oxygen* to give *2 moles of water.*

The equation always tells us how many *moles* of each *reactant* are needed and how many *moles* of each *product* are formed. Of course, to be correct the equation must be properly balanced. It is therefore essential for you to learn how to balance chemical equations as we have just described.

(b) Masses

As the equation tells us the number of moles of reactants and of products we can, by using atomic masses, calculate the masses of reactants and products.

Consider the equation:

$2HgO$	$\rightarrow$ $2Hg$	$+O_2$
2 molecules	2 atoms	1 molecule
2 moles	2 moles	1 mole

The molar mass of mercury(II) oxide is (201 + 16) g mol^{-1} = 217 g mol^{-1}. The molar mass of mercury is 201 g mol^{-1}.

(2 × 217) g mercury(II) oxide → (2 × 201) g mercury + 32 g oxygen.
i.e., 434 g mercury(II) oxide → 402 g mercury + 32 g oxygen.

4.7.4 Examples

(*a*) *Calculate the masses of each substance formed when 10 g of calcium carbonate react with excess of hydrochloric acid. (Ca = 40, C = 12, O = 16, Cl = 35·5, H = 1.)*

First a balanced equation must be produced and then the molecular masses added:

$$\underbrace{\underset{40+12+(3\times 16)}{CaCO_3}}_{100} + \underbrace{\underset{2(1+35\cdot 5)}{2HCl}}_{73} \rightarrow \underbrace{\underset{40+(2\times 35\cdot 5)}{CaCl_2}}_{111} + \underbrace{\underset{(2\times 1)+16}{H_2O}}_{18} + \underbrace{\underset{12+(2\times 16)}{CO_2}}_{44}$$

i.e., 1 mole (100 g) of calcium carbonate reacts with 2 moles (73 g) of hydrochloric acid to give 1 mole (111 g) of calcium chloride, 1 mole (18 g) of water and 1 mole (44 g) of carbon dioxide.

Therefore, by simple proportion,

10 g calcium carbonate give $\frac{111}{100} \times 10 = 11\cdot 1$ g calcium chloride,

$\frac{18}{100} \times 10 = 1\cdot 8$ g water,

and $\frac{44}{100} \times 10 = 4\cdot 4$ g carbon dioxide.

(*b*) *What mass of lead(II) nitrate must be heated to give 22·3 g of lead(II) oxide? (Pb = 207, N = 14, O = 16).*

The equation is:

$$\underbrace{\underset{2[207+2[14+(3\times 16)]]}{2Pb(NO_3)_2}}_{662} \rightarrow \underbrace{\underset{2(207+16)}{2PbO}}_{446} + \underbrace{\underset{4[14+(2\times 16)]}{4NO_2}}_{184} + \underbrace{\underset{(2\times 16)}{O_2}}_{32}$$

In this problem we are concerned only with the masses of lead(II) oxide and lead(II) nitrate. There was, therefore, no need to work out the molar masses of the other substances.

From the equation

2 moles (446 g) of lead(II) oxide are obtained from 2 moles (662 g) of lead(II) nitrate.

22·3 g of lead(II) oxide are obtained from $\frac{662}{446} \times 22{\cdot}3$ g of lead(II) nitrate

= 33·1 g of lead(II) nitrate

(*c*) *What is the maximum mass of silver chloride that can be precipitated from 8·5 g of silver nitrate by the addition of hydrochloric acid? (Ag = 108, N = 14, O = 16, Cl = 35.5).*

The equation is:

$$\underbrace{\underset{[108+14+(3\times 16)]}{AgNO_3}}_{170} + HCl \rightarrow \underbrace{\underset{(108+35{\cdot}5)}{AgCl}}_{143{\cdot}5} + HNO_3$$

From the equation:

170 g of silver nitrate yield 143·5 g of silver chloride

8·5 g of silver nitrate yield $\frac{143{\cdot}5}{170} \times 8{\cdot}5$ g of silver chloride

= 7·17 g silver chloride

(*d*) *A crystal of Iceland Spar (calcium carbonate), which is insoluble in water, has a mass of 15 g. It is placed in a solution containing 12·6 g of nitric acid. What mass of crystal will remain after reaction has ceased? (Ca = 40, C = 12, O = 16, N = 14, H = 1).*

The equation is:

$$\underbrace{\underset{[40+12+(3\times 16)]}{CaCO_3}}_{100} + \underbrace{\underset{[1+14+(3\times 16)]}{2HNO_3}}_{126} \rightarrow Ca(NO_3)_2 + H_2O + CO_2$$

From the equation:

126 g of nitric acid react with 100 g of Iceland Spar

12·6 g of nitric acid react with $\frac{12{\cdot}6}{126} \times 100$ g of Iceland Spar

= 10 g of Iceland Spar.

Mass of Iceland Spar remaining = (15 – 10) g = 5 g

Questions on chapter 4

1. Write the formula and calculate the mass of one mole of each of the following compounds: calcium oxide, magnesium chloride, aluminium carbide, carbon dioxide,

magnesium nitride, sodium phosphide, calcium carbonate, potassium sulphate, iron(III) hydroxide, iron(II) sulphate and copper(II) nitrate.

2. Calculate the percentage composition by mass of the following compounds: (*a*) magnesium oxide; (*b*) sulphuric acid; (*c*) calcium chloride; (*d*) sodium hydrogencarbonate; (*e*) copper(II) nitrate; (*f*) potassium chlorate(V); and (g) ammonium nitrate.

3. Calculate the mass of oxygen in: (*a*) 90 g of water, H_2O; (*b*) 68.5 g of red lead oxide, Pb_3O_4; (*c*) 30 g of iron(III) sulphate, $Fe_2(SO_4)_3$; (*d*) 12 g of potash alum, $KAl(SO_4)_2.12H_2O$.

4. Calculate the percentage of water of crystallization in the following compounds: (*a*) iron(II) sulphate crystals, $FeSO_4.7H_2O$; (*b*) sodium thiosulphate, $Na_2S_2O_3.5H_2O$; (*c*) sodium sulphate crystals, $Na_2SO_4.10H_2O$.

5. Calculate the empirical formulae of the following compounds from their percentage composition by weight:

(*a*) K, 56·5 %; C, 8·70 %; O, 34·8 %.
(*b*) Pb, 62·5 %; N, 8·45 %; O, 29·05 %.
(*c*) Cu, 47·3 %; Cl, 52·7 %.
(*d*) Na, 39·3 %; Cl, 60·7 %.
(*e*) H, 2·3 %; N, 97·7 %.
(*f*) H, 12·5 %; N, 87·5 %.
(*g*) Fe, 20·14 %; S, 11·53 %; O, 23·01 %; H_2O, 45·32 %.
(*h*) K, 42·4 %; Fe, 15·2 %; C, 19·6 %; N, 22·8 %.
(*i*) N, 19·3 %; Ca, 27·6 %; Cl, 49·0 %; H, 4·1 %.
(*j*) Mg, 9·8 %; S, 13·0 %; O, 26·0 %; H_2O, 51·2 %.

6. Given that the number of atoms in one mole (i.e. 12 g) of carbon is 6×10^{23}, write down

(i) the number of molecules in 1·4 g of nitrogen gas,
(ii) the number of electrons required to convert 3·2 g of sulphur to sulphide ions (S^{2-}),
(iii) the number of sodium ions in 250 cm^3 of 1·0M sodium chloride solution. (*C.*)

7. When hydrogen sulphide is passed into a solution of copper(II) sulphate, copper(II) sulphide is precipitated according to the equation:

$$CuSO_4 + H_2S \rightarrow CuS + H_2SO_4$$

(*a*) What weight of copper(II) sulphide will be precipitated from a solution containing 8 g of anhydrous copper(II) sulphate?
(*b*) What is the percentage by weight of copper in the precipitate? (*O. and C.*)

8. (*a*) Washing soda has the formula $Na_2CO_3.10H_2O$. What weight of anhydrous sodium carbonate is left when all the water of crystallization is expelled by heating 50 g of washing soda?
(*b*) When excess lead(II) nitrate solution was added to a solution of sodium sulphate, 10 g of lead(II) sulphate were precipitated. What weight of sodium sulphate was present in the original solution? (*C.*)

9. (*a*) When zinc carbonate is heated, the reaction is represented by the equation $ZnCO_3 \rightarrow ZnO + CO_2$. Calculate (1) the loss of weight when 2·5 g of zinc carbonate is heated, (2) the weight of zinc in the residue of zinc oxide.
(*b*) What weight of iron(III) sulphate can be obtained from 8 g of iron? (*C.*)

10. (*a*) Find the simplest formula of the white crystalline substance having the following composition: nitrogen 35 per cent, hydrogen 5 per cent and oxygen 60 per cent.
(*b*) A white crystalline salt ($M.xH_2O$) contained 51·2 per cent of water of crystallization. If the molecular mass of the anhydrous salt (M) is 120, find the formula of the crystals.
(*c*) Calculate the percentage gain in mass when mercury is completely converted into its red oxide. (*B.*)

11. Crystalline barium chloride has the formula $BaCl_2.xH_2O$. When 10 g of it are heated, the residue of anhydrous barium chloride weighs 8·52 g. Find x. (*C.*)

12. A solution containing 10 g of lead(II) nitrate is mixed with excess of sodium chloride solution. Calculate (1) the weight of lead(II) chloride formed, (2) the weight of sodium nitrate which would be produced. (*C.*)

13. When iron dissolves in dilute sulphuric acid, iron(II) sulphate is formed. Crystalline iron(II) sulphate has the formula $FeSO_4.7H_2O$. What mass of this salt could be obtained from 2·8 g of iron, and what volume of dilute sulphuric acid containing 98 g per litre would be required to dissolve this amount of iron? (*O. and C.*)

14. (*a*) Complete the following sentence:

The Avogadro number, $6{\cdot}02 \times 10^{23}$, is the number of . . .

(*b*) A packet of sugar contains crystals of average mass $3{\cdot}42 \times 10^{-3}$ g. How many sugar molecules will an average crystal contain? (Formula of sugar: $C_{12}H_{22}O_{11}$) (*J.M.B.*)

15. This question is concerned with an attempt to determine the molarity (number of g-molecules/litre of solution) of a saturated aqueous solution of hydrogen chloride at 20 °C.

The table below gives information about the variation in solubility of hydrogen chloride with temperature.

Temperature (°C)	0	30	40	50	60
Solubility of HCl (g/litre of water)	823	673	633	596	561

Plot the solubility curve of hydrogen chloride on graph paper. Plot temperature along the x-axis, taking 1 inch to represent 10 °C. and solubility along the y-axis, taking 1 inch to represent 40 g/litre. Begin the solubility axis at 500 g/litre.

(*a*) Use the graph to estimate the solubility of hydrogen chloride in water at 20 °C.

(*b*) Use your answer in (*a*) to calculate the percentage by mass of hydrogen chloride in a saturated solution at 20 °C (1 litre of water weighs 1000 g).

(*c*) Use your answer in (*a*) to calculate the number of g-molecules of hydrogen chloride dissolved in 1 litre of water at 20 °C (g-molecular weight of HCl is 36·5 g).

(*d*) If the assumption is made that there is no volume change when hydrogen chloride dissolves in water, calculate the molarity of saturated hydrogen chloride at 20 °C.

(*e*) There is in fact a volume change. What additional information would you require to calculate the molarity without making the assumption in (*d*)? (*L.*)

5

The Electronic Structure of Atoms and the Periodic Table

5.1 Introduction

Everybody has heard or read about the splitting of the atom, of electrons, of protons, neutrons and other sub-atomic particles.

Some of the ideas of more recent years, about the structure of the atom, are very complicated and it is not going to be possible to discuss these here. If you study chemistry later at school or at university you will be able to add to what we give in this book. However, all of this work during the last 50 years has been concerned with trying to understand exactly how the various parts of atoms and molecules move in space and how the different parts are held together.

Although the recent work means that scientists now understand more about atoms and molecules, none of this has had the slightest effect on the way that atoms and molecules react together and the properties of the many compounds formed. Atoms and molecules have been reacting in the same way since the formation of the Universe. When lightning started a forest fire millions of years before man existed the carbon burnt according to the reaction:

$$C(s) + O_2(g) \rightarrow CO_2(g)$$

When the cave-man made his first fire to cook his meat, the logs burnt according to the reaction:

$$C(s) + O_2(g) \rightarrow CO_2(g)$$

Now that we understand far more than the caveman did about exactly how the atoms of carbon and those of oxygen interact to give carbon dioxide we still have the same chemical reaction:

$$C(s) + O_2(g) \rightarrow CO_2(g)$$

5.2 The structure of atoms

5.2.1 The electron

Let us look at how our knowledge about the structure of the atom has developed since the time of John Dalton at the beginning of the last century.

In *1808 John Dalton* developed the idea that a *solid atom* was the fundamental building block of all matter. No further development of the structure of the atom occurred during most of the 19th century.

In *1897* work by *J. J. Thompson*, in which he passed electric discharges through gases led to the discovery of a small negatively charged particle. This work can be easily demonstrated. The equipment, a discharge tube (figure 5.1), contains the

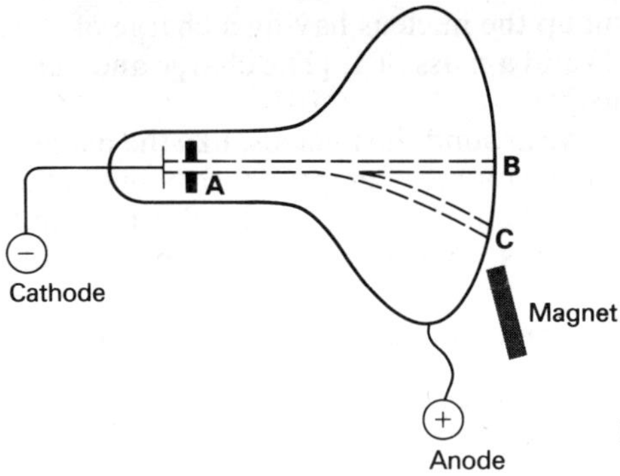

Figure 5.1 A discharge (cathode ray) tube.

gas and when the electric current is switched on, *cathode rays* are emitted from the negative plate (the cathode). The slit A enables a narrow beam to proceed. If a special chemical is painted on the glass this will glow when the rays hit it. These rays consist of a stream of negatively charged particles known as *electrons*. They can be shown to be charged by holding a magnet near the discharge tube. The magnet deflects the beam of electrons from B to C.

Experiment. Interaction of a beam of electrons with a magnetic field

Most of you have a discharge tube that you can use for this experiment. The picture-tube of a television set is a complicated discharge tube. **(You must NOT do this experiment with a COLOUR TELEVISION SET.)** The beam of electrons comes from the back and hits the chemical layer painted on the front of the tube. By the use of electronics the beam of electrons is made to travel so that it passes over the whole face of the tube 25 times every second, and when it gets to a part of the picture that is to be black the electron beam is switched off. In this way a television picture is built up.

Take a magnet and hold it in front of a **black and white** T.V. picture. The picture will be distorted, being pulled towards the magnet or pushed away from it depending on the direction of the magnetic field.

Most important note: If you have a colour television set you must not do this experiment. A colour-TV tube is much more complicated than a black and white one, and putting a magnet near it will ruin it. Once it has been so damaged, it cannot be repaired but must be replaced. Colour-TV tubes are extremely expensive.

In 1909 R. Millikan determined the charge on a single electron and its mass. Approximately 1840 of these electrons have the same mass as one hydrogen atom. Electrons in varying numbers are constituents of all atoms. They are, as we shall see, the most important atomic particle for the chemist, as it is the electrons that interact when chemical reactions take place.

5.2.2 The nucleus, protons and neutrons

In 1913 Lord Rutherford showed that an atom consists of a positively charged *nucleus* and that most of the mass of the atom was in the nucleus. He gave the name *proton* to the basic particle making up the nucleus having a charge of +1 (the electron being given a charge of −1) and a mass of 1. [The charge and mass values of 1 are arbitrary "atomic" units.]

Rutherford said that the electrons revolve around the nucleus, like the planets round the sun, and that there must be a large volume of empty space between the nucleus and the electrons. Niels Bohr worked out the orbits of the electrons and was able to calculate energies of electrons that were in agreement with experimental values.

The idea that an atom is mainly empty space can be illustrated by comparing the atom with the solar system.

Atoms have a diameter of about 10^{-10} m, and the nucleus has a diameter of 10^{-14} to 10^{-15} m.

Therefore the ratio $\dfrac{\text{diameter of atom}}{\text{diameter of nucleus}} = \dfrac{10^{-10}}{10^{-14}}$ to $\dfrac{10^{-10}}{10^{-15}}$, i.e. 10^4 to 10^5.

The diameter of the sun is 864 000 miles, and that of the orbit of the outermost planet, Pluto, is 7 400 000 000 miles (Pluto is 40 times as far from the sun as is the earth).

$$\text{Therefore the ratio } \frac{\text{diameter of Pluto's orbit}}{\text{diameter of sun}} = \frac{7\,400\,000\,000}{864\,000}$$

$$= 8565 \approx 10^4$$

Hence if the nucleus were the same size as the sun, the outer electrons would be farther from it than is Pluto—there really is a lot of empty space!

In 1932 Sir James Chadwick discovered another particle in the nucleus, with a mass similar to that of the proton but with no electric charge (see section 6.3.4). This was called the *neutron.* All nuclei (except hydrogen) contain neutrons as well as protons.

During the 1920's other ideas than those of Rutherford and Bohr were being developed. These ideas, which are not needed by us, have given us better mathematical descriptions of how the electron moves around the nucleus. What is done is to use the mathematical equations of wave-motion to describe how large collections of electrons behave.

We are usually concerned with small numbers, or individual electrons. These are quite satisfactorily regarded as particles with a charge of −1 and a mass of $\dfrac{1}{1840}$ that of the hydrogen atom.

5.2.3 The planetary atom

Hence the atom of any element is composed of two parts:

(*a*) *The nucleus.* This is at the centre of the atom and contains *protons* (*symbol "p"*) and *neutrons* (*symbol "n"*). It is minute in size compared with the atom as a whole, although it supplies nearly all the mass. The number of protons in an atom of a particular element is referred to as its *atomic number*, and the number of (protons + neutrons) as the *mass number.*

(*b*) A system of *electrons* (*symbol "e"*) rotating around the positively charged nucleus.

Since the atom is electrically neutral, the number of electrons must equal the number of protons. (See figure 5.2).

5.2.4 Symbols for atoms

It is very convenient to be able to give the *atomic number* (i.e. number of protons or electrons) and the *mass number* (i.e. number of protons + neutrons) of an atom in a concise way. The method used is very simple. The *atomic number* is written in front of the symbol for the atom and just below it, and the *mass number* is written in front of the symbol for the atom and just above it.

Thus: $^{12}_{6}C$ means a carbon atom, that has an atomic number of 6 and a mass number of 12. (6 protons and $(12-6) = 6$ neutrons.)

$^{35}_{17}Cl$ is a chlorine atom, that has 17 protons and $(35-17) = 18$ neutrons.

We can now also define an *element* as *a substance consisting entirely of atoms with the same atomic number* (i.e., with the same number of protons or electrons).

Thus the *hydrogen* atom which has an atomic number of 1 and a mass number of 1, contains 1 proton in the nucleus and has 1 electron, and the *helium* atom, with an atomic number of 2 and a mass number of 4, contains 2 protons and 2 neutrons in the nucleus and has 2 electrons.

5.2.5 Isotopes and a redefinition of relative atomic mass

In section 3.7 the relative atomic mass was defined as the *number of times one atom of a particular element is heavier than one twelfth of an atom of carbon-12.* On this basis the relative atomic mass will be the same as the mass number since protons and neutrons have the same mass and the mass of the electrons can be neglected. However, inspection of the table of relative atomic masses at the back of this book shows that this is clearly not so. The reason is that most elements have *isotopes*, i.e., *they consist of a mixture of atoms that all have the same number of protons*

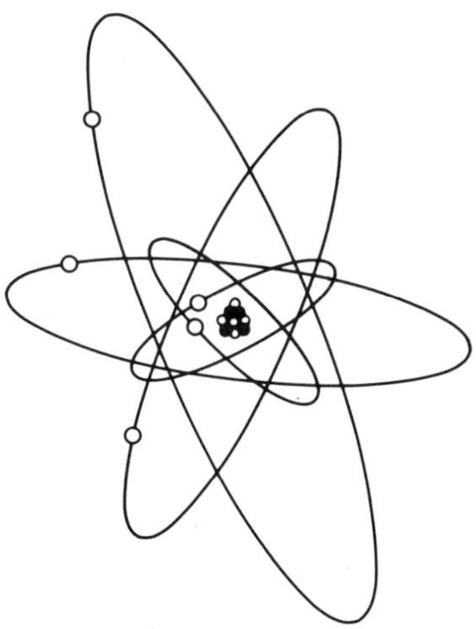

Figure 5.2 The planetary atom.

but different numbers of neutrons. Hence the atomic numbers are the same (they are, therefore, atoms of the same element) but the mass numbers are different. For example, chlorine exists as two isotopes, chlorine-35 ($^{35}_{17}Cl$) and chlorine-37 ($^{37}_{17}Cl$). The nucleus of each chlorine-35 atom contains 17 protons and 18 neutrons, and that of each chlorine-37 atom contains 17 protons and 20 neutrons (see figure 5.3). There are about 3 times as many chlorine-35 atoms in chlorine gas as there are chlorine-37 atoms. Hence the relative atomic mass of chlorine gas is $\frac{(35 \times 3) + 37}{4} = 35{\cdot}5$. This is the number in the table of relative atomic masses.

The element carbon is of importance as it is now used as the standard for relative atomic masses. Carbon has two stable isotopes, carbon-12 and carbon-13. The carbon-12 atom has 6 protons and 6 neutrons in its nucleus, and the carbon-13 atom has 6 protons and 7 neutrons in its nucleus. Naturally occurring carbon is mainly carbon-12 (98·89 %), the other 1·11 % being carbon-13. This means that the relative atomic mass of natural carbon is greater than the standard 12·000 of carbon-12. As you can see from the table of atomic masses that of carbon is 12·01. This is obtained from the percentages of the naturally occurring isotopes:

$$\text{Relative atomic mass of carbon} = \frac{(98{\cdot}89 \times 12) + (1{\cdot}11 \times 13)}{100} = 12{\cdot}01$$

Three isotopes of the hydrogen atom are known, these are hydrogen-1, hydrogen-2 and hydrogen-3. The hydrogen-1 atom contains 1 proton and 0 neutrons in its nucleus. The hydrogen-2 atom, also known as deuterium, contains 1 proton and 1 neutron in its nucleus, and the hydrogen-3 atom, also known as tritium, contains 1 proton and 2 neutrons in its nucleus (see figure 5.3). The natural concentration of deuterium is very low, and that of tritium is almost nil. Hence the relative atomic mass of hydrogen is close to 1 (1·008).

It should be clear to you that the existence of isotopes of most elements means

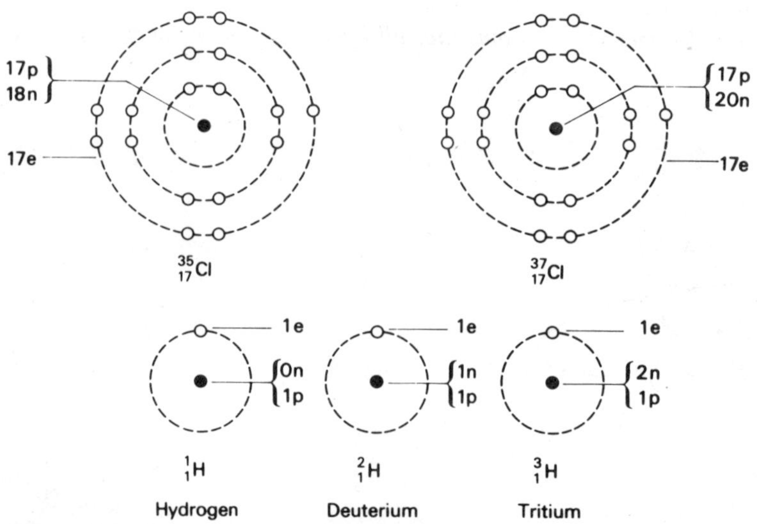

Figure 5.3 Schematic diagrams of chlorine and hydrogen isotopes showing the electron energy shells.

that our definition of *relative atomic mass* in section 3.7 is not quite correct as it refers to one atom, and for certain elements some atoms have a different mass from others (i.e. these are isotopes). What is needed in the definition is *not one particular atom* but the *"average-atom"*, the *average atom* being the mean of the naturally occurring isotopes taking into account the relative proportions in which they occur.

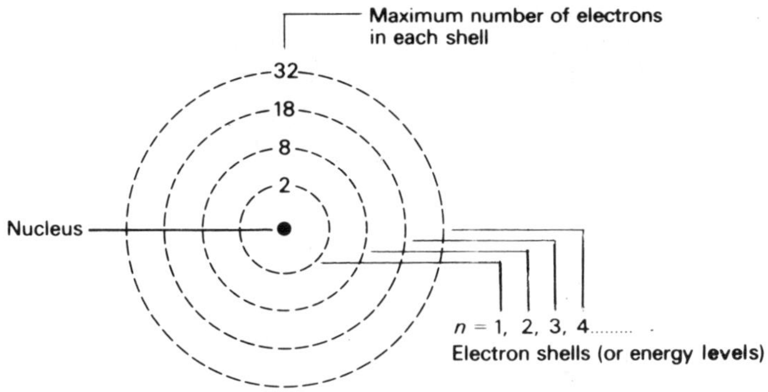

Figure 5.4 The arrangement of the electron energy shells.

5.2.6 Electron shells and energy levels

(*a*) We have not yet discussed the forces that hold the various parts of the atom together. The most important forces, so far as we are concerned, are those between the charged electrons and protons, i.e. electrostatic forces. It is the electrostatic force of attraction between the positive nucleus and the negative electrons that stops the electrons from flying away from the nucleus.

The electrons of an atom may be regarded as being in concentric rings or *shells* about the nucleus. These different *electron shells* are different distances from the nucleus. Now you probably already know that electrostatic forces *increase* as the distance between the charges *decrease*, hence the electrons that are closest to the nucleus are held most strongly. The more strongly an electron is held the *lower* is its energy, i.e. the more energy you would have to use to pull the electron away from its nucleus. Hence these *electron shells correspond to different levels of energy of the electron.*

(*b*) Now each shell can hold up to a certain number of electrons. The first shell, i.e. the one closest to the nucleus, can hold 1 or 2 electrons, the second shell can hold up to 8 electrons, the third shell can hold up to 18 electrons, and the fourth shell up to 32 electrons. (In general the maximum number of electrons in a shell is given by $2n^2$ where n is the number of the shell.) The shells have also been given symbols, these are K, L, M, N . . .

Hence:

Number of shell	1	2	3	4
Symbol of shell	K	L	M	N
$2n^2$	(2×1^2)	(2×2^2)	(2×3^2)	(2×4^2)
Number of electrons	2	8	18	32
Energy level	Lowest ———	———	———	—Increasing——→

(*c*) It is, perhaps, helpful to picture these electron shells as in figure 5.3, 5.4, 5.5 and 5.6. Remember that the diagrams tell you that the electrons in the smallest shell (i.e. nearest to the nucleus) have the lowest energy. (*Note* These pictures are helpful diagrams—they must not be interpreted to show that electrons are arranged in circular rings around the nucleus.) The electronic structures can also be written in brackets after the symbol as shown in figure 5.5.

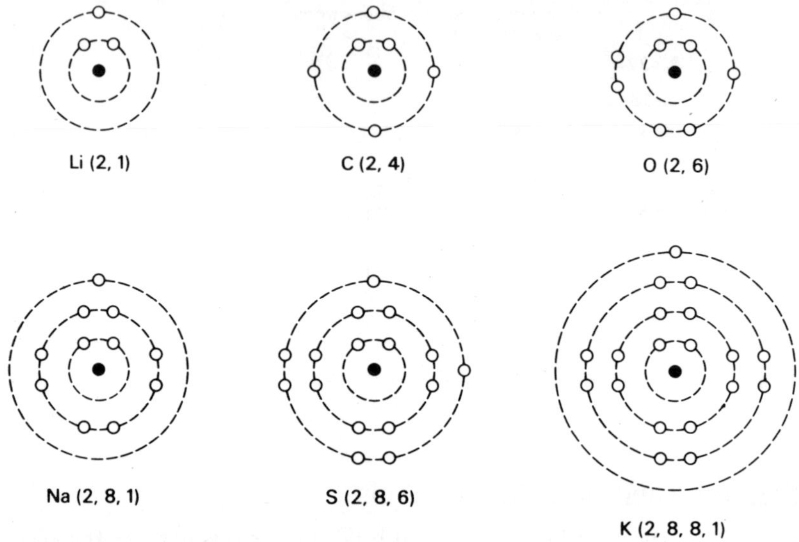

Figure 5.5 The electronic structures of some atoms.

5.2.7 The electronic structures of the elements

The electrons in an atom tend to have the lowest energy possible, so in the hydrogen atom the first electron goes into the shell closest to the nucleus ($n = 1$ or K shell). In the helium atom both electrons go into the K shell. In the lithium atom, the first two electrons go into the K shell which is then full, so the third electron goes into the next innermost shell, i.e. the next lowest in energy ($n = 2$ or L shell). All the succeeding elements fill the K shell and then the L shell, until the element neon is reached and this has 8 electrons in the L shell. When a shell is *filled*, the electrons are associated in *pairs*, hence the helium and neon atoms can be represented as in figure 5.6.

The first 20 elements have electronic structures as in the table.

Notice that although the third shell can contain up to 18 electrons, the fourth

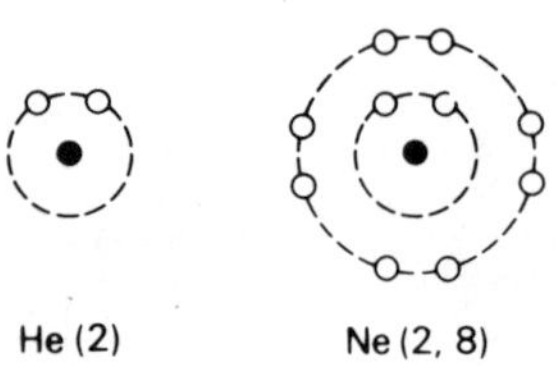

Figure 5.6 The electronic structures of helium and neon.

Atomic number	Element	Symbol	Number of electrons in shell 1 K	2 L	3 M	4 N	←Shell number ←Shell symbol
1	Hydrogen	H	1				
2	Helium	He	2				
3	Lithium	Li	2	1			
4	Beryllium	Be	2	2			
5	Boron	B	2	3			
6	Carbon	C	2	4			
7	Nitrogen	N	2	5			
8	Oxygen	O	2	6			
9	Fluorine	F	2	7			
10	Neon	Ne	2	8			
11	Sodium	Na	2	8	1		
12	Magnesium	Mg	2	8	2		
13	Aluminium	Al	2	8	3		
14	Silicon	Si	2	8	4		
15	Phosphorus	P	2	8	5		
16	Sulphur	S	2	8	6		
17	Chlorine	Cl	2	8	7		
18	Argon	Ar	2	8	8		
19	Potassium	K	2	8	8	1	
20	Calcium	Ca	2	8	8	2	

shell begins to fill when only 8 are in the third. (The element potassium has this electronic structure.) This pattern is repeated with heavier elements, and must mean that there is something special about 8 electrons in a shell.

The atoms with filled inner shells and 8 electrons in the outer shell are neon, argon, krypton, xenon and radon. These are all the noble gases and their chemistry is characterized by them all being very inert and taking part in very few (if any) chemical reactions.

Hence we say with some justification that an *octet* of electrons (i.e. 8) shows particular stability.

5.3 The periodic table

5.3.1 Introduction

Elsewhere in this book elements are classified in various ways. For example we speak of metallic elements, non-metallic elements, gaseous elements, solid elements, elements that react readily with, say, water, and elements that are inert. All of these classifications follow from the most comprehensive and fundamental method of grouping together the elements—*the periodic table.*

Before periodic classifications, chemistry was a jumble of unrelated facts. During the 18th century the ancient art of *alchemy* had developed to the point where some alchemists (who we can call early scientists) were no longer only concerned with converting other elements into gold but were making careful observations of how different substances behaved. Then in the 19th century, following John Dalton, atomic masses of the elements were being determined.

It was from attempts to relate the chemical and physical properties of the elements and their compounds to their atomic masses that the periodic table was born.

5.3.2 Early periodic classifications

(a) Döbereiner's Triads

In 1817 J. W. Döbereiner, a German chemist, noticed that certain trios (or triads) of elements had atomic masses that increased regularly, e.g.

(i) Chlorine (35·5), bromine (80) and iodine (127). The atomic mass of bromine is approximately half-way between those of chlorine and iodine. Also reference to chapter 20 will show that, both in its physical properties and in the vigour of its chemical reactions, bromine is between chlorine and iodine.

(ii) Calcium (40), strontium (88) and barium (137). Again, strontium has an atomic mass half-way between those of calcium and barium and has intermediate properties.

Some other trios of elements, such as iron, cobalt and nickel, have atomic masses that are very close to one another.

(b) Newlands' Law of Octaves

By 1863 atomic masses were more readily known, and a British chemist, John Newlands, arranged the elements that were available at that time in ascending order of their atomic masses, i.e. starting with hydrogen (atomic mass = 1). His list started thus:

H	Li	Be	B	C	N	O
F	Na	Mg	Al	Si	P	S
Cl	K	Ca	Cr	Ti	Mn	Fe

He then noticed a resemblance between each element and the one eight places farther on in the arrangement. As Newland put it: "The eighth element, starting from a given one, is a kind of repetition of the first, like the eighth note in an octave of music". For example, in the above table this *periodic* repetition of properties is obvious with lithium, sodium and potassium. It is less obvious, however, in some other cases and, as the table was continued, Newlands' rule did not hold true at all. Indeed, his ideas met with ridicule from some other scientists, one of whom suggested that it would be just as valuable to arrange the elements in alphabetical order!

5.3.3 Mendeleev's periodic table

(a) The periodic law

The *periodic law* on which the *periodic table* is based was discovered in 1869 almost at the same time and quite independently by a Russian chemist *Dmitri Ivanovich Mendeleev* (figure 5.7) and a German chemist, *Julius Lothar Meyer.*

It was Mendeleev who appreciated the important general significance of the law that, as stated by Mendeleev, "the properties of the elements are in periodic dependence upon their atomic masses". In his periodic table he left gaps where necessary to ensure that similar elements were grouped together. He also put some elements in positions not justified by the then accepted values for their atomic masses. Moreover, he altered the pattern of his table for elements that did not fit into a straightforward arrangement of the type proposed by Newlands.

(b) The periodic table

Mendeleev's periodic table is regarded by many as the beginning of modern chemistry, for as we shall see in the remainder of this chapter it brings an order to chemistry by collecting together the elements with similar properties. The periodic table illustrated in figure 5.8 is a copy of that published by Mendeleev in 1871. (The original is in German.)

You will see that the table is split into horizontal rows that Mendeleev called series (we now call these *periods* as shown on the left of figure 5.8). Vertically the table is split into *groups*.

Mendeleev collected together in a group elements with similar properties. By similar properties he meant chemical properties, in particular the ratios in which the elements combine with oxygen or hydrogen. To emphasize this he put the formula of the group oxide and hydride underneath the group number in his periodic table.

Figure 5.7 Dmitri Ivanovich Mendeleev (1834–1907). The stamp is one of four issued in the Soviet Union in 1934 on the 100th anniversary of Mendeleev's birth. The background is a periodic table. Is it Mendeleev's table? (Photo: A. Rendell.)

(c) Mendeleev's predictions

An important reason why Mendeleev was able to make a much more successful classification on the basis of atomic masses than had Newlands was that he left gaps where he thought this was necessary and was confident that elements would be discovered in the future to fill these gaps. By considering the properties of adjacent elements in the periodic table he predicted the properties of some of these undiscovered elements. Some of his predictions, e.g. for germanium (which he called eka-silicon) and gallium (eka-aluminium), were astonishingly accurate, as the table comparing Mendeleev's predictions with the properties actually found show for germanium:

Period	Series	Group 1 R_2O	Group 2 RO	Group 3 R_2O_3	Group 4 RH_4 RO_2	Group 5 RH_3 R_2O_5	Group 6 RH_2 RO_3	Group 7 RH R_2O_7	Group 8 — RO_4
1	1	H=1							
2	2	Li=7	Be=9.4	B=11	C=12	N=14	O=16	F=19	
3	3	Na=23	Mg=24	Al=27.3	Si=28	P=31	S=32	Cl=35.5	
4	4	K=39	Ca=40	—=44	Ti=48	V=51	Cr=52	Mn=55	Fe=56 Co=59 Ni=59 Cu=63
	5	(Cu=63)	Zn=65	—=68	—=72	As=75	Se=78	Br=80	
5	6	Rb=85	Sr=87	?Yt=88	Zr=90	Nb=94	Mo=96	—=100	Ru=104 Rh=104 Pd=106 Ag=108
	7	(Ag=108)	Cd=112	In=113	Sn=118	Sb=122	Te=125	I=127	
6	8	Cs=133	Ba=137	?Di=138	?Ce=140	—	—	—	— — — —
	9	(—)	—	—	—	—	—	—	
	10	—	—	?Er=178	?La=180	Ta=182	W=184	—	Os=195 Ir=197 Pt=198 Au=199
	11	(Au=199)	Hg=200	Tl=204	Pb=207	Bi=208			
7	12	—	—	—	Th=231	—	U=240	—	— — — —

Figure 5.8 Mendeleev's Periodic Table of 1871.

Property	Properties of eka-silicon (Mendeleev's predictions 1871)	Properties of Germanium (Discovered 1886)
Atomic mass	72	72·6
Density/g cm^{-3}	5·5	5·47
Specific heat	0·073	0·076
Atomic volume/cm^3	13	13·2
Colour	Grey	Grey-white
Dioxide formula	EsO_2	GeO_2
Dioxide colour	White	White
Density of dioxide/g cm^{-3}	4·7	4·7
Chloride formula	$EsCl_4$	$GeCl_4$
B. Pt of chloride/°C	Below 100	86
Density of chloride/g cm^{-3}	1·9	1·89
Ethyl compound formula	$Es(C_2H_5)_4$	$Ge(C_2H_5)_4$
B.Pt of ethyl compound/°C	160	160
Density of ethyl compound/g cm^{-3}	0·96	0·96

5.3.4 The periodic table

(a) Electronic structure

We have seen earlier in this chapter that the different elements have different numbers of electrons and that these are arranged in energy shells. The outermost shells are of highest energy, i.e. they are the most reactive. *It is these outermost electrons that take part in chemical reactions and determine the chemical properties of the elements.* It is, therefore, the *electronic structure of the atom* that is the fundamental feature on which the periodic table is based. Of course, Mendeleev did not know this, the electron not being discovered until 1897, 28 years after he had proposed his periodic table.

A periodic table of the form used these days is illustrated in figure 5.9 and at the back of the book. In this we have given the *names* and *symbols* of the elements together with their *atomic numbers* and *relative atomic masses.*

(b) The alkali metals

Within a group the elements are similar.

Let us consider the elements of group 1, lithium, sodium, potassium, rubidium and caesium. They are all very reactive metals with a valency of one. Their electronic structures are:

	Atomic number	Number of electrons in shell (n) 1	2	3	4	5	6
Lithium	3	2	1				
Sodium	11	2	8	1			
Potassium	19	2	8	8	1		
Rubidium	37	2	8	18	8	1	
Caesium	55	2	8	18	18	8	1

Thus an atom of each of these elements has one electron in the outer shell, i.e.

this electron is of higher energy, and therefore more easily removed, than the others. If it is removed the remaining species is a positively charged ion e.g.: $Na \rightarrow Na^+ + e$,
and the electronic structure now has 8 electrons in its highest energy shell. This, as we have seen, is a particularly stable number of electrons. The size of the atom increases with increase in the atomic number, hence the outer electron moves progressively further away from the nucleus on passing from lithium to caesium. Also the increasing number of inner electrons increasingly screen this outer electron. Because of these two factors it becomes progressively easier to remove the outer electron on moving from lithium to caesium, i.e. *the reactivity of the group 1 elements increases on going down the group*. Therefore potassium reacts more vigorously with water than does sodium, and caesium reacts with explosive violence.

The hydroxides of the metals in group 1 all dissolve in water to give alkaline solutions, hence their name *alkali metals* (see section (*f*) for a definition of a metal).

(c) The halogens

Fluorine, chlorine, bromine and iodine are, like the alkali metals, reactive elements with a valency of one. This, too, is expected from their electronic structures.

	Atomic number	Number of electrons in shell (n) 1	2	3	4	5
Fluorine	9	2	7			
Chlorine	17	2	8	7		
Bromine	35	2	8	18	7	
Iodine	53	2	8	18	18	7

An atom of each of these elements has seven electrons in its outer shell. They are just one short of the stable octet, and they react by gaining an electron, e.g.: $Cl + e \rightarrow Cl^-$, to give a negatively charged ion.

Like the alkali metals, and the elements in all other groups, the size of the atom increases as the group is descended. The ability to attract electrons is greatest in the small fluorine atom and least with iodine, hence *the reactivity of the group 7 elements decreases on going down the group*.

All of these elements readily form salts with the elements in the early groups (1, 2 and 3). Hence their name *halogens* (salt formers).

(d) The noble gases

Helium, neon, argon, krypton and xenon are five elements that were not known to Mendeleev. They are very unreactive and form very few compounds. This lack of reactivity and the fact that they are gases means that they are found principally in the atmosphere. (Some of these elements are also found with natural gas in certain parts of the world.) Their discovery did not take place until air could be liquefied during the early part of this century, although helium had been observed in the spectrum of the sun some years before.

Their lack of reactivity is readily understood in terms of their electronic structures as they all have a completely full outer octet.

Exercise. Write out a table of the electronic structures of the noble gases, as we have done for the alkali metals and halogens.

(e) Hydrogen

The first element in the periodic table, hydrogen, does not have properties that enable it to be placed satisfactorily in any group. This is due to two unique features. (i) The highest energy shell of a hydrogen atom can hold only two electrons, in contrast to all the others (except helium) that can hold 8 or more. (ii) When hydrogen loses its electron the ion formed, H^+, is a bare nucleus. This is very very small in contrast with a positive ion of any other element which must have some electrons still surrounding the nucleus.

(f) Metals and Non-metals

(i) We are now able to give a definition of metals and non-metals. *Metals* are those elements in the early groups of the periodic table.

Their atoms *tend to lose electrons* to form positive ions.

Non-metals are those elements in the later groups of the periodic table. Their atoms *tend to gain electrons* to form negative ions, rather than to lose electrons.

Metals and *non-metals* have other characteristics, such as metals being good conductors of electricity and non-metals being poor conductors of electricity. These features are discussed in section 21.1, where the type of bonding found in metals (the *metallic bond*) is also briefly discussed.

(ii) *Metallic characteristics increase as a group is descended.* We have already commented that the ease of forming a positive ion (a metallic characteristic) increases from lithium to caesium, and the ease of forming a negative ion (a non-metallic characteristic) decreases from fluorine to iodine. If a group near the middle of the periodic table is studied this increase in metallic character on descending group is even more striking. Consider group 4.

Carbon —definitely a non-metal.
Silicon —also a non-metal, but looks somewhat metallic.
Germanium —difficult to decide on inspection. It looks like a metal, but its structure is the same as carbon and silicon (it has a diamond structure (see section 6.2.7)).
Tin —definitely a metal.
Lead —definitely a metal.

(iii) As a period is crossed there is a gradual change in the properties of the elements from metallic to non-metallic, e.g. in period 3 (Na, Mg, Al, Si, P, S, Cl, Ar).

Sodium —a very reactive metal with a valency of 1 (e.g. NaH and Na_2O).
Magnesium —a reactive metal with a valency of 2 (e.g. MgO).
Aluminium —a less reactive metal with a valency of 3 (e.g. Al_2O_3).
Silicon —an unreactive non-metal with a valency of 4 (e.g. SiO_2, SiH_4).
Phosphorus —a reactive non-metal with valencies of 3 (e.g. PH_3) and of 5 (e.g. P_2O_5).
Sulphur —a reactive non-metal with valencies of 2 (e.g. H_2S) and 6 (e.g. SO_3).

Chlorine —a very reactive non-metal with a usual valency of 1 (e.g. HCl) but sometimes 7 (e.g. Cl_2O_7).

Argon —a noble gas with no known compounds.

Notice how the valency alters as the period is crossed, and how the valencies are connected with the number of electrons in the outer shell.

(g) The transition elements

These elements, as you can see from the periodic table (Figure 5.9), do not occur until period 4 and they are all *metals*. They show a close resemblance to one another. Generally they are hard, lustrous and have high melting points. Most of them have several valencies, and their ions in solution are coloured (in contrast to

KEY

Relative atomic mass → 63.55
Atomic number → 29Cu ← Symbol
Copper ← Name

1.008 1H Hydrogen

Group → / Period ↓	1	2							
2	6.939 3Li Lithium	9.012 4Be Beryllium							
3	22.99 11Na Sodium	24.31 12Mg Magnesium	← Transition metals—						
4	39.10 19K Potassium	40.08 20Ca Calcium	44.96 21Sc Scandium	47.90 22Ti Titanium	50.94 23V Vanadium	52.00 24Cr Chromium	54.94 25Mn Manganese	55.85 26Fe Iron	58.93 27Co Cobalt
5	85.47 37Rb Rubidium	87.62 38Sr Strontium	88.91 39Y Yttrium	91.22 40Zr Zirconium	92.91 41Nb Niobium	95.94 42Mo Molybde-num	98.91 43Tc Technetium	101.1 44Ru Ruthenium	102.9 45Rh Rhodium
6	132.9 55Cs Caesium	137.3 56Ba Barium	(see below) 57-71	178.5 72Hf Hafnium	180.9 73Ta Tantalum	183.9 74W Tungsten	186.2 75Re Rhenium	190.2 76Os Osmium	192.2 77Ir Iridium
7	(223) 87Fr Francium	226.0 88Ra Radium	(see below) 89-103						

Lanthanides →	138.9 57La Lanthanum	140.1 58Ce Cerium	140.9 59Pr Praseody-mium	144.2 60Nd Neodymium	(147) 61Pm Promethium	150.4 62Sm Samarium	152.0 63Eu Europium
Actinides →	(227) 89Ac Actinium	232.0 90Th Thorium	231.0 91Pa Protactinium	238.0 92U Uranium	237.0 93Np Neptunium	(242) 94Pu Plutonium	(243) 95Am Americium

ions of all the other elements that are colourless). Many of these metals are useful as catalysts (see section 10.4).

(h) Atomic masses, atomic numbers and the Mendeleev classification
Mendeleev, as we have already explained, did not put all the elements in the position which seemed correct according to the atomic masses, but he put them where their chemical properties best fitted. He believed that such *atomic masses* had been inaccurately determined and would, in time, be corrected. In most cases this happened, but there are still four pairs of elements (see exercise below) whose positions would need reversing to place them in order of their atomic masses. It should be clear to you by now that the *atomic mass* is not the fundamental

Figure 5.9 The Periodic Table (A list of the elements in alphabetical order is given at the back of the book).

			3	4	5	6	7	0
								4.003 $_{2}$He Helium
			10.811 $_{5}$B Boron	12.01 $_{6}$C Carbon	14.00 $_{7}$N Nitrogen	16.00 $_{8}$O Oxygen	19.00 $_{9}$F Fluorine	20.18 $_{10}$Ne Neon
			26.98 $_{13}$Al Aluminium	28.09 $_{14}$Si Silicon	30.97 $_{15}$P Phosphorus	32.06 $_{16}$S Sulphur	35.45 $_{17}$Cl Chlorine	39.95 $_{18}$Ar Argon
58.71 $_{28}$Ni Nickel	63.55 $_{29}$Cu Copper	65.37 $_{30}$Zn Zinc	69.72 $_{31}$Ga Gallium	72.59 $_{32}$Ge Germanium	74.92 $_{33}$As Arsenic	78.96 $_{34}$Se Selenium	79.90 $_{35}$Br Bromine	83.80 $_{36}$Kr Krypton
106.4 $_{46}$Pd Palladium	107.9 $_{47}$Ag Silver	112.4 $_{48}$Cd Cadmium	114.8 $_{49}$In Indium	118.7 $_{50}$Sn Tin	121.8 $_{51}$Sb Antimony	127.6 $_{52}$Te Tellurium	126.9 $_{53}$I Iodine	131.3. $_{54}$Xe Xenon
195.1 $_{78}$Pt Platinum	197.0 $_{79}$Au Gold	200.6 $_{80}$Hg Mercury	204.4 $_{81}$Ti Thallium	207.2 $_{82}$Pb Lead	209.0 $_{83}$Bi Bismuth	(210) $_{84}$Po Polonium	(210) $_{35}$At Astatine	(222) $_{86}$Rn Radon

157.3 $_{64}$Gd Gadolinium	158.9 $_{65}$Tb Terbium	162.5 $_{66}$Dy Dysprosium	164.9 $_{67}$Ho Holmium	167.3 $_{68}$Er Erbium	168.9 $_{69}$Tm Thulium	173.0 $_{70}$Yb Ytterbium	175.0 $_{71}$Lu Lutetium
(247) $_{96}$Cm Curium	(247) $_{97}$Bk Berkelium	(251) $_{98}$Cf Californium	(254) $_{99}$Es Einsteinium	(253) $_{100}$Fm Fermium	(256) $_{101}$Md Mendele-vium	(254) $_{102}$No Nobelium	(257) $_{103}$Lw Lawrencium

property on which the periodic table is based. The *fundamental property* is the *electronic structure*, or the *charge on the nucleus*. This is given by the *atomic number*. It was fortunate for Mendeleev that the order of atomic masses and atomic numbers are almost identical. Otherwise his classification would not have been possible.

(i) Physical properties of the elements in periods 1–3

At the same time that Mendeleev was working on his periodic table, the German chemist, Julius Lothar Meyer, also noticed the way in which certain physical properties of the elements varied with their atomic masses. He plotted curves of the values of various properties of the elements, e.g. boiling point and melting point, against their masses and noticed a repetition of physical properties in the same way that Mendeleev had done with both chemical and physical properties.

Figures 5.10–5.13 are a modern version of Lothar Meyer's curves. In them it can be seen that elements with similar chemical properties, e.g. lithium and sodium, occupy similar positions on the curves. Notice that: (i) The melting points and boiling points rise as the period is crossed until the element in group IV is reached. The melting points and boiling points of the elements in the other groups are very low. (ii) The densities rise to groups III and IV. After group IV they are of low value. (iii) The molar volumes fall as group III is approached. Then they rise again as the period is crossed.

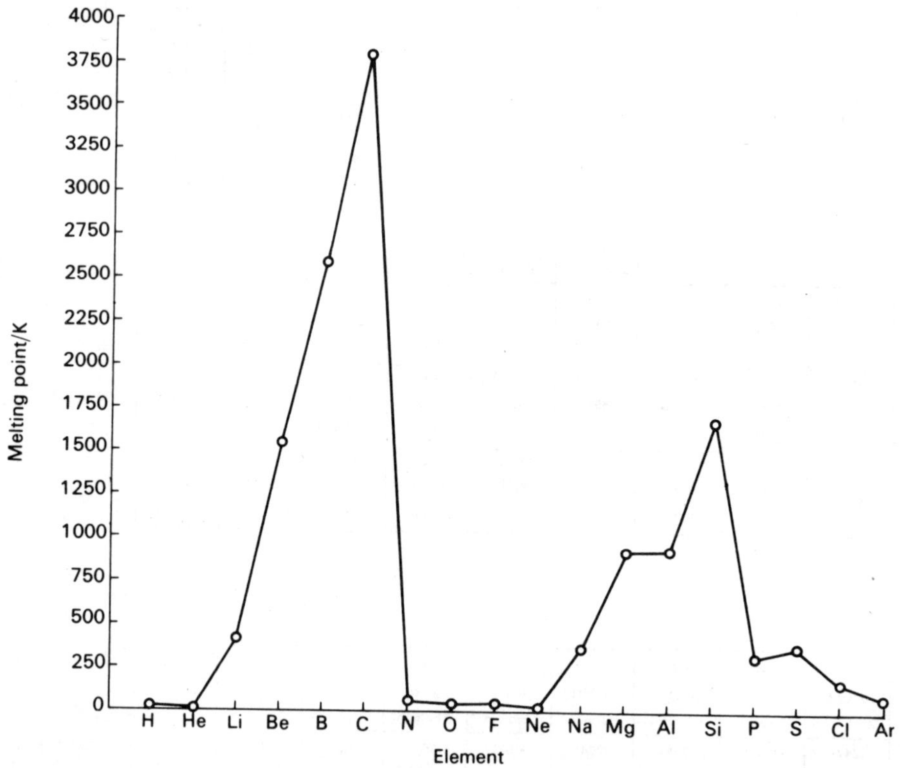

Figure 5.10 The melting points (K) of elements with atomic numbers 1–20.

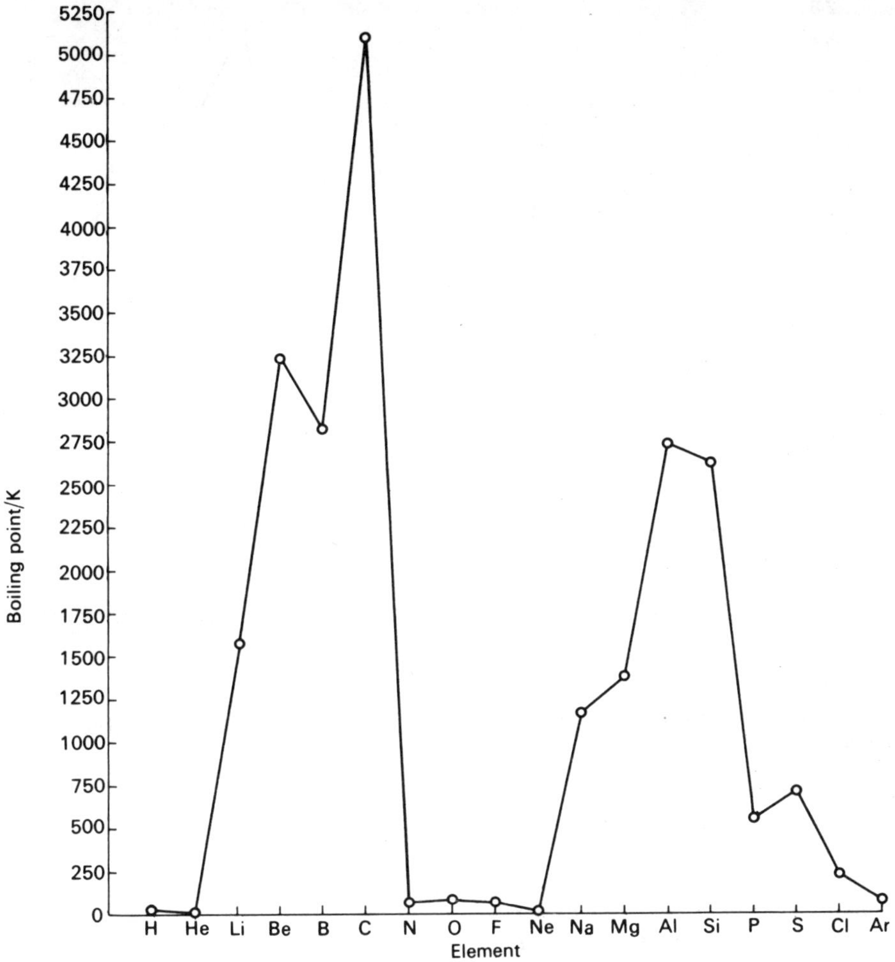

Figure 5.11 The boiling points (K) of elements with atomic numbers 1–20.

Exercise—The elements and the periodic table

(a) Using the modern periodic table give the 4 pairs of elements where the atomic masses are in the opposite order to the atomic numbers.

(b) How can this be so?

(c) Comparing the Mendeleev and modern periodic tables, how many elements were known to Mendeleev and how many have been discovered since 1871?

(d) You may wonder why the elements have the names that they do. There is a rich history connected with many of these names. Some elements have been known for many centuries and the origins of their names is lost in antiquity. Others have been named after places, people and for a variety of others reasons. Many of the names are derived from Latin or Greek words. See how many elements you can find that fit into these different groups (the number of elements in each group is given in brackets):

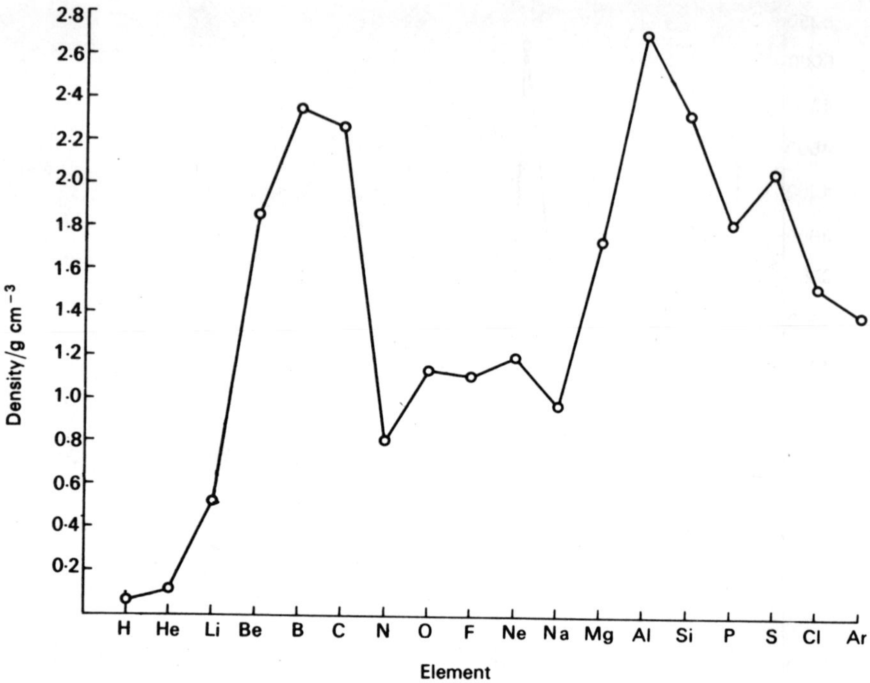

Figure 5.12 The densities (g cm^{-3}) of elements with atomic numbers 1–20. (Densities of gases given at their boiling points.)

(i) *Elements known for many centuries (16th century or earlier).* (Clue: *It is likely that their symbol will not have the same letters as their name). (12)*
(ii) *Elements named after countries, towns or other places. (23)*
(iii) *Elements named after scientists. (7)*
(iv) *Elements named after planets and other bodies in the Universe. (10)*
(v) *Elements named after colours. (13)*
(vi) *Elements named after men or women of mythology. (6)*
(vii) *(A very difficult question.) Elements that are mentioned in the Old Testament of the Bible* (Clue: *The answer to part (i) should help). (8)*

Answers

(*a*) *Argon and potassium, cobalt and nickel, tellurium and iodine, thorium and protoactinium.*

(*b*) *Because the atomic masses include the neutrons, and the first element of each pair must have, on average, more than 1 neutron more than the second. Consider the first pair.* 99.6 % *of naturally occurring argon is* $^{40}_{18}Ar$ *(i.e. Argon with 18 protons and 22 neutrons),* 93.1 % *of naturally occurring potassium is* $^{39}_{19}K$ *(i.e. Potassium with 19 protons and 20 neutrons). Hence the atomic mass of argon is more than that of potassium. Similarly for cobalt and nickel. All cobalt is* $^{59}_{27}Co$ (27p, 32n) *nickel's*

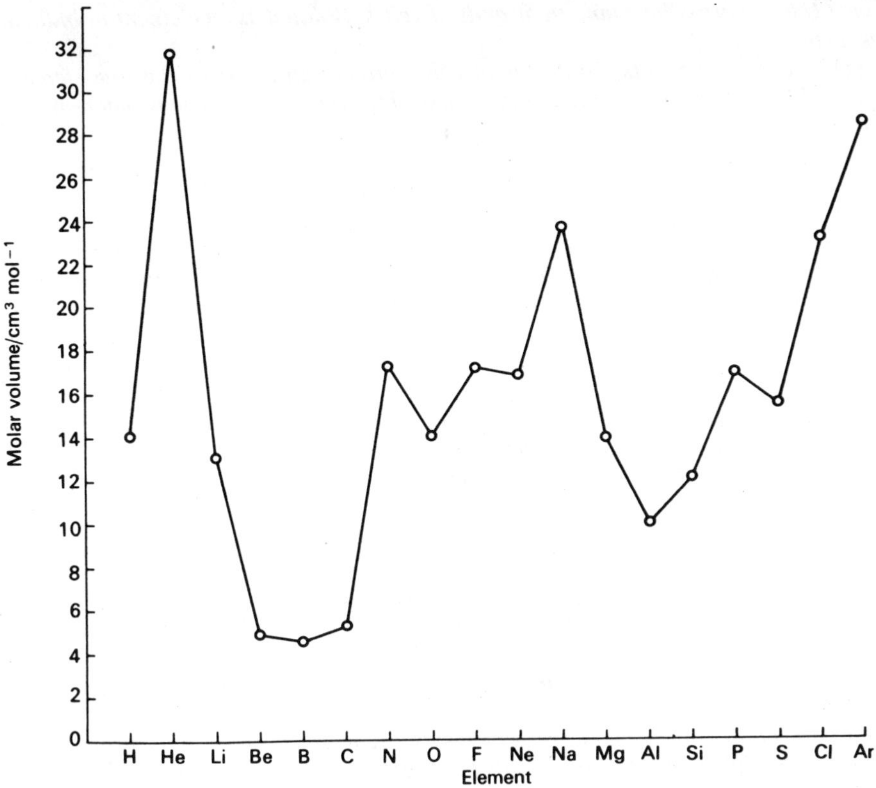

Figure 5.13 Molar volumes ($cm^3\ mol^{-1}$) of elements with atomic numbers 1–20. (Molar volumes of gases calculated at their boiling points.)

main isotopes are $^{58}_{28}Ni$ (28p, 30n) (68%) *and* $^{60}_{28}Ni$ (28p, 32n) (26%). *Hence naturally occurring cobalt has a higher* atomic mass *than does nickel.*

(*c*) *Mendeleev knew of 63 elements. The total number now known is 103, but only 89 occur naturally. (Technetium, astatine and promethium and elements 93 onwards do not.) Hence there were 26 elements still undiscovered in the earth's crust and atmosphere in Mendeleev's day.*

(*d*) (i) *Antimony (There was a book written about it in 1604 AD, probably known much earlier.)*
Arsenic (Obtained in about 1250 AD.)
Carbon (Charcoal and soot have been known since early man; diamonds were discovered about 2000 years ago in India and Borneo.)
Copper (Mined for more than 5000 years.)
Gold (4000–5000 BC, many gold objects found in Egyptian Tombs.)
Iron (Since about 3000 BC.)
Lead (about 3500 BC, used by many ancient civilizations.)
Mercury (1500 BC, in Egyptian tomb of Kurna.)
Silver (about 4000 BC.)
Sulphur (known to all ancient civilizations.)
Tin (Used for bronze since about 3000 BC, isolated as an element about 2000 BC.)

Zinc (Used in brass for many thousands of years, isolated as an element in India in the 13th century.)

(Of these 12 elements, 10 are metals that either can occur as the free element (e.g. gold and mercury) or are easily produced from their ores. The two non-metals are unique in that they can occur as the free element.)

(ii) *Americium (America), Berkelium (Berkeley, home town of the University of California where this element was first prepared), Californium (California—State in USA), Copper (Cyprus (latin)), Erbium (Ytterby, a village in Sweden), Europium (Europe), Francium (France), Gallium (France (latin)), Germanium (Germany), Hafnium (Copenhagen (latin)), Holmium (Stockholm (latin)), Lutetium (Paris (ancient name)), Magnesium (Magnesia, a district in Ancient Greece)), Manganese (Magnesia), Polonium (Poland), Rhenium (River Rhine (latin)), Ruthenium (Russia (latin)), Scandium (Scandinavia), Strontium (Strontian, a village in Scotland), Terbium (Ytterby, a village in Sweden), Thulium (Scandinavia (ancient name)), Ytterbium (Ytterby, a village in Sweden), Yttrium (Ytterby, a village in Sweden).*

(Most of these are where compounds of the element were first discovered, or where the element was first isolated or made.)

(iii) *Curium (Pierre and Marie Curie), Einsteinium (Albert Einstein), Fermium (Enrico Fermi, a famous nuclear physicist), Gadolinium (Gadolin, a Finnish chemist), Lawrencium (E. O. Lawrence, the inventor of the cyclotron), Mendelevium (Dmitri Mendeleev), Nobelium (Alfred Nobel, the inventor of dynamite).*

(All except Gadolinium are the heavy unstable elements that have been made for the first time since 1944).

(iv) *Cerium (the asteroid Ceres which was discovered in 1801 only 2 years before the element), Helium (the Sun (greek)), Mercury (planet Mercury), Neptunium (planet Neptune), Palladium (the asteroid Pallas, discovered about the same time), Phosphorus (planet Venus (ancient name)), Plutonium (planet Pluto), Selenium (the moon (greek)), Tellurium (the Earth (latin)), Uranium (planet Uranus).*

(v) *Bismuth (white mass (german)), Caesium (sky blue (latin)), Chlorine (greenish-yellow (greek)), Gold (gold (anglo-saxon)), Indium (indigo), Iodine (violet (greek)), Platinum (silver (spanish)), Praseodymium (green twin (greek)), Rhodium (rose (greek)), Rubidium (deepest-red (latin)), Silver (anglo-saxon)), Thallium (green twig (greek)), Zirconium (gold colour (arabic)).*

(The name may be the colour of the element, (e.g. chlorine and iodine), or of a compound (e.g. Zirconium after the yellow gemstone zircon), or after the colour that the element gives to a flame (e.g. Caesium and Indium). It is likely that the colours gold and silver were named after the elements and not vice versa.) (Also chromium is named after colour (greek) and iridium after rainbow (latin).)

(vi) *Niobium (Niobe, daughter of Tantalus), Promethium (Prometheus, who, according to mythology stole fire from heaven and gave it to mankind), Tantalum (Tantalus, a king in Greek mythology), Thorium (Thor, the god of war), Titanium (Titans, two men of gigantic strength), Vanadium (Vanadis, a Scandinavian goddess).*

(vii) *Carbon (as Diamond) (*Ex. *28,* 18*; 39,* 11*;* Ezek. *28,* 13*;* Jer. *17,* 1*). Copper (*Ezra *8, 27).*

Gold (many times, e.g. Gen. *2,* 11–12*;* Job *28,* 1–19*).*

Iron (many times, e.g. Gen. *4*, 22; Job *19*, 24; *28*, 2; Jer *17*, 1*)*.
*Lead (*Ex. *15*, 10; Job *19*, 24*)*.
Silver (many times, e.g. Ezra *8*, 26; Job *28*, 1; Jer. *32*, 9*)*.
*Sulphur (as Brimstone) (*Gen. *19*, 24; Job *18*, 15; Isa. *30*, 33; Ezek. *38*, 22*)*.
*Tin (*Num. *31*, 22*)*.

N.B. *If this exercise has interested you try to find the origin of the names of other elements (a Chambers Twentieth Century Dictionary will be very useful for this purpose)*.

Questions on chapter 5

1. Chlorine has an atomic number of 17 and exists mainly in two isotopic forms, A and B, of atomic masses 35 and 37 respectively. State the number of
(*a*) electrons in each atom of A,
(*b*) protons in each atom of B,
(*c*) neutrons in each atom of B,
(*d*) electrons in each $^{35}_{17}Cl^-$ ion.
Calculate the A : B ratio in ordinary chlorine gas (molecular mass = 71). (*J.M.B.*)

2. Complete the following table:

Particle	Mass No.	Atomic No.	Protons	Neutrons	Electrons
Oxygen atom	16	8			
Calcium ion		20		20	
Bromide ion	80		35		

(*J.M.B.*)

3. Strontium (Sr) is a reactive metal with a valency of 2.
Phosphorus (P) is a reactive non-metal with an atomic mass of 31.
Both strontium and phosphorus burn in oxygen to give oxides and both of these oxides react with water to give soluble compounds.
The oxide of phosphorus contains 56·3% by mass of oxygen.
Strontium phosphide—formula Sr_3P_2—is a solid with a high melting point. It reacts with water to give a gas of formula PH_3 and a strontium compound *X* as the only products.
(i) Write the formula for strontium oxide and *calculate* the formula for the oxide of phosphorus from the percentage given above.
(ii) Write equations for the reactions you would expect between water and *each* of the above oxides. Give a simple test by which you could distinguish between the solutions formed.
(iii) Write the formula you would expect for compound *X* and the equation for the reaction of compound Sr_3P_2 with water.
(iv) Calculate the volume at s.t.p. of the gas PH_3 (which is insoluble in water) formed when one mole of Sr_3P_2 reacts with an excess of water and the volume of 1·0M hydrochloric acid required to neutralize the resulting solution.
(v) Name the type of bonding you would expect in compound Sr_3P_2. Give a reason for your answer.
(vi) What valencies does phosphorus show in the above reactions? (O = 16. One mole of a gas at s.t.p. occupies 22·4 litres). (*C.*)

4. Silver is an element which exists naturally as a mixture of two isotopic forms. *A* and *B* represent atoms of these two isotopes. They occur in equal numbers.

A $^{107}_{47}Ag$ *B* $^{109}_{47}Ag$

In each case, the upper figure is the mass number and the lower figure is the atomic number.

State the number of (*a*) protons in atom *A*, (*b*) electrons in atom *B*, (*c*) neutrons in atom *A*, (*d*) neutrons in atom *B*.

What is the atomic mass of naturally occurring silver? (*J.M.B.*)

5. For the Main Groups of the periodic table, the metallic properties of the elements vary approximately with their positions as shown in the chart below.

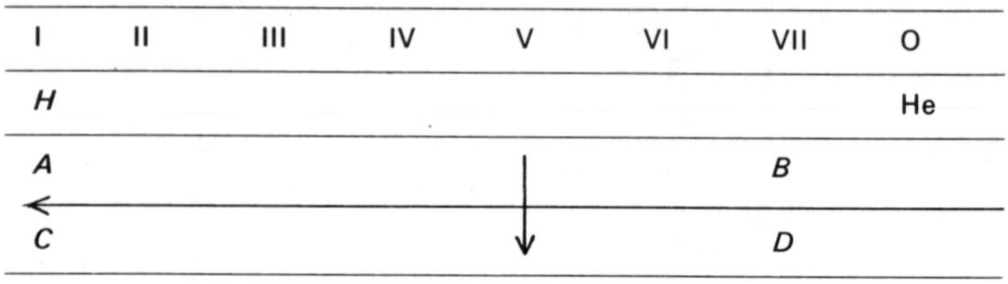

The direction of the arrows indicates an increase in metallic nature.

(*a*) Will the most metallic element be found at *A*, *B*, *C* or *D*?

(*b*) Will the most non-metallic element be found at *A*, *B*, *C* or *D*?

(*c*) The element indium has the symbol In. It is in the same group as aluminium but, whereas aluminium is the second element in the group, indium is the fourth.

(i) What is the formula for indium oxide?

(ii) Would you expect this oxide to be more basic or less basic than aluminium oxide?

(*d*) The names and symbols of the first five elements in Group II of the periodic table, reading from top to bottom, are beryllium, Be; magnesium, Mg; Calcium, Ca; strontium, Sr; barium, Ba.

(i) Which of these five elements will have the chloride of highest melting point?

(ii) Write the formula for this chloride. (*J.M.B.*)

6. Explain what is understood by the terms (*a*) atom, (*b*) atomic number, (*c*) relative atomic mass.

Discuss briefly, giving examples, the variations of chemical and physical properties to be found in (i) a period, (ii) a group of the periodic classification.

7. The elements sodium, magnesium, aluminium, silicon, phosphorus, sulphur, chlorine and argon form a series in the periodic table.

For each of *four* of these elements state:

(*a*) the electronic structure;

(*b*) the formula for its hydride and the reaction, if any, between the hydride and water,

(*c*) the formula of an oxide and whether this oxide is acidic, basic, amphoteric or neutral. (*J.M.B.*)

8. The atomic numbers of sodium and chlorine are 11 and 17 respectively.

(*a*) Give the electronic structures of the atoms of sodium and chlorine.

(*b*) Why do these elements show a valency of one in their ionic (electrovalent) compounds?

(*c*) Sketch the crystal lattice of sodium chloride.

(*d*) Explain how the crystal structure accounts for the high melting point of sodium chloride.

Describe and explain how sodium hydroxide is obtained from sodium chloride by an electrolytic process. (*A.E.B.*)

6

The Electronic Theory of Valency and Radioactivity

6.1 Introduction

This chapter consists of two topics. The first, *the Electronic Theory of Valency,* examines the various ways in which molecules can be formed from atoms. The second, *Radioactivity*, looks at the way in which reactions that involve the nucleus of atoms can take place.

6.2 The Electronic theory of valency

6.2.1 How atoms combine

Most of the chemicals that you have met in the laboratory can be divided into two classes, according to their physical properties. Members of the first class of substances are usually crystalline solids. They have high melting points, and, when melted or dissolved in water, allow the passage of an electric current. They do not dissolve in organic solvents. Sodium chloride (common salt), sodium hydroxide (caustic soda), and calcium chloride are common examples.

The other class contains solids, liquids and gases with low melting and boiling points. Only a few of them dissolve in water, and unless there is some sort of combination with the water, as with hydrogen chloride, the solutions do not conduct electricity. None of them will conduct electricity when liquefied. Such substances will often dissolve in organic solvents. Ammonia, methane, chlorine and organic compounds in general belong to this class. The atoms combine in quite different ways in these two classes, and the compounds formed are said to be *electrovalent* (*or ionic*) and *covalent* respectively.

6.2.2 Electrovalent compounds and ions

Atoms combine because by so doing they decrease their total energy, i.e. they achieve a more stable state. We saw in section 5.2.7 that the electronic structures of the noble gases (octets of electrons in the outer shell) are the most stable electronic structures and electrovalent compounds are formed when atoms gain or lose electrons to attain a *stable octet*.

Consider the combination of sodium and chlorine atoms. The sodium atom has one electron in its highest energy shell, the chlorine atom has seven (figure 6.1). To achieve a noble gas structure, the sodium atom must either gain seven electrons, or lose one. It is energetically far preferable to lose one and it then has the electronic structure of neon. Similarly the chlorine atom gains one electron rather than lose seven electrons, and it then has the electronic structure of argon.

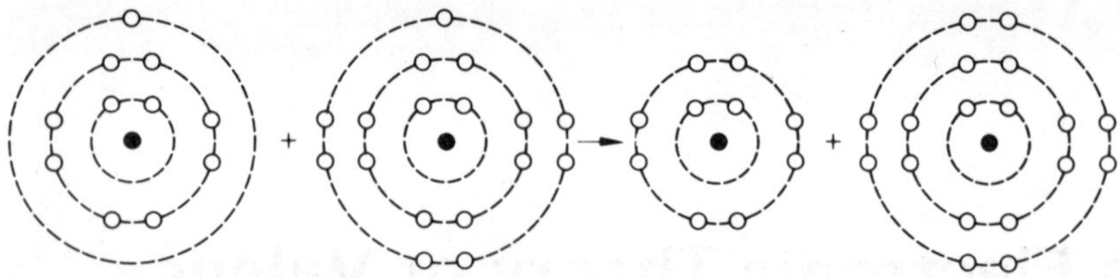

Figure 6.1 The interaction of sodium and chlorine atoms.

As the sodium atom loses one electron and the chlorine atom gains one electron they will react together in a 1 : 1 ratio, every sodium atom losing 1 electron to a chlorine atom. As both atoms are, of course, neutral the particle formed by sodium will have one more proton than electrons and it will have a net positive charge equal to the charge of a proton (i.e. Na^+). Similarly the chlorine atom will become negatively charged because of its extra electron (i.e. Cl^-).

These charged particles, formed when atoms gain or lose electrons are called *ions*. Oppositely charged ions attract one another by *electrostatic attraction*. Electrovalent compounds are, therefore, held together by electrostatic attraction between the oppositely charged ions. The high melting points of these compounds (sodium chloride will just melt in a Bunsen flame) is due to the necessity to break the strong electrostatic attractive forces. The electrostatic forces are more easily broken by water molecules and hence electrovalent compounds often dissolve in water.

Two typical *ionic crystal lattices* are those of sodium chloride (figure 6.2) and caesium chloride (figure 6.3). In both of these crystals the positive sodium or caesium ions and the negative chloride ions arrange themselves so that they are equally attracted in all directions. In sodium chloride (figure 6.2) each Na^+ ion is in the centre of 6 Cl^- ions and vice versa. In caesium chloride (figure 6.3) each Cs^+ ion is in the centre of 8 Cl^- ions and vice-versa.

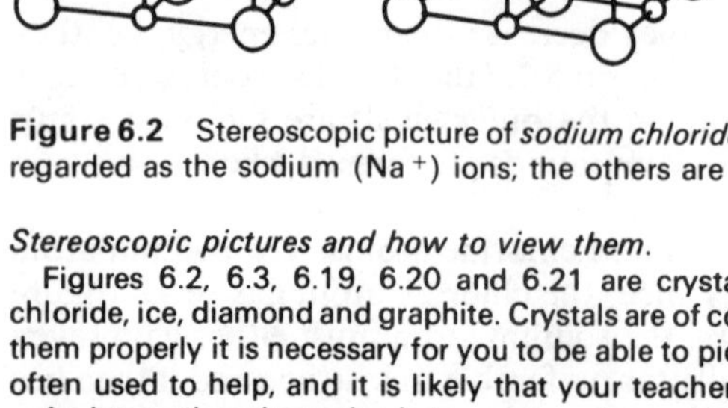

Figure 6.2 Stereoscopic picture of *sodium chloride*. Either the large or the small circles can be regarded as the sodium (Na^+) ions; the others are the chloride (Cl^-) ions.

Stereoscopic pictures and how to view them.

Figures 6.2, 6.3, 6.19, 6.20 and 6.21 are crystal structures of sodium chloride, caesium chloride, ice, diamond and graphite. Crystals are of course three-dimensional, and to understand them properly it is necessary for you to be able to picture them in three dimensions. Models are often used to help, and it is likely that your teacher has models of at least some of these.

An interesting alternative is to use two-dimensional drawings and somehow trick your brain into thinking that they are three-dimensional.

To do this we need two drawings that are slightly different. The one that the left eye sees is what the left eye would see if looking at a three dimensional object and the drawing that the right eye sees is what the right eye would see if looking at the three-dimensional object. It is not easy to work out exactly how these two drawings should differ to give the desired result. The most satisfactory way, and the way used to produce these drawings, is to have all the calculations done by a large computer. The figures here were actually drawn by the computer at the University of Hull.

To view these pictures in three dimensions it is necessary for your left eye to look at the left hand picture, and your right eye to look at the right hand picture, and to make your brain think that your two eyes are looking at the same picture. To do this is quite simple. Hold a postcard (or a piece of cardboard about 15 cm × 10 cm) with its short edge between the pictures and put your nose so that it is near the postcard. Your eyes are then about 15 cm above the picture and the postcard prevents each of your eyes from seeing the wrong picture. Hold the book so that the two pictures are evenly illuminated. Look at the pictures and allow your eyes to relax. The two images in your brain will fuse together to give one image that will be stereoscopic, i.e., some of the picture will appear to stand out from the book and some to go below the page.

Once you have discovered the knack of viewing these pictures it is easy, but at first you may have difficulty in getting the two images to fuse together in your brain. The secret is to get your eyes to relax. Do not stare at the pictures, but try to look beyond them (i.e. have them out of focus); then, when the two out-of-focus images have fused together, slowly focus your eyes on to the one stereoscopic picture.

After practice there is no need to have the postcard between the pictures. Your brain then thinks that it can see three pictures, a three-dimensional one in the middle and two two-dimensional ones on either side!

If, after trying hard, you cannot see stereoscopic pictures, then do not worry. About 10% of us cannot, usually because one eye is much more dominant than the other. Just study the pictures in two dimensions.

You will see from figures 6.2 and 6.3 that it is impossible to say that a particular sodium (or caesium) ion "belongs" to a particular chloride ion, as they are equidistant between 6 (or 8). This means that "molecules" of an electrovalent compound are not formed. The "molecule" is taken to be the atoms in their combining ratio, i.e., as given by the formula (NaCl or CsCl).

Electrovalent compounds are formed between *metals* that can attain stable electronic structures by losing one or two electrons and the *halogens* and other species (such as OH^-) that attain a stable electronic structure by having one or two excess electrons. Other ionic compounds, therefore, are magnesium fluoride ($Mg^{2+}\ 2F^-$), Calcium chloride ($Ca^{2+}\ 2Cl^-$) (figure 6.4), potassium hydroxide ($K^+\ OH^-$), Calcium Oxide ($Ca^{2+}\ O^{2-}$) and sodamide ($Na^+\ NH_2{}^-$).

Exercise: Draw diagrams to show the electronic structures of the ions in magnesium fluoride, potassium hydroxide, calcium oxide and sodamide.

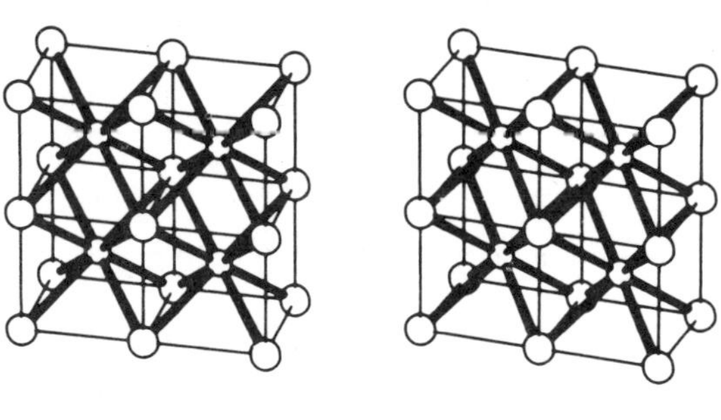

Figure 6.3 Stereoscopic picture of *caesium chloride*. Either the large or the small circles can be regarded as the caesium (Cs^+) ions; the others are the chloride (Cl^-) ions.

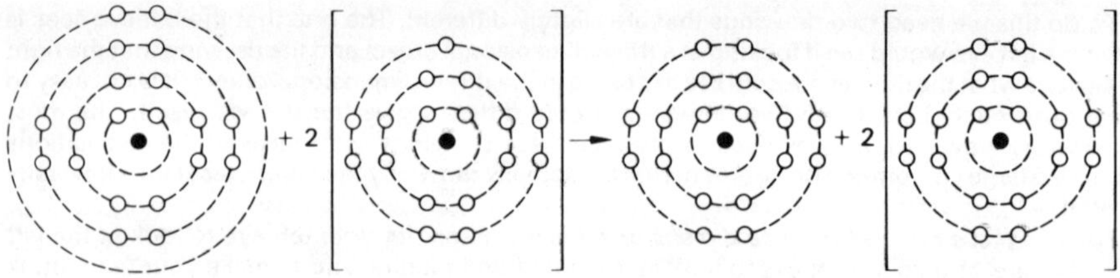

Figure 6.4 The interaction of calcium and chlorine atoms.

6.2.3 Covalent compounds

This is the second large group of compounds. When these are formed electrons on the two combining atoms are *shared.*

Thus two chlorine atoms combine as illustrated in figure 6.5, each atom contributing one electron to the pair of shared electrons. The shared electrons then belong to both atoms and in this way each chlorine atom now has eight electrons (the stable octet) in its outer shell. *It is in this way that molecules are formed.* There is little attraction *between* a molecule and its neighbours, and this accounts for the fact that most covalent compounds are gases or liquids (see section 6.2.6). Other covalent compounds are hydrogen chloride (figure 6.6), water (figure 6.7), ammonia, methane and boron trifluoride.

For hydrogen chloride (figure 6.6) and water (figure 6.7) we have drawn in all the electron shells. The hydrogen atoms in both of these molecules now have 2

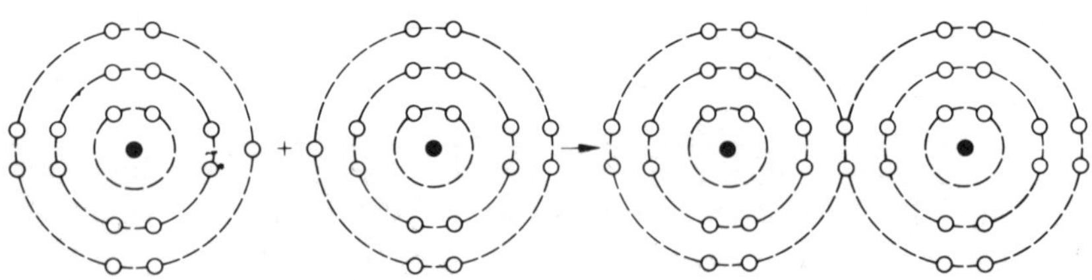

Figure 6.5 Formation of the chlorine molecule.

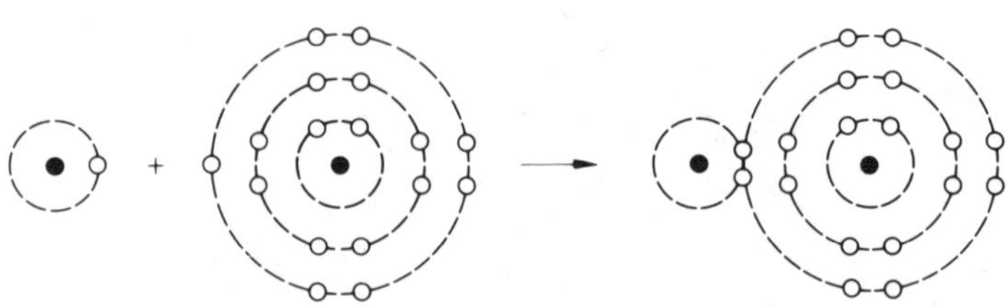

Figure 6.6 Formation of the hydrogen chloride molecule.

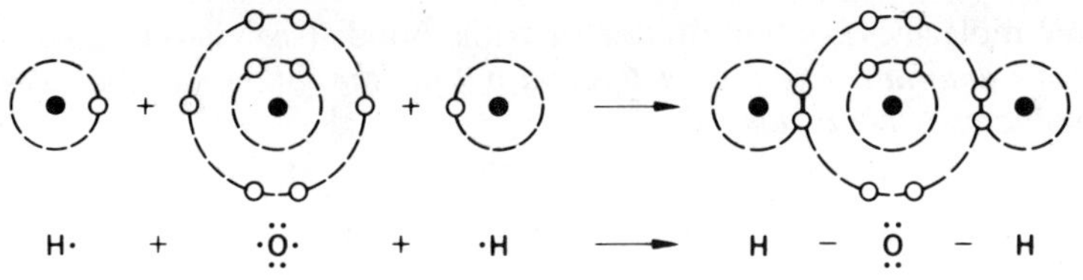

Figure 6.7 Formation of the water molecule.

electrons, i.e., they have the electronic structure of helium, and the chlorine and oxygen atoms each have the octet (8) of electrons in their outer shell. You can see from the three examples in figures 6.5, 6.6 and 6.7 that the electrons in the *inner shells* take no part in the formation of the covalent bond, and frequently these are omitted from diagrams illustrating covalent bond formation—the outer-shell electrons only are depicted surrounding the symbol for the element. When a covalent bond is formed, the electrons forming this bond are depicted as a line between the two atoms. In this way we get the diagrams as shown at the bottom of figures 6.5, 6.6 and 6.7.

Exercise: Draw diagrams like figure 6.7 for the formation of ammonia (NH_3), methane (CH_4), and boron trifluoride (BF_3) (notice that in this molecule there are only 6 electrons in the outer shell).

In all these examples above, each pair of atoms has shared 2 electrons. The type of bond formed when 2 electrons are shared is called a *single bond.* It is also possible for a pair of atoms to share 4 electrons, when a *double bond* is formed, or to share 6 electrons when a *triple bond* is formed.

An example of molecule containing double bonds is carbon dioxide (figure 6.8). Here each oxygen atom shares 4 electrons with the carbon atom and consequently all three atoms have an octet of electrons in their outer shells.

An example of a molecule containing a triple bond is the nitrogen molecule (figure 6.9). Here each nitrogen atom shares 6 electrons with the other nitrogen atom, and both atoms then have an octet of electrons.

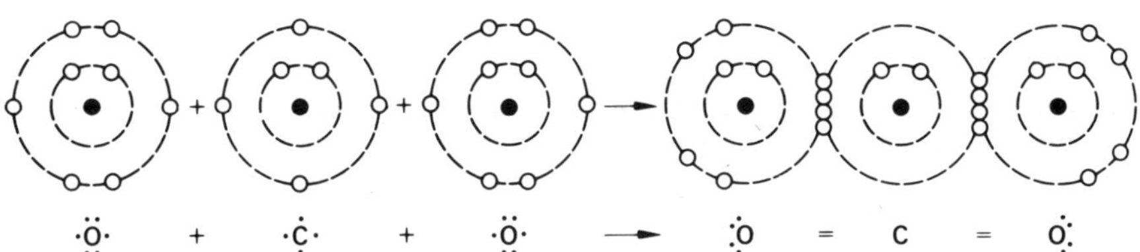

Figure 6.8 Formation of the carbon dioxide molecule.

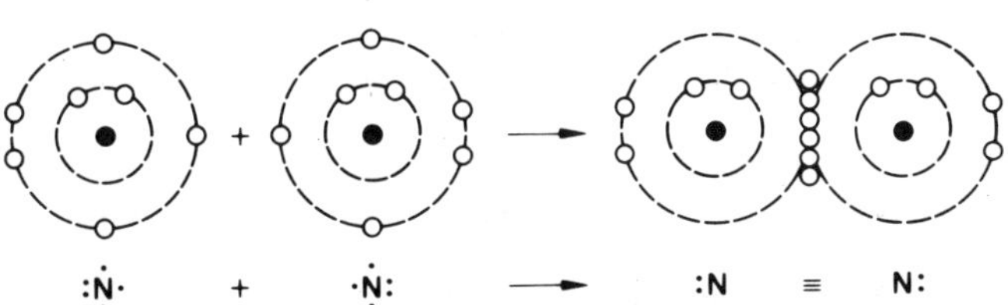

Figure 6.9 Formation of the nitrogen molecule.

Many organic molecules contain double or triple bonds (see chapter 25).

Exercise: Draw diagrams like figures 6.8 and 6.9 for the following molecules that contain double or triple bonds:

Oxygen (O = O), *hydrogen cyanide* (H—C≡N), *methanal* (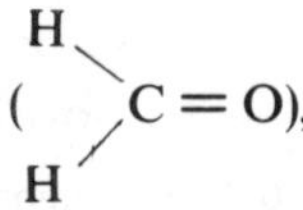),

Ethene (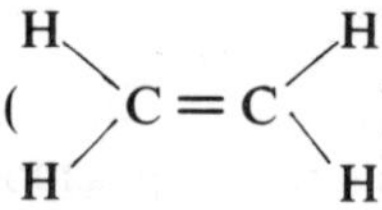), *Ethyne* (H—C≡C—H).

6.2.4 The coordinate bond

In the last section the *covalent bond* was formed by *each* of the two atoms that formed the bond *contributing an equal number of electrons to the bond.*

It is also possible to form a covalent bond where *both electrons of the bond are contributed by one atom.* This is called a *coordinate covalent bond.* Consider the two molecules ammonia and boron trifluoride (figure 6.10).

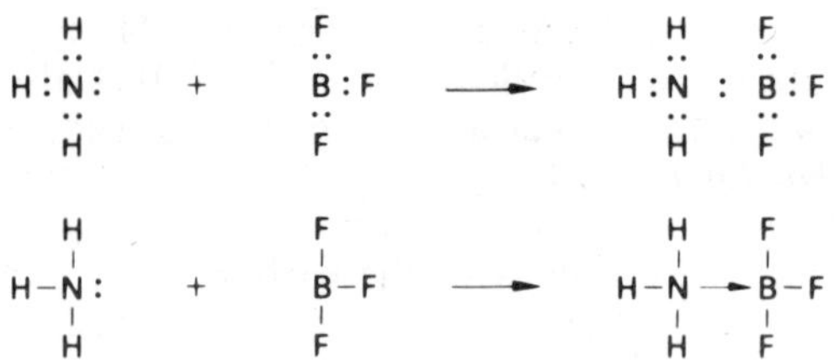

Figure 6.10 Formation of a coordinate bond.

The ammonia molecule has a pair of electrons in the outer shell of the nitrogen atom that are not being used for bonding, and the boron trifluoride molecule has only 6 electrons in the outer shell of the boron. They can, therefore, react as shown in figure 6.10 such that the spare pair of electrons on the nitrogen atom is shared with the boron atom. By so doing the nitrogen still has only the octet (8) of electrons in its outer shell, but the boron atom has now also got an octet in its outer shell. The type of bond formed in this manner is a *covalent bond*, but as one of the two atoms has supplied both electrons it is called a *coordinate bond.* The atom that supplies the electrons (here the nitrogen) is called the *donor* atom, and the atom that receives the electrons (here the boron) is called the *acceptor* atom. The coordinate bond is drawn as an arrow (→) *from the donor to the acceptor.*

Other examples of coordinate bonds are in the formation of the ammonium ion from ammonia and a proton (figure 6.11), and the formation of the beautiful deep blue $Cu(NH_3)_4{}^{2+}$ ion, when ammonia is added to a solution containing copper(II) ions (figure 6.12).

An alternative way of depicting the coordinate bond to that shown in 6.10, 6.11 and 6.12 is shown in figure 6.13.

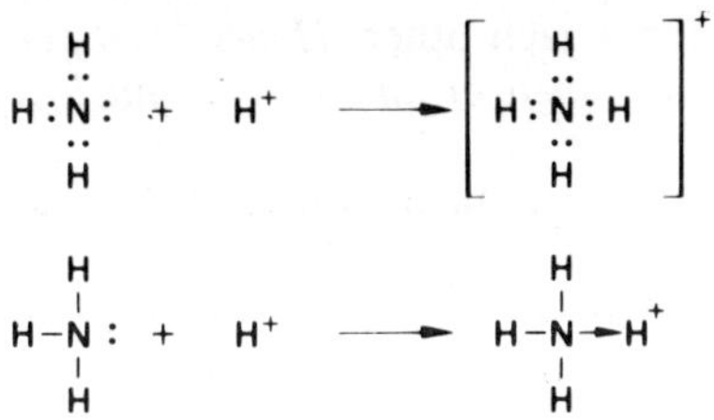

Figure 6.11 Formation of the ammonium ion.

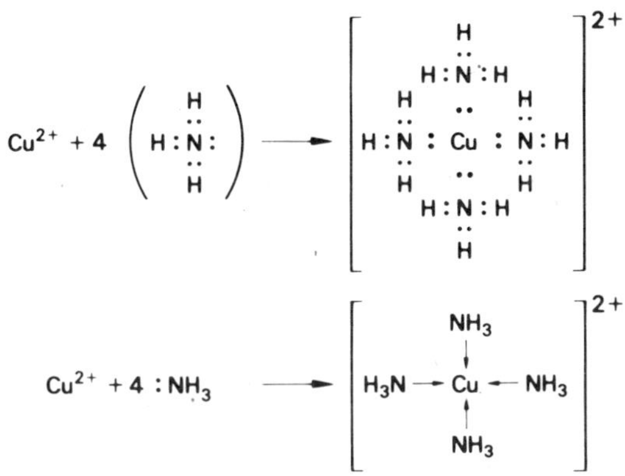

Figure 6.12 Formation of the $Cu(NH_3)_4{}^{2+}$ ion.

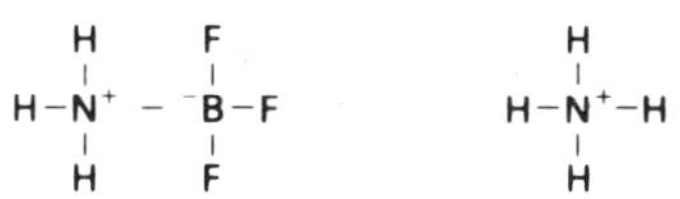

(note H was H^+, the coordinate bond puts a − ve charge on the H, the + and − charges cancel out)

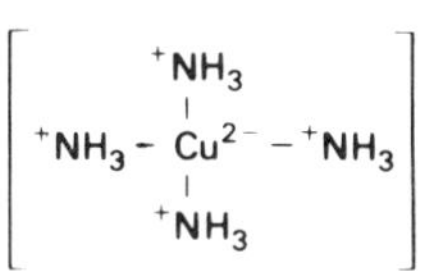

(note Cu was Cu^{2+}, the coordinate bonds put four − ve charges on the Cu, two of these cancel with the 2+ leaving Cu^{2-})

Figure 6.13 Species containing coordinate bonds.

6.2.5 The shape of covalent molecules

(a) Introduction

So far we have not discussed how the atoms in a covalent molecule are arranged in space. Our diagrams have all been 2-dimensional (on a page of the book), but the molecules that we are depicting are 3-dimensional, i.e. all the atoms do not lie in the same plane.

The shape of the simple molecules that we are discussing here can be understood in terms of one very simple idea.

We have already seen that electrons tend to pair together, and we have drawn them so in all our diagrams. The reasons why they form pairs are complicated and it is not possible for us to explain them here. These pairs of electrons are, of course, negatively charged, and like charges repel each other. *Hence each pair repels the other pairs of electrons in a particular shell of an atom as much as possible.*

The most frequently occurring number of electrons in the outer shell of an atom in a molecule is the *octet*. These 8 electrons form 4 pairs and these 4 pairs, which can be regarded as being able to move anywhere on the surface of a sphere surrounding the nucleus, repel each other. They must remain in the spherical shell, and the 4 pairs move to the position of the lowest energy. This position of lowest energy is when they are as far away from each other as possible, i.e. when *the pairs of electrons are at the 4 corners of a regular tetrahedron* (figure 6.14).

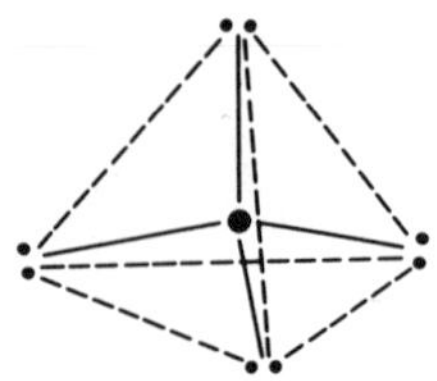

Figure 6.14 Four pairs of electrons arranged tetrahedrally.

(b) Methane

Consider, therefore, the *methane* molecule (figure 6.15(*a*)). The carbon atom has 4 pairs of electrons around it; these will be at the corners of a regular tetrahedron. The 4 hydrogen atoms are attached to these 4 pairs of electrons, hence the 4 hydrogens are at the corners of a regular tetrahedron with the carbon atom at the centre. The shape of the methane molecule is that of a *regular tetrahedron.*

(c) Ammonia

Now consider the *ammonia* molecule (figure 6.15(*b*)). The nitrogen atom has 4 pairs of electrons around it and, like those on carbon in methane, these 4 pairs of electrons will be at the corners of a regular tetrahedron. Only 3 of these electron pairs have hydrogen atoms attached to them, as in figure 6.15(*b*). The fourth pair of electrons is known as an unshared (or lone) pair. *Now the shape of the molecule is the shape given by the positions of the nuclei of the atoms making up the molecule.* Hence the shape of the ammonia molecule is that of the three hydrogen and one nitrogen nuclei, i.e. it is a triangular pyramidal molecule.

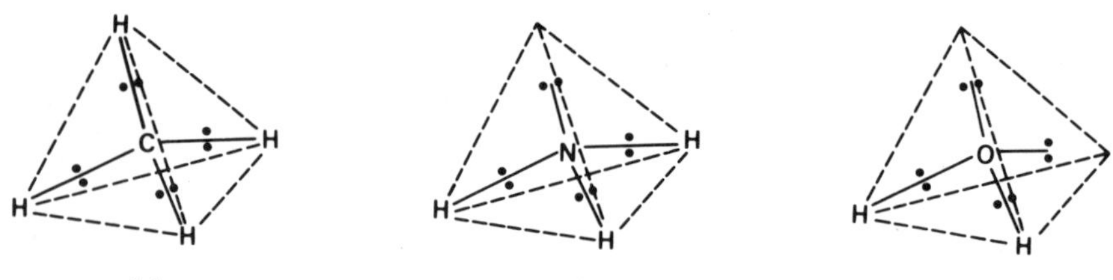

Figure 6.15 (*a*) Methane, (*b*) Ammonia, (*c*) Water.

(d) Water

Now consider the *water* molecule (figure 6.15(*c*)). The oxygen atom has 4 pairs of electrons around it and, like those of carbon and nitrogen, these 4 pairs of electrons will be at the corners of a regular tetrahedron. Now only 2 of these electron pairs have hydrogen atoms attached to them, as in figure 6.15(*c*). The water molecule has two lone pairs. The shape of the water molecule is that of the two hydrogen and oxygen nuclei, i.e. it is an *angular* (*V-shaped*) molecule.

(e) Boron trifluoride

Here, in contrast to methane, ammonia and water, there are only 6 electrons (3 pairs) in the outer shell of the boron atom. These three pairs move to the position of lowest energy (i.e., as far away from each other as possible but keeping the same distance from the nucleus). This is to the corner of a plane triangle (figure 6.16(*a*)). Each pair of electrons is shared with a fluorine atom (figure 6.16(*b*)). Hence boron trifluoride is a *plane triangular* molecule.

(f) Molecules containing double and triple bonds

Methane, ammonia, water and boron trifluoride contain only single bonds. If a molecule contains *double bonds* then the two pairs of electrons making up the double bond together repel the other pairs of electrons as much as possible. Hence if an atom forms 2 single and 1 double bond to 3 other atoms (as, for example, carbon in *methanal* ($\begin{smallmatrix}H\\H\end{smallmatrix}\!\!>C=O$)) the 3 sets of electrons (2 pairs and 1 quartet) repel one another as much as possible. This is exactly what happened in boron trifluoride (figure 6.16); the electrons move to the corners of a plane triangle. *Methanal* is, therefore, a *plane triangular* molecule (figure 6.17(*a*)).

Consider, ethene ($CH_2 = CH_2$). Here each carbon atom, like the carbon in methanal, has 2 single bonds, to the hydrogen atoms, and a double bond. Each carbon atom in ethene is, therefore at the centre of a plane triangle, i.e. the ethene molecule is planar (figure 6.17(*b*)).

Consider, now, carbon dioxide ($O = C = O$). Here the central carbon uses 4 electrons to bond each oxygen atom. The 8 electrons are, therefore, in two quartets. These two quartets of electrons repel each other as much as possible. You should easily see that they will move to opposite sides of the carbon atom, so *carbon dioxide* is a *linear* molecule (figure 6.17(*c*)).

If a *triple* bond is formed then a sextet of electrons together repel the other pairs of electrons (in an octet, of course, there can only be one other pair). This is the case in *ethyne* ($H—C{\equiv}C—H$) where the sextet of electrons forming the triple bond repel the pairs of electrons forming the C—H bonds and hence the C—H

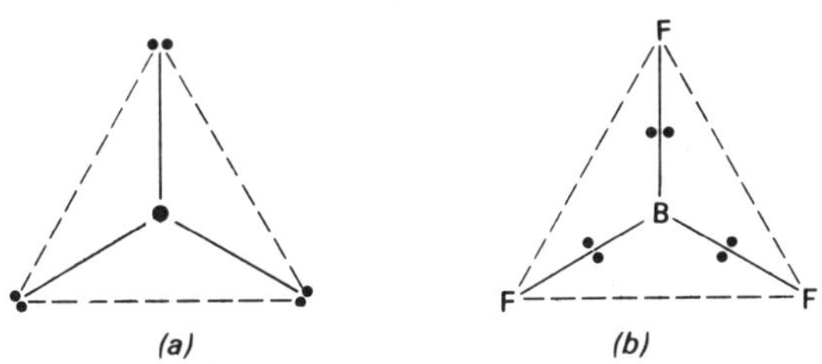

(*a*) (*b*)

Figure 6.16 (*a*) Three pairs of electrons arranged in a plane triangle, (*b*) Boron trifluoride.

and C ≡ C bonds will be on opposite sides of each carbon atom. *Ethyne* is, therefore, a *linear* molecule (figures 6.17(*d*)).

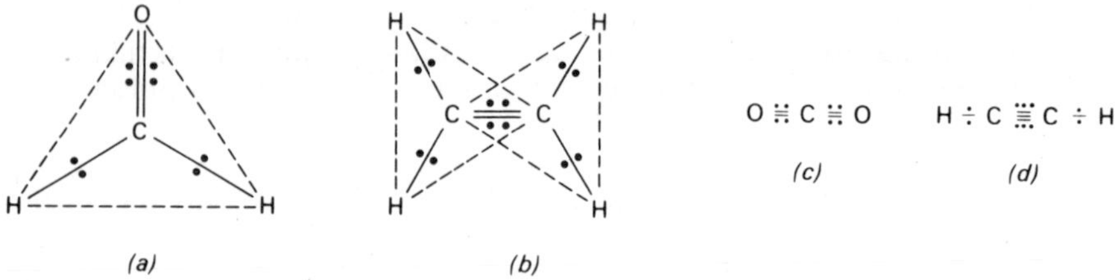

Figure 6.17 (*a*) Methanal, (*b*) Ethene, (*c*) Carbon dioxide, (*d*) Ethyne.

Experiment. To investigate the shape of molecules using balloons

An amusing experiment is to build models using balloons. The physical repulsions between the balloons, parallels the electronic repulsions between electron-pairs in molecules, hence balloons arrange themselves into the same shape as do molecules.

Modelling balloons are best for this experiment, but if these are not available, long balloons of any type can be used.

(*a*) Take two long balloons and twist each of them in the middle. Then twist the four half-balloons together. They will arrange themselves in the shape of a tetrahedron, i.e. as the 4 pairs of electrons on C in methane, N in ammonia or O in water. If 2 different coloured balloons are used, one colour represents the O—H bonds of the water molecule and the other the 2 lone pairs.

(*b*) Take a long and a small balloon, twist the long balloon in half and twist in the end of the small balloon. The three balloons will arrange themselves in a plane triangle, i.e. as the three B—F bonds in boron trifluoride. If different coloured balloons are used then the 2 half-balloons represent the C—H and the other balloon the C = O bond of methanal.

(*c*) Take 2 extra-long balloons and twist each to divide the balloon into a quarter and three-quarter. Twist them together. Then twist each long piece the same distance from the end as before, and twist together again. You will now have a pair of banana-shaped balloons with two small pieces at each end. This is the shape of the ethene molecule. The end pieces represent the four C—H bonds and the banana-shaped balloons the C = C bond. Notice that all four C—H bonds lie in one plane and this plane is at right angles to the C = C bond.

(*d*) Take two extra-long balloons, twist them in the middle, and twist them together. Twist each of the 4 half balloons as near the end as possible and twist together in pairs. You will have 2 pairs of banana-shaped balloons in a line. This is the shape of the carbon dioxide molecule. Notice that the two C = O bonds are in planes at right angles to each other.

(*e*) Try to make a model of ethyne (HC ≡ CH).

6.2.6 Molecular solids

(a) Iodine

Molecules that contain covalent bonds are small molecules as we have just described. These will often be gases, but they do form liquids and solids.

When a solid is formed from small covalent molecules, the molecules are held together in the solid by weak bonds due to the (electrostatic) attraction between the electrons and nuclei of different molecules. A good example is iodine. Solid iodine is composed of I_2 molecules. These pack together in *layers* as shown in figure 6.18. This shows the herring-bone pattern of a layer. The iodine molecules are further (by about $1\frac{1}{4}$ times) from molecules in the layers above and below them than they are from molecules in their own layer. These features give solid iodine the characteristic layer-like appearance.

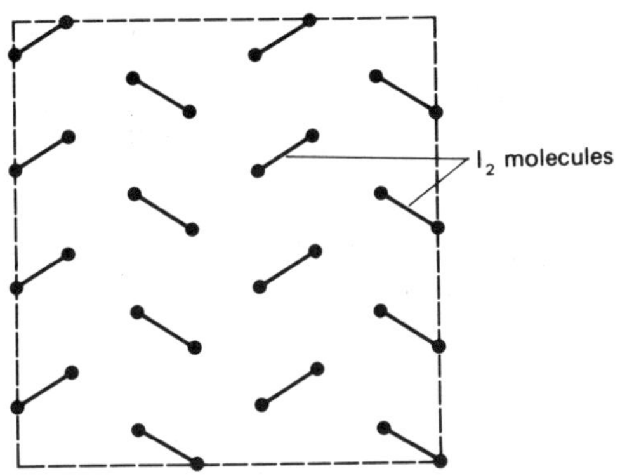

Figure 6.18 A layer of iodine molecules.

(b) Ice

The importance and widespread occurrence of water means that *ice* is a particularly common solid. When water freezes the water molecules arrange themselves in the ice crystal such that each oxygen atom is surrounded by 4 hydrogen atoms. In this way 6 membered rings of oxygen atoms, with hydrogen atoms in between them, are formed (figure 6.19). It is because ice is made up of these 6 membered rings that the snow crystals, illustrated in figure 1.1, are hexagonal in shape.

The electrons of the hydrogen-oxygen bond of the water molecule are attracted slightly more by the oxygen than the hydrogen. This means that water is a *polar molecule*, i.e. the molecule is like a little bar-magnet (except that it is electrically charged and not magnetically charged). The electrostatic attractions that can result from this polarity are the forces that orientate (or arrange) the water molecules as they are found in the ice crystal. (This electrostatic attraction involving hydrogen is in more advanced chemistry books called *the hydrogen bond.*)

One very important consequence of the orientation that takes place when water freezes to ice, is that the solid (ice) has a larger volume and hence smaller

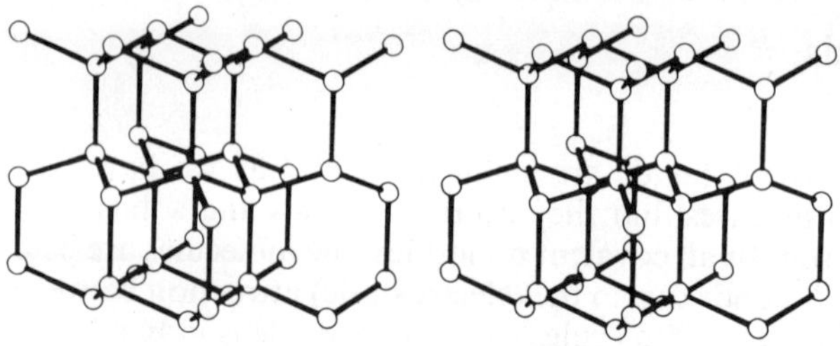

Figure 6.19 Stereoscopic picture of *ice*. The atoms shown are the oxygen atoms of the water molecules. There is a hydrogen atom in between each pair of oxygen atoms.

density than the liquid (water), i.e. *water expands on freezing*. This is most unusual; almost all other liquids contract on solidifying. The fact that water expands on freezing can be troublesome, as you will know if you have ever had a water pipe at home burst during the winter. It also means that ice floats on ponds and rivers, i.e. they freeze from the top downwards.

6.2.7 Macromolecules (giant covalent molecules)

(a) Introduction
In sections 6.2.1 and 6.2.6 we said that compounds containing covalent bonds are gases, liquids or solids with low melting and boiling points. This is usually the case, but there is also another group of molecules containing covalent bonds where the melting and boiling points can be very high. Indeed the compounds with the highest known melting and boiling points come into this class. These are the *macromolecular covalent solids*.

In a macromolecular solid the covalent bonding extends throughout space so that one can consider *the whole crystal to be one big molecule*.

(b) Diamond and graphite
Diamond is the best example of a macromolecular solid. In diamond (figure 6.20) each carbon atom is linked by strong covalent bonds to four other carbon atoms arranged tetrahedrally around it. This arrangement extends throughout the whole diamond lattice, and the structure is therefore very rigid, and the bonds are difficult to break. Hence diamond has a very high melting point. Diamonds are discussed further in section 17.2.

The other allotrope of carbon, *graphite* (figure 6.21) is, in contrast, a much softer material. This is because the strong covalent bonding is no longer in three-dimensions but only in two. There are hexagonal layers of carbon atoms held together by covalent bonds, but between the layers are much weaker bonds, and the carbon-carbon distance between layers is much bigger than in a layer. It is for this reason that graphite has a layer-like appearance, and can be split into thin sheets.

6.2.8 Metallic bonding

Another large class of substances is the metals. They contain a further type of chemical bond, the *metallic bond*. Metals are discussed in sections 5.3.4(f) and 21.1 where the metallic bond and the properties of metals are described.

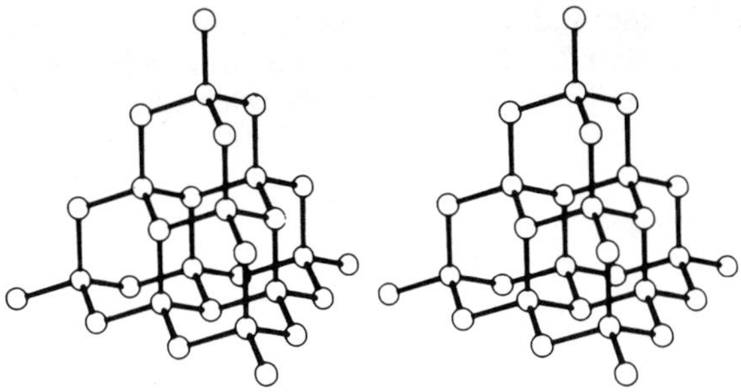

Figure 6.20 Stereoscopic picture of *diamond*.

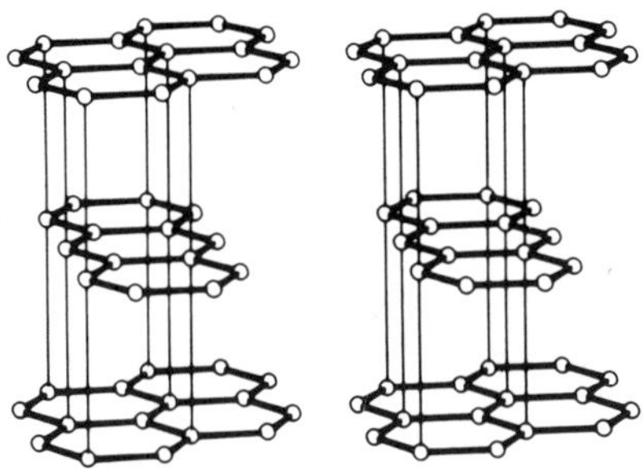

Figure 6.21 Stereoscopic picture of *graphite*.

6.3 Radioactivity and nuclear reactions

6.3.1 Nuclear physics

John Dalton thought that atoms were indivisible. We now know that this is not correct; it is possible to split up the nucleus of atoms. It should be emphasized however that during chemical reactions it is only the electrons that are involved in reactions. The reactions of nuclei, which is what this section is about, is often called *nuclear physics*.

Usually to break up nuclei you need to put in a lot of energy, but this is not always so; some atoms just break up of their own accord. This phenomenon is known as *natural radioactivity*.

6.3.2 Natural radioactivity

Atoms of certain elements, mainly some of the heavier elements such as uranium, radium and polonium, break up spontaneously and in so doing emit rays and particles. The atom *disintegrates*. This discovery was made in 1896 by the Frenchman Henri Becquerel. He found that photographic plates wrapped in black paper were darkened by uranium salts, and he suggested that the uranium salts emitted some form of rays that affected the film.

In 1898 Marie Curie (figure 6.22) found that uranium minerals such as pitchblende had more radioactivity than expected from the amount of uranium present. She therefore looked for other radioactive elements in the pitchblende and discovered polonium (named after her homeland of Poland) and radium (named after the latin word for a ray). Since then many other elements that are radioactive have been found.

Figure 6.22 Marie Sklodowska Curie (1867–1934). (This stamp is from Poland, the country of her birth). (Photo: A. Rendell).

The radioactivity in all cases originates from reactions occurring in the nuclei of atoms of the elements concerned. This is different from all *chemical reactions* as discussed elsewhere in this book. These involve only the electrons which surround the nucleus.

The rays resulting from radioactivity are of three types:

(*a*) *α (alpha) particles*. These are positively charged helium atoms, He^{2+}, i.e. helium atoms without their two electrons. As they are charged they will be deflected in a magnetic field.

(*b*) *β (beta) particles*. These are fast moving electrons. As these have a much smaller mass than α-particles and as they are negatively charged, they are deflected more in a magnetic field and in the opposite direction to α particles.

(*c*) *γ (gamma) rays*. These are a form of radiant energy. They are unaffected by a magnetic field (see figure 6.23).

When atoms of elements disintegrate to form α, β and γ rays, atoms of new elements are left behind. These may also be radioactive and themselves disintegrate to atoms of other elements. The rate at which atoms disintegrate depends on how much radioactive material is present. As less and less is left, the rate of disintegration becomes slower and slower. A sample of radium, for example, will disintegrate to half its size in 1600 years. What is left will, in its turn, disintegrate to half its size in a further 1600 years, and so on. Thus only after

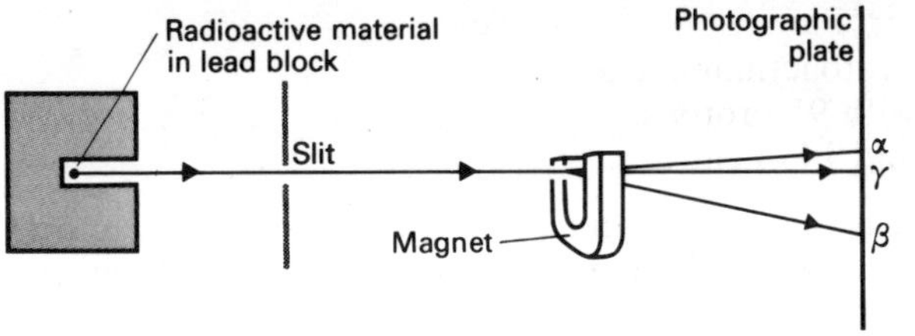

Figure 6.23 Effect of a magnetic field on α, β and γ rays. (The apparatus is assembled in a vacuum to prevent deflection of rays by gas molecules. The rays that are not emitted forwards are absorbed by the lead block. Three dark spots are produced on the plate showing that γ-rays are not deflected and that α- and β-rays are deflected to different extents in different directions).

many such periods of 1600 years will the sample of radium have completely disappeared. The graph (figure 6.24) should make this point clearer.

As the total life of a radioactive element is therefore infinite, it is usual to describe its rate of decay by means of its half-life, i.e. the time taken for half of the element to decompose. Half-lives vary considerably, e.g. radium-226 has a half-life of 1600 years, radon-222 of 3·8 days, uranium-238 of $4{\cdot}5 \times 10^9$ years and polonium-212 of 3×10^{-7} seconds.

6.3.3 Radioactive disintegrations

To understand more clearly what happens during a radioactive disintegration, we must use the symbols as described in section 5.2.4.

Thus one of the isotopes of uranium is ${}^{238}_{92}U$. This tells us that this atom has 92 protons and $(238 - 92) = 146$ neutrons in its nucleus. This particular isotope is often referred to as uranium-238, and it decays by emitting an α-particle (${}^{4}_{2}He$):

${}^{238}_{92}U$	$\rightarrow$ ${}^{234}_{90}Th$	$+$ ${}^{4}_{2}He$
i.e. a uranium atom containing 92 protons and 146 neutrons	i.e. a thorium atom containing 90 protons and 144 neutrons	i.e. a helium atom containing 2 protons and 2 neutrons.

At the same time, as with all nuclear disintegrations, γ rays are emitted.

The thorium formed is itself unstable and decays by β-particle (${}^{0}_{-1}e$) emission:

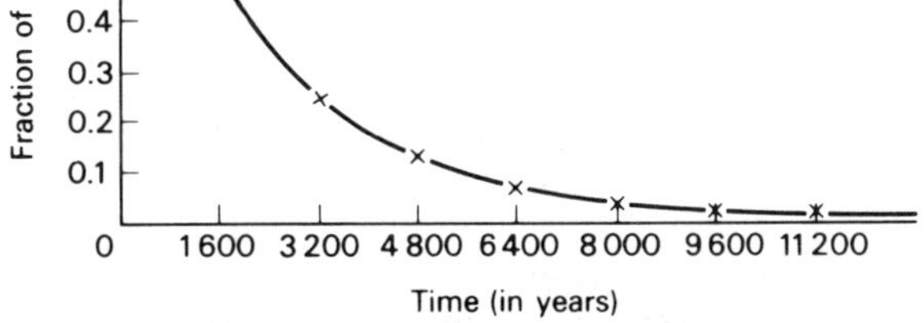

Figure 6.24 Radioactive decay curve for radium-226.

$^{234}_{90}Th$	→	$^{234}_{91}Pa$	$+ \,^{0}_{-1}e$
Thorium atom with 90 protons and 144 neutrons		Protoactinium atom with 91 protons and 143 neutrons	

What has happened here is that a neutron has changed into a proton and an electron. The electron (β-particle) is ejected and the proton increases the atomic number by one.

This protoactinium is unstable as are all the first 13 elements formed by radioactive decay of $^{238}_{92}U$. The fourteenth product is lead-206 and this is stable. Thus the total result of the successive decays of uranium-238 is:

$^{238}_{92}U$	$\rightarrow \,^{206}_{82}Pb$	$+8(^{4}_{2}He)$	$+6\,^{0}_{-1}e$
(92 protons 146 neutrons)	(82 protons 124 neutrons)	(16 protons 16 neutrons)	(6 electrons)

6.3.4 Bombardment reactions

(i) In 1919 Sir Ernest Rutherford noticed a different type of nuclear reaction when he passed a stream of α-particles from decaying radium through some nitrogen gas. Oxygen and hydrogen nuclei (protons) were produced:

$$^{14}_{7}N + ^{4}_{2}He \rightarrow ^{17}_{8}O + ^{1}_{1}H$$

The yield of oxygen and hydrogen in such a reaction is very low since there is very little chance of an α-particle hitting the very small nuclei of the nitrogen atoms. Most of the α-particles pass through the gas unchanged.

A more important bombardment reaction was made in the early 1930's by James Chadwick:

$$^{9}_{4}Be + ^{4}_{2}He \rightarrow ^{12}_{6}C + \text{neutron } (^{1}_{0}n)$$

since this led to the discovery of the neutron.

In the early days of these reactions it was only possible to react atoms of low atomic number, because the highly charged nuclei of the larger atoms repelled the α-particles. This led to methods of speeding up α-particles and protons so that they could be shot at very high speed at other nuclei. The equipment used in very large and very expensive—linear accelerators and cyclotrons. Using these, many other reactions were then found, e.g.,

$$^{39}_{19}K + ^{1}_{1}H \rightarrow ^{36}_{18}Ar + ^{4}_{2}He$$
$$^{7}_{3}Li + ^{4}_{2}He \rightarrow ^{10}_{5}B + ^{1}_{0}n$$

This second reaction is of an important type, as neutrons are formed.

These can be used to bombard other nuclei. Neutrons are particularly effective as bombarding particles because being uncharged, they are not repelled by charged nuclei.

(ii) *The 20th century Alchemists.* In section 1.7 we mentioned the predecessor of the chemist—the Alchemist. It was the dream of the Alchemist to turn other elements into gold. To do this he would have needed to carry out a nuclear-reaction. The elements that do not occur naturally have all been made since 1936. These elements are technetium, promethium, astatine and the 11 elements after uranium. They have been made by carrying out nuclear bombardment reactions on other elements. The dream of the Alchemist has been realized—one element has been converted into another. The reactions used to produce technetium, and astatine were:

$_{42}Mo + ^{2}_{1}H \rightarrow {}_{43}Tc + ^{1}_{0}n$ (a mixture of isotopes of molybdenum was used)

$_{83}Bi + ^{4}_{2}He \rightarrow {}_{85}At + ^{1}_{0}n$ (a mixture of isotopes of bismuth was used)

The elements beyond uranium in the periodic table are produced in the same way by bombardment with hydrogen nuclei, deuterium nuclei, helium nuclei (α-particles) or with neutrons. The preparation of neptunium and plutonium by neutron bombardment in a nuclear reactor is given in the next section.

Exercise on bombardment reactions

We have commented above on the Alchemists dream to convert other elements into gold.

Imagine that you could go back 200 years and take with you a supply of high energy particles to carry out bombardment reactions.

(*a*) *In order to help the Alchemist in his (until you arrived) hopeless quest, what reactions would you carry out to produce gold using as your bombarding particles* (i) *protons,* (ii) *deuterium nuclei,* (iii) *α-particles and* (iv) *neutrons. (Assume that the only product (if any) besides gold will be neutrons or electrons and that you start with iridium or platinum.)*

(*b*) *Assuming that your reactions are totally successful, do you think that you will have helped make the Alchemist a richer man?*

(*c*) *The element that the Alchemist preferred to start with was lead. Have you any comments that you can make to him about it suitability for turning into gold?*

Answer to exercise

(*a*) *Typical reactions would be*

(i) $_{78}Pt + ^{1}_{1}H \rightarrow {}_{79}Au + ^{1}_{0}n$

(ii) $_{78}Pt + ^{2}_{1}H \rightarrow {}_{79}Au + ^{1}_{0}n$

(iii) $_{77}Ir + ^{4}_{2}He \rightarrow {}_{79}Au + ^{1}_{0}n$

(iv) $^{x}_{78}Pt + ^{1}_{0}n \rightarrow {}^{x+1}_{79}Pt \rightarrow {}^{x+1}_{79}Au + {}^{0}_{-1}e$

(*b*) *You will certainly not have helped make him richer! Both platinum and iridium, (your starting materials), are still much more expensive than gold. Platinum was 8 times as expensive in 1920, and the difference was greater in the days of the alchemist! Platinum had been discovered 200 years ago (in 1741). Iridium which is more expensive still was not discovered until 1807.*

(*c*) *All bombardment reactions produce elements of* higher *atomic number. Lead has an atomic number 3* above *gold, so can only be converted into gold by a* fission *reaction. Even today that is not possible. So you could tell the Alchemist that he really was making things very difficult for himself!*

6.3.5 Nuclear energy, fission and fusion reactions

When nuclear reactions occur, enormous changes in energy can be involved, e.g.,

$^{7}_{3}Li + ^{1}_{1}H \rightarrow 2(^{4}_{2}He)$ $\qquad \Delta H = -1{\cdot}7 \times 10^{12}$ J (see chapter 9 for an explanation of ΔH).

This means that when 7 g of lithium react with 1 g of protons, $1{\cdot}7 \times 10^{12}$ joules of heat are evolved. This is because of a loss of mass that occurs. John Dalton thought that matter could neither be created or destroyed. He was almost right, and for most purposes this is so, but in the early part of this century *Albert*

Einstein showed that matter can be converted into energy, and vice versa. As matter is almost all concentrated in the nuclei of atoms clearly a nuclear reaction must be involved in any reaction creating or destroying matter. The relationship between energy and matter was shown by Einstein to be:

$$E = mc^2$$

where

E = energy liberated measured in joules;
m = loss of mass measured in kilograms;
c = velocity of light measured in meters per second ($3{\cdot}0 \times 10^8$).

Even if only a small loss of mass occurs then, because c^2 is so large, the amount of energy liberated as heat is very large.

Mass is lost, and therefore energy is evolved in two types of reaction, *fission* and *fusion*.

(a) Nuclear fission

The fission, (or splitting up) of, for example, uranium-235 or plutonium-239 by the absorption of neutrons is:

$$^{235}_{92}U + ^{1}_{0}n \rightarrow ^{103}_{42}Mo + ^{131}_{50}Sn + 2(^{1}_{0}n)$$

The uranium atom may be split in other ways, but the important point to note is that *more neutrons are produced by such a reaction than are absorbed*. Therefore, it should be possible, under suitable conditions, to arrange for the number of neutrons to increase very rapidly by one fission occurring after another. As a result of this the reaction continues at rapidly increasing rate; it is called a chain-reaction (figure 6.25).

(i) The atomic bomb

The concentration of uranium-235 in natural uranium is small, only 0·7%. This means that any neutrons released are unlikely to hit any other uranium-235 nuclei, and a *chain-reaction* does not occur.

However, in a large lump of uranium-235 the chances of neutrons (produced by fission of some uranium-235 nuclei) hitting other nuclei and splitting them is much increased, a chain-reaction does occur and an enormous amount of energy is released. This causes a temperature of millions of degrees and a terrific explosion.

If the lump of uranium-235 is below a certain size, (called the *critical mass*) a chain-reaction does not occur as most neutrons escape from the lump. For an atomic bomb, therefore, several small pairs of lumps of uranium-235 have to be rapidly brought together to give a large piece that exceeds the critical mass. An explosive chain-reaction then occurs.

(ii) The nuclear reactor

In a *nuclear reactor* (often called a *nuclear pile*) the explosive chain-reaction that occurs in the atomic bomb is not allowed to occur—the reaction is *controlled*. Naturally occurring uranium, or uranium with the concentration of uranium-235 increased, is used. This uranium, in the form of rods of the metal, is put into holes in a huge block of graphite. The fast-moving neutrons from the uranium-235 are slowed down by the graphite and are then more easily captured by the

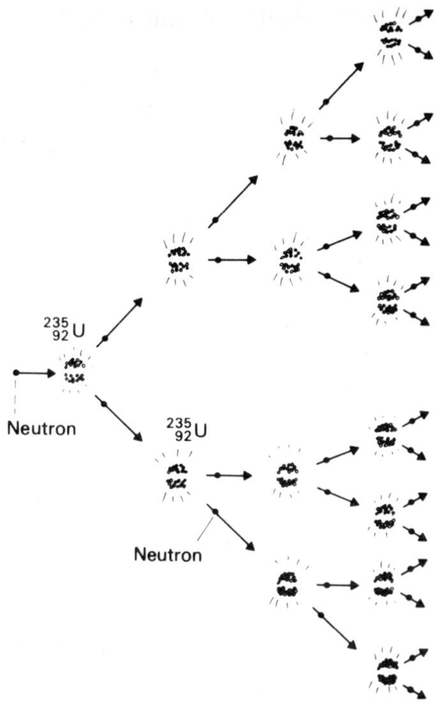

Figure 6.25 The fission of uranium-235 by neutrons —a chain reaction.

other uranium-235 atoms, and in this way a controlled chain-reaction occurs.

The graphite does not absorb any of the neutrons; it just slows them down and is therefore called a *neutron moderator*. To regulate the rate of nuclear fission reactions, by absorbing neutrons when their concentration gets too high, control-rods are used. These are made of steel with a high boron content and they very efficiently "mop up" excess neutrons.

The slow-moving neutrons are not absorbed by uranium-238, but some of the fast moving ones are. These convert uranium-238 to neptunium and then plutonium, which itself is a fissionable element:

$$^{238}_{92}U + ^{1}_{0}n \rightarrow ^{239}_{92}U \xrightarrow{-e} ^{239}_{93}Np \xrightarrow{-e} ^{239}_{94}Pu$$

Nuclear reactors, therefore, produce *energy* that can be used for driving electric generators and many are now used around the world for this purpose (see figure 28.8). They can also be used for producing *radioactive isotopes* to be used as described in the next section, and to prepare *plutonium* (by the reaction just given) that can then be used as the fuel in other nuclear reactors.

(b) Nuclear fusion

(i) In *nuclear fission* energy is released when heavy nuclei are split up. In *nuclear fusion* energy is released when small nuclei combine together.

For example the two isotopes of hydrogen, deuterium and tritium, can combine to give helium:

$$^{2}_{1}H + ^{3}_{1}H \rightarrow ^{4}_{2}He + ^{1}_{0}n + \text{energy}.$$

(ii) The source of the sun's energy is nuclear fusion reactions of this type. By successive fusion reactions inside suns, all the elements have been created starting from hydrogen.

For fusion reactions to occur very high temperatures are necessary. In our sun, although the temperature is 15 000 000 °C it is only hot enough for hydrogen atoms to fuse together to give helium.

Even hotter suns are needed to produce all the other elements and some of the stars that we see are the "factories" producing these elements.

(iii) The reaction that occurs in a *hydrogen bomb* is a fusion reaction of deuterium:

$$^{2}_{1}H + ^{2}_{1}H \rightarrow ^{3}_{2}He + ^{1}_{0}n + \text{energy}.$$

The only way to achieve the high temperatures needed (above 10 000 000 °C) is to carry out the fusion inside a reacting atomic (fission) bomb.

(iv) The light isotopes are so abundant on earth that if nuclear fusion reactions could be controlled they would provide ample sources of energy for millions of years. Many people think that fusion reactions will have to be our main energy source in centuries to come (see chapter 28). There are a lot of scientists now trying to bring about controlled fusion reactions, so far without success. The problems are enormous; how does one produce the fantastically high temperatures needed and (equally important) how do you hold materials at these high temperatures?

6.3.6 Applications of radioactive elements

Radioactive elements are produced, as we have seen, by fission. Some occur naturally, such as radium, and many others can be produced by neutron bombardment of materials put inside a nuclear reactor.

Such radioactive elements have found such numerous uses in industry, medicine, and scientific work, that it is not possible to give them all. Some examples are given below.

(a) Industrial applications

(i) Thickness and density measurements

Beta-particles can penetrate thin sheets of metal. The proportion of particles absorbed will for a given material be proportional to its thickness; or for a given thickness be proportional to the density of the material. Beta-particle thickness gauges are used for checking the thickness of metal sheets in steelmills (figure 6.26) and for checking that packets, tins and other containers are correctly filled (figure 6.27).

(ii) Tracing pipes and finding leaks

If the path of a water pipe, say, needs to be determined, or a leak detected a radioisotope can be put into the pipe and its passage followed by means of a detector above the ground.

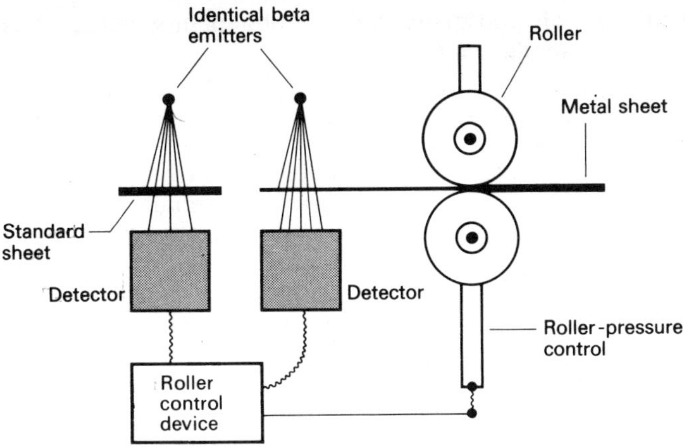

Figure 6.26 A β-particle gauge controlling the thickness of a metal sheet.

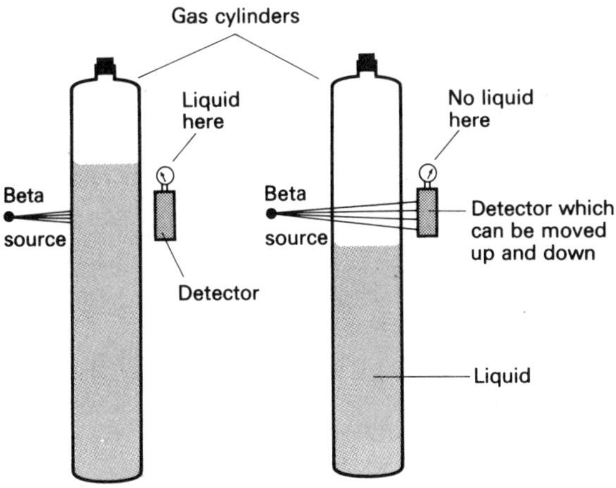

Figure 6.27 A β-particle gauge used for measuring the level of liquids in closed containers.

(b) Use in medicine

The best known use of radioactive elements is in the treatment of cancer. Some cancerous growths can be eliminated by exposure to gamma rays. Originally radium was used for this purpose, but now cobalt-60 and caesium-137 are used. Other radioactive isotopes can be implanted in cancerous growths to destroy them.

(c) Scientific applications

(i) Study of biological synthesis

To help understand photosynthesis and related processes, carbon dioxide containing radioactive carbon ($^{14}_{6}C$) is used, and the path of this carbon can be followed during the growth of the plant. The chemicals that incorporate the carbon-14 can be isolated and identified.

To study the absorption of phosphate fertilizer by plants, radioactive phosphorus can be used in a similar manner.

(ii) Dating of archaeological specimens
This is another fascinating use of radiochemical techniques. Carbon dioxide in the atmosphere contains a tiny proportion of radioactive carbon-14. In the atmosphere the proportion of carbon-14 remains approximately constant because although it is decaying, it is also being produced by cosmic-rays coming into the atmosphere from outer space. When they are growing the plants, e.g. trees, use in photosynthesis the radioactive carbon in the proportion that it occurs in the carbon dioxide. When growth stops no more carbon dioxide is used, and the radioactive carbon, that is now incorporated in the plants, decays. Carbon-14 has a half-life of 5600 years, and therefore if a sample of preserved wood has half as much carbon-14 in it as a living sample of wood, then its age is 5600 years. If it has a quarter as much its age is 11 200 years. If the carbon-14 content can be determined it is quite simple mathematically to determine the age of the sample. Many datings are now made this way, and in general they agree with those made by other methods.

Questions on chapter 6

1. Explain carefully the difference between electrovalency and covalency.
Of which of these two types of valency is ammonia an example?
By a simple electronic diagram show the essential structure of the ammonia molecule. (Give a key to your diagram to explain any symbols used.) *(J.M.B.)*

2. Explain simply the difference between electrovalency and covalency. Show by means of simple electronic diagrams the difference between (*a*) a "molecule" of sodium chloride and (*b*) a molecule of chlorine. (Give a key to your diagrams). *(J.M.B.)*

3. Use the information in the table to answer parts (*a*), (*b*) and (*c*).

Element	Carbon	Oxygen	Fluorine	Sodium	Sulphur
Symbol	C	O	F	Na	S
Atomic number	6	8	9	11	16
Mass number	12	16	19	23	32

(*a*) Draw simple diagrams to show the structures of *two* of the elements in the table.
(*b*) Give the formulae of (i) *two* covalent compounds, (ii) *two* ionic compounds, formed only from the elements in the table.
(*c*) Describe, in terms of electrons, the formation of *one* of the covalent compounds and *one* of the ionic compounds in (*b*) from their constituent atoms. *(C.)*

4. Refer to your copy of the periodic table.
(*a*) Write down the electronic structure of atoms of elements with (i) atomic number 6, (ii) atomic number 17, (iii) atomic number 20. Name a metal from these three elements. Give reasons for your choice, based on its electronic structure.
(*b*) Name and give the symbols for *three* noble (inert) gases.
(*c*) Explain the electronic structure (outer shells of electrons only) of (i) tetrachloromethane (carbon tetrachloride), (ii) calcium chloride. (iii) Select, from these two compounds in (i) and (ii), an ionic compound and state the ions present in this compound.
(*d*) An element has atomic number 17. The nuclei of some of its atoms have 18 neutrons, while other have 20 neutrons.

(i) What name is given to different atoms of the same element?
(ii) Write down the relative atomic masses of these atoms.
(iii) Why have these two atoms the same chemical properties? (*O. and C.*)

5. Write out and complete the following table.

Particle	Mass number (Atomic mass)	Number of protons	Number of neutrons	Number of electrons
Li atom		3	4	
Li^+ ion				
^{12}C isotope	12			6
^{13}C isotope	13	6		
N^{3-} ion	14	7		
Ne atom			10	10
S^{2-} ion	32			18

(*a*) Which particles, apart from the Ne atom, have the same total number of electrons as a noble gas?
(*b*) The electron configuration of the carbon atom can be written as C : 2, 4. Show in a similar way the configurations of (i) the Li atom, (ii) the S^{2-} ion.
(*c*) Write the formulae of (i) a compound containing the S^{2-} ion, (ii) a covalent compound of sulphur, (iii) a compound composed only of Li^+ and N^{3-} ions. (*C.*)

6. (*a*) An element *Q*, with non-metallic properties, has an atomic mass of 75 and is normally trivalent.
(i) Write down the equation for the complete combustion of *Q* in oxygen.
(ii) Calculate the mass of calcium that should combine with one gram of *Q*.
(iii) State whether the solution in water of the oxide of *Q* is likely to be acidic, alkaline or neutral.
(*b*) Tetrachloromethane, CCl_4, is a wholly covalent compound. Would you expect it to be: (i) volatile, (ii) miscible with water, (iii) a conductor of electricity? (Ca = 40)
(*O. and C.*)

7. The numbers of protons, neutrons and electrons in particles *A* to *F* are given in the following table:

Particle	Protons	Neutrons	Electrons
A	3	4	2
B	9	10	10
C	12	12	12
D	17	18	17
E	17	20	17
F	18	22	18

(*a*) Choose from the table the letters that represent
(i) a neutral atom of a metal,
(ii) a neutral atom of a non-metal,
(iii) an atom of a noble gas (inert gas),
(iv) a pair of isotopes,
(v) a cation (positive ion),
(vi) an anion (negative ion).
(*b*) Give the formulae of the compounds you would expect to be formed between (i) *C*

and *D*, (ii) *E* and hydrogen, (iii) *C* and oxygen. (You may use the letters *C*, *D* and *E* as symbols for the elements.)

In *each* case, say whether the compound will be an acid, a base or a salt.

(*c*) What will be the formula of a compound containing only particles *A* and *B*? Give *two* physical properties you would expect this compound to have. (*C*.)

8. In what respects has Dalton's atomic theory required modification in the light of modern knowledge?

9. What do you mean by the term "radioactivity"? Describe the three types of radiation emitted by radioactive elements. How is it possible to show that there are three types of radiation emitted?

In what ways do radioactive changes differ from ordinary chemical changes?

10.

Element	Symbol	Atomic number
Lithium	Li	3
Beryllium	Be	4
Fluorine	F	9
Magnesium	Mg	12
Silicon	Si	14

(*a*) The distribution of extra-nuclear electrons in the magnesium atom can be represented as

Mg: 2,8,2.

Show in a similar way the electron distribution in the atoms of lithium, beryllium, fluorine and silicon.

(*b*) Write formulae for the oxides you would expect lithium, beryllium and silicon to form. Which one of these oxides is most likely to be acidic?

(*c*) Write formulae for the simple ions formed by lithium, beryllium, fluorine and magnesium.

(*d*) Fluorine forms compounds with all four of the other elements in the table. Which one of the fluorides is most likely to be covalent? Give the formula of this fluoride and explain the bonding in it in terms of electrons.

(*e*) Beryllium hydroxide is a white, insoluble solid and is amphoteric. Explain what *amphoteric* means and say what you would expect to *observe* when sodium hydroxide solution is added gradually to a solution of beryllium chloride. (*C*.)

11. Balance the following equations for nuclear reactions:

$^{231}_{91}Pa \rightarrow {}^{227}Ac + {}_{2}He$

$^{216}_{84}Po \rightarrow {}_{83}Bi + {}^{4}He + e$

$^{238}_{92}U + {}^{4}_{2}He \rightarrow Pu + n$

$^{238}_{92}U + n \rightarrow U$

$^{9}_{4}Be + {}^{4}_{2}Ne \rightarrow {}^{12}_{6}C + ?$

7

The Gaseous State

7.1 Introduction

The main differences between the three states of matter were outlined in chapter 1. In this chapter we are concerned with the laws that describe the behaviour of gases. A number of these, normally studied in physics, is included here because a knowledge of them is essential to chemistry.

7.2 The diffusion of gases

If a girl using perfume enters a room the smell is rapidly detected all over the room. Less pleasantly, if some hydrogen sulphide is prepared in a laboratory its smell is immediately noticed by everyone present. In both cases the rapid spread of the "smell" is a rapid spread of molecules by a process known as *diffusion*.

7.2.1 Some experiments on gaseous diffusion

(*a*) (i) Place one drop of bromine liquid in a gas jar and cover the gas jar with a glass plate. Note that the bromine rapidly vaporizes and its brown colour becomes visible throughout the whole jar.
(ii) Remove the glass plate and invert another jar over the one containing bromine vapour. Notice that after some time the vapour seems to have diffused upwards and imparted an even colour to both jars. (N.B. Bromine vapour is over five times as dense as air.)

(*b*) Assemble the apparatus shown in figure 7.1(*a*). Invert over the porous pot a gas jar full of hydrogen which is less dense than air. A jet of liquid will shoot through the tube at *A*. Remove the gas jar. Liquid will rise up the tube *B*.

This is explained by the high rate of diffusion of hydrogen compared with air. Before the jar of hydrogen is placed over the pot there is a constant movement of air particles through the small holes in the porous pot. The movement takes place in both directions and the number of particles in the pot remains, to all intents and purposes, constant. With the jar of hydrogen in position, hydrogen particles move into the pot much more rapidly than the air particles move out. Accordingly a pressure is set up inside the pot which forces out the jet of liquid.

When the gas jar is removed the opposite process occurs and the pressure decreases.

(*c*) Hold a gas jar full of carbon dioxide, which is denser than air, over the pot as shown in figure 7.1(*b*). Water will rise up the tube *B*. What explanation

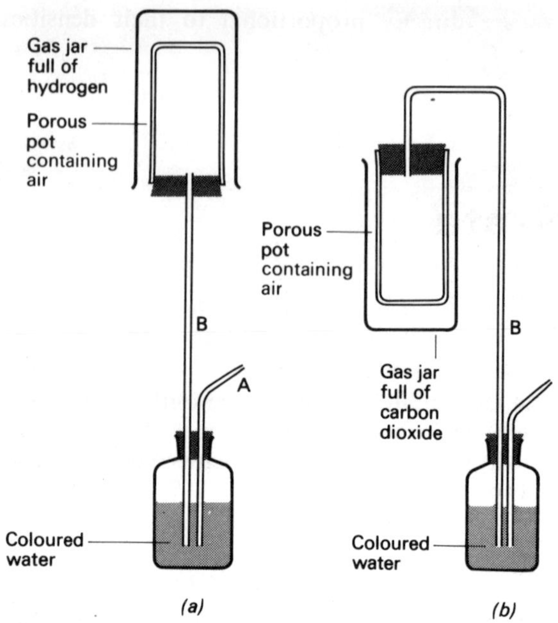

Figure 7.1 The diffusion of gases.

can you offer for this? What do you expect will happen when you remove the jar containing the carbon dioxide?

7.2.2 Graham's law of diffusion

These experiments indicate that the rate of diffusion of a gas depends on its density, i.e. gases of low density diffuse more rapidly than gases of high density. Thomas Graham investigated this relationship and formulated his *Law of Diffusion:*

At a constant temperature and pressure the rate of diffusion of a gas is inversely proportional to the square root of its density.

Using symbols, this becomes:

$$R \propto \frac{1}{\sqrt{d}} \text{ or } R = \frac{k}{\sqrt{d}}$$

(where R = rate of diffusion in $cm^3\ s^{-1}$ or other suitable units, and d = density).

If the rates of diffusion of two gases, A and B, are compared, it follows that:

$$\frac{R_A}{R_B} = \frac{\sqrt{d_B}}{\sqrt{d_A}}$$

Alternatively, the times taken for equal volumes of two different gases to diffuse, under similar conditions, may be compared. Then:

$$\frac{T_A}{T_B} = \frac{\sqrt{d_A}}{\sqrt{d_B}}$$

Molecular masses of gases are directly proportional to their densities. Therefore:

$$\frac{R_A}{R_B} = \frac{\sqrt{M_B}}{\sqrt{M_A}}$$

i.e. Graham's Law affords a means for determining molecular masses.

7.3 Boyle's law

By considering gases to be composed of small particles in constant motion, it is also possible to explain the variation of the volume of a gas with pressure.

Suppose that a certain mass of gas enclosed in a container of volume V exerts a pressure P. This pressure is caused by the particles which make up the gas hitting the walls of the container. Provided the temperature of the gas does not alter, the number of collisions with the wall in a particular length of time remains constant.

Now suppose that the same mass of gas is put into a container only half as big, i.e. of volume $\frac{V}{2}$. The number of particles in a particular volume will then be twice as great. This means that collisions with the walls will be twice as frequent and the pressure doubled, i.e. $2P$.

In other words, if the volume of a fixed mass of gas is halved, its pressure is doubled. This was first realised by Robert Boyle (figure 13.7) in 1662, and is summarised in Boyle's Law:

The pressure of a fixed mass of gas at constant temperature is inversely proportional to its volume.

i.e. Pressure × Volume = Constant

If V_1 and V_2 are the different volumes of a fixed mass of gas at pressures P_1 and P_2, then

$$P_1 V_1 = P_2 V_2$$

7.4 Charles' law

To explain the expansion of gases when they are heated, consider a fixed mass of gas. When the temperature of the gas is raised more energy is supplied to the particles, and they therefore move more rapidly. This means that particles will hit the walls of a container more frequently and, unless the volume is increased, this will result in an increase in pressure.

Charles observed that gases maintained at constant pressure expanded equally for equal rises in temperature. His law states:

A fixed mass of gas, at constant pressure, expands $\frac{1}{273}$ of its volume at 0 °C, for each degree Centigrade rise in the temperature, i.e. a gas with a volume of 273 cm^3 at 0 °C would expand by 1 cm^3 for each degree rise in temperature and would contract 1 cm^3 for each degree fall in temperature. (figure 7.2).

Theoretically gases would have a zero volume at −273 °C but in practice they all liquefy above this temperature. This temperature of −273 °C, at which a gas that did not liquefy would have no volume, is the lowest temperature possible. It is therefore known as the *absolute zero. The absolute scale of temperature* is

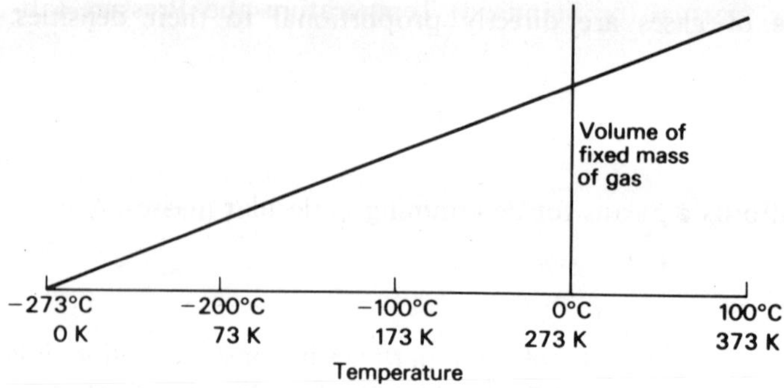

Figure 7.2 Graph illustrating Charles' Law.

obtained merely by adding 273° to the Centigrade temperatures, i.e.

$x\,^\circ\mathrm{C} = (x+273)\mathrm{K}$

Thus $-273\,^\circ$C becomes 0 K, $0\,^\circ$C becomes 273 K, etc. (K is the symbol for the *Kelvin*, the temperature unit on the absolute scale.)

Charles' Law may therefore be stated more conveniently for our purpose: *The volume of a fixed mass of gas, at constant pressure, is directly proportional to its absolute temperature.*

i.e. $$\frac{\text{Volume}}{\text{Temperature in K}} = \text{Constant}$$

If V_1 and V_2 are the different volumes of a fixed mass of gas at absolute temperatures T_1 and T_2, then:

$$\frac{V_1}{T_1} = \frac{V_2}{T_2}$$

7.5 The general gas equation

By combining the Laws of Boyle and Charles, an expression relating the temperature, pressure and volume of a fixed mass of gas is obtained:

$$\frac{P_1 V_1}{T_1} = \frac{P_2 V_2}{T_2}$$

This is a very important equation. It means, for example, that if the volume of a given mass of gas under one set of conditions is known its volume under a different set of conditions may be calculated. *Always remember that the temperature used must be on the Absolute scale.*

7.5.1 Normal (or standard) temperature and pressure

When stating the volume or density of a gas it is usual to give the temperature and pressure at which it was found. This is necessary, as these quantities vary considerably under different conditions.

For comparison purposes, standard values of temperature and pressure, namely $0\,^\circ$C (i.e. 273 K) and 760 mm mercury pressure, have been chosen. These values are known as Normal (or Standard) Temperature and Pressure (abbreviated to n.t.p. or s.t.p.).

7.5.2 Problems on the gas laws

1. *A gas occupies 420 cm³ at 17 °C and 770 mm pressure. What is its volume at s.t.p.?*

$P_1 = 770$ mm $\qquad P_2 = 760$ mm
$V_1 = 420$ cm^3 $\qquad V_2 = x$ cm^3
$T_1 = 17°C = 290$ K $\qquad T_2 = 0°C = 273$ K

$$\frac{P_1 V_1}{T_1} = \frac{P_2 V_2}{T_2}$$

$$\therefore \frac{770 \times 420}{290} = \frac{760 \times x}{273}$$

$$x = \frac{770 \times 420 \times 273}{290 \times 760} \text{ cm}^3$$

$$= 400 \text{ cm}^3$$

Volume of gas at s.t.p $= 400$ cm^3

2. *A gas occupies 36 cm³ at 27 °C and 750 mm pressure. At what temperature will the same mass of gas occupy 72 cm³ at 1000 mm pressure?*

$P_1 = 750$ mm $\qquad P_2 = 1000$ mm
$V_1 = 36$ cm^3 $\qquad V_2 = 72$ cm^3
$T_1 = 27°C = 300$ K $\qquad T_2 = x$ K

$$\frac{P_1 V_1}{T_1} = \frac{P_2 V_2}{T_2}$$

$$\therefore \frac{750 \times 36}{300} = \frac{1000 \times 72}{x}$$

$$x = \frac{1000 \times 72 \times 300}{750 \times 36} = 800 \text{ K}$$

Temperature is 800 K or (800–273)°C = 527°C

7.6 Gay-Lussac's law of combining volumes

It was known during the time of Joseph Priestley that when hydrogen burns it combines with approximately half its volume of oxygen. Henry Cavendish showed that the volumes of the combining gases in this reaction were almost exactly in the ratio 2:1. The French scientist Gay-Lussac was, however, the first to investigate systematically the combining volumes of gases. His interest in this subject arose from experiments he was doing on the percentage composition of samples of air obtained from various parts of the world, and as a result of which he established a method of gas analysis. By an experiment on the reaction of hydrogen with oxygen, he showed that 200 cm^3 of hydrogen combine with 100 cm^3 of oxygen to give 200 cm^3 of steam, or, in other words, that:

2 vols. of hydrogen + 1 vol of oxygen → 2 vols. of steam

By extending his work to other gaseous combinations, he showed that:

1 vol. of hydrogen + 1 vol. of chlorine → 2 vols. of hydrogen chloride
2 vols. of carbon monoxide + 1 vol. of oxygen → 2 vols. of carbon dioxide
1 vol. of hydrogen chloride + 1 vol. of ammonia → solid ammonium chloride
1 vol. of nitrogen + 3 vols. of hydrogen → 2 vols. of ammonia

In all the cases he investigated he noticed the simple relationship existing between the volumes of the gases involved, and in 1808 he advanced his now famous law:

When gases combine, they do so in volumes which bear a simple ratio to one another, and to the volume of the product if it is gaseous, provided all the volumes are measured at the same temperature and pressure. As volumes vary with temperature and pressure, the reason for the last part of his law should be obvious.

The law *only applies to gases.* If hydrogen and oxygen were combined under such conditions that water, and not steam, was formed the volume of the water would be negligible and would bear no simple relationship to the volume of the gases. Such is the case, also, with the formation of solid ammonium chloride and numerous other reactions, e.g.:

Solid carbon + 1 vol. of oxygen → 1 vol. of carbon dioxide
Solid sulphur + 1 vol. of oxygen → 1 vol. of sulphur dioxide

It is difficult to see at first why 2 volumes of hydrogen and 1 volume of oxygen should give only *two* volumes of steam. This is because steam is $\frac{3}{2}$ times denser than the mixture of hydrogen and oxygen, but this whole question should become clearer later in the chapter.

A certain similarity exists between Gay-Lussac's Law and one of the main points in Dalton's theory. Gay-Lussac stated that:

When volumes of gases combine they do so in numbers which bear a simple ratio to one another.

Dalton's theory stated that:

When atoms combine they do so in numbers which bear a simple ratio to one another.

This similarity led J. J. Berzelius, a Swede, to suggest that equal volumes of all gases at the same temperature and pressure contain the same number of atoms or "compound atoms". Unfortunately, deductions made from this suggestion disagreed with Dalton's theory. For example:

1 vol. of hydrogen + 1 vol. of chlorine → 2 vols. of hydrogen chloride.

If Berzelius' suggestion is correct, and 1 volume of hydrogen contains n atoms, then 1 volume of chlorine will also contain n atoms, and 2 volume of hydrogen chloride will contain $2n$ "compound atoms", i.e.

n atoms of hydrogen + n atoms of chlorine → $2n$ "compound atoms" of hydrogen chloride, or

1 atom of hydrogen + 1 atom of chlorine → 2 "compound atoms" of hydrogen chloride.

Therefore 1 "compound atom" of hydrogen chloride must contain $\frac{1}{2}$ an atom of hydrogen and $\frac{1}{2}$ an atom of chlorine. Accepting the fact that the atom is indivisible, this statement is absurd.

John Dalton was inclined to believe that this showed Gay-Lussac's experimental work to be faulty, and it was left to the Italian physicist Avogadro to reconcile Dalton's theory with Gay-Lussac's Law.

7.7 Avogadro's law

To Amedeo Avogadro (figure 7.3), the Italian, we owe the conception of the molecule, and the abandoning of the term 'compound atom'. The term 'molecule' applies to elements and compounds, and is defined as *the smallest part of an element or compound that can exist alone*. This means that elements do not necessarily exist as single atoms, but may exist as groups of two or more atoms. The number of atoms in a molecule of an element is referred to as its *atomicity*. Thus helium, which contains one atom per molecule, is monatomic; hydrogen, which contains two atoms per molecule, is diatomic and so on. In 1811 Avogadro formulated his law:

Figure 7.3 Amedeo Avogadro (1776–1856). This stamp, issued by Italy to commemorate the centenary of his death, not only has his signature on it but is also the only stamp in the world with a full written statement of a scientific law "Equal volumes of all gases at the same conditions of temperature and pressure contain the same number of molecules." (Photo: A. Rendell.)

Equal volumes of all gases at the same temperature and pressure contain the same number of molecules.

Applying this relationship to the combination between hydrogen and chlorine, we get:

1 vol. of hydrogen + 1 vol. of chlorine → 2 vols. of hydrogen chloride
n molecules of hydrogen + n molecules of chlorine → $2n$ molecules of hydrogen chloride
1 molecule of hydrogen + 1 molecule of chlorine → 2 molecules of hydrogen chloride

This means that if the molecules of hydrogen and chlorine contain an even number of atoms this deduction is not in conflict with Dalton's theory.

There is much evidence for believing that the hydrogen chloride molecule does in fact contain only one atom of hydrogen and one atom of chlorine combined together. Where the hydrogen is concerned, for example, hydrogen chloride gives rise to only one sodium compound by substitution of hydrogen with sodium, namely sodium chloride, and that compound contains no hydrogen. Similar compounds, such as sulphuric acid, which contains two hydrogen atoms, form two sodium derivatives: sodium sulphate, Na_2SO_4, and sodium hydrogensulphate, $NaHSO_4$. Since hydrogen chloride contains only one hydrogen atom, the hydrogen molecule must be diatomic. (Further evidence in support of the diatomicity of hydrogen is beyond the scope of this book.) By similar reasoning the chlorine molecule can be shown to be diatomic.

The reaction between hydrogen and chlorine now becomes clear if, instead of using the more conventional equation, we represent it in this way:

$$HH + ClCl \rightarrow HCl + HCl$$

i.e. one molecule of hydrogen containing two atoms combines with one molecule of chlorine containing two atoms, to give two molecules of hydrogen chloride each containing one atom of hydrogen combined with one atom of chlorine.

It is important to realise exactly the difference between an *atom* of an element and a *molecule* of an element. An atom is the smallest particle of an element that can take part in a chemical change. Notice that single atoms of hydrogen and chlorine are present in a molecule of hydrogen chloride. Therefore, in accordance with the definition, these atoms can be considered to have taken part in a chemical change.

A molecule of an element is the smallest part of it that can exist alone. Therefore the hydrogen and chlorine, before they have taken part in a chemical reaction, are represented as aggregates of two atoms, i.e. as molecules, in which form those elements normally exist.

7.7.1 The Avogadro constant

In section 3.10, where the *mole* was discussed, we explained that the number of particles in a mole of a substance is $6{\cdot}02 \times 10^{23}$, and that this number is known as *Avogadro's constant.*

It is possible therefore to restate *Avogadro's Law*, which refers to gases only, as:

Equal volumes of all gases at the same temperature and pressure contain the same number of moles.

7.7.2 Problems on Gay-Lussac's law and Avogadro's law

(*a*) *50 cm³ of each of the following gases are burned in excess of oxygen: carbon monoxide, propane (C_3H_8). Calculate in each case the volume of oxygen consumed and the volume of any gaseous product. Assume that all volumes are measured at the same room temperature and pressure.*

Carbon monoxide

First, the equation is needed:

$$2CO + O_2 \rightarrow 2CO_2$$

i.e. 2 molecules carbon monoxide + 1 molecule oxygen → 2 molecules carbon dioxide, or $2n$ molecules carbon monoxide + n molecules oxygen → $2n$ molecules carbon dioxide.

By Avogadro's Law, if n molecules of oxygen occupy 1 volume, then, at the same temperature and pressure, $2n$ molecules of carbon monoxide and $2n$ molecules of carbon dioxide will each occupy 2 volumes, i.e.

2 vols. carbon monoxide + 1 vol. oxygen → 2 vols, carbon dioxide
50 cm^3 carbon monoxide + 25 cm^3 oxygen → 50 cm^3 carbon dioxide,

i.e. 25 cm^3 oxygen are consumed and 50 cm^3 carbon dioxide are formed.

Propane

$C_3H_8 + 5O_2 \rightarrow 3CO_2 + 4H_2O$

i.e. 1 mol. propane + 5 mols. oxygen → 3 mols. carbon dioxide + 4 mols. water

Therefore, by Avogadro's Law

1 vol. propane + 5 vols. oxygen → 3 vols. carbon dioxide + some water

(N.B. The water will be liquid at room temperature), i.e.

50 cm^3 propane + 250 cm^3 oxygen → 150 cm^3 carbon dioxide,

i.e. 250 cm^3 oxygen are consumed and 150 cm^3 carbon dioxide are formed.

(*b*) *If all volumes are measured at normal pressure and 120 °C, state the volumes and names of the gases remaining when 100 cm^3 of methane are mixed with 300 cm^3 of oxygen and the mixture exploded.*

$CH_4 + 2O_2 \rightarrow CO_2 + 2H_2O$

i.e. 1 mol. methane + 2 mols. oxygen → 1 mol. carbon dioxide + 2 mols. steam

By Avogadro's Law,

1 vol. methane + 2 vols. oxygen → 1 vol. carbon dioxide + 2 vols. steam

100 cm^3 methane + 200 cm^3 oxygen → 100 cm^3 carbon dioxide + 200 cm^3 steam

As only 200 cm^3 of oxygen are used in the reaction, 100 cm^3 of oxygen will be left.

Also remaining will be 100 cm^3 of carbon dioxide and 200 cm^3 of steam formed by the reaction.

7.8 Relationship between relative molecular mass and vapour density

Equal volumes of all gases at the same temperature and pressure contain the same number of molecules. Therefore, if a given volume of a gas A has a mass 3 times that of the same volume of a gas B, each molecule of gas A must be 3 times as heavy as each molecule of gas B. It is therefore obvious that the molecular mass of a gas is directly proportional to its density.

The methods used in physics for expressing densities are not normally used with gases, since the densities of gases are so small that the fractional values obtained would be awkward to use. Instead, the density of hydrogen is used as a standard to give what is called the *vapour density* of a gas.

Vapour density is the number of times the density of a gas is greater than the density of hydrogen at the same temperature and pressure.

Or alternatively:

Vapour density is the number of times a given volume of gas is heavier than the same volume of hydrogen, measured at the same temperature and pressure.

These definitions are similar to those used for specific gravity or relative density, except that:

(*a*) hydrogen, not water, is used as the standard;

(*b*) since gases are concerned, the volumes of which alter greatly with temperature and pressure, temperature and pressure must be carefully specified in the definition. As with specific gravity, vapour density has no units; it is a ratio.

An exact relationship between vapour density and molecular mass may now be derived:

$$\text{Vapour density} = \frac{\text{Mass of a given volume of gas}}{\text{Mass of an equal volume of hydrogen at same temperature and pressure}}$$

Therefore, using Avogadro's Law,

$$\text{Vapour density} = \frac{\text{Mass of } n \text{ molecules of gas}}{\text{Mass of } n \text{ molecules of hydrogen}}$$

$$= \frac{\text{Mass of 1 molecule of gas}}{\text{Mass of 1 molecule of hydrogen}}$$

But the hydrogen molecule contains two atoms,

$$\therefore \text{Vapour density} = \frac{\text{Mass of 1 molecule of gas}}{\text{Mass of 2 atoms of hydrogen}} \quad \ldots \text{(A)}$$

By definition,

$$\text{Molecular mass} = \frac{\text{Mass of 1 molecule of gas}}{\text{Mass of 1 atom of hydrogen}} \ldots \text{(B)}$$

Comparing *A* and *B*, we see that:

Molecular mass = 2 × Vapour density

If the mass of the carbon-12 atom is used as the standard in the definition of molecular mass the hydrogen atom has a mass of 1·008 units. Then the relationship between molecular mass and vapour density becomes:

Molecular mass = 2 × 1·008 vapour density
i.e. Molecular mass = 2·016 vapour density

7.8.1 Problems involving vapour density

(*a*) *Calculate the molecular mass of oxygen given that at s.t.p. 333 cm^3 of oxygen has a mass of 0·48 g. 1 dm^3 (1000 cm^3) of hydrogen at s.t.p. has a mass of 0·09 g*

333 cm^3 of oxygen has a mass of 0·48g

$\therefore$ 1000 cm^3 of oxygen has a mass of $\dfrac{0{\cdot}48 \times 1000}{333}$ g = 1·44 g

$$\text{Vapour density} = \frac{\text{Mass of given volume of gas}}{\text{Mass of an equal volume of hydrogen at same temperature and pressure}}$$

$$= \frac{1{\cdot}44}{0{\cdot}09} = 16{\cdot}0$$

Molecular mass = 2 × vapour density = 2 × 16·0 = 32·0

(*b*) *Assume air to be a mixture of oxygen and nitrogen (section 13.2). Calculate its percentage composition by volume if 1000 cm^3 at s.t.p. has a mass of 1·30 g. 1000 cm^3 of hydrogen at s.t.p. has a mass of 0·09 g.*

As in example (a), vapour density of air $= \dfrac{1{\cdot}30}{0{\cdot}09} = 14{\cdot}4$

This is the average vapour density of the oxygen (vapour density = 16) and the nitrogen (vapour density = 14) that make up air. (Other gases are present in small amounts only and can be neglected here.)

Let the fraction of oxygen in the air be = x

Then the fraction of nitrogen in the air = $1-x$

$(16 \times x) + 14(1-x) = 14{\cdot}4$

$2x = 0{\cdot}4$

$x = 0{\cdot}2$

i.e. air is 20% oxygen and 80% nitrogen

Note: This value for the vapour density of air is of value since gases with vapour densities less than 14·4 are less dense than air. (If they are substantially less dense, they can be collected by downwards displacement of air.) Gases with vapour densities greater than 14·4 are denser than air. (If they are substantially denser, they can be collected by upwards displacement of air.)

7.9 The mole and molar volume

7.9.1 Introduction

The *mole* of a substance is $6{\cdot}02 \times 10^{23}$ particles of the substance, and the mass of any substance containing a mole of particles is the *molar mass*. Thus the molecular mass of methane, CH_4, is $(12+4\times 1) = 16$ and the molar mass of methane is 16 g. Similarly the hydrogen molecule, which contains two hydrogen atoms each with an atomic mass of 1, has a molar mass of 2 g. If the modern carbon-12 standard is used the atomic mass of the hydrogen atom is 1·008 and the molar mass of hydrogen molecules is 2·016 g. The density of hydrogen at s.t.p. is 0·09 g dm^{-3} (g per litre). Therefore the volume of one mole of hydrogen at s.t.p. $= \dfrac{2{\cdot}016}{0{\cdot}090}$ dm^3 = 22·4 dm^3 (litres). This volume is known as the *molar volume*. At s.t.p. it is equal to 22·4 litres for all gases. At room temperature and pressure the molar volume is approximately 24 litres for all gases.

The reason for the constancy of this figure for all gases becomes clear with an example:

Helium has a molecular mass of 4, i.e. its molecular mass is twice that of hydrogen. Density is proportional to molecular mass (see section 7.8). Therefore helium is twice as dense as hydrogen at the same temperature and pressure. As helium has twice the molecular mass *and* twice the density of hydrogen, a mole of it will occupy the same volume.

In general terms, if a gas has a molecular mass *n* times greater than hydrogen it will be *n* times denser. Therefore a mole of any gas will occupy the same volume, e.g.

	Molar mass	Volume of 1 g at s.t.p.	Volume of 1 mole at s.t.p.
Hydrogen, H_2	2 g	11 200 cm^3	22 400 cm^3
Helium, He	4 g	5 600 cm^3	22 400 cm^3
Methane, CH_4	16 g	1 400 cm^3	22 400 cm^3
Oxygen, O_2	32 g	700 cm^3	22 400 cm^3
Sulphur dioxide, SO_2	64 g	350 cm^3	22 400 cm^3

This fact enables us to have:

(*a*) an alternative method for calculating the molecular mass of a gas, the density of which is known (see problem (*a*) below).

(*b*) a method for calculating the volumes of gases produced during chemical reactions (see problems (*b*) and (*c*) below).

7.9.2 Problems involving molar volumes

(*a*) *0·16 g of methane at s.t.p. occupy 224 cm³. Calculate its molecular mass.*

The molar volume of a gas at s.t.p. is 22·4 litres. 224 cm^3 of methane has a mass of 0·16 g.

22 400 cm^3 of methane has a mass of $\dfrac{0{\cdot}16 \times 22\,400}{224}$ g = 16 g

The molecular mass of methane = 16

(*b*) *Calculate the volume of oxygen, measured at s.t.p., which is evolved when (i) 1 mole of potassium chlorate(V), (ii) 24·5 g of potassium chlorate(V), are heated to constant mass. (K = 39; Cl = 35·5; O = 16.)*

(i) The equation for the reaction is:

$$2KClO_3 \rightarrow 2KCl + 3O_2$$

$$2[\underbrace{39 + 35{\cdot}5 + (3 \times 16)}_{245}]$$

i.e. 2 moles of potassium chlorate(V) gives 3 moles of oxygen

∴ 1 mole of potassium chlorate(V) gives 1·5 moles.

But 1 mole of oxygen at s.t.p. occupies 22·4 litres

∴ 1 mole of potassium chlorate(V) gives 1·5 × 22·4 litres of oxygen
= 33·6 litres of oxygen

(ii) 245 g of potassium chlorate(V) gives 3 moles of oxygen, i.e. give $3 \times 22{\cdot}4$ litres of oxygen at s.t.p.

$\therefore$ 24·5 g of potassium chlorate(V) give $\dfrac{3 \times 22{\cdot}4 \times 24{\cdot}5}{245}$ litres at s.t.p.

= 6·72 litres at s.t.p.

(*c*) What mass of mercury(II) oxide must be heated to give 380 cm^3 of oxygen measured at 27 °C and 780 mm pressure? (Hg = 201, O = 16.).

First the volume of oxygen at s.t.p. is calculated:

$$P_1 = 780 \text{ mm} \qquad P_2 = 760 \text{ mm}$$
$$V_1 = 380 \text{ cm}^3 \qquad V_2 = x \text{ cm}^3$$
$$T_1 = (273+27)\text{K} = 300\text{K} \qquad T_2 = 273\text{K}$$

$$\frac{P_1V_1}{T_1} = \frac{P_2V_2}{T_2} \quad \therefore \quad \frac{780 \times 380}{300} = \frac{760 \times x}{273}$$

$$\therefore x = \frac{780 \times 380 \times 273}{300 \times 760}$$

Vol. at s.t.p. = 355 cm^3

The equation is: $2HgO \rightarrow 2Hg + O_2$

$$\underbrace{2(201+16)}_{434}$$

i.e. 1 mole of oxygen is obtained from 434 g of mercury(II) oxide.

22 400 cm^3 of oxygen at s.t.p, are obtained from 434 g of mercury(II) oxide.

355 cm^3 at s.t.p. are obtained from $\dfrac{355 \times 434}{22\,400}$ g of mercury(II) oxide

= 6·88 g of mercury(II) oxide

7.10 Cannizzaro's method for determining atomic masses

In 1811 Avogadro proposed his now famous law. Much more progress in theoretical chemistry would have been possible during the next fifty years if his views had been fully appreciated at the time. It was not until 1858, two years after Avogadro's death, that any real progress was made. Then a fellow Italian, Stanislao Cannizzaro, realised the importance of Avogadro's views, and deduced from them a new method for the determination of atomic masses. Many of the doubts and difficulties existing at the time in the minds of chemists were the gradually dispersed.

Cannizzaro's method for finding atomic masses can be applied only to those elements, such as carbon and nitrogen, which form a large number of gaseous or readily volatile compounds. The vapour density of each of these compounds is found. We have already proved, using Avogadro's Law, that the molecular mass is double the vapour density. So the molecular mass of each of these compounds is known.

The number of units of mass of the element under consideration, in the molecular mass of each compound, is then determined by a process of analysis. Taking carbon as an example, the following are results that might be obtained from a small number of its compounds:

Compound	Vapour density	Molar mass	% carbon in compound	Mass of carbon in one mole (i.e. molar mass) of compound = $\frac{\text{molecular mass} \times \text{percentage}}{100}$
Methane	8	16 g	75·0	12 g
Ethane	15	30 g	80·0	24 g
Propane	22	44 g	81·8	36 g
Carbon monoxide	14	28 g	42·9	12 g
Carbon dioxide	22	44 g	27·3	12 g
Ethene	17	28 g	85·7	12 g

However many compounds are analysed in this way, it is found that the number of grams of carbon in the molar mass of one of its compounds is either 12 or a multiple of 12.

Consider what this means. Of all the compounds analysed, it is practically certain that at least one of them contains only one carbon atom per molecule. Therefore the lowest value obtained for the number of grams of carbon in one mole of a compound is the atomic mass. If this is not clear, consider a compound CX_a, containing one carbon atom. Let the atomic mass of carbon $= n$ and of X $= m$. Then the molecular mass $= n + am$, i.e. the number of units by mass of carbon in the molecular mass $= n$. It is therefore apparent from the table of results that the atomic mass of carbon is 12.

Questions on chapter 7

1. Calculate the volume at s.t.p. of a gas which occupies:
(*a*) 380 cm^3 at 900 mm and 27 °C
(*b*) 580 cm^3 at 950 mm and 17 °C
(*c*) 257 cm^3 at 570 mm and −16 °C
(*d*) 190 cm^3 at 852 mm and 11 °C
(*e*) 331 cm^3 at 765 mm and 14 °C
(*f*) 284 cm^3 at 755 mm and 23 °C

2. How can you explain the pressure exerted by a gas on the walls of its containing vessel? State Graham's Law of Gaseous Diffusion.

Explain how the molecular mass of a gas may be determined by measurements of the rates of diffusion of gases. (*J.M.B.*)

3. What do you understand by the phrase "diffusion of a gas"? What evidence does diffusion provide concerning the nature of gases?

Explain how you would expect the rate of diffusion of a gas to change if its temperature was increased.

Describe and explain an experiment to show that hydrogen diffuses faster than air.

The vapour of a compound *Q* containing 87·5 % nitrogen and 12·5 % hydrogen diffuses under identical experimental conditions, at the same rate as oxygen. Deduce (*a*) the empirical formula of *Q*; (*b*) its molecular formula. Show your reasoning clearly. (H = 1; N = 14; O = 16) (*A.E.B.*)

4. Calculate the molecular masses of the following substances:
(*a*) chlorine, if 280 cm^3 at s.t.p. have a mass of 0·89 g;
(*b*) nitrogen, if 420 cm^3 at s.t.p. have a mass of 0·525 g;
(*c*) propanone, if 290 cm^3 of propanone vapour at 91 °C and 784 mm pressure have a mass of 0·58 g.

5. 1000 cm^3 of hydrogen chloride at s.t.p. were dissolved in water. Calculate the volume of a solution of sodium hydroxide containing 42 g per dm^3 required to neutralise the acid. (*O. and C.*)

6. The equation for the action of heat on sodium hydrogencarbonate is

$$2NaHCO_3 \rightarrow Na_2CO_3 + H_2O + CO_2$$

Calculate (i) the mass of the residue on heating 21 g of this substance, (ii) the volume of carbon dioxide evolved, measured at s.t.p. (*O. and C.*)

7. The action of heat on lead(II) nitrate is given by the equation

$$2Pb(NO_3)_2 \rightarrow 2PbO + 4NO_2 + O_2$$

If 16·55 g of lead(II) nitrate are heated, what is (i) the mass of the residue, (ii) the volume of oxygen liberated (measured at s.t.p.)? (*O. and C.*)

8. The following is the equation for the reaction between sodium hydrogensulphite and hydrochloric acid:

$$NaHSO_3 + HCl \rightarrow NaCl + H_2O + SO_2$$

Calculate (i) the mass, (ii) the volume (in dm^3 at s.t.p.) of the sulphur dioxide formed on treating 13 g of sodium hydrogensulphite with excess hydrochloric acid. (*O. and C.*)

9. (*a*) A compound of nitrogen and oxygen contains 46·7% by mass of nitrogen. Calculate the empirical formula for this compound. (N = 14; O = 16)

(*b*) 1·4 g of a gaseous oxide of carbon occupy 1·2 litres at room temperature and pressure.

What is (i) the molecular mass of this oxide, (ii) its likely formula? (1 mole of any gas at room temperature and pressure occupies 24 litres. C = 12; O = 16)

10. 10 cm^3 of ethene, (C_2H_4), are burned completely in oxygen. Calculate (i) the volume of oxygen used, (ii) the volume of the product. All volumes are measured at s.t.p. (*O. and C.*)

11. 10 cm^3 of each of the following gases were burned in excess of oxygen: carbon monoxide, hydrogen sulphide. Calculate the volume of oxygen consumed and the volume of the product in each case. All volumes were measured at s.t.p. (*O. and C.*)

12. If 5 g of impure calcium carbonate, containing 5 per cent of inert impurity, were subjected to strong heating, what would be the mass of the solid residue?

What mass and what volume of gas, measured at 15 °C and 800 mm, would also be evolved? (*B*)

13. State fully Gay-Lussac's Law of combining volumes. Write equations for (*a*) the explosion of hydrogen and oxygen; (*b*) the synthesis of ammonia from nitrogen and hydrogen.

If all volumes are measured at s.t.p., state the volume and name of the gas remaining when (i) 200 cm^3 of hydrogen are exploded with 500 cm^3 of oxygen, (ii) 300 cm^3 of hydrogen combine completely with 100 cm^3 of nitrogen. (*J.M.B.*)

14. State the law expressing the combination of gases by volume and describe an experiment which demonstrates the law in the case of hydrogen and chlorine.

5 cm^3 of a gas, containing *only* carbon and hydrogen, required 15 cm^3 of oxygen for complete combustion and 10 cm^3 of carbon dioxide were formed. All volumes were measured at the same temperature and pressure. What is the molecular formula of the gas? (*D.*)

15. Define *vapour density* and *molecular mass* and state how the two are related to one another.

Calculate the vapour density of carbon dioxide, chlorine, ammonia and nitrogen oxide.

Excess of hydrochloric acid is allowed to react with 10 g of anhydrous sodium carbonate until the reaction is complete. Calculate the volume of gas liberated if it is measured at n.t.p. (s.t.p.) (*J.M.B.*)

16. The chloride of a metal has a vapour density of 175. What is its molecular mass?

The chloride contains 40·6 per cent of chlorine by mass. (Atomic mass of chlorine = 35·5.)

How many atoms of chlorine are there in one molecule of the chloride?

What is the maximum value for the atomic mass of the metal? (*J.M.B.*)

17. A certain specimen of iron(II) sulphide contained 12 per cent of free iron. 10 g of this solid were treated with excess of hydrochloric acid and the gas was measured at s.t.p. What volume did it occupy? (*C.W.B.*)

18. Sketch and label the apparatus you would use to determine the loss in mass on heating a given mass of potassium chlorate(V), and the volume of oxygen evolved, in one and the same experiment.

2·45 g of potassium chlorate(V) when heated lost 0·96 g in mass and gave off 672 cm^3 of oxygen at s.t.p. From these results calculate the vapour density of oxygen. (*O.*)

19. Define the *vapour density* of a substance and state how it is related to the molecular mass.

Three compounds of a certain element contain 36·36, 50 and 60 per cent of the element, and have vapour densities of 22, 32 and 40 respectively. Calculate the masses of the element in one molecule of each of these compounds, and hence deduce a probable atomic mass for the element. Explain carefully the reasoning on which the deduction is based. (*O.*)

20. State (*a*) Gay-Lussac's Law, (*b*) Avogadro's law. Describe an experiment that illustrates Gay-Lussac's law.

X is a gaseous compound of silicon and hydrogen; its density was found to be 1·333 grams per dm^3. When 100 cm^3 of X was completely decomposed, 200 cm^3 of hydrogen was obtained. (The relative atomic mass of silicon (Si) is 28; one mole of a gas occupies 24 dm^3 at the conditions used in these determinations.)

(i) What is the molecular mass of *X*?

(ii) Show how Avogadro's law enables you to deduce the number of hydrogen atoms in one molecule of X.

(iii) What is the formula of X?

(iv) X burns readily in air. What products would you expect to obtain under these conditions? (*A.E.B.*)

21. Define (*a*) vapour density, (*b*) molecular mass.

A compound has the formula XCl_2, and the vapour density of this chloride is 40.

What is the molecular mass of the chloride?

What is the atomic mass of *X*? (*J.M.B.*)

22. Nitrogen combines with certain metals forming compounds called nitrides; two such metals are lithium (Li) and calcium. These nitrides react with water giving off ammonia and leaving a metallic hydroxide.

When 1·75 g of lithium nitride is treated with water, 1·18 litres of ammonia at 17 °C and 765 mm pressure is evolved.

(*a*) Calculate the empirical formula of lithium nitride.

(*b*) What is the valency of lithium in this compound?

(*c*) Write the equation for the action of water on lithium nitride.

(*d*) Write the formula for calcium nitride.

(Li = 7·00; N = 14·0; Molar volume of a gas is 22·4 litres at s.t.p.) (*L.*)

23. When heated in a current of nitrogen, magnesium reacts with it to form magnesium nitride, Mg_3N_2.

Magnesium nitride reacts with water to form magnesium hydroxide and ammonia.

(*a*) Write the equation for the first reaction, and calculate the volume of nitrogen, measured at s.t.p., required to react with 8·0 g of magnesium.

(*b*) Write the equation for the second reaction and calculate the volume of ammonia produced, measured at s.t.p., if the magnesium nitride formed in (*a*) reacts completely with water. (Assume that all the ammonia can be driven off and collected.) (*J.M.B.*)

8

The Activity (or Electrochemical) Series and Electrolysis

8.1 Introduction

The study of a scientific subject, particularly one where a large number of facts has to be known, is always made simpler if some form of classification is possible. Several useful classifications have already been made in this book. We have classified substances into solids, liquids and gases, and into elements, compounds and mixtures. We have classified elements according to their valency; their method of linkage with one another as covalent or electrovalent. The elements have also been classified as metals or non-metals. *The activity series (or electrochemical series) is a method of classifying the chemical elements according to their reactivity*, i.e. according to the ease with which they take part in chemical reactions, and is a great help in the study of these elements and their compounds.

There are several methods by which such a series can be obtained. For example, for metals, which are the most important members of the activity series we can

1. consider the chemical properties of metals; or
2. consider the chemical properties of compounds of the metals.

8.2 The activity series

8.2.1 Chemical properties of the metals

A very good idea of the comparative reactivity of metals can be obtained by examining their behaviour with air or oxygen, water, acids and with salts of other metals.

(a) Reactions of metals with air (see also chapter 13)
Potassium, sodium, calcium and magnesium all burn vigorously in air or oxygen. Aluminium, zinc and iron also burn readily if they are strongly heated, particularly in oxygen. Tin, lead, copper and mercury are all converted into the oxide on heating in air, but they do not burn. Silver, gold and platinum are completely unaffected by air or oxygen. (Tarnishing of silver does occur in industrial areas, but this is due to hydrogen sulphide in the atmosphere.)

(b) Reactions of metals with water (see also chapter 11)
Those metals that burn in air or oxygen do in general react with water or steam. Of all the more common metals, potassium reacts most vigorously with cold water:

$$2K(s) + 2H_2O(l) \rightarrow 2KOH(aq) + H_2(g)$$

Sodium and calcium react similarly, but less violently.

Magnesium reacts *very* slowly with cold water. Zinc and iron do not react with cold water at all. When heated, all three metals react readily with steam, e.g.,

$$3Fe(s) + 4H_2O(g) \rightleftharpoons Fe_3O_4(s) + 4H_2(g)$$

Other metals in table 8.1, which gives the activity series for common metals, are without appreciable action on either water or steam.

(c) Reactions of metals with acids

Potassium, sodium and calcium react vigorously and *dangerously* with dilute hydrochloric acid to give hydrogen. The other metals occurring above lead in the series (see table 8.1) react similarly, but with less vigour. A very slight reaction occurs with lead.

These reactions with acids indicate that all these metals form ions more readily than does hydrogen.

Let us consider as an example the reaction of zinc with hydrochloric acid. Both hydrogen ions and chloride ions are present in hydrochloric acid solution. When the zinc is added to the acid each zinc atom loses two electrons to two hydrogen ions to form two hydrogen atoms. These combine together and gaseous hydrogen is released from the solution (equation 8.1). The chloride ions are unaffected and since they appear on both sides of the equation, as can be seen if equation 8.1 is written as equation 8.2, the chloride ions can be omitted to give equation 8.3. This is known as an *ionic equation*, and can be used to represent the action of any acid on zinc (except oxidizing acids).

$$Zn + 2HCl \rightarrow ZnCl_2 + H_2 \tag{8.1}$$

$$Zn + 2H^+ + 2Cl^- \rightarrow Zn^{2+} + 2Cl^- + H_2 \tag{8.2}$$

$$Zn + 2H^+ \rightarrow Zn^{2+} + H_2 \tag{8.3}$$

Metals below lead in the activity series, such as copper, silver and gold, do not react with hydrochloric acid, indicating that these metals form ions less readily than hydrogen. For these and other reasons it is usual to include hydrogen in the activity series of metals. Summing up, we can say: *Metals above hydrogen in the activity series, and which therefore form ions more readily than hydrogen, displace hydrogen from hydrochloric acid, while metals below hydrogen in the series do not displace hydrogen from the acid.*

Certain acids, e.g. nitric acid and concentrated sulphuric acid, behave as acids *and* oxidising agents. The reactions of these acids are more complicated, and are not governed by the same rule.

(d) Reactions of metals with salts of other metals

If a metal is added to a solution of a salt of a less reactive metal, the less reactive metal is displaced from the solution. For example, zinc, when added to a solution of copper(II) sulphate, gives a deposit of copper.

$$Zn + CuSO_4 \rightarrow ZnSO_4 + Cu$$

$$\text{or } Zn + Cu^{2+} + SO_4^{\ 2-} \rightarrow Zn^{2+} + SO_4^{\ 2-} + Cu$$

$$\text{or } Zn + Cu^{2+} \rightarrow Zn^{2+} + Cu$$

This is a reaction similar to the action of an acid on a metal. The zinc forms ions more readily than copper, and therefore displaces the copper.

Experiment. A silver tree

A beautiful experiment is the displacement of silver from silver nitrate solution by copper:

$$2Ag^{+}(aq) + Cu(s) \rightarrow 2Ag(s) + Cu^{2+}(aq)$$

If a piece of copper wire is placed in 2% silver nitrate solution the silver is displaced and grows as lustrous silver crystals on the copper wire. A sheet of copper metal can be cut into the shape of a tree. After about 15 minutes in the silver nitrate solution it looks like a silver christmas tree (figure 8.1).

Although this type of reaction generally holds true for metals, a complication occurs with those which normally react with water. We can collect together the ideas that have been mentioned above in Table 8.1. The metals are arranged in order of decreasing activity.

Figure 8.1 A silver tree. (Photo: A. Rendell).

TABLE 8.1 THE ACTIVITY SERIES AND PROPERTIES OF METALS

Metal	Air or oxygen on metal	Water on metal	Acids on metal	Metals on salts of other metals
K	Burn in air or oxygen	React with cold water (with decreasing ease)	Displace hydrogen from acids which are not oxidising agents (with decreasing ease)	Each metal will displace a metal lower in the series from a solution of one of its salts
Na				
Ca				
Mg				
Al		React with steam when heated		
Zn				
Fe				
Sn	Converted into the oxide by heating in air	No reaction with water or steam		
Pb				
(H)			React only with oxidizing acids	
Cu				
Hg				
Ag	Unaffected by air or oxygen			
Au			No action with acids	
Pt				

You must not expect a table like table 8.1 to give an exact guide to the properties of metals. It does, however, give a good general guide.

8.2.2 Chemical properties of compounds of metals

It is reasonable to suppose that when a compound is formed from a reactive metal (that is, a metal that combines readily with other elements) it will be comparatively stable. Sodium and chlorine, for example, show a great affinity for one another. Therefore it is unlikely that these elements, once combined, will be readily parted. In general, we can say that compounds of metals high in the electrochemical series are less readily decomposed than those low in the series. Other properties of compounds also depend on the position in the series of any metal contained in it. This is made clear by studying certain typical compounds:

(a) Oxides

Only the oxides of the unreactive metals mercury, silver and gold decompose into the elements on heating. If heated in a stream of hydrogen the oxides of zinc and of metals above it in the series are unaffected; those of copper, lead, tin and iron are reduced to the metal: e.g.,

$CuO(s) + H_2(g) \rightarrow Cu(s) + H_2O(g)$

If heated with carbon these four metal oxides and also zinc oxide are reduced to the metal, e.g.,

$ZnO(s) + C(s) \rightarrow Zn(g) + CO(g)$

Most oxides of metals higher than zinc in the activity series cannot be reduced in these ways (e.g., K_2O, Na_2O, CaO, Al_2O_3). These metals have to be obtained from their compounds by *electrolysis*.

The oxides of the three metals of the top of the series, potassium, sodium and calcium, combine with water to give the hydroxide:

$CaO(s) + H_2O(l) \rightarrow Ca(OH)_2(s)$

A slight reaction occurs with magnesium oxide, but other oxides are unaffected.

(b) Hydroxides

Potassium and sodium hydroxides are readily soluble in water to give strongly alkaline solutions. Calcium hydroxide is slightly soluble, but other hydroxides are practically insoluble in water.

Potassium and sodium hydroxides are unaffected by heat; the others decompose to give the oxide and water:

$Cu(OH)_2(s) \rightarrow CuO(s) + H_2O(g)$

(c) Carbonates

Only sodium and potassium carbonates are soluble in water. They are also stable to heat. The others are insoluble and decompose on heating to give the oxide and carbon dioxide.

$PbCO_3(s) \rightarrow PbO(s) + CO_2(g)$

Table 8.2 summarises the properties of some important compounds of the metals, which are again arranged in order of decreasing reactivity.

(d) Nitrates

All nitrates are soluble in water.

On heating the solid nitrate, those of sodium and potassium decompose to give the nitrite and oxygen:

$2KNO_3(s) \rightarrow 2KNO_2(s) + O_2(g)$

Most other nitrates give the oxide, nitrogen dioxide and oxygen:

$2Pb(NO_3)_2(s) \rightarrow 2PbO(s) + 4NO_2(g) + O_2(g)$

In the case of the very unreactive elements, the oxides that are formed decompose further to give the metal and oxygen:

$2AgNO_3(s) \rightarrow 2Ag(s) + 2NO_2(g) + O_2(g)$

8.3 Electrolysis

8.3.1 Introduction

The following experiments serve to introduce the main ideas of electrolysis.

TABLE 8.2 THE ACTIVITY SERIES AND PROPERTIES OF COMPOUNDS

Metal	Oxides				Hydroxides		Carbonates	Nitrates
	Action of heat	Action of hydrogen	Action of carbon	Action of water	Solubility	Action of heat	Solubility and action of heat	Action of heat
K	No reaction	No reaction	No reaction	React to give hydroxide	Very soluble	No reaction	Soluble; unaffected by heat	Decompose to give nitrite and oxygen
Na								
Ca					Slightly soluble	Decompose to give oxide and water	Insoluble Decompose on heating to give oxide and carbon dioxide	Decompose to give oxide, nitrogen dioxide and oxygen
Mg				No reaction; insoluble	Insoluble			
Al								
Zn			Reduced to metal					
Fe		Reduced to metal						
Sn								
Pb								
Cu								
Hg	Decompose to give metal and oxygen				Hydroxides unstable or non-existent	Hydroxide unstable or non-existent	Carbonates unstable or non-existent	Decompose to give metal, nitrogen dioxide and oxygen
Ag								
Au								
Pt								

Experiment. Electrolysis

(*a*) Set up an electrolytic cell (figure 8.2) by connecting a lamp-holder or an ammeter and a switch in series with two copper plates and a battery or accumulator.

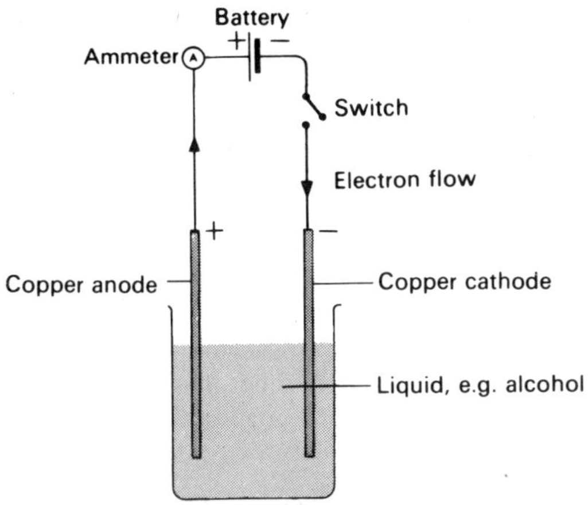

Figure 8.2 An electrolytic cell.

Immerse the copper plates in a beaker of alcohol, and the ammeter or bulb will show that no current passes. Repeat the experiment using other liquids, e.g. methylbenzene and tetrachloromethane, with the same result.

(*b*) Set up the same apparatus with distilled water in the beaker. No measurable current passes, but if a sufficiently high potential difference is applied and a sensitive ammeter is used a small reading is obtained on the ammeter. Then add small quantities of various substances, separately, to the water.

With addition of acids, alkalies or salts a current passes. Chemical reactions occur at the plates.

With addition of sugar, glucose, alcohol, etc., there is still no passage of current.

(*c*) Set up a similar circuit with the copper electrodes immersed in potassium iodide which has been melted in a crucible. A current passes and iodine vapour can be seen at one of the copper plates, indicating chemical decomposition. If the potassium iodide is allowed to cool and solidify no current then passes.

Those compounds which conduct electricity when they are dissolved in water or melted, are called electrolytes. Acids, alkalies and salts are the most important groups of electrolytes.

Other compounds, which do not allow the passage of electricity in these circumstances, are called *non-electrolytes.*

The copper plates, by which the current enters or leaves the solution of the electrolyte, are called *electrodes.*

The electrode connected to the positive pole of the battery or other source of direct current is the *anode.*

The electrode connected to the negative pole of the battery or other source of direct current is the *cathode.*

8.3.2 Explanation of electrolysis

Electrolysis is the chemical change which takes place when an electric current passes through an electrolyte.

Arrhenius in 1873 explained the conductivity of solutions of electrolytes by assuming the presence of ions in the solution. We have already discussed ions in connection with the electronic theory. All electrovalent compounds consist of positively and negatively charged ions, present in such numbers that the substance as a whole is electrically neutral. These ions are normally held together by the force that exists between particles that are positively and negatively charged, called electrostatic attraction. But the ions are free to move when the compound is melted, or when the ions become separated as they do when the compound is dissolved in water. Thus barium chloride ($BaCl_2$) consists of barium and chloride ions. Each barium ion carries a double positive charge and each chloride ion a single negative charge. There are, therefore, in barium chloride twice as many chloride ions as barium ions. The charge on each barium ion is due to the neutral barium atom losing two electrons; that on each chloride ion to the gain by the neutral chlorine atom of one electron. Remember that the number of charges on an ion is equal to the valency of the atom or radical concerned.

Some examples of ions given by various electrovalent compounds are:

Compound	Formula	Ions
Sodium chloride	$NaCl$	Na^+, Cl^-
Magnesium chloride	$MgCl_2$	Mg^{2+}, $2Cl^-$
Sodium sulphate	Na_2SO_4	$2Na^+$, SO_4^{2-}
Lead nitrate	$Pb(NO_3)_2$	Pb^{2+}, $2NO_3^-$
Sodium hydroxide	$NaOH$	Na^+, OH^-

Some *covalent* compounds, such as the mineral acids, also dissolve in water to give solutions which conduct an electric current. This is because they react with water to give ions. They are not ionised when pure. Examples of ions given by covalent compounds are:

Compound	Formula	Ions
Sulphuric acid	H_2SO_4	$2H^+$, SO_4^{2-}
Nitric acid	HNO_3	H^+, NO_3^-
Hydrochloric acid	HCl	H^+, Cl^-

All electrolytes, such as those mentioned above, which are completely ionized or almost completely ionized in solution are called *strong electrolytes.*

Some covalent compounds, such as water and ethanoic acid, only dissociate (i.e. split up) slightly to give ions. Their solutions are poor conductors of electricity and these compounds are called *weak electrolytes.*

It is convenient to remember that, for all electrolytes, metal ions, hydrogen ions and ammonium ions are positively charged. Other ions are negatively charged. The various ions are responsible for conducting the electricity through the solution.

8.3.3 Electrolysis of acids

(a) Dilute sulphuric acid

Experiment. To demonstrate this set up the apparatus shown in fig-

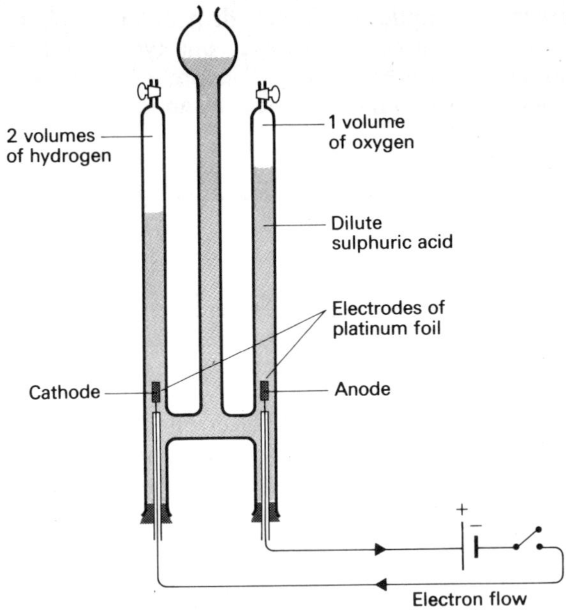

Figure 8.3 Electrolysis of dilute sulphuric acid.

ure 8.3, using electrodes of platinum foil. When an electric current is passed, two volumes of hydrogen are formed for every one volume of oxygen.

Ions are present in the solution from two sources:
Sulphuric acid yields H^+ ions and SO_4^{2-} ions.
Water yields a small number of both H^+ ions and OH^- ions.

When the switch is closed electrons from the battery flow to one of the electrodes and make it negatively charged. It is from this electrode that the electrons enter the solution. *The electrode that feeds electrons into the solution is always the cathode.* The other electrode, the *anode*, becomes positively charged by losing electrons to the battery.

The positively charged hydrogen ions are attracted by the negatively charged cathode, and start moving in that direction. The negatively charged hydroxide and sulphate ions are attracted by the positively charged anode, and start moving towards it. Instead of the normal random movement of ions in the solution, under the influence of the electric field ions of opposite charges are moving in opposite directions.

When hydrogen ions reach the cathode each one receives an electron from it and is converted into an atom:

$$H^+ + e \rightarrow H$$

Hydrogen atoms do not normally exist alone, and pairs of atoms unite to form hydrogen molecules:

$$2H \rightarrow H_2$$

The reactions at the anode are more complicated. Both hydroxide and sulphate ions arrive there, but usually under such conditions only one type of ion is discharged. In this case hydroxide ions are discharged in preference to the sulphate ions, i.e. each hydroxide ion loses an electron to the anode:

$$OH^- - e \rightarrow OH$$

The uncharged hydroxide species formed is called a radical. It is highly reactive and decomposes to give water and oxygen:

$$4OH \rightarrow 2H_2O + O_2$$

We may now sum up what happens as far as elecrons are concerned. The cathode loses electrons to the positive ions. The anode gains electrons from the negative ions. This amounts to a flow of electrons from the cathode to the anode, i.e. an electric current through the solution. The number of electrons lost by the cathode must equal the number gained by the anode. Four electrons are needed to produce one molecule of oxygen by discharge of four hydroxide ions. The same quantity of electricity is involved in the production of two molecules of hydrogen by the discharge of four hydrogen ions. Therefore two molecules of hydrogen are produced for every molecule of oxygen, and this means, applying Avogadro's Law, that two volumes of hydrogen are produced for every volume of oxygen. This is in accordance with the experimental observation (see section 10.6.7).

Other changes take place in the solution. Sulphate ions are moving away from the cathode and hydrogen ions are being discharged at the cathode. This means that the sulphuric acid concentration in the neighbourhood of the cathode decreases.

Sulphate ions accumulate at the anode. Further, the discharge of hydroxide ions at the anode causes more water to ionize (see section 10.6.7.) resulting in an increase of hydrogen-ion concentration, i.e. there is an increase of sulphuric acid concentration in the neighbourhood of the anode.

The main effects of the electrolysis of sulphuric acid are therefore:

(*a*) 1 volume of oxygen produced at the anode;
(*b*) 2 volumes of hydrogen produced at the cathode;
(*c*) an increase in sulphuric acid concentration at the anode;
(*d*) a decrease in suphuric acid concentration at the cathode.

The total amount of sulphuric acid in the solution remains constant, and the hydrogen and oxygen being produced come from the water:

$$2H_2O \rightarrow 2H_2 + O_2$$

(b) Very dilute hydrochloric acid (with carbon electrodes)

Hydrochloric acid yields H^+ and Cl^- ions.
Water yields a small number of H^+ and OH^- ions.

When an electric current is passed, hydrogen ions move to the cathode, where they are discharged by gain of electrons. Therefore gaseous hydrogen is evolved:

$$H^+ + e \rightarrow H, \quad 2H \rightarrow H_2$$

OH^- and Cl^- ions are attracted to the anode. In very dilute solutions OH^- ions are discharged in preference to the Cl^- ions:

$OH^- - e \rightarrow OH, \quad 4OH \rightarrow 2H_2O + O_2$

Therefore oxygen is evolved at the anode.

What changes in acidity will occur in the neighbourhood of the electrodes, and what will be the relative volumes of the gases evolved?

(c) Concentrated hydrochloric acid (with carbon electrodes)

Experiment. Set up the apparatus shown in figure 8.4. On passing an electric current chlorine is evolved at the anode and hydrogen at the cathode. They can be detected by the usual tests.

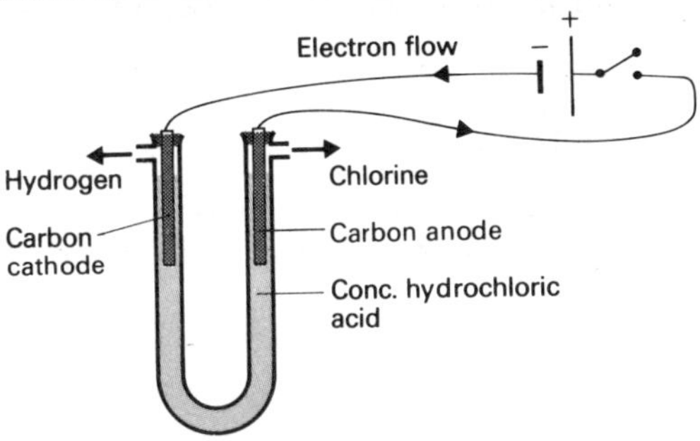

Figure 8.4 Electrolysis of concentrated hydrochloric acid.

As with dilute hydrochloric acid, hydrogen ions move to the cathode, where they are discharged to give hydrogen.

OH^- and Cl^- ions move to the anode but in this case, because the acid is concentrated, there is much higher concentration of chloride ions than in the dilute acid, and it is the chloride ions that are discharged, *not* the hydroxyl ions. The chlorine atoms immediately unite to give molecules:

$Cl^- - e \rightarrow Cl, \quad 2Cl \rightarrow Cl_2$

As two electrons are involved in the formation of (*a*) a hydrogen molecule by the discharge of two H^+ ions, and (*b*) a chlorine molecule by the discharge of two Cl^- ions, equal volumes of the two gases are evolved.

8.3.4 Electrolysis of salt solutions

(a) Sodium chloride solution (carbon electrodes)

Sodium chloride yields Na^+ and Cl^- ions.
Water yields a small number of H^+ and OH^- ions.

When the current is switched on, sodium and hydrogen ions move to the cathode. H^+ ions are discharged much more readily than Na^+ ions. Therefore hydrogen is evolved:

$H^+ + e \rightarrow H, \quad 2H \rightarrow H_2$

As with concentrated hydrochloric acid, chlorine is evolved at the anode. What difference would you expect if a very dilute solution of sodium chloride was used?

The overall result of this electrolysis is to convert sodium chloride and water into hydrogen, chlorine and sodium hydroxide:

$$2NaCl + 2H_2O \rightarrow H_2 + Cl_2 + 2Na^+ + 2OH^-$$

You will be able to deduce that the sodium hydroxide accumulates near the cathode. This is the basis for an important industrial preparation of sodium hydroxide (see section 21.2.7(*c*) and figures 21.4 and 21.5).

(b) Sodium hydroxide solution

The apparatus used for the electrolysis of dilute sulphuric acid (figure 8.3) is suitable for this investigation.

Sodium hydroxide yields Na^+ and OH^- ions.
Water yields a small number of H^+ and OH^- ions.

During electrolysis hydroxide ions are discharged at the anode, and the hydrogen ions at the cathode. Two volumes of hydrogen and one volume of oxygen are obtained:

$$H^+ + e \rightarrow H, \qquad 2H \rightarrow H_2$$
$$OH^- - e \rightarrow OH, \qquad 4OH \rightarrow 2H_2O + O_2.$$

The overall result of the electrolysis is that the amount of sodium hydroxide in the solution remains constant. As with sulphuric acid the compound being used up is the water:

$$2H_2O \rightarrow 2H_2 + O_2.$$

(c) Copper(II) sulphate solution

(i) Using a copper cathode and a platinum anode.

Copper(II) sulphate yields Cu^{2+} and $SO_4^{\ 2-}$ ions.
Water yields a small number of H^+ and OH^- ions.

When the current is switched on, Cu^{2+} and H^+ ions move to the cathode. Copper ions are more readily discharged than hydrogen ions, and metallic copper is deposited on the cathode:

$$Cu^{2+} + 2e \rightarrow Cu$$

At the anode the hydroxide ions are discharged, yielding water and oxygen. The overall reaction is: $2CuSO_4 + 2H_2O \rightarrow 2Cu + O_2 + 4H^+ + 2SO_4^{\ 2-}$

(ii) Using copper electrodes.
If both electrodes are made of copper the same reactions occur at the cathode.
At the anode *neither* $SO_4^{\ 2-}$ *nor* OH^- *ions are discharged.* Instead the anode gains its electrons by conversion of copper of the anode into ions which go into solution:

$$Cu \rightarrow Cu^{2+} + 2e \quad \text{(these electrons are left on the anode)}$$

This is known as electrode ionization. During the course of electrolysis the total concentration of copper(II) sulphate in the solution remains unchanged, and the cathode "grows" at the expense of the anode. The use which is made of this fact in the electrolytic purification of copper is discussed in section 24.3.2.

8.3.5 Preferential electrode reactions

When more than one type of ion migrates to an electrode during electrolysis only one type of ion is discharged. The main factor in deciding which ion is discharged is its position in the electrochemical series. Thus, those metals which are most reactive form ions which are not readily discharged.

Just as metals can be arranged in order of their activity, so can non-metals and non-metallic radicals. The following groups of metallic and non-metallic ions are arranged in order of their ease of discharge:

Metallic ions		Non-metallic ions
K^+	Least readily discharged	SO_4^{2-}
Na^+		NO_3^-
Ca^{2+}	↓	Cl^-
Mg^{2+}		Br^-
Zn^{2+}		I^-
H^+	Most readily discharged	OH^-
Cu^{2+}		

There are, however, two other factors which influence the course of electrolysis.

(a) Ionic concentration

According to its position in the series given above, the hydroxide ions should be discharged more readily than the chloride ion. Indeed, this is so with the electrolysis of a solution containing a very low concentration of chloride ions, e.g. very dilute hydrochloric acid. It must be remembered, however, that the hydroxide-ion concentration is extremely low in water, so that when concentrated hydrochloric acid is electrolysed it is the chloride ions which are discharged.

(b) Material of the electrodes

When copper(II) sulphate solution is electrolysed using a copper anode neither hydroxide nor sulphate ions are discharged, but electrode ionization occurs.

A more important example of the material of an electrode influencing the course of an electrolytic reaction is with the electrolysis of sodium chloride solution using a mercury cathode (see section 21.2.7(*c*)). Here sodium ions are discharged in preference to hydrogen ions, despite the positions of these two elements in the electrochemical series. This is because the liberated sodium forms an amalgam with the mercury of the cathode.

Table 8.3 will be of help in deciding which are the preferred electrode reactions in the more important examples of electrolysis.

The material of the electrodes is given in this table only where it affects the course of electrolysis, but it should be remembered that an electrode must be made of a conductor, but it must not be attacked by a chemical produced, e.g. if chlorine is evolved the electrode should not be made of a metal.

Exercise. Use the data given in the table, and deduce what happens when solutions of the following substances are electrolysed:

(*i*) Na_2SO_4;
(*ii*) *KOH*
(*iii*) *KCl (in dilute and concentrated solutions).*

Find, where more than one gas is evolved, the relative volumes of those gases.

8.3.6 Electroplating

Electroplating is often used to coat one metal with another. This is done either as a protection against corrosion or to improve appearance. Thus silver plating is used for many domestic articles. Chromium plating is used for preventing iron from rusting (see also section 24.2.8). (For the best cover this involves plating the iron first with nickel and then with chromium.) Some galvanized (zinc-coated)

TABLE 8.3

Ions migrating to electrode	Ion discharged and preferred electrode reactions
SO_4^{2-} and OH^-	OH^- ($OH^- - e \rightarrow OH$, $4OH \rightarrow 2H_2O + O_2$)
SO_4^{2-} and OH^- (copper anode)	Neither ion is discharged, but Cu atoms of anode ionise ($Cu \rightarrow Cu^{2+} + 2e$)
Cl^- and OH^- (low Cl^- concentration)	OH^- ($OH^- - e \rightarrow OH$, $4OH \rightarrow 2H_2O + O_2$)
Cl^- and OH^- (high Cl^- concentration)	Cl^- ($Cl^- - e \rightarrow Cl$, $2Cl \rightarrow Cl_2$)
Na^+ and H^-	H^+ ($H^+ + e \rightarrow H$, $2H \rightarrow H_2$)
Na^+ and H^+ (mercury cathode)	Na^+ ($Na^+ + e \rightarrow Na$, $Na + Hg \rightarrow Na/Hg$)
K^+ and H^+	H^+ ($H^+ + e \rightarrow H$, $2H \rightarrow H_2$)

iron and much tin-coated iron is also produced by electroplating rather than by a mechanical process.

An object which is to be electroplated is made the cathode of an electrolytic cell.

Experiment. The coating of copper by nickel

Set up the apparatus shown in figure 8.5.
Both electrodes in this experiment must be scrupulously cleaned.

Nickel(II) ions are discharged at the cathode and coat it.

$Ni^{2+} + 2e \rightarrow Ni$

At the anode electrode ionization occurs (see section 8.3.4(*c*)) so that the overall concentration of the solution remains unchanged. Zinc plating may be performed similarly using a zinc anode and a slightly acidified solution of zinc sulphate as the electrolyte.

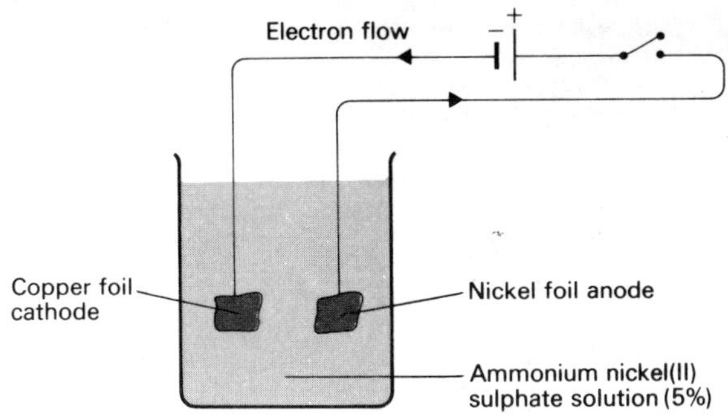

Figure 8.5 Nickel plating of copper.

8.4 Faraday's laws of electrolysis

Michael Faraday (figure 8.6) in 1830 investigated the masses of substances produced by varying quantities of electricity during the electrolysis of solutions. He decided that:

1. *The mass of a substance liberated during electrolysis is directly proportional to the quantity of electricity passed through the solution.*

Quantity of electricity is measured in coulombs. A coulomb is equivalent to a current of one ampere passing for one second, i.e. if a current of x amperes passes for y seconds, the quantity of electricity that passes equals xy coulombs.

This law is to be expected from the foregoing discussion, although Faraday discovered it by experiment. An electric current is a flow of electrons. It produces chemical change by causing ions in solution to gain or lose electrons. When a silver ion, for example, is discharged an electron is gained.

$$Ag^+ + e \rightarrow Ag$$

Two electrons would discharge two ions and so on, i.e. the mass of silver discharged is proportional to the number of electrons which arrive at the cathode, i.e. is proportional to the quantity of electricity which passes.

2. *The quantities of electricity required to liberate 1 mole of different substances bear a simple relationship to one another.*

To liberate 1 mole of silver, it is found that 96 500 coulombs of electricity are needed. This quantity of electricity is known as a *Faraday. A Faraday is defined as the quantity of electricity required to deposit one mole of silver.* As there are $6{\cdot}02 \times 10^{23}$ atoms in a mole of silver and each atom of silver is formed from one ion by gaining one electron it is seen that one Faraday corresponds to $6{\cdot}02 \times 10^{23}$ electrons, i.e. *a Faraday is a mole of electrons.*

Experiment. To determine the quantity of electricity required to liberate a mole of copper

Set up the apparatus shown in figure 8.7.

Beaker *A* contains dilute copper(II) sulphate solution and copper foil electrodes and beaker *B* dilute silver nitrate and silver foil electrodes. Weigh each cathode before the experiment. Connect up the circuit and pass a small

Figure 8.6 Michael Faraday (1791–1867). (Photo: The Science Museum, London).

electric current for about an hour. During this period silver is deposited on the silver cathode and copper on the copper cathode. At both anodes, electrode ionization occurs so that the overall concentrations of both solutions remains constant.

After switching off the current, dry each electrode and wash by dipping in water followed by ethanol and then propanone. Allow to dry and reweigh. (N.B. Silver crystals sometimes fall from the cathode. If so, these must be washed, dried and weighed with the cathode.)

Suppose 0·108 g of silver (Atomic mass 108) and 0·031 75 g of copper (Atomic mass 63·5) are deposited.

One Faraday (96 500 coulombs) liberates one mole of silver, i.e. 108 g.

Therefore 0·108 g of silver is liberated by $\dfrac{96\,500 \times 0{\cdot}108}{108}$ coulombs

= 96·5 coulombs.

The same quantity of electricity deposited the copper.

Therefore 0·031 75 g of copper is deposited by 96·5 coulombs

Therefore 63·5 g of copper (one mole) is liberated by $\dfrac{96{\cdot}5 \times 63{\cdot}5}{0{\cdot}031\,75}$ coulombs

= 193 000 coulombs or two Faradays i.e., two moles of electrons.

(In this experiment the silver voltameter has been used to measure the quantity of electricity which passed through the copper(II) sulphate solution. A less accurate result may be obtained by using the copper voltameter only and measuring the current registered by the ammeter and the time in seconds for which it passes. The product of these quantities gives the number of coulombs of electricity which has passed. Hence the number of coulombs required to liberate a mole is easily calculated.)

This result might be expected as two electrons are required to discharge each copper ion whereas only one is required to discharge each silver ion.

$Cu^{2+} + 2e \rightarrow Cu$

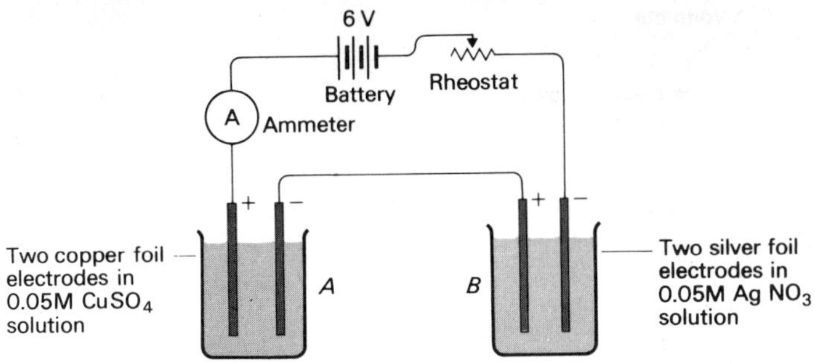

Figure 8.7 Determination of the quantity of electricity required to liberate 1 mole of copper.

In general it should be obvious that the number of moles of electrons (Faradays) required to discharge one mole of a substance is the number of charges carried by its ion.

8.5 Simple voltaic cells

8.5.1 Introduction

So far in this chapter we have been studying *electrolytic cells*. In these the electric current is passed into the cell and this causes the electrolyte solution to be decomposed and form gases etc. at the electrodes.

We now want to look at cells in which chemicals react and thereby produce an electric current. These cells are called *Voltaic* or *Galvanic* cells after Alessandro Volta and Luigi Galvani who in 1801 discovered such cells.

For the electrodes of his cell Volta used copper and zinc. If, like Volta, we place a plate of zinc and a plate of copper in a beaker of dilute sulphuric acid then the copper is unaffected but the zinc begins to react and hydrogen gas bubbles are formed on it. This is because, as we saw in section 8.2.1, zinc reacts with sulphuric acid:

$$Zn + H_2SO_4 \rightarrow ZnSO_4 + H_2,$$
$$\text{or } Zn + 2H^+ \rightarrow Zn^{2+} + H_2.$$

Here the zinc is losing electrons and forming ions ($Zn \rightarrow Zn^{2+} + 2e$) and the hydrogen ions are gaining electrons to form hydrogen gas ($2H^+ + 2e \rightarrow H_2$). These last two equations, in brackets, are called *half-equations* as they describe half of a reaction. The half-reaction for zinc in which electrons are formed is called an *oxidation*, i.e., the zinc is being *oxidized*, and the half-reaction for the hydrogen ions in which electrons are being used is called a *reduction*, i.e. the hydrogen ions are being *reduced*. (See also section 15.8).

This reaction between zinc and sulphuric acid, during which, as we have just seen, electrons move into the solution from the zinc plate and combine with the hydrogen ions from the acid to give hydrogen gas is known as a *local action*, and can be suppressed by rubbing a little mercury on the zinc plate.

If now the treated zinc plate and the copper plate are connected by a wire through a voltmeter, as in figure 8.8, the voltmeter registers a voltage (of about 1 volt), the zinc begins to react and bubbles of hydrogen begin to appear on the

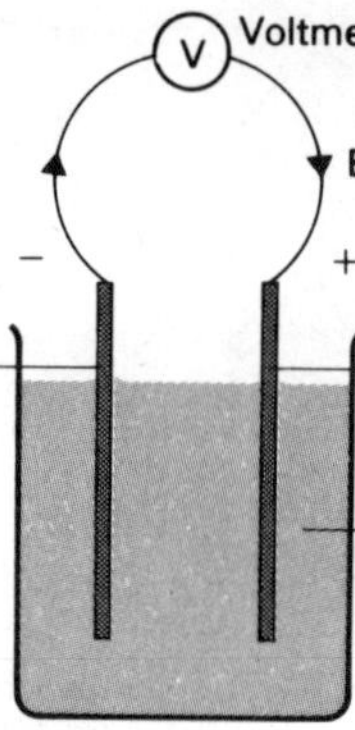

Figure 8.8 A simple Voltaic Cell.

copper electrode. Why is this so? It must be because electrons are now able to flow from the zinc electrode through the wire to the copper electrode. The electrons are generated by the zinc electrode dissolving: $Zn \rightarrow Zn^{2+} + 2e$ and they are used up at the copper electrode in converting hydrogen ions into hydrogen gas: $2H^{+} + 2e \rightarrow H_2$.

It is at the zinc electrode that *oxidation* occurs, and from which electrons leave this voltaic cell. Hence this electrode is defined as the *anode*. It is at the *copper* electrode that electrons enter the cell and a *reduction* occurs. Hence this electrode is defined as the *cathode*. Electrons are negatively charged and so they flow from the negative to positive electrode through the wire connecting them. So in this *Voltaic cell* the *zinc anode* is the *negative* electrode and the *copper cathode* is the *positive* electrode.

Notice that this relationship between anode and cathode and positive and negative electrodes is reversed in the *electrolytic cell* that we discussed earlier, since in these cells the electrode where the electrons enter the cell and a reduction occurs is again the *cathode* (by definition), and the electrode where an oxidation occurs and electrons leave the cell is again the *anode* (by definition). But as the electrons are being put into the cell from an external source (from a battery for example) the electrode where the electrons enter the cell will be more negative than the electrode where the electrons leave, i.e., the *cathode* will be the *negative* electrode, and the *anode* the *positive* electrode.

A troublesome feature of a simple Voltaic cell such as that illustrated in figure 8.8 is that the voltage generated quickly drops as hydrogen bubbles cover the copper plate. This is known as *polarisation*, and if the cell is to continue working the hydrogen must be removed by either shaking the electrode or adding an oxidizing agent to the electrolyte in the cell so that some other preferential electrode reaction occurs and hydrogen gas is not formed.

The voltage recorded on the voltmeter is known as the *electromotive force* (*e.m.f.*) of the cell.

A very simple and practical Voltaic cell is the *Daniell* cell (figure 8.9). Here an electrode of zinc in zinc sulphate solution is placed in a porous pot. This pot is surrounded by saturated copper(II) sulphate solution in which is an electrode of copper. Zinc has a greater tendency to lose electrons than has copper. If the

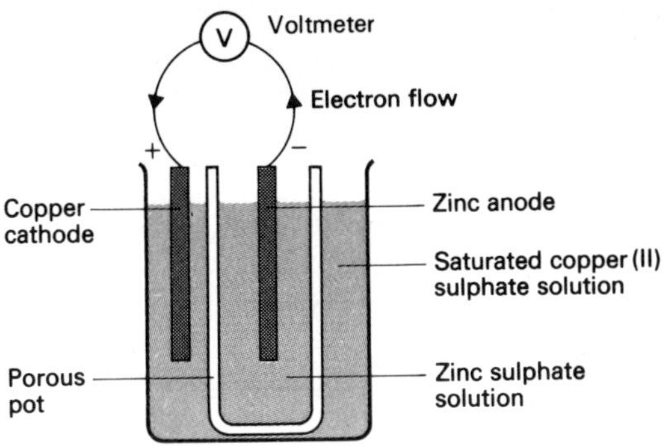

Figure 8.9 The Daniell Cell.

electrodes are connected together through a voltmeter then electrons released by the zinc atoms flow through the wire to the copper and zinc passes into solution as ions:

$$Zn \rightarrow Zn^{2+} + 2e$$

Electrons collect on the copper and there react with copper(II) ions from the solution:

$$Cu^{2+} + 2e \rightarrow Cu$$

The copper atoms formed adhere to the copper electrode which increases in size; at the same time the solution becomes paler as copper ions are removed from solution unless there are copper(II) sulphate crystals present that dissolve as the cell is used. The total reaction which occurs in this cell is that which occurs when zinc is placed in copper(II) sulphate solution:

$$Cu^{2+} + Zn \rightarrow Zn^{2+} + Cu$$

The difference is that in the cell the electrons from the zinc reach the copper ions through the external circuit thereby causing a current to flow.

8.5.2 Simple cells and the electrochemical series

By using a cell with copper as one electrode and using other metals for the other electrode it is possible to arrange the metals in order of the electromotive force they produce, i.e. the voltage recorded on the voltmeter.

It is normal practice however to use a *hydrogen* electrode as this *standard electrode*. This electrode is made by bubbling hydrogen over a platinum electrode coated with finely divided platinum immersed in an electrolyte of 1 M hydrochloric acid. The platinum adsorbs hydrogen on to its surface so that in effect there is a hydrogen electrode in a solution containing hydrogen ions. Consider the cell in figure 8.10. One half of the cell consists of a hydrogen electrode in a 1 M solution of hydrogen ions. The other half is a zinc electrode in a 1 M solution of zinc ions. The solutions are connected by a "bridge" containing saturated potassium chloride solution. This salt "bridge" gives a highly conducting link between the two solutions and replaces the porous pot in the Daniell cell described earlier. The zinc is found to be the negative electrode of this cell and the

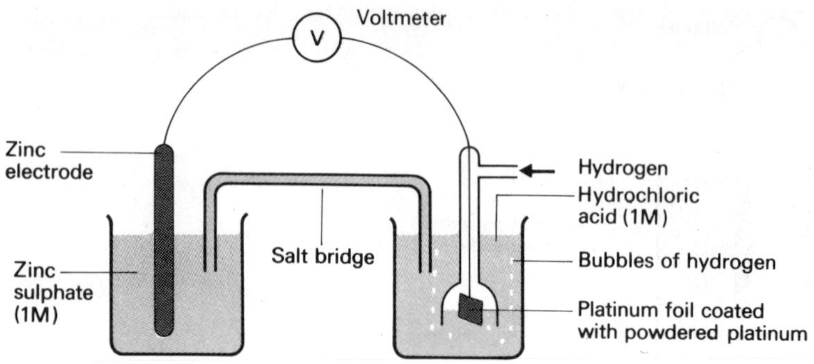

Figure 8.10 Determination of electrode potential.

voltage of the cell is found to be 0·76 volts. So that a definite value can be assigned to each electrode from the voltages of such cells it is necessary to assign a value of zero to one electrode. Chemists have chosen the hydrogen electrode and hence we say this has a zero electrode potential. Hence zinc is said to have an electrode potential of −0·76 volts.

If the zinc electrode is replaced by a copper electrode immersed in a 1 M solution of copper ions, the hydrogen electrode is the negative and copper the positive electrode. The voltmeter needle is deflected in the opposite direction and registers 0·34 volts. Copper is said to have an electrode potential of +0·34 volts. By a series of experiments of this type the electrode potentials of most metals may be determined. (The hydrogen electrode is inconvenient to use and may be replaced by a different one that has been standardised. An electrode known as a calomel electrode, because it contains mercury(I) chloride whose old name was calomel, is frequently used.)

TABLE 8.4 ELECTRODE POTENTIALS OF COMMON METALS.

K/K^{+}	−2·92 volts	Sn/Sn^{2+}	−0·14 volts
Na/Na^{+}	−2·71 volts	Pb/Pb^{2+}	−0·13 volts
Mg/Mg^{2+}	−2·36 volts	$(H/H^{+}$	0·00 volts)
Al/Al^{3+}	−1·67 volts	Cu/Cu^{2+}	+0·34 volts
Zn/Zn^{2+}	−0·76 volts	Ag/Ag^{+}	+0·80 volts
Fe/Fe^{2+}	−0·44 volts	Hg/Hg_2^{+}	+0·85 volts
Ni/Ni^{2+}	−0·25 volts	Au/Au^{+}	+1·68 volts

In table 8.4 are given the electrode potentials of the common metals. Also given is the metal ion that is formed when the metal electrode goes into solution. The electromotive force for any particular cell can be calculated from these figures, it being the difference between the electrode potentials of the two metals concerned. Thus a Daniell cell which has copper and zinc electrodes, will have an electromotive force of 0·34−(−0·76) volts = 1·10 volts.

The metals listed in the order above constitute the *electrochemical series*. It is a list of the metals in order of their ease of releasing electrons. The ease of electron loss, (i.e., ease of oxidation), decreases down the table. Earlier in the chapter this series was derived from chemical evidence and there we called it the *activity series*. With very few exceptions the activity and electrochemical series are identical—the ease with which a metal loses electrons determines both its reactivity and its electrode potential.

8.5.3 Other cells

(a) The dry battery (figure 8.11)
In this cell the zinc case is the negative electrode. Manganese(IV) oxide, mixed with carbon to improve its conductivity, is the positive electrode. The reactions which occur are complex but are basically these:

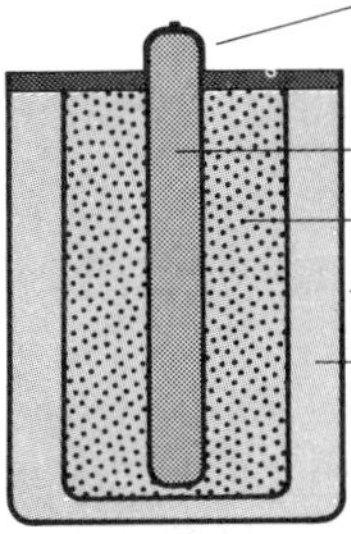

Figure 8.11 A dry battery.

(i) Negative electrode

$Zn \rightarrow Zn^{2+} + 2e$

These electrons pass through an external circuit to the positive electrode.

(ii) Positive electrode

$2MnO_2 + H_2O + 2e \rightarrow Mn_2O_3 + 2OH^-$

i.e. the electrons are taken up.

A single cell of this type has a voltage of 1·5. Batteries of them in series are used to obtain higher voltages.

(b) The lead accumulator
The lead accumulator with which you are most familiar, i.e. a car battery, usually consists of six cells connected in series. Each cell contains a negative electrode made of lead and a positive electrode of lead(IV) oxide. They are immersed in fairly concentrated sulphuric acid (figure 8.12).

When the starter motor, or some other electrical equipment, in the car is used the reactions that occur are, at the negative electrode:

$Pb + SO_4^{2-} \rightarrow PbSO_4 + 2e$

and at the positive electrode:

$PbO_2 + 4H^+ + SO_4^{2-} + 2e \rightarrow PbSO_4 + 2H_2O$

The electrons flow in the direction indicated, both electrodes become coated in

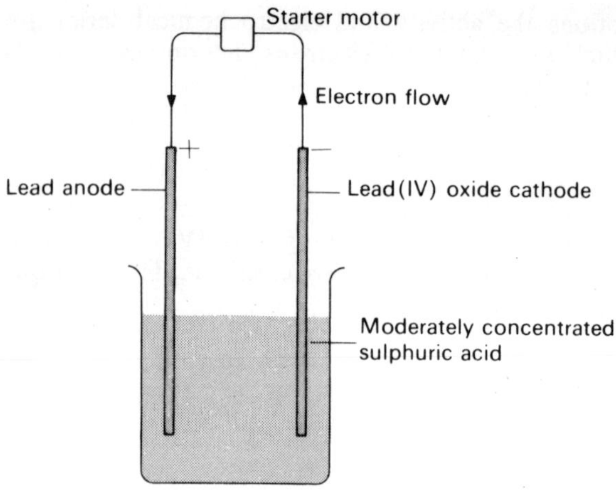

Figure 8.12 The lead accumulator.

lead(II) sulphate and the sulphuric acid electrolyte becomes diluted. As discharge proceeds the specific gravity of the acid decreases from 1.25 to 1.16 or lower. It is usual to check on the condition of car battery using a hydrometer to measure the specific gravity of the acid.

Fortunately the accumulator can be recharged by an outside source of electrical energy, e.g. by the generator in a car or, if necessary, by using a 'battery charger'. An electric current passes through the accumulator in the opposite direction to the flow during discharge.
The reactions given above are then reversed, i.e.

$$PbSO_4 + 2e \rightarrow Pb + SO_4^{2-}$$

$$PbSO_4 + 2H_2O \rightarrow PbO_2 + 4H^+ + SO_4^{2-} + 2e$$

The electromotive force given by a single cell is 2.0 volts. Therefore the normal car battery of six cells will have a voltage of 12 volts.

Questions on chapter 8

1. What is meant by the term *electrolyte*, and what three classes of compound are electrolytes?

Describe what happens in the electrolysis of a copper(II) sulphate solution between platinum plates, assuming the presence in the solution of copper(II), sulphate, hydrogen and hydroxyl ions.

How is crude copper purified by an electrolytic process? (*J.M.B.*)

2. Give *two* simple electronic diagrams to show and to explain the difference between (*a*) a sodium atom, (*b*) a sodium ion. (Give a key to your diagrams to explain the symbols you use.)

Assume that in a dilute solution of sodium chloride there are ions of four kinds: (*a*) hydrogen ions, (*b*) sodium ions, (*c*) hydroxyl ions, (*d*) chloride ions. State what product is liberated during electrolysis of *dilute* sodium chloride solution with platinum electrodes at (i) the anode, (ii) the cathode, and give a brief explanation to account for these products at their respective electrodes. Give a labelled diagram of the apparatus you would set up to perform this experiment. (*J.M.B.*)

3. State only, and give the equation of the reaction (if any) of (*a*) water (or steam), (*b*) oxygen, (*c*) dilute hydrochloric acid on zinc, copper, iron and calcium. Arrange the foregoing metals in descending order of chemical activity, and compare this order with that of the same metals in the electromotive series. (*B.*)

4. This question concerns the metals lead, iron, mercury and aluminium.

(*a*) Place the four metals in order of reactivity, putting the most reactive first.

(*b*) State which of these metals would react with warm, dilute hydrochloric acid and give equations for the reactions which occur.

(*c*) Describe what you would see if a strip of lead were placed in a solution of silver nitrate ($AgNO_3$) and left for a few hours. Give an equation for the reaction occurring.

(*d*) One of the above metals is obtained from its oxide by electrolysis, another by heating its oxide, and the other two by heating their oxides with carbon monoxide. State which metals are obtained by each of the three processes.

(*e*) What is the connection between the way in which the metal is obtained from its oxide and your order of reactivity given in (*a*)? Give a reason for your answer. (*J.M.B.*)

5. An element *X* has an atomic mass of 88. When a current of 0·5 amp. was passed through the fused chloride of *X* for 32 minutes 10 seconds, 0·44 g of *X* was deposited at the cathode. (1 Faraday = 96 500 coulombs).

(*a*) Calculate the number of Faradays to liberate 1 mole of *X*.

(*b*) Write the formula for the *X* ion.

(*c*) Write the formula for the hydroxide of *X*. (*J.M.B.*)

6. When dilute sulphuric acid is electrolysed using platinum electrodes, two gases are evolved,

(*a*) (i) Name the gas formed at the anode, (ii) Name the gas formed at the cathode.

(*b*) Write equations for the reactions at the anode and the cathode. In each case you should write either one complete equation or a set of partial equations.

(*c*) Although the volumes of gases collected are expected to be in a 1 : 2 ratio, the volume of one gas actually collected during the early part of the electrolysis is always slightly less than half that of the other.

(i) Why is a 1 : 2 ratio expected? (ii) Suggest a reason why it is not obtained in practice.

(*d*) How does the method of conduction of electricity through dilute sulphuric acid differ from that through a copper wire? (*J.M.B.*)

7. Explain what happens when a direct current is passed (*a*) through a *very dilute* solution of common salt between platinum electrodes, (*b*) a concentrated solution of common salt between carbon electrodes.

A silver voltameter is connected in series with a copper voltameter. What mass of silver will be deposited in one voltameter whilst 0·16 g of copper is deposited in the other? (Atomic masses: Cu = 64, Ag = 108) (*L*)

8. The relative positions of the elements rubidium (Rb), beryllium (Be) and bismuth (Bi) in the electrochemical or metal activity series are given below, with the positions of some more familiar elements. The formulae of three chlorides are also given.

Position in series	*Chloride*
rubidium	RbCl
sodium	
magnesium	
beryllium	$BeCl_2$
iron	
lead	
hydrogen	
bismuth	$BiCl_3$
copper	

Make use of this information to answer the following questions about rubidium, beryllium and bismuth. You may assume that these elements do not show variable valency.

(*a*) Write formulae for the hydroxides of these three elements. Which one of these hydroxides will be soluble in water?
(*b*) Give the formulae of the solid compounds you would expect to remain when samples of the three hydroxides in (*a*) are heated.
(*c*) How, and under what conditions, would you expect rubidium and beryllium to react with water? Write equations for these reactions. State, with reasons, whether any reaction between bismuth and water is to be expected.
(*d*) What would you expect to happen when beryllium and bismuth are separately treated with dilute hydrochloric acid?
(*e*) Which one of the three elements could readily be obtained by heating its oxide in a stream of hydrogen? (*C.*)

9. A steady current of 0·2 A is passed for 4825 s through the cells *P* and *Q* illustrated below.
(*a*) Name the gases formed at (i) the positive electrode, (ii) the negative electrode, in cell *P*.

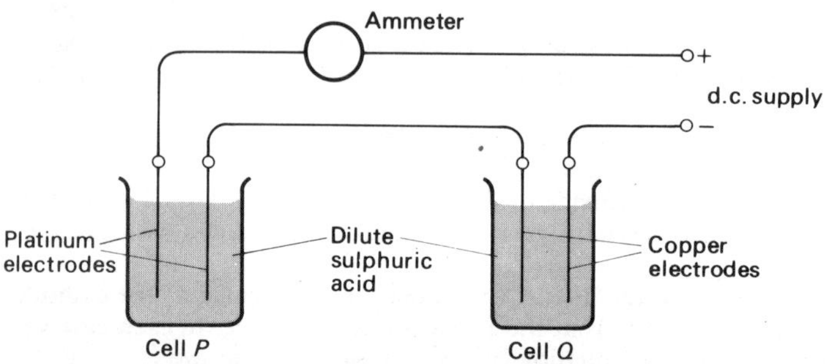

(*b*) Explain the chemical changes that occur during electrolysis at (i) the positive electrode in cell *P*, (ii) the positive electrode in cell *Q*.
(*c*) Calculate (i) the quantity of electricity in coulombs that flows through the circuit, (ii) the mass of gas liberated at the negative electrode in cell *P*, (iii) the change in mass of the positive electrode in cell *Q*.

Experiment	Cell *P*	Cell *Q*
I	Ethanol	A solution of sodium chloride in water
II	Molten potassium chloride	Mercury
III	A solution of ammonia in water	A solution of hydrogen chloride in water
IV	A solution of sugar in water	Molten paraffin wax

(*d*) In separate experiments, the dilute sulphuric acid is replaced by the above liquids.
State for *each* experiment whether or not you would expect to see any chemical change (i) in cell *P*, (ii) in cell *Q*. (The Faraday constant = 96 500 c mol^{-1}; H = 1; O = 16; Cu = 63·5.) (*C.*)

10. (*a*) Draw a diagram of an apparatus suitable for the electrolysis of copper(II) sulphate solution using platinum electrodes, and for the collection of the products. Give the names and polarities of the electrodes, the names of the products and the equations for the electrode reactions.
After passing a current for, say, ten minutes, what would be the effect of reversing the current and passing it in the opposite direction for about twenty minutes?

(*b*) In the example given above, electricity is used to bring about a chemical reaction. Describe a way in which a chemical change can be used to release electrical energy. Carefully indicate the reactions which occur and where they take place. (*J.M.B.*)

11. Explain the following observations:

(*a*) When a piece of clean magnesium ribbon is placed in a small volume of copper(II) sulphate solution, the blue colour of the solution fades, and a brown deposit and bubbles of a gas form on the surface of the magnesium.

(*b*) When a zinc plate and a copper plate are partially immersed in dilute sulphuric acid, bubbles appear slowly at the surface of the zinc plate if the plates are not in contact, but they appear rapidly at the copper plate if the plates are touching. (*J.M.B.*)

12. (*a*) A steady current of 1 amp is passed between copper electrodes through a solution of copper(II) sulphate for 1 hour. The cathode gains in mass by 1·19 g and the anode loses mass by the same amount.

In another experiment, a steady current of 2 amps is passed for 1 hour between copper electrodes in a warm alkaline solution of sodium chloride. The anode is found to lose in mass by 4·76 g.

Calculate the number of Faradays of electricity required to cause 1 mole of copper (63·6 g) to dissolve from the anode in each case, and comment on the answer.

(1 Faraday = 96 500 coulombs = 26·8 amp hours; 1 coulomb = 1 amp for 1 sec).

(*b*) Why is it inadvisable to examine the acid levels in a car battery by the light of a burning candle?

(*c*) The elements magnesium, nickel, hydrogen, copper and silver stand in that order in the electrochemical series (magnesium most active).

(i) What would you expect to happen at the cathode if an aqueous solution of nickel(II) nitrate and silver nitrate is electrolysed with platinum electrodes?

(ii) Magnesium is placed in an aqueous solution of copper(II) sulphate in one beaker. Copper is placed in an aqueous solution of magnesium sulphate in another beaker. What would you expect to happen in each beaker? (*L.*)

13. A copper rod stands in a dilute solution of copper(II) sulphate in a beaker. A zinc rod stands in a second beaker containing a dilute solution of zinc sulphate.

(*a*) Draw a labelled diagram of this apparatus, adding to it any items necessary to enable it to act as a voltaic cell.

(*b*) Show, in your diagram, the direction of flow of electrons when this cell is connected to an external circuit.

(*c*) Describe carefully all changes that will take place in each beaker while current flows. Explain these changes.

(*d*) Give two reasons why the flow of electrons will, after a time, diminish.

(*O. and C.*)

14. (*a*) Explain why a solution of sugar is a non-conductor of electricity but a solution of sodium chloride is a good conductor. What is formed at each electrode when a current is passed through a solution of sodium chloride by means of carbon electrodes? Explain how you would demonstrate the presence of each product.

(*b*) A current of 2A is passed for 2 hours through (i) a solution of copper(II) sulphate (using copper electrodes) and (ii) a dilute solution of sulphuric acid (using platinum electrodes), the two solutions being connected in series.

In (i), calculate the changes in mass of the cathode and anode, stating whether a gain or loss.

In (ii), calculate the volumes at s.t.p. of the gases liberated at the cathode and anode. Name the gases. (Cu = 63·5; 1 faraday = 96 500 coulombs; molar volume of a gas at s.t.p. = 22·4 dm^3.) (*C.*)

15. Name the particles present in the crystal lattices of (*a*) metallic sodium, (*b*) sulphur, (*c*) sodium chloride.

Explain the following in terms of the structures of the substances involved.

(*d*) Sodium conducts electricity but sulphur is a non-conductor.

(*e*) Solid sodium chloride does not conduct electricity, but it will conduct when molten and is then decomposed by the electric current.

Cadmium is an element (symbol Cd) of relative atomic mass 112. Suppose you are asked to determine the quantity of electricity required to deposit one mole of cadmium atoms during the electrolysis of aqueous cadmium sulphate using cadmium electrodes.

(i) Sketch the electrical circuit you would use.
(ii) State the measurements you would make.
(iii) Indicate how you would calculate the required result.

In experiments of this kind it is found that the quantity of electricity required to deposit one mole of cadmium atoms is twice that required to deposit one mole of silver atoms from aqueous silver nitrate. How do you explain these results? (*A.E.B.*)

9

Energy and Chemistry

9.1 What is energy?

Energy is that which makes things move. It can exist in a number of forms and these forms may be changed from one to another.

Suppose that you are digging the garden. The energy to move the soil is supplied by your body. This source of energy has to be refuelled by eating food.

If you use a cultivator to dig the garden, it too needs refuelling by using petrol or connecting to the electricity supply. To generate the electricity supply, coal or oil may be burned at a power station.

If a substance is being used to provide energy it is called a *fuel* (see section 17.6). *Fuel*, in one form or another, *is needed to do any useful job that involves moving*. Thus

(a) Fuels store *chemical energy*.

(b) The *chemical energy* in coal is changed to *heat energy* when it is burnt in a power station.

(c) The *heat energy* turns water into steam which drives turbines in a dynamo.

(d) The spinning dynamo has *kinetic energy* which it converts into *electrical energy*.

(e) If *electrical energy* is used to raise a lift in a building, the lift gains *potential energy*, energy that it has by virtue of its position.

In this chapter we are concerned with chemical energy. When a chemical reaction takes place, some of this energy is usually changed to heat energy (occasionally light energy is produced). In fact whenever a chemical reaction occurs either *chemical energy is changed to heat energy* or *heat energy is changed to chemical energy*. The study of such changes is called *thermochemistry*.

If, when a chemical reaction occurs, *heat is evolved*, the reaction is said to be *exothermic*. If during the chemical reaction *heat is absorbed* then the reaction is said to be *endothermic*.

9.2 Energy units

All energy is measured in *joules* (symbol used being J). A *joule* is defined as *the quantity of energy transferred when a force of 1 newton acts through a distance of 1 metre*. A joule is also equal to the energy change when 1 coulomb of electricity flows through a potential difference of 1 volt, or when 1 watt of electrical power flows for 1 second.

Heat energy used to be measured in calories, a calorie being the quantity of heat required to raise the temperature of 1 g of water by 1 °C. 4·18 joules are needed to raise the temperature of 1 g of water by 1 °C, hence 4·18 joules are equal to one calorie. The joule is such a small amount of energy that it is often convenient to use *kilojoules* (symbol: "kJ") (1000 J = 1 kJ).

Calories are still often used. You will, for example, find the "calorie content" of many foods given on their packets. This is a measure of the energy content of the food when used as a fuel by you, i.e. when you eat it. You might like to work out how much energy you eat when you have, say, a dish of breakfast cereal. A source of confusion is that for food the calorie used is always the kilocalorie, i.e. 1000 calories. This is sometimes written as Calorie i.e. with a capital C, but unfortunately it is sometimes written with a small c.

9.3 The source of chemical energy

You already know that atoms of the elements combine together to form chemical compounds. In so doing bonds are formed between the elements, and whenever a chemical bond is formed energy is released. If we were able to make chemical compounds from free atoms (that is from atoms that were not associated with any other atoms, either as a solid or liquid element or as a molecule) then we would be able to measure the decrease in the energy content of the atoms as they come together and form a molecule and release energy. But this is usually not possible as atoms are always found with other atoms as the solid or liquid element, or combined with other atoms in molecules (except for the noble gases that do occur as free gaseous atoms). What we have to measure, therefore, when a chemical reaction occurs, is *the difference in the energy between the reactants* (the chemicals that we start with), *and the products* (the chemicals formed).
We can help to explain these ideas with simple energy diagrams.

In figure 9.1(a) the higher horizontal line represents the total energy in some substances which react together. The lower line represents the total energy obtained from these reactants. The energy content of the products is *less* than that of the reactants so that when the reactants change to the products, *heat is evolved*, i.e. figure 9.1(a) represents an *exothermic reaction*.

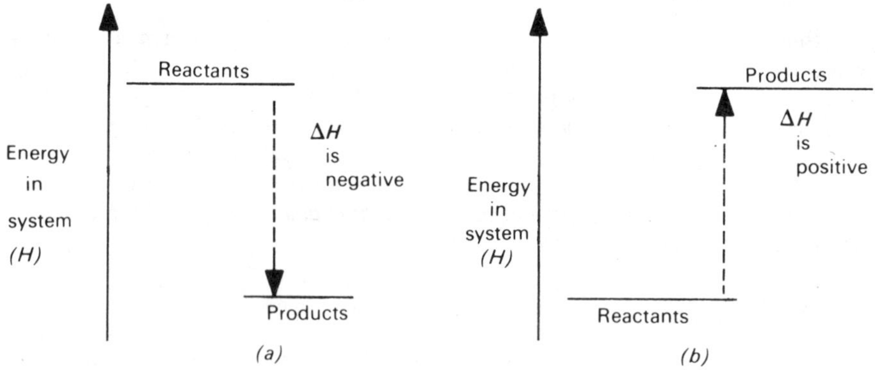

Figure 9.1 Energy content of chemical systems (a) for an exothermic reaction (b) for an endothermic reaction.

The energy content of a system (i.e. either the reactants or the products) is usually represented by the letter H. The *difference* between the energy content of the reactants and products is represented by ΔH (Δ is the Greek letter "delta" and means "change of"), i.e. the heat evolved in the reaction represented in figure 9.1(a) is ΔH.

$$H_{\text{products}} - H_{\text{reactants}} = \Delta H$$

As H_{products} is less than $H_{\text{reactants}}$, the difference ΔH *will be negative.*

In figure 9.1(b) the lower horizontal line represents the total energy in the substances which react together. The higher line represents the total energy in the products. Here the energy content of the products is *more* than that of the reactants so that when the reactants change to the products, *heat is absorbed,* i.e. figure 9.1(b) represents an *endothermic reaction.* As before ΔH is obtained by subtracting the energy content of the reactants from the energy content of the products:

$$H_{\text{products}} - H_{\text{reactants}} = \Delta H$$

Here ΔH will be positive.

For an exothermic reaction ΔH is always negative and for an endothermic reaction ΔH is always positive.

Other names for the energy content of a system, H, are *heat content* or *enthalpy.* By convention values of ΔH given in this and other text-books always refer to the quantity of heat evolved not by atoms and molecules but by moles of the substances involved and to emphasise this fact units are expressed as "per mole" (written mol^{-1}).

9.4 Energy changes on melting and boiling

In physics you may have studied the heat changes that occur when a liquid is changed into its vapour or when a solid is melted. Both of these changes involve the particles composing the liquid or solid gaining more freedom, i.e. bonds between the particles are broken and therefore the changes require energy.

The energy required to melt a solid is known as the *latent heat of fusion* and it is *the heat required to convert 1 g of a solid into liquid at the same temperature.* A definition involving moles is better for the chemist. The *molar heat of fusion* of a solid is *the quantity of heat required to convert 1 mole of a solid into liquid at the same temperature.*

The molar heat of fusion of ice is 6·0 kJ mol^{-1}, i.e.,

$$H_2O(s) \rightarrow H_2O(l) \qquad \Delta H = +6{\cdot}0\,\text{kJ}\,\text{mol}^{-1}.$$

Similarly the energy required to vaporize a liquid is known as the *latent heat of vaporization* and it is *the heat required to convert 1 g of a liquid into vapour at the same temperature.* Again, chemists prefer moles to grams, hence *the molar heat of vaporization* of a liquid is *the quantity of heat required to convert 1 mole of a liquid into vapour at the same temperature.*

The molar heat of vaporization of water is 44·0 kJ mol^{-1}. This quantity of heat converts 18 g of water (i.e. 1 mole) at 100 °C into steam, i.e.

$$H_2O(l) \rightarrow H_2O(g) \qquad \Delta H = +44{\cdot}0\,\text{kJ}\,\text{mol}^{-1}.$$

For fusion and vaporization, ΔH is positive. As is common experience you have to put heat into ice to melt it, or into water to boil it. These values say, therefore, that 1 mole of steam contains 44·0 kJ of energy more than 1 mole of water, and 1 mole of water contains 6·0 kJ of energy more than 1 mole of ice. The value for fusion is much less than for vaporization, because when a solid melts much of the attraction between the atoms or molecules remains, but when a liquid vaporizes all the attractions must be overcome as in the gas the molecules are widely separated.

If the processes are reversed, e.g., if we consider the condensation of steam to water, an equivalent amount of heat is liberated. i.e.,

$$H_2O(g) \rightarrow H_2O(l) \qquad \Delta H = -44{\cdot}0\,\text{kJ mol}^{-1}$$

Latent heats of vaporization are a rough measure of the attractions existing between molecules in a liquid. For example, for the halogens the molar heats of vaporization are:

fluorine = $+6{\cdot}7\,\text{kJ mol}^{-1}$
chlorine = $+17{\cdot}5\,\text{kJ mol}^{-1}$
bromine = $+30{\cdot}9\,\text{kJ mol}^{-1}$
iodine = $+43{\cdot}5\,\text{kJ mol}^{-1}$

These values show that the attractive forces between the molecules must increase as the mass of the molecules increases, because more energy is required to separate them. This statement is generally true but there are exceptions, e.g. water has a much higher latent heat of vaporization than hydrogen sulphide although the hydrogen sulphide molecule is heavier. Why do you think this is so? (*Answer*: It must be that there are some additional bonds between water molecules, that have to be broken, that do not exist in hydrogen sulphide.)

9.5 Energy changes on burning

The most common type of chemical reaction in which heat is evolved is burning. The heat change on burning is called the *heat of combustion* and it is *the quantity of heat evolved when 1 mole of a substance is completely burned in oxygen.* As heat is evolved, the compound or compounds formed have a total energy that is less than the starting materials. This means that the ΔH values for combustion are always negative.

Examples are:

$$Mg(s) + \tfrac{1}{2}O_2(g) \rightarrow MgO(s) \qquad \Delta H = -601\ \text{kJ mol}^{-1}$$
$$C(s) + O_2(g) \rightarrow CO_2(g) \qquad \Delta H = -393\ \text{kJ mol}^{-1}$$
$$CO(g) + \tfrac{1}{2}O_2(g) \rightarrow CO_2(g) \qquad \Delta H = -283\ \text{kJ mol}^{-1}$$
$$CH_4(g) + 2O_2(g) \rightarrow CO_2(g) + 2H_2O(l) \qquad \Delta H = -890\ \text{kJ mol}^{-1}$$

To agree with the definition of heat of combustion, all of the above equations are written with *1 mole* of the *substance concerned* reacting with the appropriate amount of oxygen. The first equation, for example, tells us that 1 mole of magnesium combines with $\frac{1}{2}$ a mole of oxygen gas to form one mole of magnesium oxide, and 601 kJ of energy are released for every mole (24 g) of magnesium burnt.

If water is the product of combustion, the *total heat* evolved will depend on whether steam or liquid water is formed. As steam has a higher energy content than liquid water (because of the latent heat of vaporization) more heat is evolved if liquid water is the product.

Considering the reaction between hydrogen gas and oxygen gas:

$$H_2 + \frac{1}{2}O_2(g) \rightarrow H_2O(g) \qquad \Delta H = -242 \text{ kJ mol}^{-1}$$

we see that 242 kJ of heat are evolved for every mole (2 g) of hydrogen gas that is burned, when the product is steam. We have already seen that for $H_2O(g) \rightarrow H_2O(l)$, $\Delta H = -44$ kJ mol^{-1}. If we add together these two equations, the resulting equation is for the burning of hydrogen to give liquid water:

$$H_2(g) + \frac{1}{2}O_2(g) \rightarrow H_2O(l) \qquad \Delta H = -286 \text{ kJ mol}^{-1}$$

The heat change of -286 kJ mol^{-1} is obtained by adding together the two ΔH values.

To measure the heat of combustion of a substance a *calorimeter* is used. In an accurate calorimeter the substance is burned in a container that is immersed in a known mass of water. It is very difficult to accurately measure heats of combustion. Rough values can however be obtained using very simple apparatus.

Experiment. To determine the heat of combustion of some alcohols

Set up the apparatus shown in figure 9.2.

The tin should be of 250–500 cm^3 capacity and about 200 g of water is weighed or carefully measured into it. Clamp the tin just above where the flame from the burner will be, and read the temperature of the water when it becomes steady. Fill the burner with an alcohol (methanol or ethanol). Fit the wick through a glass tube in a cork, and weigh the burner with contents. Light the burner and place it under the tin and put the apparatus in a large metal container. Stir the water with the thermometer and when the temperature has risen about 15 °C extinguish the flame, continue stirring the water and read the maximum temperature that it reaches. When the burner has cooled reweigh it.

Repeat the experiment using a different alcohol, and calculate the heats of combustion of the two alcohols. This is done as follows (using some results for methanol):

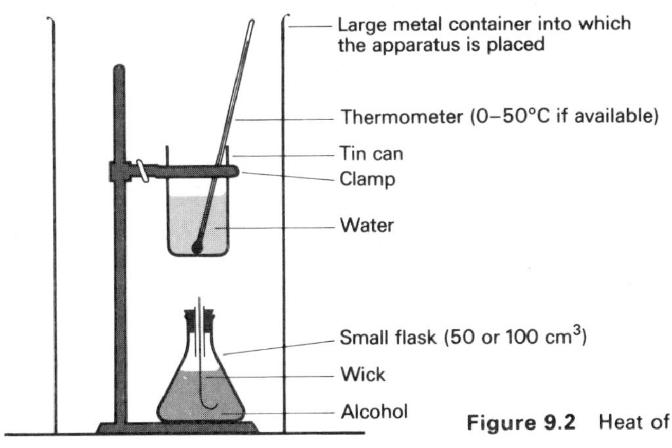

Figure 9.2 Heat of combustion of alcohols.

Mass of water in the container	= 200 g
Initial temperature of the water	= 23·2 °C
Final temperature of the water	= 37·4 °C
Mass of burner initially	= 101·58 g
Mass of burner finally	= 100·82 g
∴ Rise in temperature of water	= 14·2 °C
Mass of methanol burnt	= 0·76 g

∴ Total heat evolved = 4·18 × 200 × 14·2 J (because, as we saw in section 9.2, 4·18 J are needed to raise the temperature of 1 g of water by 1 °C)
= 11 871 J = 11·87 kJ
∴ 0·76 g of methanol give 11·87 kJ of heat

Hence 1 mole of methanol (32 g) gives $\frac{11{\cdot}87 \times 32}{0{\cdot}76}$ kJ of heat = 500 kJ mol^{-1}.

The heat of combustion of methanol (ΔH) = − 500 kJ mol^{-1}.

In a similar experiment using ethanol we found ΔH to be = −908 kJ mol^{-1}. Why do you think the amount of heat obtained by burning a mole of ethanol is more than that obtained by burning a mole of methanol?

The values that are obtained by this experiment are considerably lower than the values found using better types of calorimeters. (ΔH(methanol) = −726 kJ mol^{-1}, ΔH(ethanol) = −1367 kJ mol^{-1}). Why do you think they are lower?

9.6 Energy changes on dissolving

Energy is usually evolved or absorbed when a substance dissolves in water. This heat change is called the *heat of solution* and it is *the quantity of heat evolved or absorbed when 1 mole of a substance is dissolved in sufficient water so that no further heat change occurs on dilution.* There are several reasons why heat changes occur when a substance dissolves. If, for example, an ionic compound is dissolved in water the ions must be separated. To do this requires energy and dissolving is an *endothermic* process. On the other hand many ions, particularly positive ones, when they dissolve in water become hydrated, i.e., water molecules attach themselves to the ions. Hydration liberates energy and if on dissolving hydration is important the reaction is an *exothermic* process.

Experiment. To investigate heats of solution

1. Set up the apparatus shown in figure 9.3. Half fill the bottle with water and take the reading on the thermometer when the temperature is steady. Remove the cork and add about 3 g of sodium hydroxide. Replace the cork, gently shake the bottle and read the thermometer again.
2. Repeat the experiment with ammonium nitrate.
3. Repeat the experiment with sodium chloride.

You will have found that in experiment (1) the temperature of the solution increased, i.e., the reaction $NaOH(s) + H_2O(l) \rightarrow Na^+(aq) + OH^-(aq)$ is *exothermic*. This heat came from the water molecules attaching themselves to the sodium ions. In experiment (2) the temperature of the solution decreased, i.e., the reaction $NH_4NO_3(s) + H_2O(l) \rightarrow NH_4^+(aq) + NO_3^-(aq)$ is *endothermic*. This

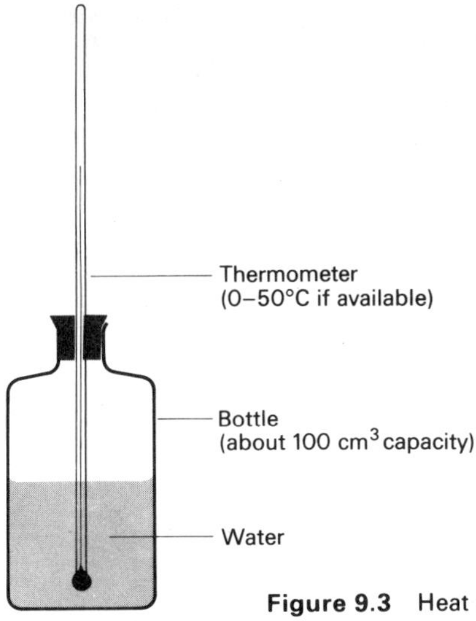

Figure 9.3 Heat of solution apparatus.

heat change resulted from energy being required to separate the ammonium ions and nitrate ions in the solid when it dissolved. Finally in experiment (3) the temperature of the solution remained almost the same, i.e. the reaction $NaCl(s) + H_2O(l) \rightarrow Na^+(aq) + Cl^-(aq)$ is said to be *thermoneutral*; this is because the heat required to break up the solid sodium chloride is balanced almost exactly by the heat liberated by water molecules attaching themselves to the sodium ions.

Experiment. To determine the heats of solution of anhydrous and hydrated calcium chloride

Set up the apparatus shown in figure 9.4. Measure 50 cm^3 of water into the drinking cup and take its temperature when steady. Add a known mass of about 1 g of anhydrous calcium chloride and with the thermometer stir the solution. Read the maximum temperature that the solution attains.

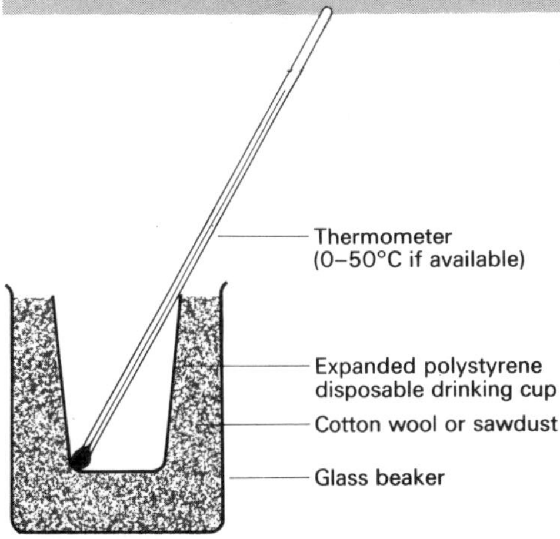

Figure 9.4 A simple calorimeter.

Repeat the experiment with a known mass of about 2 g of calcium chloride-6-water.

Calculate the heats of solution as follows:

Initial temperature of the water	= 20·9 °C
Final temperature of the water	= 24·2 °C
Mass of anhydrous calcium chloride used	= 1·10 g

Total mass of solution = (50 + 1·1)g = 51·1 g. Assume that, as for water, 4·18 J are needed to raise the temperature of 1 g of solution by 1 °C.
Heat gained by solution = 51·1 × 3·3 × 4·18 J = 705 J.

$$\therefore \text{Heat liberated by 1 mole (111 g) of } CaCl_2 = \frac{705 \times 111}{1{\cdot}1} = 71\,100 \text{ J}$$

$\therefore \Delta H$(of solution) = -71 kJ mol^{-1} (within limits of experimental accuracy).

The negative heat of solution, for this anhydrous salt, is because the energy liberated by hydration is greater than that used in separating the calcium and chloride ions.

For hydrated calcium chloride you will have found that the temperature will have fallen because the ions are hydrated before the salt dissolves and the fact that energy is needed to separate the ions now shows itself.

This simple method neglects the heat capacity of the container and also any heat that may be lost to or gained from the surroundings. Do you think that these are very serious errors? Because these are neglected, are the values that you obtain too large or too small?

9.7 Energy changes when acids and alkalis react together

Note: **For sections 9.7, 9.8 and for a number of experiments in chapter 10 a knowledge of volumetric analysis (section 12.6) is required.**
The energy change that occurs when an acid is neutralized by an alkali is known as the *heat of neutralization* and it is *the quantity of heat evolved when 1 mole of hydrogen ions (i. e., 1 g) reacts with 1 mole of hydroxide ions (i. e., 17 g), to form 1 mole of water (i.e., 18 g):*

$$H^+(aq) + OH^-(aq) \rightarrow H_2O(l).$$

Experiment. To measure the heat of neutralization of hydrochloric acid by sodium hydroxide

Set up the apparatus shown in figure 9.4. Measure 50 cm^3 of 1M sodium hydroxide solution into the drinking cup and read its temperature when it becomes steady. Add a measured 50 cm^3 of 1M hydrochloric acid solution and stir the solution with the thermometer. Read the highest temperature that the solution reaches.

Calculate the heat of neutralization of the acid by the alkali as follows:

Total volume of solution	= 100 cm^3
Initial temperature	= 22·10 °C
Final temperature	= 28·95 °C
∴ Temperature rise	= 6·85 °C
Total heat of change of the solution	= 100 × 6·85 × 4·18 J = 2863 J

Now 50 cm³ of 1M HCl contains $\frac{1 \times 50}{1000}$ mole of HCl = 0·05 mol and similarly 50 cm³ of 1M NaOH contains 0·05 mol, so the two solutions do neutralize each other exactly, and the heat liberated if 1 mole of each was used would be:

$$\frac{2863 \times 1}{0{\cdot}05} = 5726 \text{ J} = 57{\cdot}3 \text{ kJ}$$

Hence ΔH (neutralization) = $-57{\cdot}3$ kJ mol^{-1}

The value obtained in this example of the calculation is the accepted value for the heat of neutralization of an acid by an alkali, i.e.:

$$HCl(aq) + NaOH(aq) \rightarrow Na^+(aq) + Cl^-(aq) + H_2O(l) \quad \Delta H = -57{\cdot}3 \text{ kJ mol}^{-1}$$

If, when you carry out this experiment, you give to your teacher the results that we have just used in the sample calculation he would know that you had cheated and not recorded correctly the temperatures from your own experiment. Why?

The experiment could be repeated using various combinations of strong acids and alkalies. Similar values for the heat of neutralization would be obtained: For example:

$$HNO_3(aq) + NaOH(aq) \rightarrow Na^+(aq) + NO_3^-(aq) + H_2O(l) \quad \Delta H = -57{\cdot}3 \text{ kJ mol}^{-1}$$

$$\tfrac{1}{2}H_2SO_4(aq) + KOH(aq) \rightarrow K^+(aq) + \tfrac{1}{2}SO_4^{2-}(aq) + H_2O(l) \quad \Delta H = -57{\cdot}3 \text{ kJ mol}^{-1}$$

$$HCl(aq) + \tfrac{1}{2}Ca(OH)_2(aq) \rightarrow \tfrac{1}{2}Ca^{2+}(aq) + Cl^-(aq) + H_2O(l) \quad \Delta H = -57{\cdot}3 \text{ kJ mol}^{-1}$$

This is because the fundamental reaction is in every case the same, i.e., the reaction

$$H^+(aq) + OH^-(aq) \rightarrow H_2O(l) \quad \Delta H = -57{\cdot}3 \text{ kJ mol}^{-1}$$

1 mole + 1 mole → 1 mole

When weak acids or weak alkalies are used, the heat of neutralization is less:

$$NH_4OH(aq) + HCl(aq) \rightarrow NH_4^+(aq) + Cl^-(aq) + H_2O(l) \quad \Delta H = -51{\cdot}4 \text{ kJ mol}^{-1}$$

$$CH_3COOH(aq) + NaOH(aq) \rightarrow CH_3COO^-(aq) + Na^+(aq) + H_2O(l) \quad \Delta H = -56{\cdot}1 \text{ kJ mol}^{-1}$$

This is because weak acids, such as ethanoic acid, or weak alkalis, such as ammonium hydroxide, are only partly ionized and energy is used up as they ionize before neutralization.

9.8 Other heats of reaction

The heat changes that we have just been considering, those that occur on combustion, on solution, and on neutralization are all heats of chemical reactions.

We can define a *heat of reaction* as *the quantity of heat evolved or absorbed when the number of moles represented by the normal chemical equation react together.*

Experiment. To determine the heat of precipitation of silver chloride

Again the apparatus shown in figure 9.4 is used. Measure 50 cm^3 of 0·5 *M* silver nitrate solution into the drinking cup and take its temperature when it becomes steady. Add 50 cm^3 of 0·5 *M* sodium chloride solution and stir the solution with the thermometer. Read the highest temperature recorded by the thermometer.

Calculate the heat of precipitation of silver chloride as follows:

Total volume of solution	= 100 cm^3
Initial temperature	= 23·9 °C
Final temperature	= 27·8 °C
∴ Temperature rise	= 3·9 °C
Total heat change of the solution	= 100 × 3·9 × 4·18 J
	= 1630 J

Now 50 cm^3 of 0·5 M silver nitrate contains $\frac{0{\cdot}5 \times 50}{1000}$ mole

= 0·025 mol of $AgNO_3$

Similarly 50 cm^3 of 0·5 M sodium chloride contains

0·025 mol of NaCl

Therefore the heat change for the reaction:

$$AgNO_3 + NaCl \rightarrow AgCl + NaNO_3$$

or $Ag^+(aq) + Cl^-(aq) \rightarrow AgCl(s)$

1 mole 1 mole 1 mole

is $\frac{1630 \times 1}{0{\cdot}025} = 65\,200\,J = 65{\cdot}2\,kJ$

∴ $\underline{\Delta H = -65{\cdot}2\ kJ\,mol^{-1}}$

(The correct value which, as we have seen earlier, will always be a larger number (ΔH more negative), than you obtain by carrying out the above experiment, is $-65{\cdot}7$ kJ mol^{-1}).

Many other heats of reaction can be measured using the simple calorimeter illustrated in figure 9.4. For example the heat of the reaction:

$$Zn(s) + CuSO_4(aq) \rightarrow Cu(s) + ZnSO_4(aq)$$

is obtained by adding a large excess of zinc powder to 50 cm^3 of 0·2M copper(II) sulphate solution. If the temperature rise is 10° C, calculate the heat of this reaction (*Answer*: $\Delta H = -209$ kJ mol^{-1}).

9.9 The law of constant heat summation

This law states that *the total change in heat content during a chemical change is independent of the intermediate stages, and depends only on the chemical nature and physical state of the starting materials and products.* This law, which was discovered in the early part of the 19th century by G. H. Hess, is often called Hess's law. *The law of conservation of energy,* says that *energy can neither be created nor destroyed; but can only be converted from one form into another.* Thus Hess's law is another statement of this law as it applies to chemical change.

To illustrate the law of constant heat summation consider the formation of carbon dioxide from carbon. This can be done directly, or by forming carbon monoxide first and then burning it:

Either	$C(s) + O_2(g)$	$\rightarrow CO_2(g)$	$\Delta H = -393\ \text{kJ mol}^{-1}$
or	$C(s) + \frac{1}{2}O_2(g)$	$\rightarrow CO(g)$	$\Delta H = -110\ \text{kJ mol}^{-1}$
followed by	$CO(g) + \frac{1}{2}O_2(g)$	$\rightarrow CO_2(g)$	$\Delta H = -283\ \text{kJ mol}^{-1}$

The heats of reaction for the two-stage process equal the heat of reaction of the one-stage process. With this law it is obviously possible to calculate one of the heats or reaction knowing only the other two.

9.10 Heat of formation

The heat of formation of a compound is the quantity of heat evolved when one mole of a compound is formed from its elements.
If heat is evolved when a compound is formed, e.g.,

$$Ca(s) + \tfrac{1}{2}O_2(g) \rightarrow CaO(s) \quad \Delta H = -635\ \text{kJ mol}^{-1}.$$

the compound is said to be an *exothermic compound.*
If heat is absorbed when a compound is formed, e.g.,

$$2C(s) + H_2(g) \rightarrow C_2H_2(g) \quad \Delta H = +229\ \text{kJ mol}^{-1}$$

the compound is said to be an *endothermic compound.*

Many heats of formation cannot be experimentally determined and they must be calculated using the law of constant heat summation.

As an example let us find the heat of formation of methane, knowing the heats of combustion of methane, carbon and hydrogen.

We want to know ΔH for

$$C(s) + 2H_2(g) \rightarrow CH_4(g) \quad \Delta H = ? \quad (1)$$

$$CH_4(g) + 2O_2(g) \rightarrow CO_2(g) + 2H_2O(l) \quad \Delta H = -890\ \text{kJ mol}^{-1} \quad (2)$$

$$C(s) + O_2(g) \rightarrow CO_2(g) \quad \Delta H = -393\ \text{kJ mol}^{-1} \quad (3)$$

$$H_2(g) + \tfrac{1}{2}O_2(g) \rightarrow H_2O(l) \quad \Delta H = -286\ \text{kJ mol}^{-1} \quad (4)$$

If we add together equations (3) and twice equation (4) then we have equation (5)

$$C(s) + 2H_2(g) + 2O_2(g) \rightarrow CO_2(g) + 2H_2O(l) \quad \Delta H = -393 + 2 \times (-286)\ \text{kJ mol}^{-1} \quad (5)$$

Subtracting equation (2) from equation (5) gives equation (6) and this is equation (1)

$$C(s) + 2H_2(g) \rightarrow CH_4(g) \quad \Delta H = -393 + 2 \times (-286) - (-890) \quad (6)$$

$$\text{i.e. } \Delta H = -75\ \text{kJ mol}^{-1}$$

Hence methane is formed from its elements by an exothermic reaction.

Heats of formation are a guide to the stability of compounds. Endothermic compounds, i.e., compounds that have a higher energy content than the elements from which they are obtained, are unstable; when they decompose into their elements energy is released. The heats of formation of metallic oxides bear a close relationship to the positions of metals in the electrochemical series (see table 8.1):

Oxide	ΔH (formation)
CaO	$-635\ kJ\ mol^{-1}$
MgO	$-602\ kJ\ mol^{-1}$
ZnO	$-348\ kJ\ mol^{-1}$
FeO	$-272\ kJ\ mol^{-1}$
PbO	$-219\ kJ\ mol^{-1}$
CuO	$-157\ kJ\ mol^{-1}$

Of these calcium oxide is the most stable and copper(II) oxide the least stable.

9.11 Some thermochemical problems

The only way to understand thermochemical problems is to do them. Here are five problems for you to do. The answers are given at the end of the problems, but do not look at these answers until you have tried the problems for yourself. Look back through this chapter to get any information that you need.

1. *How many joules are required to raise the temperature of* 25 g *of water from* 10 °C *to* 30 °C?

2. *How many kilojoules are required to convert* 36 g *of ice at* 0 °C *into steam at* 100 °C?

3. *When* 0·5 mole *of a given salt (molecular mass* = 74*) was dissolved in* 1500 cm³ *of water the temperature fell by* 1·5 °C. *Calculate* ΔH *for the reaction.*

salt (s) + water → salt (aq).

4. *When* 100 g *of water is heated with a small bunsen flame, the water temperature rises from* 20 °C *to* 60 °C *in* 2 minutes.

The same flame is used to evaporate a liquid, of molecular mass 50. *When this liquid is boiled for* 10 *minutes, it loses* 20 g *in mass. Calculate the molar heat of vaporization of the liquid.*

5. *Given the following heats of reaction:*

(*a*) $H_2(g) \rightarrow 2H(g)$ $\Delta H = +436\ kJ\ mol^{-1}$

(*b*) $Br_2(g) \rightarrow 2Br(g)$ $\Delta H = +224\ kJ\ mol^{-1}$

(*c*) $HBr(g) \rightarrow H(g) + Br(g)$ $\Delta H = +366\ kJ\ mol^{-1}$

Calculate the heat of the reaction:

$H_2(g) + Br_2(g) \rightarrow 2HBr(g)$.

Is this reaction exothermic or endothermic?

Answers

1. *To raise the temperature of* 1 g *of water by* 1 °C *requires* 4·18 J
   ∴ *To raise the temperature of* 25 g *of water by* 20 °C requires
   25 × 20 × 4·18 J = 2090 J

2. *Molecular mass of water* = 18
Molar heat of fusion of ice = +6·0 kJ mol^{-1} (*section* 9.4)
∴ *To melt* 18 g *of ice requires* 6·0 kJ
∴ *To melt* 36 g *of ice requires* 12·0 kJ

To raise the temperature of 36 g *of water from* 0 °C *to* 100 °C *requires*
$4{\cdot}18 \times 36 \times 100\,\text{J} = 15\,048\,\text{J} = 15{\cdot}05\,\text{kJ}$

Molar heat of vaporization of water = +44·0 kJ mol^{-1} (*section 9.4*)
∴ *To convert* 18 g *of water at* 100 °C *into steam requires* 44·0 kJ
∴ *To convert* 36 g *of water at* 100 °C *into steam requires* 88·0 kJ
∴ *Total heat required* = 12·0 + 15·05 + 88·0 = 115·0(5) kJ

3. *Mass of solution* = (1500 + 37)g = 1537 g
∴ *Total heat lost by* 0·5 mole *of salt* = 1537 × 1·5 × 4·18 J = 9637 J = 9·6 kJ.
∴ *Heat lost by* 1 mole = 2 × 9·6 kJ = 19·2 kJ (i.e. ΔH = 19·2 kJ mol^{-1})

4. *Heat supplied by the flame in heating the water* = 100 × 40 × 4·18 J = 16·72 kJ

∴ *Heat supplied per minute* $= \dfrac{16{\cdot}72}{2} = 8{\cdot}36\,\text{kJ}$

∴ *In* 10 minutes *the flame supplies* 83·6 kJ *of heat*
20 g *of the liquid is evaporated by* 83·6 kJ

∴ 50 g (*i.e.*, 1 mole) *is evaporated by* $\dfrac{83{\cdot}6 \times 50\,\text{kJ}}{20} = +209\,\text{kJ}$

or for liquid → *vapour* $\Delta H = +209\,\text{kJ mol}^{-1}$

5. *To obtain the equation* $H_2(g) + Br_2(g) \rightarrow 2HBr(g)$ *from the* 3 *equations given it is necessary to add together equations* (*a*) *and* (*b*) *to give*

$H_2(g) + Br_2(g) \rightarrow 2H(g) + 2Br(g) \quad \Delta H = +660\,\text{kJ mol}^{-1}$
(ΔH from +436 + 224 = +660)

If equation (*c*) *is multiplied by* 2 *it becomes:*

$2HBr(g) \rightarrow 2H(g) + 2Br(g) \quad \Delta H = +732\,\text{kJ mol}^{-1}$
(ΔH from +366 × 2 = 732)

Subtracting this last equation from the one above it gives the equation we want.

$H_2(g) + Br_2(g) \rightarrow 2HBr(g) \quad \Delta H = -72\,\text{kJ mol}^{-1}$

(ΔH from +660 − (+732) = −72)

i.e., 72 kJ *of heat are given out for every mole of gases that react; the reaction is exothermic.*

Questions on chapter 9

1. Calculate the quantity of heat required to heat:
(*a*) 50 g of water from 10 °C to 60 °C.
(*b*) 200 g of water from 40 °C to 100 °C.
(*c*) 500 g of water from 70 °C to 80 °C.
(*d*) 1000 g of water from 20 °C to 25 °C.

2. Calculate the quantity of heat required to convert:
(*a*) 18 g of ice at 0 °C to water at 0 °C.
(*b*) 36 g of ice at 0 °C to water at 0 °C.
(*c*) 24 g of ice at 0 °C to water at 20 °C.
(*d*) 90 g of ice at 0 °C to water at 100 °C.
(Molar heat of fusion of ice = 6·0 kJ mol^{-1}).

3. Calculate the quantity of heat required to convert:
(*a*) 9 g of water at 100 °C to steam at 100 °C.
(*b*) 18 g of water at 80 °C to steam at 100 °C.
(*c*) 36 g of water at 0 °C to steam at 100 °C.
(*d*) 180 g of ice at 0 °C to steam at 100 °C.
(Molar heat of vaporization of water = 44 kJ mol^{-1}).

4. When 1 g of carbon is burnt in excess of oxygen it raises the temperature of 400 g of water by 19·5 °C. Calculate the molar heat of combustion of carbon. (C = 12)

5. A certain fuel has a molecular mass of 46. If 5 g of this fuel is found to raise the temperature of 500 g of water from 15 °C to 80 °C, calculate the heat of combustion of the fuel.

6. 20 g of ammonium nitrate (molecular mass = 80) is dissolved in sufficient water to give 250 cm^3 of solution. The temperature drops from 20 °C to 14 °C. Assuming that 4·18 kJ raises the temperature of one litre of solution by 1 °C, calculate the heat of solution of ammonium nitrate.

7. The heat of solution of sulphuric acid in water is −71·0 kJ mol^{-1}. Express this by means of a thermochemical equation. Given that the molecular mass of sulphuric acid is 98, calculate the rise in temperature when 4·9 g of the acid are dissolved in sufficient water to give 250 cm^3 of solution.

8. Describe briefly an experiment which could be used to determine the heat of the reaction:

$$H^+(aq) + OH^-(aq) \rightarrow H_2O(l)$$

Briefly state how you would modify your experiment to determine the volume change in the production of one mole of water. Make clear what measurements you would make. (*L*.)

9. When 500 cm^3 of 2M hydrochloric acid is mixed with 500 cm^3 of 2M sodium hydroxide solution, there is a rise in temperature of 13·7 °C. Calculate the heat of neutralization.

Why would a lower value be obtained if ammonium hydroxide were substituted for the sodium hydroxide?

10. 100 cm^3 of a molar solution of an acid are mixed with (*a*) 100 cm^3 M NaOH, (*b*) 100 cm^3 2M NaOH, (*c*) 100 cm^3 3M NaOH. The temperature rises obtained are (*a*) 6·8 °C, (*b*) 13·6 °C, (*c*) 13·6 °C. Explain these figures and deduce the basicity of the acid.

11. What information is given by the following expression:

$$H_2(g) + Cl_2(g) \rightarrow 2HCl(g); \quad \Delta H = -184 \text{ kJ mol}^{-1}.$$

Give an explanation of the energy change in the light of the chemical bonds involved.

Explain how the energy change is consistent with the Law of Conservation of Energy. (*L*.)

12. The thermochemical equation for the decomposition of hydrogen peroxide is

$$2H_2O_2 \rightarrow 2H_2O + O_2 \quad \Delta H = -190 \text{kJ mol}^{-1} \text{ of oxygen produced}$$

A solution containing 8·5 g of hydrogen peroxide in 100 cm^3 of solution is placed in an insulated container of negligible heat capacity. Some manganese(IV) oxide is added to catalyse its decomposition. What rise in temperature occurs?

13. The molar heats of formation of aluminum oxide and chromium(III) oxide are −1590 kJ mol^{-1} and −1130 kJ mol^{-1} respectively. What can you say about the relative stability of these two oxides?

Calculate the heat of the reaction:

$$2Al + Cr_2O_3 \rightarrow Al_2O_3 + 2Cr$$

10

The Rate of Chemical Change and Equilibria

10.1 Introduction

Many of the reactions studied in this book take place very rapidly.

If silver nitrate solution is added to potassium chloride solution a white precipitate of silver chloride is formed at once:

$$\left[\begin{array}{l} AgNO_3 + KCl \rightarrow AgCl\downarrow + KNO_3 \\ Ag^+(aq) + Cl^-(aq) \rightarrow AgCl(s) \end{array}\right.$$

All other precipitation reactions seem to occur immediately. The neutralization of an acid by an alkali is also very rapid. Other chemical reactions are very slow, e.g., the rusting of iron, while others, such as zinc dissolving in acid, occur at a moderate speed. We are going to examine in this chapter the factors that can affect how rapidly a particular reaction occurs (this is called its *rate*), and to see how it is possible to *alter the rates* of chemical reactions.

You, or your teacher, may have already used one of a number of methods to alter the rate of a reaction, e.g.,

(*a*) If in the reaction of zinc with dilute sulphuric acid the bubbles of hydrogen gas are being formed rather slowly, more concentrated sulphuric acid can be carefully added to increase *the concentration* of the acid.

(*b*) If you have a mixture of chemicals that is reacting slowly, it is common practice to heat the mixture to speed up the reaction. When a reaction becomes too vigorous it is usual to cool the reaction flask under a tap. *The temperature* is affecting the rate.

(*c*) If oxygen is being prepared from either potassium chlorate(V) or hydrogen peroxide, a little manganese(IV) oxide is added as a *catalyst* to speed up the reaction.

(*d*) If hydrogen is mixed with chlorine in a room lit with artificial light no reaction occurs, but in bright *sunlight* there is a rapid reaction.

(*e*) If powdered chalk reacts with hydrochloric acid, the reaction is much more rapid than if lumps of marble are used. The *size of the solid particles* affects the rate of a reaction.

There are, in fact, three factors that can influence the rate of a reaction. These are:

(*a*) *The concentration of the reagents.* For solutions this means the amount of the reactant in the solution, for gases it means the pressure of the gas and for solids it means the size of the solid particles.

(*b*) *The temperature of the reaction.*
(*c*) *The presence of a catalyst, or in some cases light.*

In the following experiments we investigate these three factors.

10.2 The influence of concentration on the rates of chemical reactions

Experiment. (*a*) The reaction between dilute hydrochloric acid and sodium thiosulphate

Assemble a boiling tube rack with 6 tubes and corks as shown in figure 10.1(*a*), and place behind the rack a white card on which a black line has been drawn about 3 cm from the bottom of the card.

Prepare a solution of about 30 g l^{-1} sodium thiosulphate in water (about 105 cm^3 required for each experiment) and a 1 M solution of hydrochloric acid (about 75 cm^3 required for each experiment). Put into the tubes, from left to right, 30, 25, 20, 15, 10 and 5 cm^3 of the sodium thiosulphate solution and put 0,

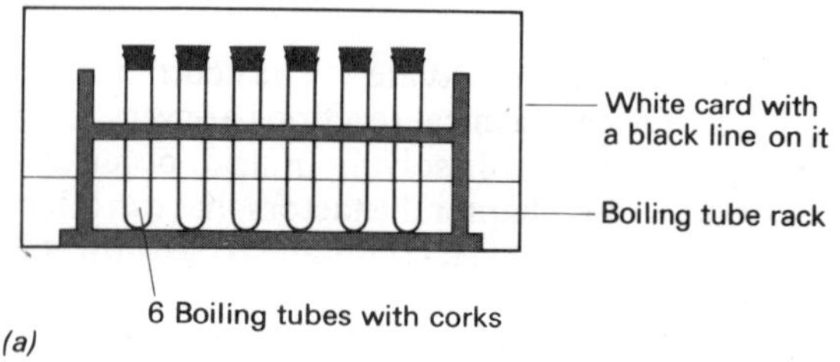

(*a*)

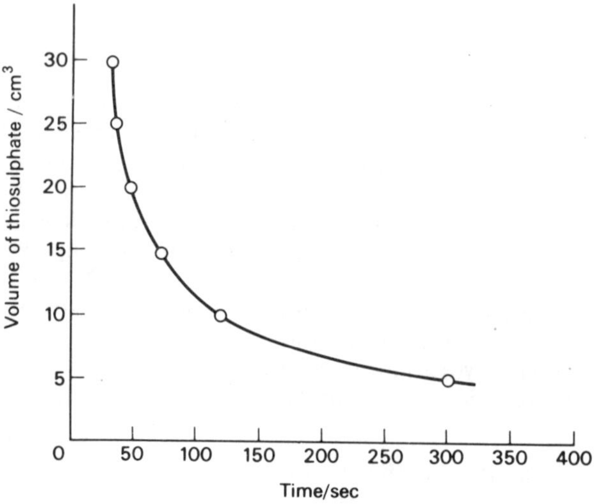

(*b*)

Figure 10.1 (*a*) Apparatus for studying the rates of chemical reactions. (*b*) Concentration-time curve for the reaction of hydrochloric acid with sodium thiosulphate.

5, 10, 15, 20 and 25 cm^3 of water into each tube respectively. This gives a total volume in each tube of 30 cm^3.

To the first tube add 10 cm^3 of the hydrochloric acid and at the same time start a stop clock. Cork the tube, shake it and put it in the rack. The solution will go milky as solid sulphur particles are formed by the reaction:

$$\left[\begin{array}{lll} Na_2S_2O_3 & +2HCl & \rightarrow 2NaCl + H_2O + SO_2 + S \\ S_2O_3{}^{2-}(aq) & +2H^+(aq) & \rightarrow H_2O(l) + SO_2(aq) + S(s) \end{array}\right.$$

Watch the black line through the solution and stop the stop clock when the line behind the boiling tube just disappears. Record the time. Repeat the experiment 5 times by adding 10 cm^3 of hydrochloric acid to successive tubes.

Plot a graph of concentration of thiosulphate (that can be calculated or just use the volume of sodium thiosulphate used as this is proportional to concentration) against time. The graph should be as in figure 10.1(*b*).

You will see that this experiment shows that the time taken for the same amount of reaction to occur (in this case for sufficient sulphur to be formed to make the solution opaque) increases as the solutions become more dilute.

Experiment. (*b*) The reaction between iron(III) ions and iodide ions

You again use the boiling tubes and corks as shown in figure 10.1(*a*). The reaction studied here is:

$$2Fe^{3+}(aq) + 2I^-(aq) \rightarrow 2Fe^{2+}(aq) + I_2(aq)$$

The following solutions are required

1 $g l^{-1}$ sodium thiosulphate solution
4 $g l^{-1}$ potassium iodide solution
25 $g l^{-1}$ ammonium iron(III) sulphate-12-water made up in 0·5M sulphuric acid
A freshly prepared 1 % starch solution.

Put the sodium thiosulphate solution in a burette and put 1 cm^3 into each of 6 boiling tubes.

Put the iron(III) solution in a second burette and add 10, 8, 6, 4, 2 and 1 cm^3 respectively to each of the boiling tubes followed by 3 drops of starch solution. Then add 0, 2, 4, 6, 8 and 9 cm^3 of water to the 6 tubes respectively. This makes the total volume in each tube equal to 11 cm^3, but the concentration of the iron(III) decreases progressively from tube to tube.

To the first tube add 5 cm^3 of the potassium iodide solution, (from a burette or a measuring cylinder), and at the same time start a stop-clock. Cork the tube, quickly shake it and put it in the rack. Stop the clock as soon as a blue colouration appears in the tube. Record the time.

Repeat the experiment 5 times by adding 5 cm^3 of the potassium iodide solution to successive tubes.

Plot a graph of concentration of iron(III) (this can be calculated, or just use the volumes of iron(III) solution used) against time. It should be the same shape as the room temperature curve in figure 10.5.

This experiment uses a clever method to enable you to measure how long the reaction takes. When you add the potassium iodide solution to the mixture of iron(III) ions, sodium thiosulphate and starch, the iron(III) ions react with the iodide to give iodine:

$$2Fe^{3+}(aq) + 2I^-(aq) \rightarrow 2Fe^{2+}(aq) + I_2(aq)$$

But this iodine immediately reacts with the sodium thiosulphate:

$$I_2(aq) + S_2O_3{}^{2-}(aq) \rightarrow 2I^-(aq) + S_4O_6{}^{2-}(aq),$$

and this continues to happen until all the thiosulphate has been used up. Then the iodine reacts with the starch to give the blue colour. Mixtures like this, that react for some time and then signal that they have reached a certain stage, are called "clock reactions". Having carried out this experiment you will understand why.

As in the previous experiment you see that the time taken for the same amount of reaction to occur (in this case until the blue colour appears) increases as the solutions become more dilute.

In both of these reactions you have carried out a reaction between two solutions. Often a chemical reaction occurs between a solid and a liquid. In this next, very simple, experiment you will see how the rate of reaction between pieces of marble (calcium carbonate) and hydrochloric acid changes, *as the acid becomes more diluted* by the reaction:

$$\left[\begin{array}{llllll} CaCO_3 & + \; 2HCl & \rightarrow CaCl_2 & + \; H_2O & + \; CO_2 \\ CaCO_3(s) & + \; 2H^+(aq) & \rightarrow Ca^{2+}(aq) & + \; H_2O(l) & + \; CO_2(g) \end{array}\right.$$

Experiment. (*c*) The reaction between marble and hydrochloric acid

Put about 15 g of marble chips into a 100 cm³ conical flask and fit a cork, *with a V-groove cut down one side*, into the flask. Weigh the flask and contents. Remove from the balance and remove the cork and add about 30 cm³ of about 1 M hydrochloric acid, and start a stop-clock. Replace the cork and reweigh (remembering the weight will be about 30 g more than before adding the acid.) Weigh the flask and contents at regular intervals, say every 30 seconds. (A direct reading balance is very convenient for this experiment.)

Plot a graph of mass of flask and contents against time. It should be similar to figure 10.2. The rate of the reaction, as measured by the loss in mass of CO_2 gas, decreases as the acid is used up, i.e., as its concentration decreases.

Experiment. (*d*) The reaction between magnesium and hydrochloric acid.

Set up the apparatus as shown in figure 10.3(*a*). In the boiling tube place 20 cm³ of 0·5 M hydrochloric acid solution. Remove the rubber bung, tip in between 0·05 g and 0·1 g magnesium ribbon cut into small pieces, and immediately replace the bung. Read at 30 second intervals the volume of hydrogen collected in the gas syringe. Record your results in graphical form as

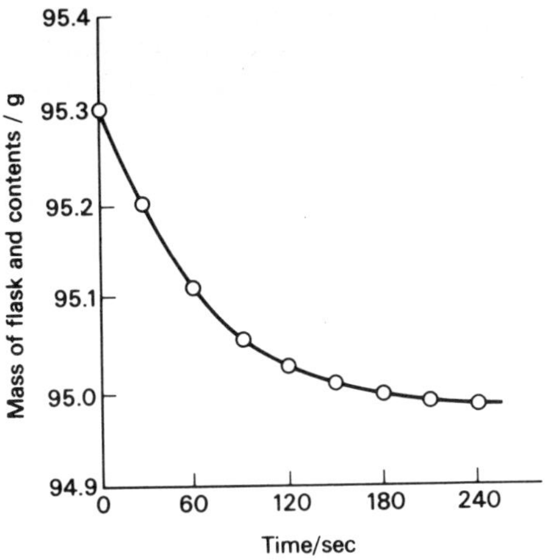

Figure 10.2 Concentration-time curve for the reaction of marble with hydrochloric acid.

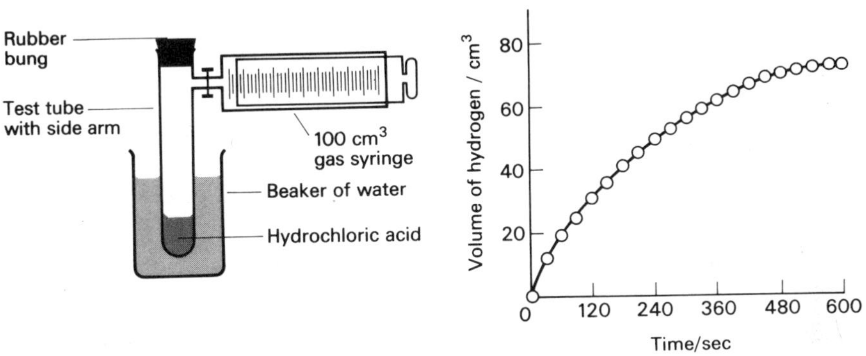

(a) (b)

Figure 10.3 (*a*) Apparatus for studying the rate of a chemical reaction that has a gaseous product. (*b*) Product-time curve for the reaction of magnesium with hydrochloric acid.

in figure 10.3(*b*). Here again the reaction rate decreases as the acid in the solution, and the magnesium, are used up.

It is not difficult to understand why the rate of reaction should depend upon concentration. Let us consider a simple reaction between two gases A and B:

$$A + B \rightarrow C + D$$

If there are, say, 1000 molecules of each gas present in a closed container they will move about in a random manner. Whenever an atom of A hits an atom of B there is a chance of them reacting with one another. If now the number of molecules of A is doubled to 2000 and the size of the container is kept the same then the chances of a collision between a molecule of A and a molecule of B is doubled, i.e., the rate of reaction is doubled. If the number of molecules of both A

and B is increased to 2000 the chances of collision are increased four times.

If the pressure of a gaseous mixture is increased the concentration of both gases will be increased by compression, and therefore the rate of reaction will also increase.

Another way in which the concentration of a *solid* reagent can be changed is by changing its *surface area*. The finer the particles the bigger the surface area and therefore the higher the amount of the solid next to the other, liquid, reagent, i.e., the higher the "concentration" of the solid.

If a lump of marble is placed in dilute hydrochloric acid, only the surface is reacting with the acid and the reaction is slow. If the marble is crushed to a powder the surface area is greatly increased and so is the rate of reaction. For this reason powdered metals are very reactive, e.g., powdered lead will ignite if it is sprinkled into oxygen. Also coal-dust in coal mines is very hazardous, and a spark may cause it to burn explosively (see section 17.8 and figure 17.10).

10.3 The influence of temperature on the rates of chemical reactions

The effect of heat on the rate of a reaction is already very familiar to you. You heat chemical reactions to make them go faster.

Experiment. (*a*) The influence of heat on the reaction between dilute hydrochloric acid and sodium thiosulphate.

This is the experiment from the previous section repeated at different temperatures. In the previous section the reaction took place at room temperature. For this experiment put the boiling tubes containing the sodium thiosulphate and water (each carefully labelled) into a beaker of water, that also contains a thermometer, maintained at 30 °C over a carefully adjusted Bunsen burner, and also put the six 10 cm^3 quantities of hydrochloric acid into 6 more tubes in the beaker.

Take the first thiosulphate tube from the beaker and add the solution to one portion of hydrochloric acid, starting the stop-clock. Cork and shake the tube and put it in the tube-rack. Stop the clock when the line disappears as in the previous experiment.

Repeat with the 5 other tubes.

Repeat the whole experiment with the water in the beaker at 40 °C.

Plot the graphs of thiosulphate volume against time. They will be as in figure 10.4; the higher the temperature the faster the reaction takes place.

Experiment. (*b*) The influence of heat on the reaction between iron(III) and iodide ions.

This is the experiment in the previous section repeated at different temperatures. The changes to be made in the experiment are as for the thiosulphate-hydrochloric acid reaction first described. The graphs of volume of iron(III) ions against time are as figure 10.5. Notice that the rate of this reaction

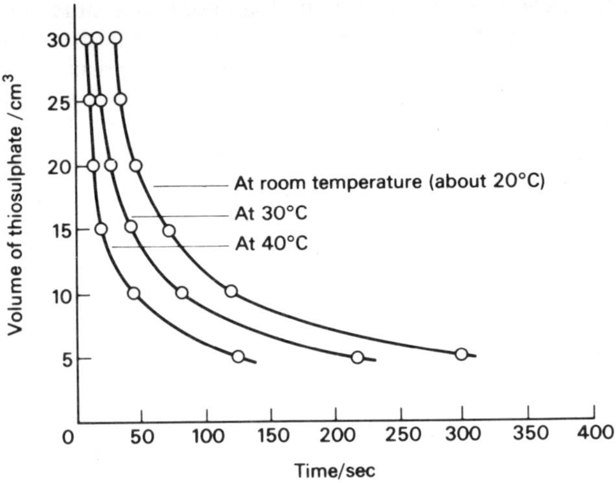

Figure 10.4 Concentration-time curves for the reaction of hydrochloric acid with sodium thiosulphate at different temperatures.

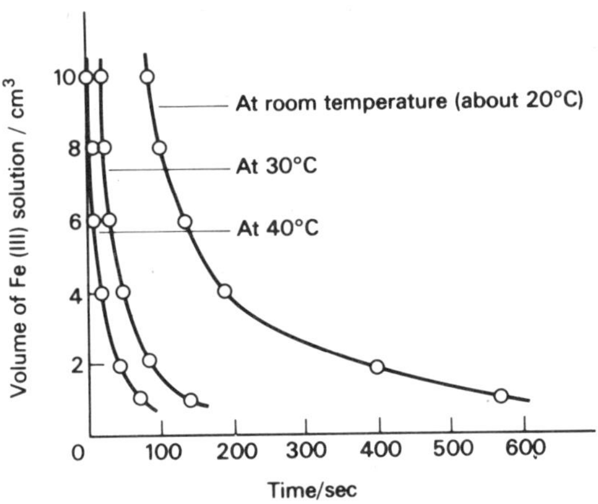

Figure 10.5 Concentration-time curves for the reaction of iron(III) ions with iodide ions at different temperatures.

increases much more with change in temperature than the rate of the hydrochloric acid-thiosulphate reaction.

The effect of heat on a chemical reaction is that the rate of a reaction is approximately doubled for every 10 °C increase in temperature, i.e., an increase in temperature of 30 °C would cause an increase in rate of about $2 \times 2 \times 2 = 8$ times. These increases in rate are far greater than one would expect from the increased number of collisions between molecules because they are moving faster at the higher temperature. The reason for this is that collisions between molecules result in a chemical reaction only if more than a certain amount of energy is involved.

Thus two molecules that between them possess a small amount of energy can collide without reaction. If between them they have a large total amount of energy they react when they collide. An increase in temperature greatly increases the number of molecules having the necessary energy for reaction. In other words, increasing the temperature does increase the number of collisions between molecules, but it also increases to a much greater extent the number of *fruitful* collisions.

10.4 The influence of catalysis on the rates of chemical reactions

10.4.1 Catalysts

A *catalyst* is a substance that alters the rate of a chemical reaction, but is not itself chemically changed at the end of the reaction.

Catalysts are very important indeed for making slow chemical reactions go faster. Indeed some chemical reactions are so slow without a catalyst that no reaction appears to have taken place over periods of not only days, but years and centuries!

The chemical industry depends on catalysts for many of its more important industrial processes. In this book we discuss the catalysis of many of these reactions such as:

(*a*) The manufacture of hydrogen with an iron(III) oxide catalyst (section 14.3.1);
(*b*) the manufacture of carbon monoxide and hydrogen with a nickel catalyst (section 14.3.2);
(*c*) the manufacture of ammonia with an iron-molybdenum catalyst (section 16.3.2);
(*d*) the manufacture of nitric acid with platinum catalyst (section 16.6.3);
(*e*) the manufacture of sulphuric acid with a platinum or vanadium(V) oxide catalyst (section 19.7.3(*b*));
(*f*) the manufacture of methanol with a zinc oxide catalyst (section 25.6.1);
(*g*) the manufacture of ethanol with a phosphoric acid on silica catalyst (section 25.6.2(*b*));
(*h*) the manufacture of edible fats using a nickel catalyst (section 25.4.2(*b*)).

Usually a very small amount of catalyst is all that is required to cause the reaction to take place. Catalysts are usually *selective*, i.e., they are usually effective for one reaction only. In a reversible reaction (see section 10.6) they do not affect the yield of products, but just the rate at which the products are obtained. Catalysis of chemical reactions can be readily demonstrated.

Experiment. The catalysis of the reaction of magnesium with nitric acid.

Put 20 cm^3 of 1M nitric acid into each of 2 boiling tubes. To one of them add a spatulaful of cadmium sulphate. To each tube add a piece of magnesium ribbon about 1 cm long. Note how long it takes for the magnesium ribbon to dissolve in each tube.

You will have found that the magnesium dissolves more rapidly with the cadmium sulphate present—the cadmium sulphate is a *catalyst.*

Experiment. The catalysis of the decomposition of hydrogen peroxide.

Put 10 cm^3 of 20 volume hydrogen peroxide into 4 boiling tubes. To one of the tubes add a spatulaful of manganese(IV) oxide. Watch and test for the evolution of oxygen. Repeat with a second tube and, instead of manganese(IV) oxide, add 2 cm^3 of 2M sodium hydroxide solution. To the third tube add 2 cm^3 of 2M sulphuric acid. Examine the fourth tube to see if any oxygen is being liberated. Then place the tubes into a beaker of hot water and again test for oxygen evolution.

You will have found that although the hydrogen peroxide alone may give off a little oxygen, addition of manganese(IV) oxide will much increase the rate of oxygen evolution (much oxygen is evolved at room temperature). The sodium hydroxide also speeds up the reaction, oxygen being evolved on heating. By contrast, in the tube with acid added no oxygen is evolved—here the reaction has been suppressed.

10.4.2 Enzymes

A further large group of catalysts ensures that the chemical reactions in your body take place at the right speed. These catalysts that occur in living matter are called *enzymes.* There are specific enzymes in your body to catalyse the many different reactions that take place. One enzyme called *amylase* is used to catalyse the conversion of the starch that we eat into sugars. This enzyme is in saliva and an experiment that uses saliva to catalyse starch hydrolysis is given in section 26.3.

Enzymes are now to be found in washing powders. Are they catalysts there, and what reactions do they catalyse? (*Answer*: Yes. They catalyse the decomposition of protein molecules that make up the dirt on the clothes—e.g., grass and egg stains).

10.4.3 How do catalysts work?

There are two ways in which a substance can affect the course of a chemical reaction and itself remain unaffected at the end.

The most important way, and the way in which all solid catalysts work, is by *adsorption.*

(*a*) *Adsorption* is the name given to the process of a gas or liquid molecule attaching itself to the surface of a solid. (You must be very careful not to confuse *adsorption* with *absorption* where the gas or liquid goes into the solid like water into a sponge). In the oxidation of sulphur dioxide to sulphur(VI) oxide using a platinum catalyst (section 19.7.3(*b*)) the sulphur dioxide and oxygen are both adsorbed onto the platinum surfaces. They are very close to one another, and the interaction with the metal also weakens the bond in the oxygen molecule. The sulphur dioxide and oxygen are able to react together more easily and they leave the surface as sulphur(VI) oxide. Fresh molecules of sulphur dioxide and oxygen then take their place.

All the manufacturing processes given above use *adsorption catalysts.*

Exercise. Sketch a periodic table (as in figure 5.9) and mark on it the metals used as industrial catalysts in the list in section 10.4.

You will see that the metals that can act as adsorption catalysts all occur together in group VIII of the periodic table, or in groups nearby. This is because these particular metals adsorb gases most strongly.

(*b*) The second way in which a catalyst can participate in a chemical reaction is by forming an intermediate compound that then decomposes to give the reaction products and at the same time regenerates the catalyst unchanged.

Suppose that a catalyst C is effective for the slow reaction:

$$A+B \rightarrow AB$$

The intermediate is formed and decomposes in two rapid reactions that take place one immediately after the other:

$$A+C \rightarrow AC$$

$$AC+B \rightarrow AB+C$$

The catalyst is reformed in the second stage and appears to have taken no chemical part in the reaction of A with B.

Because the reaction that destroys the intermediate compound follows immediately after its formation, it is usually not possible to isolate or directly examine an intermediate compound. Only in very few special cases is it possible; the next experiment is one such case.

Experiment. The catalysis by cobalt(II) ions of the reaction of potassium sodium tartrate and hydrogen peroxide.

Put into a boiling tube $\frac{1}{2}$g of potassium sodium tartrate, 5 cm³ of 20-volume hydrogen peroxide solution and 10 cm³ of water. Heat the tube in a hot water bath (60–80 °C) and note whether any gas is being formed.

Add about 0·1 g of cobalt(II) chloride, sufficient so that when it dissolves the solution is pink in colour. After a short time the colour of the solution will change to green and a gas will be evolved. Keep the tube in the water bath when suddenly the green colour will fade and the solution becomes pink again.

Here a green intermediate compound of cobalt has been formed, and when all the reactants were used up the original pink colour caused by cobalt(II) ions was reformed.

So far all the catalysts we have examined, except sulphuric acid with hydrogen peroxide, have speeded up the chemical reaction. They are *positive catalysts* or *promotors*. Sometimes a catalyst is wanted that slows down a chemical reaction. Such a catalyst is called a *negative catalyst* or an *inhibitor*. Hydrochloric acid is a negative catalyst for the decomposition of hydrogen peroxide.

Inhibitors are used to prevent rusting of iron, e.g., antifreeze solutions contain them, and to prevent oxidation by the air of plastics. Chemicals that are put into bread to keep it fresh for several days are inhibiting the complex chemical reaction that occur when bread goes stale, and chemicals that are put into petrol to prevent "knocking" are there to inhibit the complex chemical reactions that take place in a motor-car engine.

10.5 The influence of light on the rates of chemical reactions

The influence of light on a chemical reaction is called *photocatalysis*.

Light is a form of energy and when light shines onto certain mixtures of chemicals it starts off a chemical reaction. It is an *initiator*. In some cases the reaction, once started, continues in the absence of light, in others the light must be a continual source of energy to keep the reaction going.

Experiment. The effect of light on silver halides.

Put dilute solutions of sodium chloride, potassium bromide and potassium iodide into 6 test tubes (each solution into 2 tubes). Add silver nitrate solution to each tube, when silver halides are precipitated. Immediately put one tube of each silver halide into a darkened cupboard and put the other tube near a window. Examine and note what has happened to each tube after about 15 minutes.

You will have found that the precipitates in the tubes near the window have darkened, but those in the cupboard are their original white, cream or yellow colour. In the tubes near the window *photocatalysis* has caused the formation of black particles of silver metal. This is exactly what happens in photography. Where the light falls on the negative in the camera a small number of atoms of silver are formed. Developing the film causes more silver to be deposited on the silver already there from the exposure. In this way a negative is formed. It is easy to see why it is a negative, because the black part of it, i.e., the silver, is the light part of the picture being taken.

Other examples of photocatalysis are the reaction between hydrogen and chlorine (section 14.4.6), the fading of dyes in sunlight, and photosynthesis (section 18.6.2(*a*)).

10.6 Reversible reactions

10.6.1 What are reversible reactions?

Most of the reactions that we study in this book are of the form:

Reactants $\rightarrow$ Products.

A reasonable question that might be asked is "can the products of a reaction react together to give the reactants?" The answer is yes, and there are examples of this scattered throughout the book. Here are some of them.

(*a*) When hydrated copper sulphate is heated it loses water:

$CuSO_4.5H_2O(s) \rightarrow CuSO_4(s) + 5H_2O(l)$

and when water is added to anhydrous copper sulphate the hydrated salt is reformed:

$$CuSO_4(s) + 5H_2O(l) \rightarrow CuSO_4.5H_2O(s)$$

(*b*) When hydrogen sulphide is passed into a solution of iron(II) sulphate a black precipitate or iron(II) sulphide is obtained:

$$FeSO_4(aq) + H_2S(g) \rightarrow FeS(s) + H_2SO_4(aq),$$

and when dilute sulphuric acid is added to the mixture of products, the precipitate disappears and hydrogen sulphide is evolved:

$$FeS(s) + H_2SO_4(aq) \rightarrow FeSO_4(aq) + H_2S(g)$$

(*c*) When water in streams and rivers that contain carbon dioxide passes through limestone, the limestone dissolves and the water becomes "hard" with calcium hydrogencarbonate:

$$H_2O(l) + CO_2(aq) + CaCO_3(s) \rightarrow Ca(HCO_3)_2(aq)$$

But this hardness is only temporary and the calcium is reprecipitated in caves as stalactites or stalagmites (figure 22.5) or it can be removed by boiling:

$$Ca(HCO_3)_2(aq) \rightarrow H_2O(l) + CO_2(aq) + CaCO_3(s)$$

(*d*) If steam is passed over red-hot iron filings, iron(II) diiron(III) oxide and hydrogen are formed:

$$3Fe(s) + 4H_2O(g) \rightarrow Fe_3O_4(s) + 4H_2(g)$$

and if hydrogen is passed over red-hot iron(II) diiron (III) oxide, iron and steam are formed:

$$Fe_3O_4(s) + 4H_2(g) \rightarrow 3Fe(s) + 4H_2O(g)$$

(*e*) If acid is added to a yellow solution of potassium chromate(VI) an orange solution of potassium dichromate (VI) is formed:

$$2K_2CrO_4(aq) + 2HCl(aq) \rightarrow K_2Cr_2O_7(aq) + 2KCl(aq) + H_2O(l)$$

and if alkali is added to the orange dichromate(VI) solution then yellow potassium chromate(VI) solution is formed:

$$K_2Cr_2O_7(aq) + 2KCl(aq) + H_2O(l) \rightarrow 2K_2CrO_4(aq) + 2HCl(aq)$$

(*f*) If ammonium chloride is heated, ammonia and hydrogen chloride are formed:

$$NH_4Cl(s) \rightarrow NH_3(g) + HCl(g)$$

When these two gases are allowed to cool in contact with one another, ammonium chloride is reformed:

$$NH_3(g) + HCl(g) \rightarrow NH_4Cl(s)$$

Such reactions, *that can be made to go in either direction by changing the conditions, are known as reversible reactions.*

It is usual to indicate that a reaction is reversible by using double arrows in the equation thus:

(*a*) $CuSO_4 5H_2O(s) \rightleftharpoons CuSO_4(s) + 5H_2O(l)$
(*b*) $FeSO_4(aq) + H_2S(g) \rightleftharpoons FeS(s) + H_2SO_4(aq)$
(*c*) $H_2O(l) + CO_2(aq) + CaCO_3(s) \rightleftharpoons Ca(HCO_3)_2(aq)$
(*d*) $3Fe(s) + 4H_2O(g) \rightleftharpoons Fe_3O_4(s) + 4H_2(g)$
(*e*) $2K_2CrO_4(aq) + 2HCl(aq) \rightleftharpoons K_2Cr_2O_7(aq) + 2KCl(aq) + H_2O(l)$
(*f*) $NH_4Cl(s) \rightleftharpoons NH_3(g) + HCl(g)$

It is usual to call the reaction with the reactants written on the left and the products written on the right the *forward reaction*, and the reaction with the reactants written on the right and the products written on the left the *backward* or *reverse reaction*.

10.6.2 The state of equilibrium

If it is possible to have both the reaction from the compounds on the left of the $\rightleftharpoons$ to give those on the right, and the reaction of the compounds on the right of the $\rightleftharpoons$ to give those on the left, can both reactions take place at the same time?

Consider the example (*d*):

$$3Fe(s) + 4H_2O(g) \rightleftharpoons Fe_3O_4(s) + 4H_2(g)$$

If the steam is passed over the iron and the hydrogen removed as it is formed, as in figure 10.6(*a*), then there is no possibility of the hydrogen and iron oxide reacting together. Similarly if the hydrogen is passed over the iron(II) diiron(III) oxide in the same apparatus (figure 10.6(*b*)) so that the steam is swept out as it is formed, it cannot react with the iron.

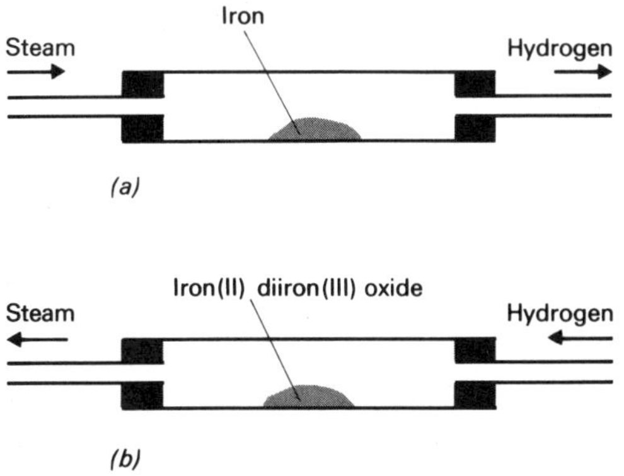

Figure 10.6 (*a*) The reaction of iron with steam. (*b*) The reaction of iron(II) and diiron(III) oxide with hydrogen.

However, if the steam and iron are heated in a closed vessel, hydrogen and iron(II) diiron(III) oxide will again be formed but now neither of these products can escape. *Therefore as soon as they are formed they will start to react to give back the starting materials.* When the forward reaction has just begun, its rate will be high because the steam and iron will be there in high concentration, and at this time the reverse reaction will be slow because the hydrogen and iron(II) diiron(III) oxide will be there in very low concentration. As time goes on the concentration of iron and steam will decrease and their reaction rate will decrease, but the concentration of iron(II) diiron(III) oxide and hydrogen will increase and their reaction rate will increase. Eventually a state will be reached when the rate at which the iron and steam react together will be the same as the rate at which the iron(II) diiron(III) oxide and hydrogen react together, and exactly as much of the iron and steam will be being reformed from the iron(II) diiron(III) oxide and hydrogen as are being lost forming the iron(II) diiron(III) oxide and hydrogen. *A state of equilibrium* will be set up. *A state of equilibrium* is achieved when *the rates of the forward and backward reactions are the same.*

10.6.3 The factors that affect the state of equilibrium

The state of equilibrium exists when the rates of the forward and reverse reactions are the same. If we can do something to the system that affects one of these rates and leaves the other unchanged (or changes it but not to the same extent) then the *position of the equilibrium* will be changed.

This is best illustrated by some simple experiments.

Experiment. Precipitation of iron(II) sulphide.

Pass some hydrogen sulphide gas through a solution of iron(II) sulphate in water; black iron(II) sulphide is precipitated. To the mixture add some sulphuric acid (50 % by volume); the black sulphide redissolves. The reaction ((*b*) above) is:

$$FeSO_4(aq) + H_2S(g) \rightleftharpoons FeS(s) + H_2SO_4(aq).$$

When the acid is added it *increases the concentration* of one of the compounds on the right hand side and the reverse reaction takes place to form more of the compounds on the left hand side.

Experiment. The chromate(VI)-dichromate(VI) equilibrium.

Dissolve some potassium chromate(VI) in water; a yellow solution will be obtained. Add, from a dropper, some 2M hydrochloric acid and the solution will go orange. Add, with another dropper some 2M sodium hydroxide and the solution will go yellow again. The alternate addition of acid and alkali can be repeated many times, the colour changing from yellow to orange and back to yellow.

The reaction ((*e*) above) is:

$$\underset{\text{yellow}}{2CrO_4^{2-}(aq)} + 2H^+(aq) \rightleftharpoons \underset{\text{orange}}{Cr_2O_7^{2-}(aq)} + H_2O(l)$$

On adding acid the concentration of H^+ is increased and the *forward reaction* takes place to form an excess of the orange dichromate(VI) ions. When alkali is added this removes the H^+ ions and so the reaction responds by the *reverse reaction* taking place to give an excess of the yellow chromate(VI) ions. Here again it is a change in the *concentration of a reactant* that is affecting the position of the equilibrium.

Experiment. Sublimation of ammonium chloride.

Heat some ammonium chloride in an evaporating dish with an inverted funnel over it (figure 1.14, with ammonium chloride only in the evaporating dish). The ammonium chloride will reform as beautiful crystals on the funnel. The reaction is (f) in section 10.6.1.

Ammonium chloride exists as a stable compound, (you got some out of a bottle!) so at room temperature the equilibrium above lies on the left hand side, i.e., it is all ammonium chloride. However, when the ammonium chloride is heated the forward reaction produces the two gases, ammonia and hydrogen chloride i.e., the equilibrium position now favours the compounds on the right hand side of the equation. These are cooled again by the cold funnel and the backward reaction takes place to reform ammonium chloride crystals on the funnel. The factor that has determined the extent to which the forward and reverse reactions take place, and hence the position of equilibrium, is the *temperature*.

Experiment. Equilibrium of cobalt and chloride ions

In this experiment you can observe the effect of the *concentration of a reactant* and of *temperature* on an equilibrium.

Dissolve a little of a cobalt(II) salt by heating it in 2–3 cm^3 of water in a boiling tube. Cool under a tap. The solution will be pink or red. Add concentrated hydrochloric acid from a dropper until the solution becomes deep blue in colour. Add water from a dropper until the solution becomes pink or red again.

Heat the solution in a Bunsen flame and the pink solution will turn blue. Cool the solution under a tap and the blue solution will turn pink. The heating and cooling may be repeated many times. The reaction is:

$$Co^{2+}(aq) + 4\,Cl^{-}(aq) \rightleftharpoons CoCl_4^{2-}(aq)$$

When the cobalt(II) is surrounded by water molecules (i.e. on the left hand side of the equation) it is pink in colour. When the cobalt(II) is surrounded by chloride ions (i.e. on the right hand side of the equation in what is called a complex ion) it is blue. When adding hydrochloric acid to the pink solution in water the concentration of Cl^- ions is increased and the *forward reaction* takes place to form mainly the blue complex ions. When water is added this decreases the concentration of chloride ions again and so the reaction responds by the *reverse reaction* taking place to give mainly the pink cobalt(II) (aq) ions. Here it is the change in the *concentration of a reagent* that is affecting the position of the equilibrium.

For this equilibrium the blue complex is more stable in hot solutions than in cold and hence when the solution is heated more of the blue complex is formed and the solution turns from pink to blue. On cooling, the reverse reaction takes place as the cobalt(II) (aq) ions become the most stable. It is the *temperature* that is affecting the position of the equilibrium.

Pressure, too, influences the position of equilibrium in some reactions, particularly those involving gases. Consider, for example, a mixture of dinitrogen tetraoxide and nitrogen dioxide in equilibrium at a particular temperature:

$2\,NO_2(g)$ $\qquad$ $N_2O_4(g)$.
dark brown $\qquad$ pale brown

Increasing the pressure on the mixture causes more dinitrogen tetraoxide to be present. Reducing the pressure has the opposite effect. The changes can be followed by changes in the colour of the mixture. An experiment using these gases to demonstrate this that is often given in textbooks (including the previous edition of this one!) is not valid because when pressures are changed temperature changes also occur (you have probably held your finger over the end of a bicycle pump and felt it get hot on pumping), and these cause complications. (You should read an article of ours in the School Science Review (1978, 211, **60**, 309) about this if you would like to know more.)

If there are the *same number* of molecules on each side of the equilibrium *then pressure has no effect* on the position of the equilibrium. Thus in the reaction:

$$H_2(g) + I_2(g) \rightleftharpoons 2HI(g)$$

the application of pressure does not affect the equilibrium position.

10.6.4 Le Chatelier's principle

The effects of changing the various conditions, that we have just been examining, were summarized by Henry le Chatelier in 1888:

If any of the conditions of a system in equilibrium is changed, the system will adjust itself in order to lessen the change.

We have just examined the most important conditions. They are (*a*) concentration, (*b*) temperature and (*c*) pressure.

10.6.5 Industrial processes

Many important chemicals are manufactured by means of reversible reactions. It is, therefore, most important to choose the right conditions for obtaining the best possible yields of the required product.

Let us look at 3 such processes.

(a) Manufacture of ammonia.

Ammonia is synthesized from its elements (section 16.3.2):

$$N_2 + 3H_2 \rightleftharpoons 2NH_3 \qquad \Delta H = -92\ \text{kJ mol}^{-1}$$

Effect of pressure. The equation shows that 4 molecules of the starting materials give 2 molecules of the products. Therefore, by Avogadro's Hypothesis, there is a decrease in volume when ammonia is formed. This means that high pressure will favour the yield of ammonia because the reduction in volume will tend to reduce the pressure. In practice pressures between 250 and 1000 atmospheres are used.

Effect of temperature. The reaction is exothermic. If heat is applied the reaction will move so that heat is absorbed. Therefore a high temperature in this example will give a lower yield of ammonia. In practice it is found that at very low temperatures, even in the presence of a catalyst, the rate of reaction is low, i.e. a high yield is obtained but too slowly to be of any practicable use. This means that an optimum temperature of about 500 °C, together with a catalyst, is used. This gives a reasonable yield (about 10 %) very rapidly.

(b) Manufacture of nitrogen oxide.

Nitrogen oxide can be manufactured from its elements:

$$N_2 + O_2 \rightleftharpoons 2NO \qquad \Delta H = +180 \text{ kJ mol}^{-1}$$

Effect of pressure. 2 molecules of the starting materials gives 2 molecules of the product, i.e. there is no change in volume when this reaction takes place. Pressure, therefore, is without effect on the position of equilibrium.

Effect of temperature. The reaction is endothermic. Raising the temperature will therefore increase the yield of nitrogen oxide because in this way heat will be absorbed. At a high temperature the rate of reaction is also highest and no catalyst is necessary.

(c) Manufacture of sulphur(VI) oxide

Sulphur(VI) oxide is manufactured from sulphur dioxide and oxygen (section 19.7.3(b)):

$$2SO_2 + O_2 \rightleftharpoons 2SO_3 \qquad \Delta H = -189 \text{ kJ mol}^{-1}$$

Effect of pressure. When sulphur(VI) oxide is formed there is a decrease in volume so that a high yield is favoured by a high pressure. In practice a good enough yield is obtained at atmospheric pressure and the cost of using a high pressure is not worthwhile.

Effect of temperature. The reaction is exothermic and therefore a high temperature favours the backwards reaction. In practice a temperature of 500 °C and a catalyst are used; this ensures that equilibrium is reached quickly. At a lower temperature a higher yield would be obtained but it would be obtained more slowly.

Effect of concentration. A large excess of air is used to ensure the maximum yield from the expensive sulphur dioxide.

10.6.6 The effect of a catalyst on the position of equilibrium.

In the reactions (*a*) and (*c*) just discussed a catalyst is used to help the reactions to occur. It must be noted that the catalyst has no effect on the yield. i.e. *The position of equilibrium is unaffected by the presence of a catalyst.* The catalyst increases the rate of the forward and reverse reactions and therefore the speed at which the equilibrium condition is reached.

10.6.7 The dissociation of water

A very important equilibrium system is that of water.

Water is in equilibrium with hydrogen ions and hydroxyl ions:

$$H_2O(l) \rightleftharpoons H^+(aq) + OH^-(aq)$$

This equilibrium is of great importance in the chemistry of aqueous solutions as it enables water to give or take hydrogen or hydroxyl ions as required by other substances dissolved in water.

In pure water the equilibrium lies well over to the left. In every 1 gram of pure water there are only 0·000 000 000 1 g of H^+ ions and 0·000 000 001 7 g of OH^- ions.

The reverse reaction:

$H^+(aq) + OH^-(aq) \rightarrow H_2O(l)$

is the reaction:

acid + base → water,

and it is when a substance dissolves in water that requires H^+ or OH^- that the equilibrium position of water changes and the appropriate H^+ or OH^- ions are produced.

A number of examples of this will be found throughout this book (for example in the section on salt hydrolysis (section 12.5)).

Often the equation for the dissociation of water is not written as:

$H_2O \rightleftharpoons H^+ + OH^-$

but as:

$2H_2O \rightleftharpoons H_3O^+ + OH^-$

This second equation is more correct than the first. What it is saying is that free H^+ ions do not exist in aqueous solution but are always firmly attached to water molecules. This is true. The H^+ ion is a very very small species, as it consists of a nucleus (of the hydrogen atom) without any electrons. It is readily attracted by electrons in the water molecule and attaches itself to them.

However, so far as its chemistry is concerned it is an H^+, and in this book we have not complicated our equations by putting in the extra H_2O molecule.

If you, or your chemistry teacher, prefers to write H_3O^+ instead of H^+ then all you have to do is add H_2O to all the H^+s in this book!

Questions on chapter 10

1. A chemist attempts to prepare some hydrogen by allowing sulphuric acid to react with zinc. Using lumps of zinc of about 0·5 cm diameter and 0·2 *M* acid he finds that gas is given off but the reaction is too slow. State and explain as far as you can, three ways in which he could alter the conditions of reaction so that the gas would be formed more rapidly. Which of these would you recommend as the most suitable, and why? If the chemist wanted to produce 36 litres of hydrogen, at room temperature and pressure, how much zinc (to the nearest 5 g) would he need? (The volume of one mole of gas at 1 atm pressure and room temperature is approximately 24 litres. Atomic masses: Zn = 65·4, H = 1·00.)

2. In a series of experiments in which magnesium ribbon of uniform width reacted with excess dilute hydrochloric acid, the rate of evolution of hydrogen was found to be as follows:

Length of ribbon (cm)	1·0	2·0	3·0	4·0	5·0	6·0	7·0
Rate of evolution of hydrogen (cm^3/min)	1·1	1·8	2·7	3·6	4·6	5·4	6·4

Draw a graph and state what conclusion you reach. Calculate the rate of evolution of hydrogen from a piece of magnesium ribbon 12 cm long under the same conditions.

Describe, with experimental details, how these measurements could be made. (*L.*)

3. $2SO_2 + O_2 \rightleftharpoons 2SO_3 + \text{heat evolved.} \qquad \Delta H = -189 \text{ kJ mol}^{-1}$

The equation represents a system of equilibrium. State the changes in the equilibrium concentration of sulphur(VI) oxide which would be caused by (*a*) adding oxygen, (*b*) heating the mixture, (*c*) increasing the pressure. (*J.M.B.*)

4. Phosphorus pentachloride undergoes dissociation on heating:

$PCl_5(\text{gas}) \rightleftharpoons PCl_3(\text{gas}) + Cl_2(\text{gas})$

What would be the effect on the equilibrium concentration of phosphorus pentachloride of (*a*) increasing the pressure, (*b*) adding chlorine? (*c*) The equilibrium concentration of chlorine increases as the temperature is increased. What can be deduced about the evolution or absorption of heat as phosphorus pentachloride dissociates? (*J.M.B.*)

5. (*a*) The formation of methanol from hydrogen and carbon monoxide can be represented by

$CO + 2H_2 \rightleftharpoons CH_3OH \qquad \Delta H = +90{\cdot}54 \text{ kJ mol}^{-1}$

What weight of hydrogen would react to cause a heat change of 90·54 kJ?

(*b*) What would be the effect on the equilibrium concentration of methyl alcohol in this endothermic reaction if (i) the temperature was increased, (ii) the pressure was increased, (iii) the hydrogen concentration was increased? (*J.M.B.*)

6. (*a*) The equation for the dissociation of calcium carbonate is

$CaCO_3 \rightleftharpoons CaO + CO_2 \qquad \Delta H = +178 \text{ kJ mol}^{-1}$ (heat taken in)

What will be the effect on the proportion of calcium carbonate in the equilibrium mixture of (i) increasing the temperature, (ii) increasing the pressure?

What conditions would be suitable for manufacturing calcium oxide from calcium carbonate on a large scale?

(*b*) At 400 °C, the three gases, hydrogen, iodine and hydrogen iodide exist together in equilibrium. The equation for the reaction is

$H_2 + I_2 \rightleftharpoons 2HI$

What effect will an increase in pressure have on (i) the rate of reaction of hydrogen with iodine and (ii) the position of equilibrium?

Name the product formed when hydrogen iodide is dissolved in water and describe what would happen if chlorine were passed into the solution.

(*c*) Write an equation for a reaction which can be catalysed by manganese(IV) oxide (manganese dioxide). What would be the effect on the rate of this reaction of (i) adding more manganese(IV) oxide, (ii) using the same weight of catalyst but increasing the particle size?

(*d*) The rate of reaction between hydrogen and chlorine depends on the intensity of light falling on the reagents. For another reaction whose rate also depends on the intensity of the light, (i) name the reagents and (ii) either name the products or state the importance or use of the reaction. (*J.M.B.*)

7. (*a*) When 1·0g of powdered manganese(IV) oxide (manganese dioxide) was shaken with 25 cm^3 of hydrogen peroxide solution, a reaction took place which was completed in 1 minute. State three different ways in which the reaction could be slowed down.

(*b*) It is said that copper(II) oxide will catalyse the decomposition of hydrogen peroxide. Outline the experiments which you would make to test this statement. Include in your answer details of the measurements you would make and the results that you would expect if the oxide were a catalyst.

(*c*) Explain why catalysts are important in manufacturing processes. Illustrate your answer with two examples from industrial chemistry, naming the catalyst in, and writing an equation for, each reaction. (*J.M.B.*)

8. (*a*) Explain briefly what you understand by *photosynthesis*.

(*b*) What is a catalyst? Write equations for two catalysed reactions of industrial importance and for each reaction name the catalyst used. Describe experiments you would carry out to compare the effectiveness of two given metallic oxides when used as catalysts in the decomposition of hydrogen peroxide solution. (*C.*)

9. A crystal of pure calcium carbonate weighing 7·50 g was placed in a flask with 50 cm^3 of dilute hydrochloric acid. The flask was kept at constant temperature and the carbon dioxide evolved was collected in a graduated vessel. The volume of carbon dioxide was recorded at 20 minute intervals and corrected to s.t.p. Some of the calcium carbonate remained undissolved at the end of the experiment. The results of the experiment are given in the following table:

Time from start of reaction (min.)	Volume of carbon dioxide corrected to s.t.p. (cm^3)
20	655
40	910
60	1065
80	1100
100	1120
120	1120

(*a*) Write the equation for the reaction between calcium carbonate and hydrochloric acid.

(*b*) 655 cm^3 of gas were evolved during the first 20-minute interval. Write down the volumes of gas evolved (i) during the second 20-minute interval (20 to 40 minutes), (ii) during the third 20-minute interval (40 to 60 minutes), (iii) during the fourth 20-minute interval (60 to 80 minutes). Explain the variations in the volumes of gas formed during these intervals.

(*c*) Why is there no increase in volume of gas after 100 minutes?

(*d*) Plot the graph of the volume of carbon dioxide evolved against the time. How much gas was evolved after (i) 30 minutes, (ii) 50 minutes? (*C.*)

10. (*a*) Describe and write equations for (i) one reaction of industrial importance in which heat is absorbed, (ii) one reaction of industrial importance in which heat is produced, (iii) two reactions in which light energy is absorbed.

(*b*) What is meant by the calorific value of a fuel? A fuel such as coal and a food such as bread are used in very different ways. Explain why the calorific value is an important characteristic of both the fuel and the food. (*C.*)

11. A small flask was connected to a gas syringe by means of a stopper and delivery tube. 30 cm^3 of water and 0·5 g of manganese(IV) oxide were placed in the flask. 5 cm^3 of hydrogen peroxide were added, the flask was quickly stoppered and readings of the volume of gas in the syringe were recorded every 10 seconds.

Time	0	10	20	30	40	50	60	70	80
Volume (cm^3)	0	18	30	40	48	53	57	58	58

(*a*) On graph paper plot a graph of the volume shown on the syringe against the time in seconds. Label this curve "*A*". When the reaction was complete, the stopper was removed and the syringe was emptied of gas. Without emptying the flask, another 10 cm^3 of water and 5 cm^3 of hydrogen peroxide were added, and the experiment repeated exactly as before. On the same sheet of graph paper and using the same axes as were used for the previous graph, sketch a second graph to show how the volume of gas collected would vary with time in this second experiment. Label this second curve "*B*". Explain the reason for choosing the line you have drawn.

(*b*) In this reaction, the manganese(IV) oxide acts as a catalyst. Explain fully how you would attempt to prove this.

(*c*) Write an equation and state the conditions for another reaction of manganese(IV) oxide in which it does not act as a catalyst. What part is played by the oxide in this reaction? (*J.M.B.*)

12. (*a*) Describe an experiment that you have either seen or performed to show how temperature affects the rate of *either* the reaction of a metal with a dilute acid *or* the decomposition of hydrogen peroxide solution. Include in your description the apparatus used, the experimental procedure and the measurements made.

(*b*) An excess of dilute hydrochloric acid is added to 10 g of marble (calcium carbonate) and left at constant temperature until the reaction has stopped.

(i) Calculate the maximum volume of carbon dioxide at s.t.p. formed when the reaction has stopped.

(ii) Draw a graph to show approximately how you would expect the volume of carbon dioxide formed to vary with *time* during the experiment. (*C.*)

11

Water

11.1 Introduction

Water is such an important chemical and is involved in so many chemical reactions that you should have some general idea of its properties before we proceed further. Its importance to life, as we know it, is common knowledge; in this book we are concerned with it from a narrower point of view.

There are many natural sources of water, but it is never found completely pure.

Rain water is the purest form of natural water, formed as it is by a process resembling distillation. Water is, however, a solvent for gases, so that rain water will inevitably contain small quantities of those present in the atmosphere, notably carbon dioxide, oxygen and nitrogen. An experiment which demonstrates the presence of such gases is described in section 1.16. In our larger towns, rain water passing through heavily polluted atmospheres has a more dubious content.

Sea water forms the largest proportion of water on the earth. It contains as much as 3·6 per cent of dissolved solids, mainly common salt. The sea is now regarded as a "chemical store" for many other substances besides salt, and already bromine and magnesium are extracted from it in large quantities (see sections 20.2.3 and 22.1.3). Some idea of the vast quantities of the elements that are present are given in table 11.1.

River water is of widely varying purity, but always contains dissolved solids in addition to gases. The actual solids present depend on the soil over which the river passes; if the river has run over insoluble materials, such as granite, the solid content will be small, but if it has passed over chalk the water will contain calcium hydrogencarbonate (bicarbonate). Much river water is used for domestic purposes and the substances dissolved in such water are fully discussed in section 22.2.

Spring water has travelled downwards through soil and consequently the solid impurities depend on the various strata through which the water has passed. In some areas mineral springs contain very high proportions of dissolved substances. At Epsom spring water contains magnesium sulphate (Epsom salts), while the much recommended Harrogate water contains hydrogen sulphide!

If chemically pure water is required then the various substances dissolved in it must be removed. The most important methods of doing this are *distillation* (section 1.7 and 22.2.4(b)) and *ion-exchange* (section 22.2.4(e)).

Water as one of the earth's natural resources is discussed further in section 28.4.

11.2 Physical properties of water

Water is a colourless, odourless, tasteless liquid. It is such a common liquid that use is made of its physical constants for such purposes as calibrating thermo-

TABLE 11.1 ESTIMATED QUANTITIES OF SOME ELEMENTS IN ONE CUBIC MILE OF SEA-WATER

	Tonnes
Chlorine (as Cl^- ions)	91 000 000
Sodium (as Na^+ ions)	50 000 000
Magnesium (as Mg^{2+} ions)	6 200 000
Sulphur (as SO_4^{2-} ions)	4 300 000
Calcium (as Ca^{2+} ions)	2 000 000
Potassium (as K^+ ions)	1 800 000
Bromine (as Br^- ions)	350 000
Carbon (as HCO_3^- ions)	72 000
Strontium (as Sr^{2+} ions)	37 000
Boron (as HBO_3^{2-} ions)	18 000
Fluorine (as F^- ions)	4 700
Iodine (as I^- ions)	100–12 000
Arsenic	45–350
Silver	up to 45
Gold	23

(*Note.* The first 11 elements are the major constituents. The last 4 in the table illustrate the large quantities of some minor constituents.)

meters, and in defining the gram. *The boiling point of pure water at 760 mm of mercury pressure is called 100 °C.* At pressures higher than normal, as in a pressure cooker, the boiling point is above 100 °C; at pressures below normal, as at the top of Mount Everest, it is below 100 °C.

The presence of dissolved solids also raises the boiling point of water.

The freezing point of pure water at 760 mm of mercury is called 0 °C. Increased pressure lowers the freezing point slightly. The presence of dissolved solids lowers the freezing point of water appreciably, the fact that sea water freezes at below 0 °C being well known.

The density of water under a pressure of 760 mm at 4 °C is 1 g per cm³. Pressure has such a slight effect on the density that it is quite common to omit mention of it in stating the density.

The physical constants listed above are often used to distinguish water from other liquids.

Water is a good solvent for many substances. It is because water is so universal in its solvent action that absolutely pure water is difficult to obtain. Many substances which are said to be insoluble in water do in fact dissolve slightly.

Many chemical reactions will proceed only in solution. *This is particularly true of reactions between ions,* for example the reaction which takes place when silver nitrate solution is added to sodium chloride solution does not occur when the two substances are dry.

$$AgNO_3 + NaCl \rightarrow AgCl + NaNO_3$$

$$Ag^+(aq) + Cl^-(aq) \rightarrow AgCl(s)$$

11.3 Chemical properties of water

Water is a stable compound in the sense that it is not easy to decompose it directly into its elements. It does, however, enter into a wide variety of chemical reactions.

The following list is by no means comprehensive; many of the reactions are discussed more fully elsewhere in this book.

11.3.1 Action with metals

As mentioned in section 8.2.1(*b*), the action of a metal on water is very much influenced by its position in the electrochemical series. This is clearly shown by the behaviour of the following metals;

(a) Potassium

When a small piece of potassium (no larger than a pea) is dropped into a trough full of water it melts to a silvery ball and vigorously shoots about over the surface of the water. It effervesces, and the gas evolved burns with a violet flame. After a short time the reaction is complete and the potassium disappears.

The solution left is alkaline to litmus and contains potassium hydroxide. The inflammable gas is hydrogen, the colour being imparted to the flame by the potassium.

$$2K + 2H_2O \rightarrow \underset{\text{potassium hydroxide}}{2KOH} + H_2$$

$$2K(s) + 2H_2O(l) \rightarrow 2K^+(aq) + 2OH^-(aq) + H_2(g)$$

(b) Sodium

Sodium is less vigorous in its action, but, like potassium, a small piece of sodium melts and shoots about on the surface of the water, with a ready effervescence as hydrogen is evolved. However, there is insufficient heat to ignite the hydrogen. If the sodium is placed on a piece of filter paper that is floating on the water then the hydrogen will ignite. Again an alkaline solution is produced, this time containing sodium hydroxide:

$$2Na + 2H_2O \rightarrow \underset{\text{sodium hydroxide}}{2NaOH} + H_2$$

$$2Na(s) + 2H_2O(l) \rightarrow 2Na^+(aq) + 2OH^-(aq) + H_2(g)$$

It is possible to collect the gas from this reaction by wrapping a piece of clean sodium in dry copper gauze to make it sink, and placing an inverted test-tube full of water above the bubbles (figure 11.1). The gas can then be shown to burn.

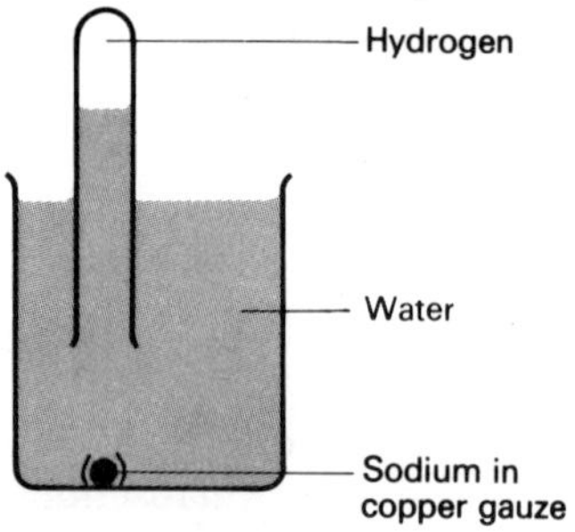

Figure 11.1 The reaction of sodium with water.

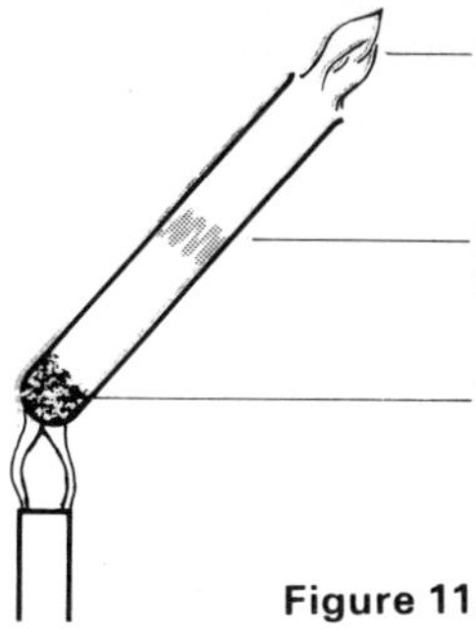

Figure 11.2 The reaction of magnesium with steam.

(N.B. Some teachers recommend the use of the less reactive sodium-lead alloy or lithium to demonstrate this type of reaction. Several unexplained explosions have resulted from the use of sodium.)

(c) Calcium

When calcium is added to water it sinks and bubbles of gas are formed. These can be collected, as with sodium, and shown to be hydrogen.

An alkaline solution is obtained, but as the reaction proceeds the liquid becomes milky. This is because the alkali, calcium hydroxide, which is produced, is only slightly soluble, and therefore small white particles of it become suspended in the liquid.

$$Ca + 2H_2O \rightarrow \underset{\text{calcium hydroxide}}{Ca(OH)_2} + H_2$$

$$Ca(s) + 2H_2O(l) \rightarrow Ca^{2+}(aq) + 2OH^-(aq) + H_2(g)$$

(d) Magnesium

There is little reaction between magnesium and cold distilled water. Its action with steam can be shown as follows:

A few cubic centimetres of water are placed in a hard glass test-tube, and sufficient silica wool is added to soak up the water. Some folded magnesium ribbon is inserted half-way up the tube as shown (figure 11.2). The tube is clamped at an angle of 45° and is heated strongly just below the magnesium. The wet silica wool is then gently heated and the magnesium ignites (further heating of the magnesium may be necessary); a hydrogen flame is seen at the mouth of the tube:

$$Mg(s) + H_2O(g) \rightarrow \underset{\text{magnesium oxide}}{MgO(s)} + H_2(g)$$

(e) Iron

When steam is passed over strongly heated iron filings or steel wool, hydrogen and iron(II) diiron(III) oxide are formed. (For experimental details, see section 10.6.2).

$$3Fe(s) + 4H_2O(g) \rightleftharpoons Fe_3O_4(s) + 4H_2(g)$$

(f) Copper

Copper and other metals below hydrogen in the electrochemical series do not react with water or steam.

11.3.2 Reactions with non-metals

In general, water does not react with non-metals. If, however, steam is passed over white-hot coke a mixture of carbon monoxide and hydrogen, known as water gas, is obtained:

$$C(s) + H_2O(g) \rightarrow \underbrace{CO(g) + H_2(g)}_{\text{water gas}}$$

Cold water combines with chlorine to give a mixture of hydrochloric and chloric(I) acids:

$$Cl_2(g) + H_2O(l) \rightarrow \underset{\text{hydrochloric acid}}{HCl(aq)} + \underset{\text{chloric(I) acid}}{HClO(aq)}$$

11.3.3 Reactions with oxides

Many oxides of non-metals combine with water to give acids:

$$\underset{\text{sulphur dioxide}}{SO_2(g)} + H_2O(l) \rightleftharpoons \underset{\text{sulphurous acid}}{H_2SO_3(aq)}$$

$$\underset{\text{carbon dioxide}}{CO_2(g)} + H_2O(l) \rightleftharpoons \underset{\text{carbonic acid}}{H_2CO_3(aq)}$$

Oxides of metals do not react with water except for those of metals high in the electrochemical series. With these, alkalies are obtained:

$$\underset{\text{calcium oxide}}{CaO(s)} + H_2O(l) \rightarrow \underset{\text{calcium hydroxide}}{Ca(OH)_2(aq)}$$

11.4 The composition of water

Until the end of the eighteenth century water was believed to be an element. The first evidence to the contrary resulted from an experiment by Joseph Priestley (figure 13.8) in 1781, when he exploded a mixture of hydrogen and air to amuse, so it is related, some of his friends. A dew formed inside the glass vessel being used, but Priestley thought that this was due to water vapour present in the air at the beginning of the experiment. Henry Cavendish repeated these experiments and managed to collect a small quantity of a liquid assumed to be water. He also exploded a mixture of hydrogen and oxygen and obtained water, but his product was acidic due to impurities in the oxygen used, and he therefore delayed publication of his results so that he could investigate the cause of the acidity.

Meanwhile the French scientist, Antoine Lavoisier (figure 13.10), who had also been investigating the burning of hydrogen, heard of Cavendish's experiments, and realised that water must be a compound of hydrogen and oxygen. He published this information without acknowledging his debt to the work of Cavendish. An experiment describing the preparation of water from hydrogen is given in chapter 14.

Various other experiments also indicate the composition of water. Thus, Lavoisier obtained hydrogen by passing steam over red-hot iron. Assuming iron to be an element, the hydrogen must have come from the water. The other product was iron(II) diiron(III) oxide, which is also obtained by burning iron in oxygen. Therefore water must also contain oxygen.

The reaction of steam with magnesium also gives a clear indication that water is composed of hydrogen and oxygen.

11.4.1 The formula of water

We have often used the formula H_2O for water, but so far we have only shown that it is a compound of hydrogen and oxygen. The formula can be proved by determining the volume composition of steam.

Experiment. Determination of the formula of water

The apparatus (figure 11.3) consists of a stout glass calibrated tube. On top of the glass tube is a three-way tap; two platinum wires, with a small gap between them, are fixed into the tube. The other ends of the wires are connected to a sparking-coil. A wide glass jacket surrounds the tube with inlet and outlet tubes fitted as shown in the figure.

The three-way tap is set so that B is connected to C and the mercury level is raised until the mercury just touches the tap. The tap is then turned to connect A to B. Hot water is put into the outer jacket and steam passed through it, to heat the apparatus to about 100 °C. The temperature of the apparatus can be read on the thermometer.

Approximately equal volumes of the gases hydrogen and oxygen are introduced into the apparatus by connecting each gas supply, in turn, to tube A. After removing the air from the connecting tube via B, the tap is turned to connect A to C and the mercury is run out of the tap until a measured volume of each gas has been put into the apparatus. The volumes of the gases must not be measured until they reach the temperature of the apparatus (i.e. until the gas in the apparatus has stopped expanding). Volumes are measured at atmospheric pressure by equalizing the level of the mercury in the two limbs of the apparatus.

When a spark is passed through the wires, a mild explosion occurs and steam is formed. It is advisable to put a cork in the open end of the apparatus before exploding the gas mixture, so that there is no chance of mercury being blown out of it. The mercury levels are again equalized and the volume of steam plus excess reactant gas is noted. Cold water is then passed into the outer jacket and as the apparatus cools the steam produced by the explosion condenses. The mercury levels are again equalized and the volume of excess

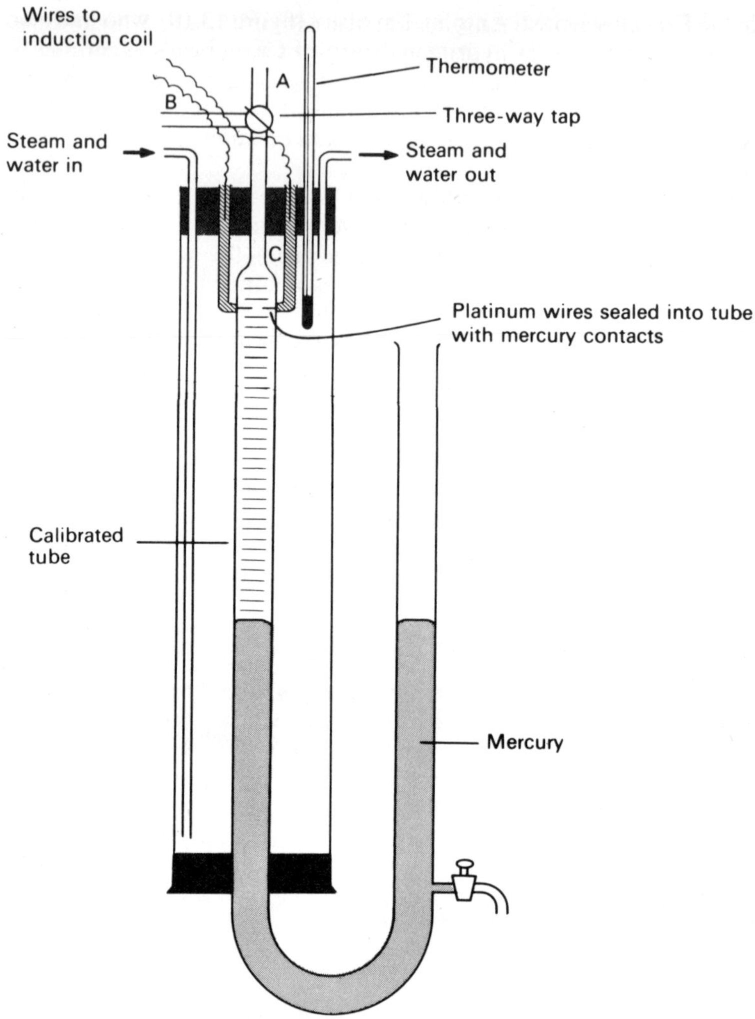

Figure 11.3 The formula of water.

reactant gas, and the temperature are noted. This excess gas can be shown to be oxygen by collecting it, via exit A or B. A glowing wooden splint reignites in the gas, confirming that it is oxygen.

Typical results would be:

Volume of hydrogen (at 100 °C)	= 20 cm^3
Volume of oxygen (at 100 °C)	= 22·5 cm^3
Volume of steam + excess oxygen (at 100 °C)	= 32·5 cm^3
Volume of excess oxygen (at 20 °C)	= 9·8 cm^3

All volumes are required at the same temperature (say 100 °C). Therefore the reading taken at 20 °C must be corrected to 100 °C by using Charles' Law:

$$\frac{V_1}{T_1} = \frac{V_2}{T_2} \quad \text{(section 7.4).}$$

$$\frac{9 \cdot 8}{293} = \frac{x}{373} \qquad \therefore\ x = 12 \cdot 5\ \text{cm}^3$$

i.e. Volume of excess oxygen (at 100 °C) = 12·5 cm^3
and volume of oxygen which has reacted = 22·5 − 12·5 = 10·0 cm^3.
Also volume of steam (at 100 °C) = 32·5 − 12·5 = 20 cm^3.
∴ 20 cm^3 of hydrogen + 10 cm^3 of oxygen → 20 cm^3 of steam,
i.e. 2 volumes of hydrogen + 1 volume of oxygen → 2 volumes of steam.

Therefore by Avogadro's Law,
2 molecules of hydrogen + 1 molecule of oxygen → 2 molecules of steam.
Therefore 1 molecule of water is made from 1 molecule of hydrogen and $\frac{1}{2}$ molecule of oxygen.

Therefore, as hydrogen and oxygen are both diatomic, *the formula of water is* H_2O.

(*Note:* The method given here differs from that given elsewhere in that the combining ratio of 2 : 1 is not assumed and equal volumes of hydrogen and oxygen gases are used. Under these conditions the steam formed does not condense until well below 100 °C. Therefore there is no need to use, as is done elsewhere, an inflammable high boiling alcohol vapour in the outer jacket. Hot water can be used and it is much more convenient.)

11.4.2 The composition of water by mass

The first determination of the composition of water by mass was made by Jean Dumas. For his experiment he prepared hydrogen, and took elaborate precautions to purify it. The pure, dry hydrogen was passed over a weighed quantity of heated copper(II) oxide, with consequent production of water. If the mass of water formed was also measured its percentage composition could be calculated.

An adaptation of Dumas' experiment, giving approximately correct results, can be carried out in the school laboratory.

Experiment. The composition of water

Set up the apparatus shown in figure 11.4.

Pass hydrogen through the U-tube containing anhydrous calcium chloride to dry it and then over the large boat containing copper(II) oxide. Lead the steam formed into the pair of calcium chloride tubes, to absorb it. Weigh the boat containing copper(II) oxide and also the calcium chloride tubes before and after the experiment. An extra calcium chloride tube must be present to prevent entry of moisture from the atmosphere.
Work out the results as follows:

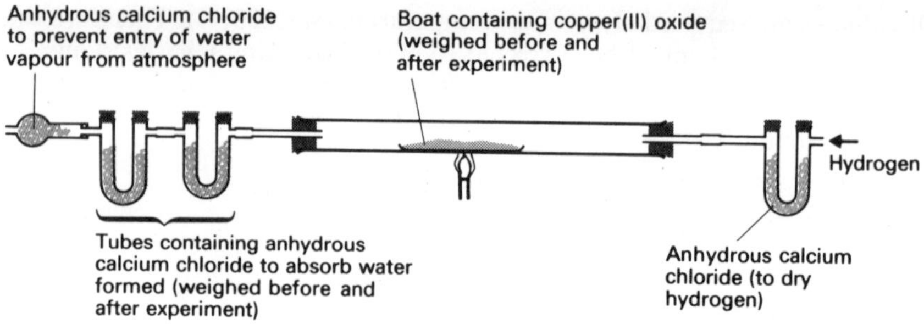

Figure 11.4 The composition of water by mass.

Mass of calcium chloride tubes at end of experiment $= w_1$ g
Mass of calcium chloride tubes at start of experiment $= w_2$ g
$\therefore$ Gain in mass of tubes
= mass of water formed $= w_1 - w_2$ g

Mass of copper(II) oxide boat at beginning $= w_3$ g
Mass of copper(II) oxide boat at end $= w_4$ g
Loss in mass
= mass of oxygen in water formed $= w_3 - w_4$ g

$$\therefore \text{Percentage of oxygen in water} = \frac{w_3 - w_4}{w_1 - w_2} \times 100\,\%$$

$$\text{and Percentage of hydrogen in water} = 100 - \left(\frac{w_3 - w_4}{w_1 - w_2} \times 100\right)\%$$

Dumas' results gave an average of 11·136 per cent hydrogen, 88·864 per cent oxygen, i.e. a ratio by weight of hydrogen to oxygen of 1 : 7·98. The accepted value for this ratio is now 1 : 7·94.

In the laboratory experiment outlined above it is unlikely that you will get a result very near to this. Can you suggest reasons for inaccuracy?

11.5 Water of crystallization

When some dry copper(II) sulphate crystals are heated in a test-tube steam is formed, and the blue crystals change to a white powder. The water must be combined in the copper(II) sulphate crystals, otherwise they would necessarily be damp. Furthermore, copper(II) sulphate crystals have a definite composition by mass, indicating that the water forms part of a compound. This combined water is called *water of crystallization*, as the shape of the crystals is dependent on this water. *Water of crystallization is defined as a fixed amount of water with which some crystals are associated, when they are obtained from a solution in water.*

The formula of copper(II) sulphate crystals is $CuSO_4.5H_2O$. It is said that each molecule of copper(II) sulphate crystals contains five molecules of water of crystallization. That the water is only loosely combined is obvious from the ease with which it is given up on heating.

Many other salts contain water of crystallization. Other examples are:

Sodium carbonate crystals (Washing soda)	$Na_2CO_3.10H_2O$
Iron(II) sulphate crystals (Green Vitriol)	$FeSO_4.7H_2O$
Sodium sulphate crystals (Glauber's salts)	$Na_2SO_4.10H_2O$
Sodium thiosulphate crystals (Hypo)	$Na_2S_2O_3.5H_2O$
Aluminium potassium sulphate (Alum)	$AlK(SO_4)_2.12H_2O$

Crystalline compounds containing water are said to be *hydrated*; without their water of crystallization they are said to be *anhydrous*.

i.e. $CuSO_4.5H_2O \xrightarrow{\text{heat}} CuSO_4 + 5H_2O$

hydrated copper(II) sulphate → anhydrous copper(II) sulphate

If water is added to anhydrous copper(II) sulphate the above change is reversed. This is often used as a test to discover whether a liquid contains water or not. A few drops of the liquid are added to white anhydrous copper(II) sulphate. The development of a blue colour indicates the presence of water, although not necessarily of pure water.

It should be mentioned that not all crystalline substances contain combined water of crystallization. For example, crystals of potassium nitrate, sodium chloride (common salt), ammonium sulphate and potassium chlorate(V) are all anhydrous.

Experiment. Determination of the percentage of water of crystallization in Epsom salts (hydrated magnesium sulphate)

Weigh a clean dry crucible and lid empty and then containing a few grams of Epsom salts. Heat the crucible, at first gently and then more strongly. During the initial heating the lid should almost cover the crucible to prevent loss of solid by "spitting".

When most of the water has been expelled as steam remove the lid completely. After about ten minutes' heating cool the crucible, lid and contents, preferably in a desiccator, and reweigh.

To ensure that no water of crystallization is left, heat for a further five minutes' and weigh again. If the last two weighings are the same it can be assumed that no water is left.

Then the percentage of water of crystallization

$$= \frac{\text{Loss in mass of Epsom salts}}{\text{Initial mass of Epsom salts}} \times 100$$

You should be able to make your own table of weighings for this experiment.

The results can be used to determine the actual number of molecules of water of crystallization in the Epsom salts:

Suppose that the loss in mass is found to be 51.2 per cent.

Let the formula of Epsom salts be $MgSO_4xH_2O$, where x is the number of molecules of water of crystallization. ($Mg = 24$, $S = 32$, $O = 16$, $H = 1$.)

$$\begin{array}{ll} & \text{Mg} \quad \text{S} \quad\quad \text{O}_4 \quad\quad x\text{H}_2\text{O} \\ \text{Molecular mass of Epsom salts} & = 24 + 32 + (4 \times 16) + x(2 + 16) \\ & = 120 + 18x \end{array}$$

Of these $(120 + 18x)$, $18x$ are due to water of crystallisation.

∴ Percentage of water of crystallisation

$$= \frac{18x}{120 + 18x} \times 100$$

$$= 51{\cdot}2 \text{ (experimental result)}$$

$$\therefore\ 18x = 0{\cdot}512\ (120 + 18x)$$

$$= 61{\cdot}4 + 9{\cdot}22x$$

$$\therefore\ 8{\cdot}78x = 61{\cdot}4 \text{ and } x = \frac{61{\cdot}4}{8{\cdot}74} = 7 \text{ (nearest whole number)}$$

Formula of Epsom salts = $MgSO_4.7H_2O$

The method outlined can be used with any hydrated compound, provided that it is stable to heat in its anhydrous form. If not, other methods, which we are not able to discuss here, have to be used.

11.6 Efflorescence, deliquescence and hygroscopy

Sodium carbonate crystals (washing soda) are given the formula $Na_2CO_3.10H_2O$, yet in an experiment to find the number of molecules of water of crystallisation present the result is usually less than ten. The discrepancy is particularly great if the crystals have been exposed to the air for a considerable time. This is because the crystals are *efflorescent, i.e. they give up some of their water of crystallization to the atmosphere.* For washing soda the reaction is:

$$Na_2CO_3.10H_2O(s) \rightarrow Na_2CO_3.H_2O(s) + 9H_2O(g)$$

The monohydrate is visible as a white powder on the surface of the crystals.

With some substances, not necessarily hydrates, a process opposite to efflorescence occurs. Thus, if some calcium chloride is exposed to the atmosphere it absorbs water to such an extent that a concentrated solution of calcium chloride is eventually obtained. This process is called *deliquescence.* Other deliquescent substances are sodium hydroxide and phosphorus(V) oxide. All three of these substances are used as drying agents.

Some substances, such as copper(II) oxide, absorb water vapour, *but do not form solutions;* some liquids, such as concentrated sulphuric acid, also readily absorb moisture. Both these classes of substances are said to be *hygroscopic.*

Questions on chapter 11

1. Explain carefully what is meant by the following terms: hygroscopy, deliquescence, efflorescence, water of crystallization.

Illustrate your answers with examples.

2. Explain why anhydrous copper(II) sulphate can be used to detect the presence of water.

What further tests have to be performed to prove that a liquid is pure water?

3. Give a fully labelled diagram to show how the composition of water by weight may be found by preparing dry hydrogen and allowing it to act on dry copper(II) oxide.

State the weighings you would make and indicate how you would calculate the result. (*J.M.B.*)

4. Explain the formation of water (*a*) on the cool sides of a test-tube in which sodium hydrogencarbonate is being heated, (*b*) at the exhaust of a motor-car when the engine is first started from cold.

Write equations for one reaction in each case in which either water or steam reacts with (i) a metal, (ii) a non-metal, (iii) a metallic oxide. State the conditions under which each reaction occurs. Diagrams are not required.

4·76 g of the hydrated chloride of a divalent metal *M*, of atomic mass 59, contain 2·60 g of the anhydrous salt. Find the number of molecules of water of crystallisation in each molecule of the hydrated salt. (*C.W.B.*)

5. Describe a method to determine the composition of water by volume.

How and under what conditions does water react with (*a*) ammonia, (*b*) calcium, (*c*) charcoal?

In the case of (*c*), how would you establish that the reaction proceeds as you state? (*D.*)

6. Explain what is meant by water of crystallization. What are the changes that are observed when crystals of copper(II) sulphate are heated to a moderate temperature? How would you determine the percentage amount of water of crystallization in these crystals?

If in your experiment the loss of water on heating the copper(II) sulphate crystals was 36 per cent, how many molecules of water of crystallization does this represent? (*C.W.B.*)

7. State Avogadro's Law.

What are the experimental facts and theoretical reasoning which lead to the view that the chlorine molecule contains at least two atoms?

Draw a clear, labelled diagram of the apparatus used to demonstrate that two volumes of hydrogen react with one volume of oxygen to produce two volumes of steam. Describe how the experiment is carried out and state briefly the importance of the result. (*C.W.B.*)

8. State fully and completely (*a*) Gay-Lussac's Law of Combining Volumes, (*b*) Avogadro's Hypothesis.

In a suitable apparatus at 130 °C, 100 cm^3 of hydrogen were exploded with 100 cm^3 of oxygen. The volume was then found to be 150 cm^3.

The apparatus was then cooled to room temperature when the water vapour condensed and it was found that the remaining gas was oxygen and that its volume was 50 cm^3. (Assume that all volumes have been corrected to s.t.p.)

Assuming that both hydrogen and oxygen molecules are diatomic deduce *from these data* the formula for steam.

The atmosphere is said to contain water vapour. How would you obtain and collect a sample of water *from the atmosphere*? How would you prove this liquid to be pure? (*J.M.B.*)

9. State Gay-Lussac's Law. Describe the experiment you would carry out to demonstrate that this law is obeyed by the reaction in which steam is formed from hydrogen and oxygen.

How would you show that washing soda contains water of crystallization?

Give the names and formulae of *two* other salt hydrates. (*O. and C.*)

10. What do you understand by water of crystallization or hydration? How would you account for the fact that transparent sodium carbonate-10-water crystals turn to a white powder on exposure to the air?

Describe what happens to (*a*) calcium chloride, and (*b*) phosphorus(V) oxide when each is used to dry a gas.

Outline a method to determine the percentage of water of crystallization in a salt. (*D.*)

12

Acids, Bases, and Salts; Volumetric Analysis

12.1 Acids

12.1.1 Properties of acids

The word "acid" is commonly used in everyday life, and an acid is usually regarded as a highly corrosive, burning liquid. Perhaps, therefore, it should be said straight away that an acid need not be corrosive, and is usually not a liquid. To summarize the properties of acids we can say:

(*a*) *Solutions of acids have a sour taste.* This is the reason for the sour taste of vinegar, which is a dilute solution of ethanoic (acetic) acid; of fruits, which contain 2-hydroxypropane-1,2,3-tricarboxylic (citric) and other acids; and of sour milk, which contains 2-hydroxypropanoic (lactic) acid. The common mineral acids, hydrochloric, sulphuric and nitric, have a similar taste.

(*b*) *Solutions of acids turn blue litmus red.* Many vegetable and synthetic dyes are changed in colour by the action of acids. You can try the effect of acids and alkalies (see later in the chapter) on beetroot juice, flower petals, black-currant juice or any other sort of dye, often with interesting results. Litmus is a dye, obtained from certain lichens, and its solution is often used in the laboratory to determine whether a solution is acidic or not.

(*c*) *Solutions of acids usually yield hydrogen when treated with the more reactive metals (i.e. with metals occurring above hydrogen in the activity series).* Thus, when dilute sulphuric acid is added to zinc, zinc sulphate and hydrogen are formed:

$$Zn(s) + H_2SO_4(aq) \rightarrow ZnSO_4(aq) + H_2(g)$$

The ionic equation is more general and can be taken to represent the action of any dilute acid on zinc:

$$Zn(s) + 2H^+(aq) \rightarrow Zn^{2+}(aq) + H_2(g)$$

The zinc sulphate formed is called a *salt*, and this is a general name for a substance formed when all or part of the hydrogen of an acid is replaced by a metal.

Dilute nitric acid is peculiar in its behaviour and rarely gives hydrogen with a metal. This is because, in addition to being an acid, it is also an oxidising agent (see section 16.6.4(c)), and this causes a complication.

(*d*) *Solutions of acids liberate carbon dioxide from carbonates.* This is probably the best test for an acid, as only a few very weak acids are exceptions to this rule. Thus if nitric acid is added to sodium carbonate there is an effervescence, and a salt, water and carbon dioxide are formed.

$$\left[\begin{array}{lll} Na_2CO_3 + 2HNO_3 & \rightarrow & 2NaNO_3 + H_2O + CO_2 \\ & & \text{sodium nitrate} \\ Na_2CO_3(s) + 2H^+(aq) & \rightarrow & 2Na^+(aq) + H_2O(l) + CO_2(g) \end{array}\right.$$

Again the ionic equation represents the general reaction between an acid and a carbonate.

Other examples of this type of reaction are:

$CaCO_3(s)$	$+2HCl(aq)$	$\rightarrow$	$CaCl_2(aq)$	$+H_2O(l)+CO_2(g)$
calcium carbonate	hydrochloric acid		calcium chloride	
$CuCO_3(s)$	$+H_2SO_4(aq)$	$\rightarrow$	$CuSO_4(aq)$	$+H_2O(l)+CO_2(g)$
copper(II) carbonate	sulphuric acid		copper(II) sulphate	

(*e*) *Acids react with oxides or hydroxides of metals to yield salts.* If dilute sulphuric acid is warmed with copper(II) oxide, copper(II) sulphate and water are formed:

$$\left[\begin{array}{l} CuO \quad + H_2SO_4 \rightarrow CuSO_4 \quad + H_2O \\ CuO(s) + 2H^+(aq) \rightarrow Cu^{2+}(aq) + H_2O(l) \end{array}\right.$$

With sulphuric acid and sodium hydroxide there are two possible reactions, either all or part of the hydrogen of the acid being replaced by the sodium:

$$\left[\begin{array}{llll} NaOH & + H_2SO_4 & \rightarrow NaHSO_4 & + H_2O \\ & & \quad \text{sodium hydrogensulphate} & \\ OH^-(aq) & + 2H^+(aq) + SO_4{}^{2-}(aq) & \rightarrow HSO_4{}^-(aq) & + H_2O(l) \end{array}\right.$$

$$\left[\begin{array}{llll} 2NaOH & + H_2SO_4 & \rightarrow Na_2SO_4 & + 2H_2O \\ & & \quad \text{sodium sulphate} & \\ OH^-(aq) & + H^+(aq) & \rightarrow H_2O(1) & \end{array}\right.$$

An acid is often defined on the basis of the above properties; in the reactions with metals, and with carbonates, oxides and hydroxides of metals, the hydrogen present in all acids is wholly or partially replaced by a metal with the formation of a salt. Therefore:

An acid is a substance which turns blue litmus red and contains hydrogen, all or part of which can be replaced by a metal, directly or indirectly, with the formation of a salt.

This statement is true for all the common acids, but unfortunately it also holds true for some substances which are not acids. It is the hydrogen ions obtained from all acids *when they are dissolved in water* which account for their general similarity in chemical properties. Therefore a simpler and better definition of an acid is:

An acid is a substance which when dissolved in water yields hydrogen ions as the only positive ions. (See section 10.6.7 for a discussion of H^+ compared with H_3O^+.)

Ions yielded by some common acids are:

Sulphuric acid (H_2SO_4)	$2H^+$	SO_4^{2-}
Nitric acid (HNO_3)	H^+	NO_3^-
Hydrochloric acid (HCl)	H^+	Cl^-
Ethanoic (acetic) acid ($CH_3.COOH$)	H^+	$CH_3.COO^-$
Carbonic acid (H_2CO_3)	$2H^+$	CO_3^{2-}
Phosphoric acid (H_3PO_4)	$3H^+$	PO_4^{3-}

Ethanoic and carbonic acids are weak, i.e. they are only partially ionized in solution, and exist very largely as undissociated molecules.

12.1.2 Basicity of acids

Nitric, hydrochloric and ehanoic acid are only capable of giving one hydrogen ion for each molecule of acid. Such acids are said to be *monobasic*. Sulphuric acid and carbonic acid, which are capable of giving two hydrogen ions for each molecule of acid, are *dibasic*. Phosphoric acid, which yields three hydrogen ions per molecule, is *tribasic*.

The basicity of an acid is the number of hydrogen ions which can be given by one molecule of that acid.

or alternatively:

The basicity of an acid is the number of hydrogen atoms in one molecule of that acid, which can be replaced by a metal.

12.2 Bases

12.2.1 Properties of bases

(*a*) *Compounds which react with acids to form a salt and water* ONLY, *are called bases.*

In general, these are the oxides or hydroxides of metals or the ammonium group. So bases contain oxide or hydroxide ions as the only negative ions. The following equations illustrate the reactions of some bases with acids:

CuO copper(II) oxide	$+H_2SO_4$	$\rightarrow$	$CuSO_4$ copper(II) sulphate	$+H_2O$
$CuO(s)$	$+H^+(aq)$	$\rightarrow$	$Cu^{2+}(aq)$	$+H_2O(l)$
$Cu(OH)_2$ copper(II) hydroxide	$+H_2SO_4$	$\rightarrow$	$CuSO_4$ copper(II) sulphate	$+2H_2O$
$Cu(OH)_2(s)$	$+2H^+(aq)$	$\rightarrow$	$Cu^{2+}(aq)$	$+2H_2O(l)$
NH_4OH ammonium hydroxide	$+HCl$	$\rightarrow$	NH_4Cl ammonium chloride	$+H_2O$
$OH^-(aq)$	$+H^+(aq)$	$\rightarrow$	$H_2O(l)$	
KOH potassium hydroxide	$+HNO_3$	$\rightarrow$	KNO_3 potassium nitrate	$+H_2O$
$OH^-(aq)$	$+H^+(aq)$	$\rightarrow$	$H_2O(l)$	

Very few *bases* dissolve in water; those that do are called *alkalis*, and they are a most important group of compounds.

(*b*) *An alkali is a soluble base* (i.e. all alkalis are also bases). There are only four common alkalis. These are:

Sodium hydroxide, "caustic soda"	$NaOH$
Potassium hydroxide, "caustic potash"	KOH
Calcium hydroxide, "slaked lime"	$Ca(OH)_2$
Ammonium hydroxide, "ammonia"	NH_4OH

12.2.2 Properties of Alkalis

(*a*) *Solutions of alkalis have a bitter taste*, and they feel "soapy" to the touch.

(*b*) *Solutions of alkalis turn red litmus blue*. They also produce colour changes with other natural and synthetic dyestuffs.

(*c*) *Solutions of alkalis neutralize acids to form salts*. This reaction has already been mentioned. Let us consider the ionic equation for the reaction between an acid and an alkali, say hydrochloric acid and sodium hydroxide:

$$\left[\begin{array}{l} Na^+(aq)+OH^-(aq)+H^+(aq)+Cl^-(aq) \rightarrow Na^+(aq)+Cl^-(aq)+H_2O(l) \\ OH^-(aq)+H^+(aq) \rightarrow H_2O(l) \end{array}\right.$$

We see that neutralization consists of a union of hydrogen and hydroxide ions to form water. The acid provides the hydrogen ions, the alkali the hydroxide ions and other ions present remain unchanged.

(*d*) *Solutions of alkalis will precipitate insoluble metallic hydroxides when added to solutions of salts of many metals*. If caustic soda solution is added to a solution of copper(II) sulphate a pale blue deposit of copper(II) hydroxide is immediately formed:

$$\left[\begin{array}{lll} CuSO_4 & +2NaOH & \rightarrow Cu(OH)_2+Na_2SO_4 \\ Cu^{2+}(aq) & +2OH^-(aq) & \rightarrow Cu(OH)_2(s) \end{array}\right.$$

Similarly, if ammonium hydroxide solution is added to iron(III) chloride solution, a brick-red deposit of iron(III) hydroxide is formed.

$$\left[\begin{array}{lll} FeCl_3 & +3NH_4OH & \rightarrow Fe(OH)_3+3NH_4Cl \\ Fe^{3+}(aq)+3OH^-(aq) & & \rightarrow Fe(OH)_3(s) \end{array}\right.$$

It is the hydroxide ions yielded by all alkalis which are responsible for their general similarity of behaviour in the reactions listed above. Therefore we can say that an *alkali is a compound soluble in water, that gives hydroxide ions as the only negative ions*.

Ammonium hydroxide is the only common weak alkali; in solution it is only slightly split up into ammonium and hydroxide ions.

12.3 The strength of acids and alkalis

12.3.1 The pH scale

The strength of different acids and alkalis may be compared by using a mixture of different dyes known as *Universal Indicator*. The colour of this mixture depends upon the concentration of hydrogen ions (which determine acidity) in a solution. It has been found convenient to express acidity by using a scale of numbers known as the pH scale. Water has a pH of 7 and is described as neutral. Solutions which have a pH of greater value than 7 are called alkaline. Solutions with a pH of lower value than 7 are called acidic. The lower the pH value the more acidic is a solution.

Colours of Universal Indicator

pH	— 4	5	6	7	8	9	10
Colour	—Red	Orange	Yellow	Green	Blue	Indigo	Violet
Acidity/Alkalinity	—Most acidic	------	------	Neutral	--------	--------	Most alkaline.

(A key to these colours is usually given on bottles containing the indicator or on packets of indicator paper. Notice that they are in the same order as the colours in the spectrum (e.g. in a rainbow).)

Experiment. Acidity of solutions using Universal Indicator

(*a*) Prepare solutions of hydrochloric acid which are 10^{-2}M, 10^{-3}M, 10^{-4}M, 10^{-5}M and 10^{-6}M. To half a test-tube full of each solution add two drops of Universal Indicator and shake. The colours obtained will show the acidity, i.e. hydrogen ion concentration, of the various solutions. The hydrogen ion concentration decreases as the solution is diluted and therefore the pH value gets larger.

(*b*) Prepare solutions of ethanoic (acetic) acid which are 10^{-2}M, 10^{-3}M, 10^{-4}M, 10^{-5}M and 10^{-6}M. Again add two drops of Universal Indicator to half a test-tube full of each solution. The colours indicate in each case that the diluted ethanoic acid is less acidic than the corresponding hydrochloric acid solution.

If both hydrochloric and ethanoic acids were of the same strength, the same pH values would be obtained for solutions of the same concentration. In fact hydrochloric acid is a *strong acid*, i.e. it is completely ionized, whereas ethanoic acid is a *weak acid*, i.e. it is only partially ionized.

Similar experiments to these could be performed using alkalis. These would show that ammonium hydroxide is a weak alkali, i.e. is only partially split up into ammonium and hydroxide ions.

12.3.2 The role of water in acidity

An acid develops acidic properties only when it is dissolved in water. In fact a reaction occurs with the water to produce hydroxonium ions:

$$HCl(aq) + H_2O(l) \rightarrow H_3O^+(aq) + Cl^-(aq)$$

Strictly speaking it is hydroxonium ions and not hydrogen ions which cause acidity. In some books H^+ in water will always be written as H_3O^+. We have chosen not to adopt the unnecessary complication and, in our equations, do not indicate that the hydrogen ions are attached to water molecules. We represent the reaction thus:

$$HCl(aq) \rightarrow H^+(aq) + Cl^-(aq)$$

(See section 10.6.6 for a more complete discussion about H^+ and H_3O^+.)

As indicated above, a solution of hydrogen chloride in water, known as hydrochloric acid, shows all the usual reactions of an acid. This is because the hydrogen chloride reacts with water according to the above equation (or "splits up into hydrogen and chloride ions in water").

If, however, a solution of *dry* hydrogen chloride in *dry* toluene is prepared, no acidic properties are shown, e.g. the solution is without effect on Universal Indicator paper and has no reaction with metals or carbonates.

Similarly if the acid is a solid (as many organic acids are) it will not show its acidity until dissolved in water. This feature is used, for example, in the preparation of sherbet and health salts (e.g. Eno's or Andrew's) which are inert powders until mixed with water. Then the acid causes the typical fizzing (see section 18.3(a)).

12.4 Salts

12.4.1 Properties of salts

A salt is a compound obtained when all or part of the replaceable hydrogen of an acid is replaced by a metal (or the ammonium ion).

The following table gives the formulae of some acids, and of some salts derived from those acids:

Acids	Salts
Hydrochloric acid (HCl)	NaCl, sodium chloride
Nitric acid (HNO_3)	$NaNO_3$, sodium nitrate
Sulphuric acid (H_2SO_4)	Na_2SO_4, sodium sulphate
	$NaHSO_4$, sodium hydrogensulphate (bisulphate)
Carbonic acid (H_2CO_3)	Na_2CO_3, sodium carbonate
	$NaHCO_3$, sodium hydrogencarbonate (bicarbonate)
Phosphoric acid (H_3PO_4)	Na_3PO_4, tri-sodium phosphate
	Na_2HPO_4, di-sodium hydrogenphosphate
	NaH_2PO_4, mono-sodium di-hydrogenphosphate
Ethanoic (acetic) acid ($CH_3.COOH$)	$CH_3.COONa$, sodium ethanoate (acetate)

With the salts sodium chloride, sodium nitrate and sodium sulphate, all the hydrogen of the acid has been replaced by a metal; such salts are called *normal salts.*

With sodium hydrogensulphate and sodium hydrogencarbonate, only part of the replaceable hydrogen has been replaced. Sodium hydrogensulphate yields sodium ions, hydrogen ions and sulphate ions in solution, and is therefore called an *acid salt*. One of us was once triumphantly presented with some solid "sulphuric acid" from a boy's chemistry set. It proved to be sodium hydrogensulphate, which, because of the hydrogen ions and sulphate ions in its solution, does indeed give many of the more harmless reactions of sulphuric acid.

With phosphoric acid, which contains three replaceable hydrogen atoms per molecule, there are three sodium salts with formulae as given above. It should be obvious that the basicity of an acid determines how many salts can be obtained from it with a particular metal.

Although ethanoic acid contains four hydrogen atoms per molecule, only one sodium salt is formed. This is because ethanoic acid is mono-basic, only one of its hydrogen atoms being replaceable by metal.

12.4.2 Preparation of salts

The method of preparation of a salt depends partly on whether or not it is soluble in water. Certain features concerning solubility of salts can be summarized:

(*a*) All nitrates are soluble.
(*b*) All the common chlorides are soluble except those of silver and lead.
(*c*) All the common sulphates are soluble except those of barium and lead. Calcium sulphate is only slightly soluble.
(*d*) All the common carbonates are insoluble except those of sodium, potassium and ammonium.
(*e*) All common salts of sodium, potassium and ammonium are soluble.

(The word insoluble as applied to a salt is, strictly speaking, incorrect. All salts are soluble to some extent, but in some cases the solubility is so slight that for practical purposes it can be neglected.)

The main methods for preparing *soluble* salts are:

(*a*) by the action of a dilute acid on a metal;
(*b*) by the action of a dilute acid on an insoluble base;
(*c*) by the neutralization of a dilute acid with a solution of an alkali;
(*d*) by the action of a dilute acid on a carbonate.

Experiment. Preparation of zinc sulphate crystals (Method (*a*))

The reagents needed are dilute sulphuric acid and zinc. It is inconvenient to measure the exact quantities of these chemicals as required by the equation:

$$Zn(s) + H_2SO_4(aq) \rightarrow ZnSO_4(aq) + H_2(g)$$

Therefore in the preparation an excess of the metal, which is readily removed by filtration, is normally used.

Add several pieces of zinc to approximately 50 cm^3 of dilute sulphuric acid in a beaker. Warm the mixture until evolution of hydrogen has nearly stopped. Filter off excess zinc to give a filtrate of zinc sulphate solution.

To obtain zinc sulphate crystals evaporate the solution in an evaporating dish until it is so concentrated that crystals separate on cooling. At this stage the solution is said to be at its "point of crystallization". If bench sulphuric acid (1 M) is used the point of crystallization is reached after evaporation to half bulk. Or it is quite easy to test whether the solution has evaporated sufficiently by rapidly cooling a test portion in a dry test-tube to see if crystals form.

Once the point of crystallization has been reached allow the solution to cool.

Needle-like crystals of zinc sulphate-7-water ($ZnSO_4.7H_2O$) will form. Filter these off from the remaining solution and dry them on a porous plate or between filter papers.

This method can be used for preparing various other crystals, such as those of magnesium sulphate ($MgSO_4.7H_2O$), iron(II) sulphate ($FeSO_4.7H_2O$), and lead(II) nitrate ($Pb(NO_3)_2$). It should be remembered, however, that dilute hydrochloric and sulphuric acids are without action on the less reactive metals, and that it would be positively dangerous to attempt to prepare, for example, sodium chloride by the action of hydrochloric acid on sodium.

Can you suggest why diluted acids are almost invariably used in this and other methods for preparing salts?

Experiment. Preparation of copper(II) sulphate crystals (Method (*b*))

The reagents needed are dilute sulphuric acid and copper(II) oxide. As in the previous experiment, an excess of the insoluble reagent is used:

$$CuO(s) + H_2SO_4(aq) \rightarrow CuSO_4(aq) + H_2O(l)$$

Warm approximately 50 cm^3 of dilute sulphuric acid in a beaker and add copper(II) oxide, a little at a time. The black oxide will combine with the sulphuric acid producing a blue solution of copper(II) sulphate. As soon as an excess of the oxide is present, filter it off and evaporate the filtrate of copper(II) sulphate solution to the point of crystallization. Cool, when crystals of copper(II) sulphate-5-water will form. Filter off the crystals and dry them.

Many *soluble* salts can be prepared by this method, e.g.

$ZnO(s)$ zinc oxide	$+H_2SO_4(aq)$	$\rightarrow$	$ZnSO_4(aq)$ zinc sulphate	$+H_2O(l)$
$PbO(s)$ lead(II) oxide	$+2HNO_3(aq)$	$\rightarrow$	$Pb(NO_3)_2(aq)$ lead(II) nitrate	$+H_2O(l)$
$Mg(OH)_2(s)$ magnesium hydroxide	$+H_2SO_4(aq)$	$\rightarrow$	$MgSO_4(aq)$ magnesium sulphate	$+2H_2O(l)$

Experiment. Preparation of potassium nitrate crystals (Method (*c*))

The reagents needed are dilute nitric acid and potassium hydroxide (caustic potash). Both these chemicals are soluble in water, so that an excess of one of the reagents cannot be removed by filtration. The exact quantities required by the equation:

$$KOH(aq) + HNO_3(aq) \rightarrow KNO_3(aq) + H_2O(l)$$

must therefore be used. This is made simple by the use of indicators. *Indicators are substances which have different colours in acidic and alkaline solutions.*

There are three very common indicators:

Indicator	*Colour in acidic solution*	*Colour in alkaline solution*
Litmus	Red	Blue
Phenolphthalein	Colourless	Red
Methyl orange	Pink	Yellow

If a salt which gives a neutral solution is being prepared, any one of these indicators may be used. The preparation consists of two stages:

(i) finding the exact quantities of the solutions required;
(ii) using these quantities to prepare the salt.

(i) Pipette 25 cm^3 of a solution of potassium hydroxide into a conical flask and add a few drops of indicator. Slowly run in dilute nitric acid from a burette until the indicator just changes colour. One drop of acid in excess is quite sufficient to cause this colour change. Note the volume of acid added. The solution of potassium nitrate obtained by this method will be contaminated with indicator, but the volume of solutions required to give the potassium nitrate will be known.

(ii) Now mix the same quantities of acid and alkali without using the indicator. Evaporate the solution obtained to the point of crystallization and isolate the crystals of potassium nitrate by allowing the solution to cool and then filtering.

Salts which can be prepared by this method include the chlorides, nitrates and sulphates of sodium, potassium and the ammonium ion e.g.

$NaOH(aq)$	$+HCl(aq)$	$\rightarrow$	$NaCl(aq)$	$+H_2O(l)$
sodium hydroxide	hydrochloric acid		sodium chloride	
$2NH_4OH(aq)$	$+H_2SO_4(aq)$	$\rightarrow$	$(NH_4)_2SO_4(aq)$	$+2H_2O(l)$
ammonium hydroxide	sulphuric acid		ammonium sulphate	

Because of the slight solubility of calcium hydroxide, this neutralization method should not be used for the preparation of calcium salts.

Experiment. Preparation of lead(II) nitrate crystals (Method (*d*))

Lead(II) carbonate, which is insoluble, and dilute nitric acid are required.

$$PbCO_3(s) + 2HNO_3(aq) \rightarrow Pb(NO_3)_2(aq) + H_2O(l) + CO_2(g)$$

Add lead(II) carbonate little by little to 50 cm^3 of warm dilute nitric acid. Carbon dioxide will be evolved. When further addition of lead(II) carbonate produces no effervescence, filter the solution and obtain the crystals from the filtrate as in the previous experiment.

This method is suitable for many other salts but the solubility of sodium and potassium carbonates render it unsuitable for salts of these metals. As solutions of sodium and potassium carbonates are alkaline (see 12.5), the neutralization method can be used with methyl orange as indicator.

12.4.3 Preparation of insoluble salts

The most important method for preparing insoluble salts is by precipitation (or direct union of ions).

A precipitate is an insoluble substance obtained when solutions are mixed. Fundamentally the method consists of mixing two solutions which between them contain the ions making up the insoluble compound. Thus, if a solution of lead(II) nitrate is added to a solution of sodium sulphate, both lead(II) and sulphate ions are present in the resultant mixture. Therefore insoluble lead(II) sulphate is precipitated:

$$Pb(NO_3)_2 + Na_2SO_4 \rightarrow 2NaNO_3 + PbSO_4$$

In terms of ions the reaction is:

$$Pb^{2+}(aq) + SO_4{}^{2-}(aq) \rightarrow PbSO_4(s)$$

The sodium and nitrate ions remain in solution unchanged during the reaction.

The ionic equation expresses the general truth that lead(II) ions from any source (e.g. from lead(II) nitrate or lead(II) ethanoate solutions) unite with sulphate ions from any source (e.g. from sodium sulphate, magnesium sulphate or sulphuric acid solutions) to give lead(II) sulphate.

After precipitation, pure lead(II) sulphate can be obtained by filtration. The residue on the filter paper is washed by pouring a little water over it, and then dried.

Other salts which can be prepared by precipitation include silver chloride and barium sulphate:

$AgNO_3$ silver nitrate	$+ NaCl$ sodium chloride	$\rightarrow$	$AgCl$ silver chloride	$+ NaNO_3$ sodium nitrate
$Ag^+(aq)$	$+ Cl^-(aq)$	$\rightarrow$	$AgCl(s)$	
$BaCl_2$ barium chloride	$+ H_2SO_4$ sulphuric acid	$\rightarrow$	$BaSO_4$ barium sulphate	$+ 2HCl$ hydrochloric acid
$Ba^{2+}(aq)$	$+ SO_4{}^{2-}(aq)$	$\rightarrow$	$BaSO_4(s)$	

Precipitation can be used for preparing other insoluble compounds besides salts e.g., the insoluble metallic hydroxides.

If you are required to prepare an insoluble salt, starting from an insoluble substance (a common examination question!), it is important to remember that this cannot be done satisfactorily in one stage. For example, there is very little reaction between lead and sulphuric acid, and between lead(II) carbonate and sulphuric acid. To prepare lead(II) sulphate from one of these substances as starting material, a solution of lead(II) nitrate should first be prepared, and then the lead(II) sulphate by a precipitation method, i.e.

Stage I : $\underset{\text{insoluble}}{PbCO_3} + 2HNO_3 \rightarrow \underset{\text{soluble}}{Pb(NO_3)_2} + H_2O + CO_2$

Stage II : $Pb(NO_3)_2 + H_2SO_4 \rightarrow \underset{\text{insoluble}}{PbSO_4} + 2HNO_3$

12.4.4 Direct combination

Certain salts—both soluble and insoluble—can be made directly from their elements. One example, already discussed (section 2.5.1), is the preparation of iron(II) sulphide by the action of heat on a mixture of iron filings and sulphur. Iron(II) sulphide is manufactured by this method.

Anhydrous iron(III) and aluminium chlorides are normally prepared by passing dry chlorine over the heated metal.

$$2Fe(s) + 3Cl_2(g) \rightarrow 2FeCl_3(s)$$

$$2Al(s) + 3Cl_2(g) \rightarrow 2AlCl_3(s)$$

Mercury(II) iodide is readily prepared by warming together in a test tube a mixture of 2 g of mercury and 2·5 g of iodine. (N.B.: Mercury vapour is very poisonous.)

$$Hg(l) + I_2(s) \rightarrow HgI_2(s)$$

12.4.5 Preparation of acid salts

An acid salt is formed when only part of the replaceable hydrogen of an acid is replaced by a metal or ammonium. Thus caustic soda can react with sulphuric acid in two ways, depending on the proportions of the reagents present.

$$\underset{}{NaOH(aq)} + H_2SO_4(aq) \rightarrow \underset{\text{sodium hydrogensulphate}}{NaHSO_4(aq)} + H_2O(l)$$

$$2NaOH(aq) + H_2SO_4(aq) \rightarrow \underset{\text{sodium sulphate}}{Na_2SO_4(aq)} + 2H_2O(l)$$

Sodium hydrogensulphate is the acid salt, sodium sulphate the normal salt. From the equations it is clear that for a certain amount of acid, twice as much alkali is needed to prepare the normal salt as the acid salt. This fact indicates the method of preparation.

Experiment. Preparation of sodium hydrogensulphate

Place 25 cm^3 of sodium hydroxide solution in a conical flask and determine the volume of dilute sulphuric acid required to neutralize it using litmus as an indicator. Suppose that x cm^3 are required. Then mix 25 cm^3 of sodium hydroxide solution with $2x$ cm^3 of dilute sulphuric acid. Evaporate the resulting solution to the point of crystallization. Allow to cool when crystals of sodium hydrogensulphate will form.

Hydrogencarbonates, hydrogensulphites and hydrogensulphides cannot be prepared by this simple method. Details of their preparation are given elsewhere in this book.

12.5 Salt hydrolysis

Not all salts give a neutral solution, quite apart from the so-called acid salts. Sodium and potassium carbonates, as already mentioned, give alkaline solutions. In fact, the first substance ever to be called an alkali was potassium carbonate which had been isolated from wood ashes. An explanation often given of the alkalinity of these salts is that they combine with or are hydrolysed by water to some extent, giving a mixture of a strong alkali and a weak acid:

$$K_2CO_3 + 2H_2O \rightleftharpoons \underset{\substack{\text{strong}\\\text{alkali}}}{2KOH} + \underset{\substack{\text{weak}\\\text{acid}}}{H_2CO_3}$$

The solution is therefore alkaline. This is an unsatisfactory explanation.

It is better to say that carbonate ions combine to some extent with hydrogen ions from the water, forming molecules of carbonic acid. There is, therefore, a greater concentration of hydroxide ions than hydrogen ions in the resulting solution, which is therefore alkaline:

$$2K^+ + CO_3^{2-} + \underbrace{2H^+ + 2OH^-}_{\text{from water}} \rightleftharpoons 2K^+ + 2OH^- + H_2CO_3$$

Similarly, salts such as ammonium chloride, which are made from a weak alkali and a strong acid, give an acidic reaction. The equivalent equations to those above are:

$$NH_4Cl + H_2O \rightleftharpoons \underset{\substack{\text{weak}\\\text{alkali}}}{NH_4OH} + \underset{\substack{\text{strong}\\\text{acid}}}{HCl}$$

$$\text{or } NH_4^+ + Cl^- + \underbrace{H^+ + OH^-}_{\text{from water}} \rightleftharpoons NH_4OH + H^+ + Cl^-$$

In this case the acidity is due to the high concentration of hydrogen ions compared with hydroxide ions.

Concerning normal salts, we can say that: (*a*) if they are derived from a weak acid and a strong alkali they give alkaline solutions, (*b*) if they are derived from a strong acid and a weak alkali they give acidic solutions. This process where molecules of the weak acid or weak alkali are reformed making the solution basic or acidic is called *salt hydrolysis*. Salts that are not hydrolysed give neutral solutions. In no circumstances is it correct, according to our modern definitions, to refer to salts as acids or alkalies, even though they may give acidic or alkaline solutions.

A simple way to discover whether a salt is hydrolysed by water is to add a few drops of Universal Indicator to a solution of the salt. The colour will indicate immediately the approximate pH of the solution. If the solution is not neutral some hydrolysis has occurred.

12.6 Volumetric analysis

12.6.1 Standard solutions

A standard solution is a solution of known concentration, i.e. a solution known to contain 8·35 grams per dm^3 of sodium hydroxide (i.e. a solution known to have a concentration of 8·35 g dm^{-3}), would be a standard solution. For purposes of calculation it is often convenient to express *concentrations* of substances in *moles per cubic decimetre* (*mol dm*$^{-3}$) ≡ *moles per litre* (mol l^{-1}). The word *MOLARITY* is frequently used for moles per dm^3 and the symbol "M" is used to signify the molarity of a solution.

A molar solution contains 1 mole of a substance dissolved in 1 cubic decimetre (1 litre) of solution (i.e. has a concentration of 1 mol dm^{-3}). Thus molar sulphuric acid (molecular mass = 98) contains 98 g of sulphuric acid in 1 dm^3 of solution. A solution containing one-tenth of a mole per dm^3 is said to be *decimolar*, i.e. decimolar sulphuric acid (0·1M H_2SO_4) contains 9·8 g of the acid in each cubic decimetre of solution.

Similarly:

2M H_2SO_4 contains 2 mole of sulphuric acid per dm^3 of solution.

0·05M H_2SO_4 contains 0·05 mole of sulphuric acid per dm^3 of solution.

(N.B. Remember that the unit is the *mole*. The abbreviation for this unit is *mol*. Hence moles per cubic decimetre is written mol dm^{-3}, and moles per litre is written mol l^{-1}).

Since a 1M solution contains 1 mole in 1 cubic decimetre (1 litre), we are able to calculate the number of moles in any volume of a solution of any molarity.

Hence if 1 dm^3 (i.e. 1000 cm^3) of a 1M solution contains 1 mole

1 cm^3 of a 1M solution must contain $\frac{1}{1000}$ mole

V cm^3 of a 1 M solution must contain $\frac{V}{1000}$ mole

V cm^3 of an x M solution must contain $\frac{Vx}{1000}$ mole.

Exercise 1: Calculate the mass of hydrochloric acid (molecular mass = 36·5) in each of the following solutions:

(i) 1 dm^3 of 1M HCl
(ii) 1 dm^3 of 0·5M HCl
(iii) 3 dm^3 of 1M HCl
(iv) 250 cm^3 of 0·5M HCl
(v) 100 cm^3 of 0·1M HCl

(Answers at end of this chapter)

12.6.2 Volumetric calculations

(a) Consider this equation:

$NaOH + HCl \rightarrow NaCl + H_2O$

Here 1 mole of sodium hydroxide reacts with 1 mole of hydrochloric acid for, in chemical reactions, *the equation* (properly balanced) *tells us* how many moles of each substance react together. Since molar solutions of all substances contain 1 mole per dm^3, it follows that

1000 cm^3 of 1M NaOH will react with 1000 cm^3 1M HCl
or 25 cm^3 of 0·1M NaOH will react with 25 cm^3 0·1M HCl

(i.e. since equal volumes of 1 molar sodium hydroxide solution and 1 molar hydrochloric acid contain equal numbers of "molecules", and one "molecule" of sodium hydroxide reacts with one "molecule" of hydrochloric acid, equal volumes of the two solutions react with each other.)

Often the molarities of two solutions that react together are not the same but *the equation tells us* that whatever number of moles of sodium hydroxide is used, the same number of moles of hydrochloric acid will be required.

Thus to calculate the volume of 0·1M HCl that will react with 50 cm³ of 0·4M NaOH, we may proceed thus:
1 dm³ of 1M NaOH contains 1 mol.

$$\therefore 50\ \text{cm}^3 \text{ of } 0{\cdot}4\text{M NaOH contain } \frac{50 \times 0{\cdot}4}{1000} \text{ mol.}$$

The equation tells us that an equal number of moles of HCl is required.

$$\therefore \text{No. of moles of HCl required} = \frac{50 \times 0{\cdot}4}{1000}$$

Suppose x cm³ is the volume of 0·1 M HCl required

$$\text{No. of moles in this volume} = \frac{x \times 0{\cdot}1}{1000}$$

$$\text{Hence } \frac{x \times 0{\cdot}1}{1000} = \frac{50 \times 0{\cdot}4}{1000}$$

$$x = \frac{50 \times 0{\cdot}4}{0{\cdot}1} = \underline{200\ \text{cm}^3}$$

In this way unknown volumes and molarities can be calculated.

Exercise 2. For the reaction between hydrochloric acid and sodium hydroxide calculate

(i) *the volume of* 0·1M HCl *that reacts with* 50 cm³ *of* 0·1M NaOH
(ii) *the volume of* 0·1M HCl *that reacts with* 50 cm³ *of* 0·2M NaOH
(iii) *the volume of* 0·1M HCl *that reacts with* 25 cm³ *of* 0·05M NaOH
(iv) *the molarity of the* HCl *if* 25 cm³ *reacts with* 25 cm³ *of* 0·2M NaOH
(v) *the molarity of the* HCl *if* 25 cm³ *reacts with* 100 cm³ *of* 0·05M NaOH

(Answers at the end of the chapter)

(b) Consider this equation:

$2NaOH + H_2SO_4 \rightarrow Na_2SO_4 + 2H_2O$

Here 2 moles of sodium hydroxide react with 1 mole of sulphuric acid. Since 1 molar solutions of all substances contain 1 mole per dm³, it follows that:

2000 cm³ of 1M NaOH will react with 1000 cm³ 1M H_2SO_4
or 50 cm³ of 1M NaOH will react with 25 cm³ 1M H_2SO_4
or 25 cm³ of 0·1M NaOH will react with 25 cm³ 0·05M H_2SO_4

To make sure that you understand fully, try the following exercise:

Exercise 3. For the reaction between sulphuric acid and sodium hydroxide calculate:

(i) *the volume of* 0·1M NaOH *that reacts with* 25 cm^3 *of* 0·1M H_2SO_4
(ii) *the volume of* 0·1M H_2SO_4 *that reacts with* 25 cm^3 *of* 0·1M NaOH
(iii) *the volume of* 0·1M NaOH *that reacts with* 25 cm^3 *of* 0·2M H_2SO_4
(iv) *the volume of* 0·1M H_2SO_4 *that reacts with* 25 cm^3 *of* 0·2M NaOH
(v) *the molarity of the* NaOH *if* 25 cm^3 *reacts with* 25 cm^3 *of* 0·2M H_2SO_4
(vi) *the molarity of the* H_2SO_4 *if* 50 cm^3 *reacts with* 50 cm^3 *of* 0·5M NaOH
(vii) *the molarity of the* NaOH *if* 25 cm^3 *reacts with* 50 cm^3 *of* 0·1M H_2SO_4
(viii) *the molarity of the* H_2SO_4 *if* 50 cm^3 *reacts with* 25 cm^3 *of* 0·4M NaOH.
(Answers at the end of this chapter)

By carrying out chemical experiments the volumes of a standard solution of acid which react with a known volume of an alkaline solution can be determined, and it is possible to calculate the concentration of the alkaline solution or vice versa. Determinations of this type, involving the measurement of volumes of solutions which will react with one another, are examples of *volumetric analyses*.

12.6.3 The apparatus used in analysis

Three pieces of apparatus are used for measuring accurately the volume of a liquid. These are the pipette, the burette and the graduated flask (figures 12.1, 12.2, 12.3).

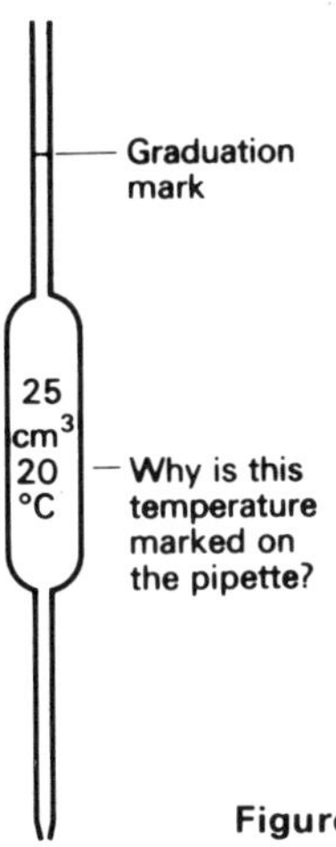

Figure 12.1 The pipette.

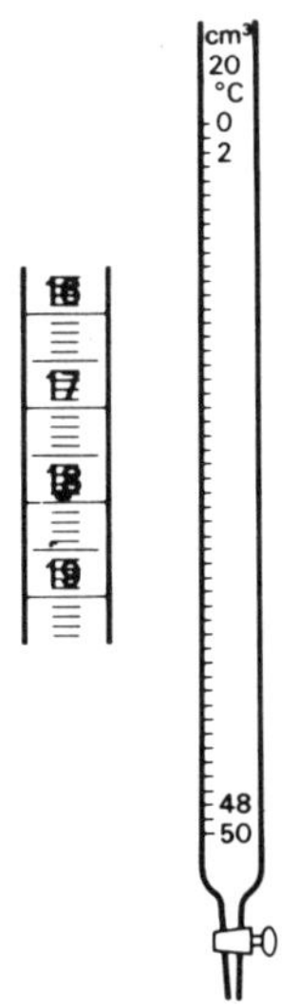

Figure 12.2 The burette.

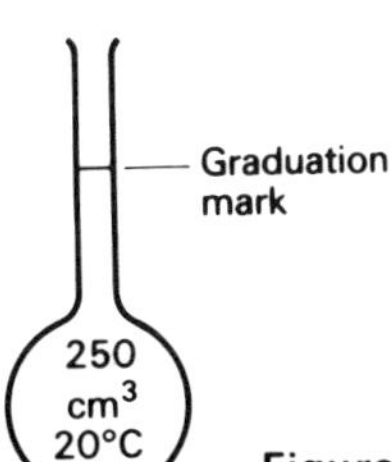

Figure 12.3 The graduated flask.

The pipette is made to deliver a given volume of liquid, usually 20 or 25 cm^3. It has a circular graduation mark etched on its surface as shown. To use the pipette, its tip is dipped well under the surface of the liquid required. Liquid is sucked up a little past the graduation mark and is held there by the tip of the first finger. (Beginners often use the thumb for this purpose, but with practice much better control can be obtained with the finger.) The liquid level is allowed to fall by easing the finger gently until the bottom of the meniscus, viewed at eye-level, is exactly level with the mark.

The liquid is then *allowed to run* into the required vessel by removing the finger. The tip of the pipette is then touched on the side of the conical flask, when a little more liquid comes out. Even after this a small amount of liquid remains in the pipette, but this is allowed for in its graduation. There are, therefore, two reasons why one should not be tempted to blow liquid out of the pipette:

(*a*) it is likely that the drop of liquid which should remain in the tip of the pipette will be blown out;
(*b*) no time has been allowed for liquid to drain down the sides of the pipette.

When pipetting poisonous liquids (and ideally at all times) do not suck on the pipette with your mouth. Use instead a rubber pipette filler.

The burette may be used to measure any volume of liquid up to 50 cm^3 and is graduated in 0·1 cm^3. Liquid is poured into the burette through a small funnel. The funnel is removed immediately, so that drops of liquid will not drain from it. A little liquid is then run out by means of the tap. This ensures that the space below the tap is full.

The initial reading of the burette is noted, having the eye level with the bottom of the meniscus. After liquid has been run out the level is again noted, the difference in two readings giving the volume used.

The graduated flask is made to hold a given volume of liquid, usually 100 cm^3, 250 cm^3, 500 cm^3 or 1000 cm^3 (1 dm^3). It must be filled to the graduation mark, the meniscus being looked at at eye-level. The volume of the liquid after being poured out is slightly less than that marked on the flask. Why is this?

12.6.4 Preparation of standard solutions

(a) 0·05M sodium carbonate

Many standard solutions cannot be prepared by weighing the solid solute. Often the substance cannot be obtained sufficiently pure; it may be a liquid and therefore too volatile, or it may be efflorescent or deliquescent. Sodium carbonate is the only common alkaline substance whose solution may be made up directly. The procedure for making 1 dm^3 of 0·05M solution is as follows:

First the molar mass of sodium carbonate must be calculated.

$$\begin{aligned} & \quad Na_2 \quad\; + C + \; O_3 \\ \text{Molar mass} &= (2 \times 23) + 12 + (3 \times 16) \\ &= \quad 46 \quad + 12 + \quad 48 \qquad = 106 \text{ g mol}^{-1} \end{aligned}$$

Molar sodium carbonate solution contains 1 mole, i.e. 106 g per dm^3 of solution.

∴ 0·05M Na_2CO_3 contains 0·05 × 106 g, i.e. 5·3 g per dm^3 of solution.

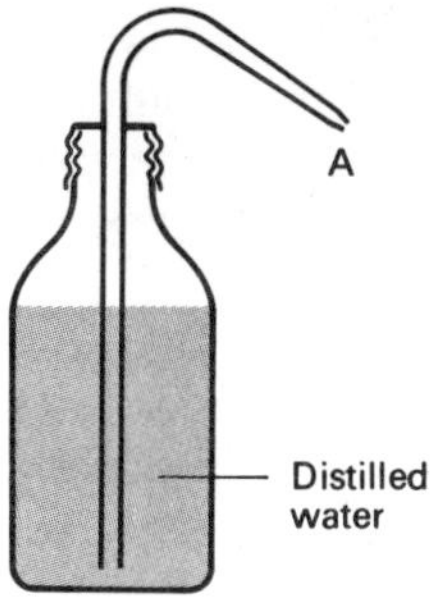

Figure 12.4 The wash-bottle. A fine jet of distilled water is obtained by squeezing the plastic bottle.

Place pure, anhydrous sodium carbonate (of analytical reagent quality) in a clean evaporating dish and heat it gently for a few minutes. This drives off any moisture which may be present. Allow the carbonate to cool, preferably in a desiccator.

Meanwhile weigh accurately a clean dry watchglass and then add 5·30 g to the pan containing the weights. Using a spatula add anhydrous sodium carbonate to the watchglass until the pans balance once more. When the weighing is complete tip the solid into a small beaker containing warm distilled water. (The water must be warm because sodium carbonate "cakes" in cold water and then requires a long time to dissolve.) Wash traces of solid adhering to the watchglass into the beaker with a jet of distilled water from a wash bottle (figure 12.4).

Stir the liquid until solution is complete and, when cool, pour it through a funnel into a litre graduated flask. The cooling is very important, as graduated flasks only hold the stated volume of liquid at one particular temperature (usually 20 °C). Permanent harm might be done to the flask, because if warmed it does not contract entirely to its original volume. Wash out the inside of the beaker several times and add the washings to the flask. Also rinse traces of sodium carbonate solution in the funnel into the flask. Finally, add distilled water to fill the flask to its graduation marking using the wash bottle for the last few drops. Shake the solution well to ensure an even mixture.

(b) Standard hydrochloric acid

It is not possible to make this by direct weighing, but use may be made of a standard solution of sodium carbonate prepared by the method described in the previous experiment. Bench dilute hydrochloric acid is approximately 2M. If a standard solution which is approximately 0·1M is required, some of this acid is diluted twenty times. The solution is then standardized, i.e. its exact concentration is determined, using standard sodium carbonate solution. Adopt the following procedure:

Wash a burette with distilled water and then with the diluted acid. Then add the acid to fill the burette. The washing with acid ensures that no water is present to dilute further the acid which has to be standardized.

Wash a pipette with water and then with 0·05M sodium carbonate solution. Then pipette 25 cm^3 of the carbonate solution into a clean conical flask. (The conical flask must be washed with water only. Why?)

Add two drops of screened methyl orange solution indicator to the sodium carbonate solution. It will turn green. Run acid into the flask from the burette until the indicator turns a neutral grey colour. This will indicate that an equivalent amount of acid has been added. The slightest excess of acid present will turn the indicator pink.

The operation of finding the volume of one solution required to react with a certain volume of another is known as a titration. The first titration is usually a rough one as you do not know where you will find the equivalence point. Carry out two or three titrations which should be within 0·1 cm^3 of one another. Enter your results in your practical book in this way:
25 cm^3 of 0·05M Na_2CO_3 in flask. Screened methyl orange indicator.

Unknown HCl in burette

	Rough	1	2	3
	cm	cm	cm	cm
2nd burette reading	26·0	24·2	48·3	24·9
1st burette reading	1·5	0·2	24·2	0.9
Volume of acid used	24·5	24·0	24·1	24·0

Average volume of acid used = 24·0 cm^3 of acid.

Therefore 25·0 cm^3 of 0·05M Na_2CO_3 react with 24·0 cm^3 of acid.

From the equation

$$Na_2CO_3 + 2HCl \rightarrow 2NaCl + H_2O + CO_2$$

it is seen that
1 mole Na_2CO_3 reacts with 2 moles HCl

Now you have used $\left(\frac{25 \times 0{\cdot}05}{1000}\right)$ mol of Na_2CO_3

The number of moles of HCl $= 2 \times \left(\frac{25 \times 0{\cdot}05}{1000}\right)$ mol, and this number of moles is in 24 cm^3.

The molarity of HCl (x M) is x in the equation

$$\frac{x \times 24}{1000} = \frac{2 \times 25 \times 0{\cdot}05}{1000}$$

$$x = \frac{2 \times 25 \times 0{\cdot}05}{24} = 0{\cdot}104\text{M}$$

A molar solution of hydrochloric acid contains 1 mole, i.e. 36·5 g of acid per dm^3. Therefore 0·104M HCl contains

$0{\cdot}104 \times 36{\cdot}5 = 3{\cdot}80$ g per dm^3.

(Screened methyl orange is made by dissolving 1·0 g of methyl orange and 1·4 g of xylene cyanol FF in 500 cm^3 of 50 per cent ethanol or may be purchased from laboratory suppliers. It is much better than ordinary methyl orange, which gives the colour change yellow to red. This colour change is difficult to distinguish accurately in artificial light or by colour blind people.)

(c) Decimolar hydrochloric acid
If an exactly decimolar (0·1M) solution of hydrochloric acid is required, it is usual to prepare and standardize a solution of higher molarity and then to dilute it. Thus the solution standardized in the experiment described above was 0·104M. Suppose that there are 900 cm³ of this solution left.

900 cm³ of 0·104M hydrochloric acid will contain $\left(\frac{900 \times 0{\cdot}104}{1000}\right) = 0{\cdot}0936$ moles. But 0·0936 moles in 0·936 dm³ would give a decimolar solution. Therefore to give a decimolar solution 36 cm³ of water would have to be mixed with the 900 cm³ of 0·104M solution.

(d) Standard sodium hydroxide solution
Sodium hydroxide is a deliquescent solid, and standard solutions cannot therefore be prepared by direct weighing. It is usual to prepare a solution of approximately the concentration required and then to standardize it as in experiment (b).

12.6.5 Choice of indicator

Indicators other than screened methyl orange must sometimes be used.

Methyl orange or *screened methyl orange* is used for the titration of a strong acid with a weak alkali or with sodium carbonate.

Phenolphthalein (colourless in acid but red in alkaline solutions) is used for the titration of a weak acid, such as acetic acid, with a strong alkali.

Any of these indicators may be used for the titration of a strong acid with a strong alkali.

12.6.6 Some problems on volumetric analysis

In all the examples so far in this chapter we have been looking at the reactions between two solutions and, by using the chemical equation to find the combining ratios of the two reacting compounds, we have been able to calculate volumes or molarities.

Sometimes a known mass of one compound is reacted with a solution. Let us look at two calculations involving masses:

Example 1
0·212 g of sodium carbonate (molecular mass = 106) was dissolved in water and was titrated with 0·1M hydrochloric acid using screened methyl orange as the indicator. What volume of acid was required?

The equation for the reaction is:

$$Na_2CO_3 + 2HCl \rightarrow 2NaCl + H_2O + CO_2$$

i.e. 1 mole of Na_2CO_3 reacts with 2 moles of HCl. . . . . (A)
molar mass of $Na_2CO_3 = 106\ g\ mol^{-1}$

$0{\cdot}212$ g of $Na_2CO_3 = \dfrac{0{\cdot}212}{106}$ mol

From (A) this would react with $\dfrac{2 \times 0{\cdot}212}{106}$ mol HCl . . . . (B)

Let volume of 0·1M HCl used in titration $= x$ cm^3

This would contain $\dfrac{x \times 0{\cdot}1}{1000}$ mol HCl . . . . (C)

The values (B) and (C) must obviously be equal

$$\frac{2 \times 0{\cdot}212}{106} = \frac{x \times 0{\cdot}1}{1000}$$

$$x = \frac{2 \times 0{\cdot}212 \times 1000}{106 \times 0{\cdot}1} = 40 \text{ cm}^3$$

Volume of acid used = 40 cm^3

Example 2
0·485 g of a divalent metal, Y, were dissolved in 25·0 cm^3 of 1 M sulphuric acid. The resulting solution needed 10·0 cm^3 of 1M sodium hydroxide to neutralize it. Calculate the atomic mass of the metal. What metal is it?

This problem is a little more complicated as we have to consider two reactions. First the metal reacts with some of the sulphuric acid to give a salt and hydrogen. As the metal is divalent the equation for this reaction must be:

$$Y + H_2SO_4 \rightarrow YSO_4 + H_2$$

The excess sulphuric acid reacts with the sodium hydroxide solution thus:

$$H_2SO_4 + 2NaOH \rightarrow Na_2SO_4 + 2H_2O$$

Now the total number of moles of sulphuric acid $= \dfrac{25 \times 1}{1000} = 0{\cdot}025$ mol

The number of moles of sodium hydroxide used $= \dfrac{10 \times 1}{1000} = 0{\cdot}01$ mol

From the equation 2 mol of sodium hydroxide react with 1 mol of sulphuric acid.
∴ 0·01 mol of sodium hydroxide reacts with 0·005 mol of sulphuric acid.
∴ The metal must have reacted with (0·025 − 0·005) = 0·020 mol of sulphuric acid.

The equation for the reaction between the metal and sulphuric acid tells us that 1 mol of sulphuric acid reacts with 1 mol of the metal.

∴ There are 0·020 mol of metal.

The metal sample has a mass of 0·485 g. Hence there are 0·02 mol in 0·485 g.

The atomic mass of the metal is the mass in 1 mol $= \dfrac{0{\cdot}485}{0{\cdot}02} = 24{\cdot}25$

Atomic mass of metal = 24·25

Inspection of the table of atomic masses (see back of book) shows that the metal is magnesium (atomic mass = 24·3).

12.6.7 Revision

You will see that throughout this chapter, and the rest of this book, we have carried out our calculations in terms of moles of substances. *This is because moles are the fundamental unit for the amount of a substance*, and calculations done this way are in their simplest possible form.

The important points to be remembered to help you with such calculations are

1. *The chemical equation is of fundamental importance.* It alone tells you the number of moles (or molar proportions) of the substances undergoing reaction.

Obviously the chemical equation *must be correctly balanced.* If it is not then you do not know a most vital piece of information about the reaction.

2. The number of moles in a certain *volume* of a certain molar solution is $\left(\frac{\text{volume} \times \text{molarity}}{1000}\right)$.

3. The number of moles in a certain *mass* of an element or compound is $\frac{\text{mass}}{\text{atomic mass}}$ for an element, or $\frac{\text{mass}}{\text{molecular mass}}$ for a compound.

Answers to exercises in this chapter

Exercise 1. (i) 36·5 g, (ii) 18·25 g, (iii) 109·5 g, (iv) 4·56 g, (v) 0·365 g.
Exercise 2. (i) 50 cm^3, (ii) 100 cm^3, (iii) 12·5 cm^3, (iv) 0·2M, (v) 0·2M.
Exercise 3. (i) 50 cm^3, (ii) 12·5 cm^3, (iii) 100 cm^3, (iv) 25 cm^3, (v) 0·4M, (vi) 0·25M, (vii) 0·4M, (viii) 0·1M.

Questions on chapter 12

1. Calculate the volumes of
(*a*) 0·1M HCl which will react exactly with:
(i) 25 cm^3 0·1M NaOH
(ii) 50 cm^3 0·2M Na_2CO_3
(iii) 30 cm^3 0·3M KOH
(*b*) 0·2M Na_2CO_3 which will react exactly with:
(i) 20 cm^3 0·1M HCl
(ii) 30 cm^3 0·2M H_2SO_4
(iii) 40 cm^3 0·3M H_3PO_4
2. Calculate the molarity of a solution of
(*a*) hydrochloric acid, 25 cm^3 of which reacts exactly with:
(i) 25 cm^3 0·2M Na_2CO_3
(ii) 30 cm^3 0·1M KOH
(iii) 10 cm^3 0·2M NH_4OH
(*b*) potassium hydroxide, 25 cm^3 of which reacts exactly with:
(i) 15 cm^3 0·5M HCl
(ii) 10 cm^3 1·2M H_2SO_4
(iii) 25 cm^3 0·125M HNO_3
3. Calculate the concentration in g per dm^3 of a solution of sulphuric acid 30 cm^3 of which reacts with:

(i) 30 cm^3 0·1M Na_2CO_3
(ii) 20 cm^3 0·1M NaOH
(iii) 25 cm^3 0·2M KOH

4. Calculate the concentration in g per dm^3 of a solution of sodium carbonate 50 cm^3 of which reacts with:
(i) 20 cm^3 0·2M HCl
(ii) 20 cm^3 0·2M HNO_3
(iii) 10 cm^3 0·2M H_2SO_4

5. Calculate the mass of magnesium carbonate (Molecular mass = 84) which will react exactly with:
(i) 50 cm^3M HCl
(ii) 50 cm^3M HNO_3
(iii) 50 cm^3M CH_3COOH
(iv) 50 cm^3M H_2SO_4

6. Explain the term "basicity of an acid".
Give the names and formulae of three acids of different basicities.
50 cm^3 of a solution of an acid containing 30·0 grams per dm^3 were found to need for complete neutralization 32·6 cm^3 of a solution of sodium hydroxide which contained 40 grams per dm^3. If the molecular mass of the acid is 92·0, calculate the basicity of the acid. *(L.)*

7. 20 cm^3 of a solution containing 0·06 g of a metal hydroxide (XOH) were exactly neutralized by 25 cm^3 of 0·1M hydrochloric acid.
(*a*) What is the molecular mass of the metal hydroxide?
(*b*) What is the atomic mass of X?
(*c*) Write an ionic equation for the reaction. *(J.M.B.)*

8. It is required to determine the concentration of a sodium hydroxide solution by titration against 0·115M hydrochloric acid. A suitable indicator is provided, but it is one whose colour change you do not know. Describe carefully how you would investigate the colour change of the indicator, perform the titration and use the results to determine the concentration of the alkali solution.
Suppose that you found that 25·00 cm^3 of alkali were neutralized by 20·50 cm^3 of the hydrochloric acid. Calculate the concentration of the alkali solution. *(J.M.B.)*

9. A solution is made by dissolving 7·5 g of sodium hydroxide, containing an inert impurity, in water and making up to 250 cm^3 of solution. If 20 cm^3 of this solution is exactly neutralized by 13 cm^3 of 1M HNO_3, calculate the percentage purity of the sodium hydroxide.

10. (*a*) If you were provided with a decimolar solution of sodium hydroxide, briefly describe how you would proceed to find the molarity of some acid that is known to be approximately 10M.
(*b*) If 20 cm^3 of 0·08M sodium carbonate solution required 25 cm^3 of a nitric acid solution for neutralization, calculate the molarity of the acid, and its concentration in g per dm^3.

11. A certain carbonate reacts with an acid according to the equation:

$$XCO_3 + 2HCl \rightarrow XCl_2 + H_2O + CO_2$$

If 1 g of this carbonate is dissolved in 50 cm^3 of molar hydrochloric acid, the remaining acid requires 30 cm^3 of molar sodium hydroxide for neutralization. Calculate the atomic mass of X.

12. (*a*) What is meant by (i) an acid, (ii) a molar solution of an acid?
(*b*) Ethanoic acid, CH_3COOH, is a monobasic acid.
(i) What ions are present in an aqueous solution of ethanoic acid?
(ii) How many grams of ethanoic acid are present in 500 cm^3 of a molar solution?
(iii) A molar solution of ethanoic acid conducts electricity but not so easily as a molar solution of hydrochloric acid. What explanation can you offer for this fact?
(*c*) Calculate the minimum volume of molar hydrochloric acid required to react with 1

gram of calcium carbonate and the volume of gas which would be evolved at standard temperature and pressure. (H = 1, C = 12, O = 16, Ca = 40; one mole of gas occupies 22·4 dm^3 at s.t.p.) *(A.E.B.)*

13. Which ion causes the acidic character shown by aqueous solutions of hydrochloric and ethanoic (acetic) acid, CH_3COOH? Explain briefly why one of these solutions behaves as a "strong" acid, the other as a "weak" acid.

You are provided with solutions of hydrochloric acid and sodium hydroxide. Describe how you would determine, as accurately as possible, the volume of hydrochloric acid required to react exactly with one dm^3 of the sodium hydroxide solution.

A water-soluble compound $MHCO_3$ reacts with hydrochloric acid according to the equation:

$$MHCO_3 + HCl \rightarrow MCl + H_2O + CO_2$$

It was found that 25·0 cm^3 of a solution containing 21·6 grams per dm^3 of $MHCO_3$ reacted with 27·0 cm^3 of a solution of hydrochloric acid containing 7·30 grams per dm^3. Calculate the following and justify your working:

(i) the molarity of the hydrochloric acid;
(ii) the molarity of the solution containing $MHCO_3$;
(iii) the mass of one mole of $MHCO_3$;
(iv) the relative atomic mass of M. *(A.E.B.)*

14. State briefly *four* methods ordinarily used in the laboratory for the preparation of salts.

Illustrate your answer by describing with full practical details how you would obtain pure, dry specimens of (*a*) copper(II) sulphate from copper(II) oxide, (*b*) calcium sulphate from calcium carbonate. *(J.M.B.)*

15. What is a salt? State very briefly *four* different methods which may be used for the preparation of salts.

Describe, with full practical details, how you would prepare (*a*) a dry crystalline specimen of a soluble salt such as copper(II) sulphate or zinc sulphate *and* (*b*) a dry specimen of an insoluble salt such as calcium sulphate or lead(II) sulphate. *(J.M.B.)*

16. *A* is pure liquid ethanoic (acetic) acid.
B is a solution of ethanoic (acetic) acid in water.
C is a solution of hydrogen chloride in toluene.
D is a solution of hydrogen chloride in water.
E is a solution of sodium hydroxide in water.
F is a solution of ammonia in water.
G is a solution of sodium nitrate in water.

Solutions *B* to *G* all contain the same number of moles of solute per dm^3.

(*a*) Choose from the above list:
(i) *two* liquids that will not conduct electricity;
(ii) the *three* liquids that are the best conductors of electricity;
(iii) a *solution* that contains molecules but no ions.

(*b*) Name an aqueous solution not in the above list that will *not* conduct electricity.

(*c*) How would you prepare, from appropriate liquids in the above list, a neutral solution of sodium chloride?

(*d*) What would happen if equal volumes of liquids *C* and *E* were shaken together and then left to settle? (Toluene is insoluble in water and has a density of 0·87 g/cm^3.)

(*e*) Give the formulae of the ions present in solution *E* and state the products you would expect to be liberated at platinum electrodes if solution *E* were electrolyzed. *(C.)*

17. What do you understand by the terms (*a*) acid, (*b*) acid salt, (*c*) normal salt? Name *one* example of each.

If you were given supplies of sodium hydroxide solution, copper(II) oxide and dilute sulphuric acid, describe with full experimental detail, how you would prepare crystalline samples of (*a*) an acid salt of sodium, (*b*) a normal salt of copper. What is the action of heat on the product obtained in (*b*)? *(L.)*

18. The element strontium (Sr) resembles calcium and barium; its position in the electrochemical series is shown in the following list

Na, Ba, Sr, Ca, Mg, Al, Zn . . . . . .

Describe with essential practical details how you would prepare from strontium carbonate:

(*a*) about 10 grams of strontium nitrate crystals, $Sr(NO_3)_2.4H_2O$. (Molecular mass 284);

(*b*) a pure dry specimen of strontium sulphate ($SrSO_4$). (Atomic masses: Sr = 88·0; N = 14·0; O = 16·0; H = 1·0; S = 32·0; C = 12·0) (*L.*)

13

Air

13.1 Burning

Burning is a common enough phenomenon but the examples which you know best such as the burning of wood are not the simplest to investigate. Indeed, for many years a consideration of what apparently happens in such reactions led to the acceptance of false ideas concerning the nature of burning. These we discuss briefly later in this chapter. It is better to see first what happens to certain elements when they are burned.

Experiment (1) (For demonstration only)

Place a small piece of dry white phosphorus in an evaporating dish floating on water in a trough. Place an open bell jar over the dish as shown (figure 13.1) and push a hot glass rod through the top to touch and ignite the phosphorus. Withdraw the rod immediately and place the stopper in the jar.

The phosphorus will burn for a short while, emitting clouds of white smoke which dissolve in the water.

When the apparatus has cooled down note that the water level will have risen approximately one fifth of the way up the bell jar.

This experiment shows that a portion, but not all, of the air is used up during burning.

Experiment (2)

Place a small quantity of magnesium ribbon in a crucible. Weigh the crucible, lid and contents. Then heat the crucible, at first gently and then more strongly, with its lid tilted slightly to permit entry of air. When the magnesium has finished burning allow the crucible, lid and contents to cool. Reweigh.

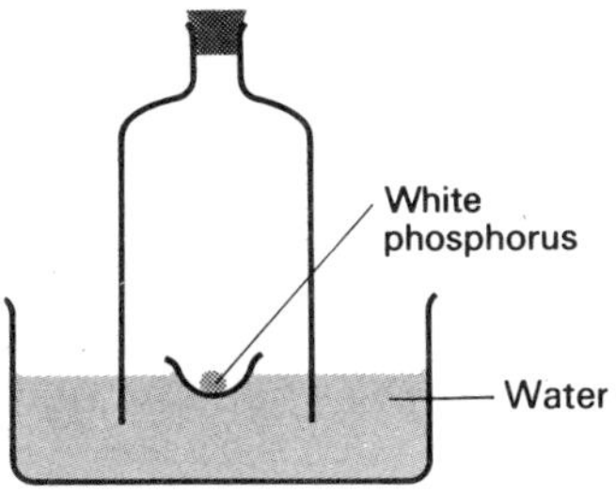

Figure 13.1 Phosphorus burning in air.

You will find that the crucible and contents has increased in mass when the magnesium burns.

Experiment (3)

Repeat experiment (2) but with the magnesium well immersed in dry sand.

This time there is no change in mass. The exclusion of air in the last experiment prevents any gain in mass of magnesium. It is obvious, therefore, that the gain in mass in experiment (2) must be caused by the magnesium combining with some substance present in the air. The first experiment indicates that only a proportion of air is used up during burning.

The gas responsible for burning is called *oxygen*. Magnesium and phosphorus combine with it to form oxides, with a consequent gain in mass. Most of the gas which is not consumed is *nitrogen*.

Reactions similar to those mentioned above take place with many other elements. Considering the reactions of some of the metals with oxygen, it is clear that the position of the metal in the activity series largely determines the ease with which it burns or combines with oxygen, e.g.:

(a) Sodium
Sodium, when heated, burns with a bright yellow flame to form sodium peroxide:

$$2Na(s) + O_2(g) \rightarrow Na_2O_2(s)$$

(b) Magnesium
Magnesium, when heated, burns readily with a dazzling flame, forming white magnesium oxide:

$$2Mg(s) + O_2(g) \rightarrow 2MgO(s)$$

(c) Zinc
Zinc is ignited if it is heated strongly in a blow-pipe flame. It burns with a bluish flame forming zinc oxide:

$$2Zn(s) + O_2(g) \rightarrow 2ZnO(s)$$

(d) Iron
If iron wire is heated in a Bunsen flame it does not burn, but it becomes coated on the surface with a bluish-black oxide of iron, iron(II) diiron(III) oxide:

$$3Fe(s) + 2O_2(g) \rightarrow Fe_3O_4(s)$$

(e) Lead
A small piece of lead heated in a crucible melts and slowly becomes coated with yellow lead(II) oxide:

$$2Pb(s) + O_2(g) \rightarrow 2PbO(s)$$

(f) Copper
Heated copper becomes black on the surface, due to formation of copper(II) oxide:

$$2Cu(s) + O_2(g) \rightarrow 2CuO(s)$$

(g) Mercury

If mercury is heated just below its boiling point for a long time and air is drawn slowly through or over it (an apparatus similar to that in figure 13.4 would suffice) then red mercury(II) oxide forms on the surface. While hot the oxide appears black and the red colour is noted only on cooling:

$$2Hg(l) + O_2(g) \rightarrow 2HgO(s)$$

(h) Silver, gold and platinum

These undergo no chemical change on heating in air.

With all the examples listed above it is comparatively easy to show that there is a gain in mass on burning, but with many other substances the gain is not so apparent. A burning candle seems to disappear almost completely on burning. With coal and wood the solid residue is slight while sulphur leaves no residue. In such cases the main products are gaseous, so that to determine whether there is a gain in mass special methods must be applied.

Experiment. To determine the mass of a candle before and after burning

Assemble the apparatus shown (figure 13.2). Weigh the whole apparatus including the watch glass. Turn on the filter pump so that air is drawn slowly through the apparatus. Light the candle. It will burn to form carbon dioxide and water vapour both of which will be absorbed by the soda-lime. After 3–4 minutes extinguish the candle. A further weighing should show a gain in mass.

Perform a control experiment without lighting the candle by drawing air through the apparatus for the same time at the same rate. This will indicate how much of the gain in mass in the first experiment was due to carbon dioxide and water vapour normally present in air.

Similar experiments could be used to show that whatever substance is used as starting material, there is always a gain in mass on burning, due to combination of the substance with oxygen in the air.

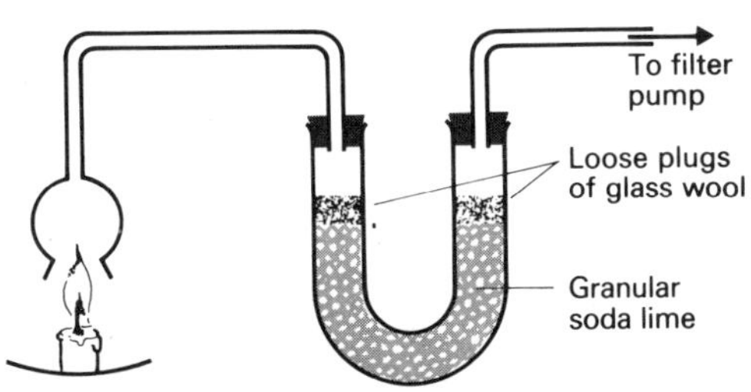

Figure 13.2 Burning of a candle.

13.2 Gases in the atmosphere

As already mentioned, the two most important gases present in the air are nitrogen and oxygen. Other gases are also present to a lesser extent. The approximate percentage composition of air by volume is:

Nitrogen	78 %
Oxygen	21 %
Argon and other noble gases	1 %
Carbon dioxide	0·03 %
Water vapour up to	4 %, obviously depending on the dampness of the atmosphere.

Other gases, such as sulphur dioxide and hydrogen sulphide, are present in small quantities, particularly where the air is polluted.

13.3 Experiments on air

Although the burning of phosphorus in a bell jar shows that approximately one-fifth of the air is used up during burning, a more refined apparatus must be used to give an accurate result. The slow combustion or smouldering of white phosphorus at room temperature can be used as in the following experiment.

Experiment. Determination of the percentage of oxygen in air

Invert a graduated tube or burette over water in a gas jar so that the water levels both inside and outside the tube are the same. Note the volume of enclosed air.

Melt a little white phosphorus under water in a test-tube and push into it the end of a piece of fairly stiff wire. Allow it to cool. Remove the wire with the bead of phosphorus on its end. Insert it in the tube as shown in figure 13.3.

The phosphorus will smoulder, combining with the oxygen present, and the water level in the tube will rise slowly. After a few hours there will be no further action, indicating that all the oxygen present has been used up. Then remove the phosphorus and the wire, and equalise again the water levels inside and

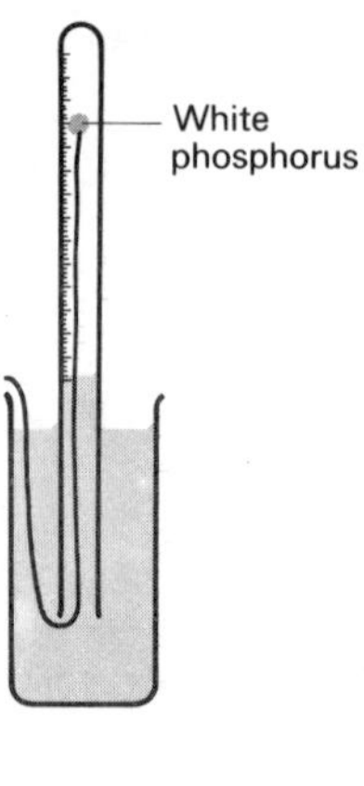

Figure 13.3 The percentage of oxygen in air.

outside the tube. This will ensure that the enclosed gas is at atmospheric pressure. Note the volume of the gas. Hence the percentage of the oxygen consumed can be readily calculated.

If during the course of the experiment there is any appreciable change of temperature and pressure in the laboratory, the volume which the remaining gas would have occupied under the original conditions must be calculated.

Experiment. Detection of carbon dioxide and water-vapour in air

Draw a stream of air through some lime water in the apparatus shown in figure 13.4. After a short time the lime water will turn milky. This proves the presence of carbon dioxide. Alternatively leave some lime water overnight in an open beaker. Again a milkiness will develop.

The condensation of moisture on a cool surface shows that a vapour must be present in air. To prove that *water* vapour is present, expose some anhydrous calcium chloride to the atmosphere. Calcium chloride is deliquescent, and in time its solution in water will be obtained. Distillation of this will yield pure water.

13.4 Separation of air into oxygen and nitrogen

If air which has been highly compressed is allowed to expand, it becomes colder. This fact is used in the production of liquid air (figure 13.5).

Air, freed from carbon dioxide and water vapour, is highly compressed, and then passed through a coiled copper pipe. As it passes through the jet at the bottom of the pipe it expands and cools down. The cooled air passing upward through the container surrounding the coil cools the oncoming compressed air, so that it in turn becomes cooler still. Eventually the air becomes so cool that it liquefies. The whole apparatus must be well lagged to prevent entry of heat from outside.

The liquid air obtained is a pale blue colour and is a mixture of colourless liquid nitrogen (b.p. = − 196 °C) and pale blue liquid oxygen (b.p. = − 183 °C).

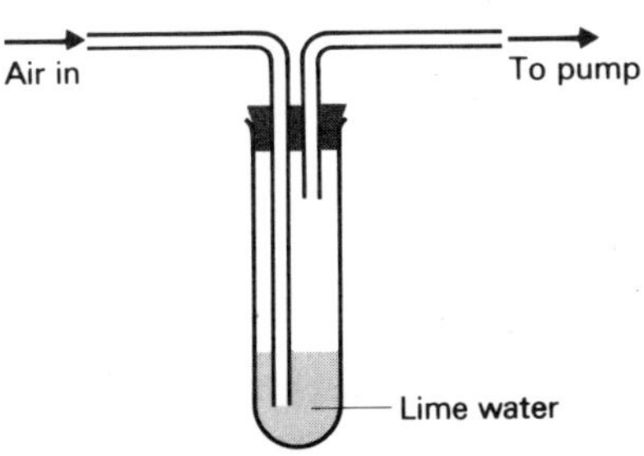

Figure 13.4 Carbon dioxide in air.

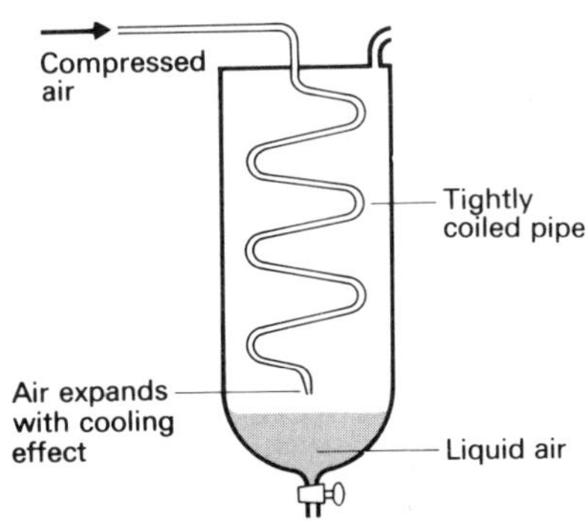

Figure 13.5 Manufacture of liquid air.

On standing, the lower-boiling nitrogen escapes first and the liquid becomes bluer in colour. In practice, the gases are separated by fractional distillation (see section 1.8), which gives a more perfect separation.

13.5 Solubility of air in water

Air is quite soluble in water.

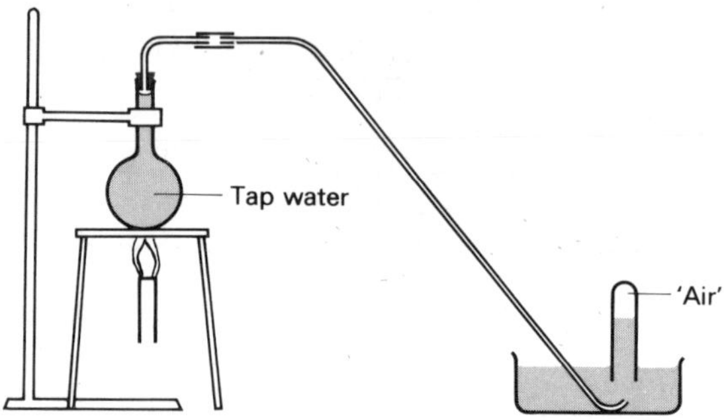

Figure 13.6 Air in water.

Experiment. The removal of air from water

Fill a large flask and delivery tube with water and set up the apparatus shown (figure 13.6). Warm the flask and collect the gas evolved in the test-tube. This will be air which was dissolved in the water. Place a glowing splint in the air; it will reignite showing that this air is rich in oxygen. This must mean that oxygen is more soluble in water than nitrogen (see also section 1.16).

13.6 Proof that air is a mixture

There has in the past been speculation as to whether air is a mixture or a compound. The following evidence proves conclusively that it is a mixture:

1. Samples of air from different parts of the world have slightly different compositions. A compound has an invariable composition.

2. Air can be separated into its components by physical means—

   (*a*) by heating water containing dissolved air, when the gas evolved has a higher proportion of oxygen than normal;

   (*b*) by the fractional distillation of liquid air, which enables oxygen, nitrogen and the noble gases to be separated.

3. "Air" can be made by mixing oxygen, nitrogen, etc., in the required proportions, without any of the energy changes which characterise the formation of compounds.

4. No simple formula can be assigned to a compound having the percentage composition by mass of air.

5. Air has the properties which would be expected from a mixture of gases. It supports combustion, but not so well as does pure oxygen. Its vapour density is between those of oxygen and nitrogen, and the carbon dioxide present responds to the normal tests for that compound.

13.7 Early ideas on burning

It is natural that the causes of burning have aroused interest since the earliest times, but it was not until the seventeenth century that the vital role of air was realised.

The early chemists (twelfth to sixteenth centuries) referred to the powder which a metal gives on heating in air, as a "calx" (i.e. a calx is what we now call an oxide). The process by which a metal was converted into its calx was called calcination. *Jean Ray* (1630) showed that the calx of tin was heavier than the tin from which it was obtained, and supposed that air attached itself in some way to the metal. *Robert Boyle* (figure 13.7) (1673) confirmed this gain in mass and then proceeded to show that sulphur would not burn in a vacuum. This could have led him to deduce the correct nature of burning, but he also showed that gunpowder, a mixture of potassium nitrate, charcoal and sulphur, burnt in the complete absence of air. He concluded therefore that air was not essential for burning.

Robert Hooke, who was once an assistant of Boyle's, came much nearer to the truth by assuming that the potassium nitrate in gunpowder contained a constitutent of air, which he called "nitre air". In support of this he showed that charcoal and sulphur readily ignited when added to molten potassium nitrate. A few years later *John Mayow* concluded that there were two gases in air—one responsible for burning and breathing, and another which supported neither of these. To prove this he showed that a burning candle in an enclosed volume of air goes out before all of the air is used up. His apparatus was similar to that used in the bell-jar experiment described earlier in the chapter, and, as in that experiment, there was a contraction in volume. He also showed that the air diminished in volume when a mouse was confined in a similar apparatus, and that the mouse soon died. A candle did not burn, and a mouse died immediately if placed in the residual gas from either of these experiments.

Mayow also performed experiments with gunpowder similar to those of Hooke, and seems to have been on the verge of discovering oxygen. Had he heated the potassium nitrate present in the gunpowder, he would have been the first to isolate oxygen.

This failure to isolate oxygen was most unfortunate for the history of chemistry. It gave rise to a theory of combustion which severely hampered the development of chemical ideas. This theory, known as the *phlogiston theory*, was first suggested by the German chemist *Johann Becher* (1699), and was developed by his pupil, *George Stahl*. They imagined that all substances capable of burning contained an element which they called phlogiston. During burning the phlogiston escaped, often giving the appearance of a flame, e.g.

Lead $\rightarrow$ Lead calx + Phlogiston

Carbon $\rightarrow$ Carbon calx + Phlogiston

With carbon very little calx was left (if it had been pure there would have been no residue). It was therefore assumed that carbon was almost pure phlogiston, but that lead contained very little phlogiston. The theory was supposedly verified by heating lead calx (i.e. lead minus its phlogiston) with carbon (almost pure phlogiston). Sure enough, metallic lead was reformed by the calx regaining its phlogiston.

Figure 13.8 Joseph Priestley (1733–1804). (Photo: The Science Museum, London).

Figure 13.7 (below) Robert Boyle (1627–1691). (Photo: The Science Museum, London)

Lead calx + Phlogiston → Lead

The Phlogistonists assumed that air was necessary for burning to absorb the phlogiston released during burning. When the air became saturated with phlogiston no further burning was possible. The theory had, therefore, a certain plausibility. It is nevertheless difficult to understand its general acceptance when it was known that burning was often associated with a gain in mass. This fact was either conveniently ignored because of the general neglect at that time of results involving measurements of mass, or else the explanation was given that phlogiston had a negative mass, i.e. a loss of phlogiston from a substance meant a gain in mass.

The work of *Carl Wilhelm Scheele* and *Joseph Priestley* (figure 13.8), both firm believers in the phlogiston theory, eventually provided sufficient evidence to bring about its downfall. Scheele, a Swedish chemist (1772), and Priestley, an English nonconformist pastor (1774) both succeeded in isolating the gas we now call oxygen.

13.8 The isolation of oxygen

Joseph Priestley placed mercury(II) oxide on the surface of mercury in the apparatus shown in figure 13.9. To heat the mercury(II) oxide the sun's rays were focused on to it with the aid of a large magnifying glass. The mercury(II) oxide decomposed to give oxygen, which displaced mercury from the flask into the trough. To Priestley's surprise, a candle burned in the gas formed with a "remarkably vigorous flame". He assumed that this was because the gas was capable of absorbing more phlogiston than ordinary air, i.e. that it was free from phlogiston to begin with. Accordingly, he named the oxygen "dephlogisticated air".

Some time after carrying out this experiment, when visiting Paris, Priestley revealed the details to Antoine Lavoisier (figure 13.10). Lavoisier, who doubted the phlogiston theory, immediately realised the significance of the results; that

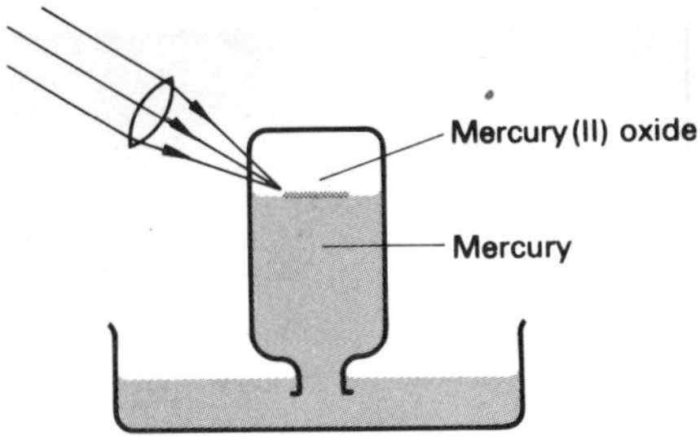

Figure 13.9 Priestley's isolation of oxygen.

"dephlogisticated air" was in fact the active constituent of ordinary air, and that it combined with other substances when they were burned in it.

To confirm this, Lavoisier placed mercury in a matrass (figure 13.11), so that it could be heated by a furnace, and assembled the apparatus shown. The mercury was then heated for twelve days at a temperature just below its boiling point. The volume of air enclosed in the apparatus then decreased to 84% of its original value. As the gas remaining would not support combustion, he gave it the name "*azote*" (from two Greek words meaning "without life").

The red powder which had formed on the surface of the mercury was then carefully collected. On heating this powder, almost the same volume of gas as had been lost was formed. This gas supported both respiration and combustion. Mercury was also formed.

Figure 13.10 Antoine L. Lavoisier (1743–1794). The stamp was issued in France in 1943 to commemorate the 200th anniversary of Lavoisier's birth. (Photo: A. Rendell).

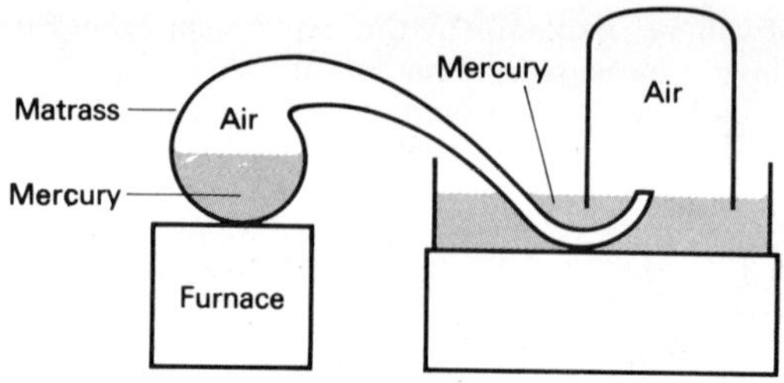

Figure 13.11 Lavoisier's isolation of oxygen from the air.

In other words, the mercury on heating combined with 16% of the air, yielding a red powder. This red powder, on stronger heating, reformed mercury and the gas with which it had combined. Lavoisier named this gas "*oxygen*", from two Greek words meaning "acid producer".

Full credit must be given to Lavoisier as the founder of a correct theory of combustion. It can also be said that he was the first chemist to realise the importance of accurate measurements and, therefore, deserves the name "Father of Modern Chemistry" often given to him. Lavoisier was unfortunately concerned with the collection of taxes in France, and suffered the fate of many others during the French Revolution, when he was beheaded at the age of 51. It is ironical that Priestley, who openly sympathised with the aims of the French Revolution, was at the same time hounded out of this country and had to settle in America, perhaps narrowly escaping a similar end.

There are several excellent biographies of both Priestley and Lavoisier. If you are interested in chemistry you should enjoy reading them. If you look you should find copies in your school or public library.

Questions on chapter 13

1. Write down the approximate composition of the air by volume. (Neglect the "noble gases".)

Write down the equation which represents the explosion of hydrogen with oxygen.

By boiling some water in a suitable apparatus 30 cm^3 of dissolved air were expelled and collected in a eudiometer (an apparatus as in figure 11.3). 50 cm^3 of hydrogen were added and after explosion the volume of the gas was found to be 50 cm^3. (Assume all volumes to be measured at the same laboratory temperature and pressure.) Calculate the percentage by volume of oxygen in this sample of air.

Explain how this result confirms that air is a mixture (*J.M.B.*)

2. Give a brief account of an experiment by which it can be shown that oxygen is a constituent of air.

Describe how you would measure accurately the percentage by volume of oxygen in the air.

Name the other gases which are found in the air and indicate *briefly* how you would demonstrate the presence of *two* of them. (*C.W.B.*)

3. A current of air is bubbled through sodium hydroxide solution and then into distilled water. Describe how you would collect a sample of the "dissolved air" and how you would find the percentage of oxygen in it.

How would you expect the result to compare with the percentage of oxygen in the atmosphere? (*C.*)

4. Describe carefully how you would determine, to within about 1 per cent, the percentage by volume of oxygen in a sample of ordinary air. Draw a diagram to show how

you would take your measurements of volume. Explain why this experiment cannot be done satisfactorily by burning a candle in a measured volume of air.

Explain briefly how oxygen is prepared from air on a large scale. (*C.*)

5. Give *two* reasons, in each case, why you consider that (*a*) water is a chemical compound, (*b*) air is a mixture.

What changes occur when the following substances are exposed to the air for some time: (i) concentrated sulphuric acid; (ii) solid sodium hydroxide; (iii) sodium carbonate-10-water crystals? What explanation can you give of each change you describe? (*C.*)

6. Give *three* reasons for believing that the nitrogen and oxygen in the air are mixed, and not combined.

Describe briefly (*a*) an experiment to find the approximate percentage by volume of oxygen in a sample of air, (*b*) an experiment to show that air contains carbon dioxide.

What use do plants make of the carbon dioxide in the air? (*O. and C.*)

14

Hydrogen

14.1 Introduction

Henry Cavendish (1766) was the first scientist to investigate the properties of "inflammable air" as he called it. The name hydrogen was applied by Lavoisier (figure 13.10), since on burning it forms water. (Hydrogen is derived from two Greek words meaning "water producer".)

Very little hydrogen occurs in the uncombined state on the earth, but its compounds are abundant. Therefore both commercially and in the laboratory it must always be obtained from compounds which contain hydrogen.

14.2 Preparation of hydrogen

14.2.1 From acids

In the laboratory, hydrogen is often prepared by the action of acids on metals. The acid used must not be an oxidising agent (so nitric acid is not suitable), and the metal must occur above hydrogen in the activity series. Dilute sulphuric acid or dilute hydrochloric acid and zinc are the usual reagents:

$$\left[\begin{array}{llll} Zn & +H_2SO_4 & \rightarrow ZnSO_4 & +H_2 \\ Zn & +2HCl & \rightarrow ZnCl_2 & +H_2 \\ Zn(s) & +2H^+(aq) & \rightarrow Zn^{2+}(aq)+H_2(g) & \end{array}\right.$$

These equations show that the action consists of a displacement by the zinc of the less active hydrogen ions.

If sulphuric acid is used, zinc and water are placed in the flask (figure 14.1), and

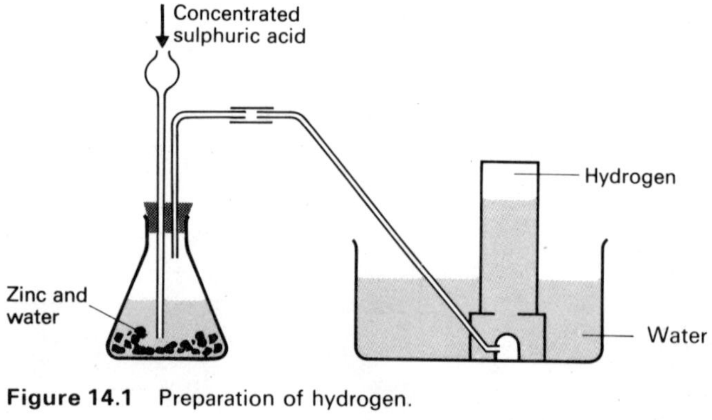

Figure 14.1 Preparation of hydrogen.

concentrated sulphuric acid is added carefully through the thistle funnel until there is a steady evolution of hydrogen. The water dilutes the acid and thereby warms the mixture. The hydrogen is then collected in gas jars over water, the first jar being neglected as it will contain air. *On no account must the flask, or any flask used in the generation of hydrogen, be warmed with a naked flame.*

Should the hydrogen be required dry, it can be bubbled through concentrated sulphuric acid (figure 14.2), and, as it is the least dense gas known, be collected in an inverted gas jar by downwards displacement of air. It could also be collected over mercury.

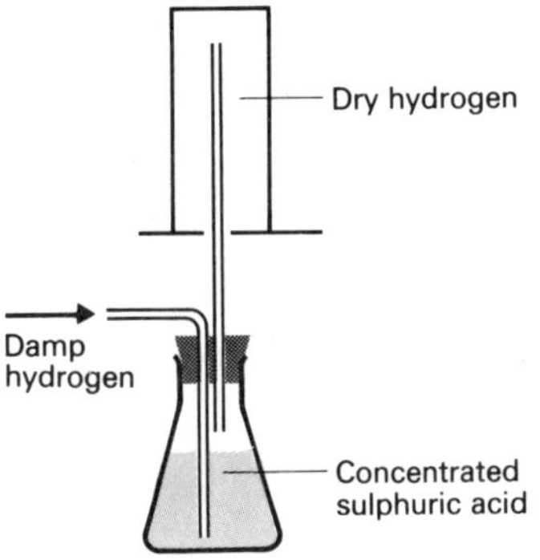

Figure 14.2 Drying of hydrogen.

When pure zinc is used in this preparation the evolution of hydrogen is slow. A few drops of copper(II) sulphate solution should then be added with the acid, when a little copper is deposited on the zinc. This makes hydrogen form readily.

14.2.2 From water

Hydrogen can also be prepared by the passage of steam over red-hot iron (figure 14.3). Steam from boiling water is passed through an iron or silica tube. This tube is loosely packed with iron filings or steel wool, and is heated by several Bunsen

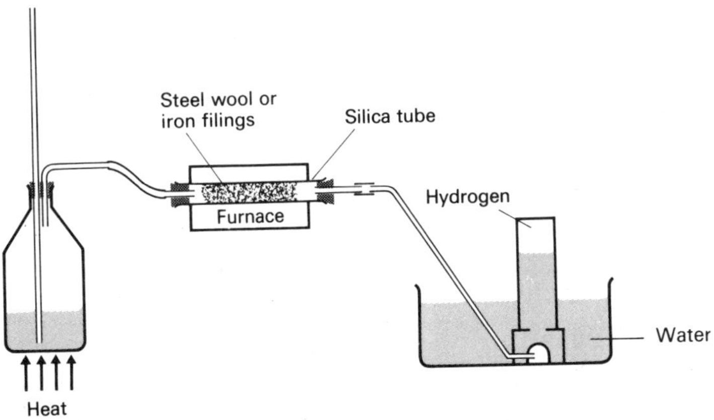

Figure 14.3 Preparation of hydrogen from steam.

burners or in a small furnace. Iron(II) diiron(III) oxide and hydrogen are formed by the following reversible reaction:

$$3Fe(s) + 4H_2O(g) \rightleftharpoons Fe_3O_4(s) + 4H_2(g)$$

The hydrogen is then collected over water, which condenses any remaining steam. (See section 10.6.2 for a further discussion of this reaction.)

Many chemical reactions are reversible. In this case, if hydrogen is passed through a tube containing iron(II) diiron(III) oxide, iron and steam are the products. *A reversible reaction is one which can proceed in either direction depending on the conditions.*

14.2.3 From alkalis

A hot solution of an alkali (or a carbonate of an alkali metal) reacts readily with zinc or aluminium to yield hydrogen:

$$\left[\begin{array}{l} \underset{}{Zn} + \underset{\text{sodium hydroxide}}{2NaOH} \rightarrow \underset{\text{sodium zincate}}{Na_2ZnO_2} + H_2 \\ Zn(s) + 2OH^-(aq) \rightarrow ZnO_2^{2-}(aq) + H_2(g) \end{array}\right.$$

$$\left[\begin{array}{l} 2Al + 2NaOH + 2H_2O \rightarrow \underset{\text{sodium aluminate}}{2NaAlO_2} + 3H_2 \\ 2Al(s) + 2OH^-(aq) + 2H_2O(l) \rightarrow 2AlO_2^-(aq) + 3H_2(g) \end{array}\right.$$

This method is rarely used as a preparation.

14.3 Manufacture of hydrogen

Hydrogen is of increasing importance, and most of it is now manufactured from water gas, from hydrocarbons or by electrolysis.

14.3.1 From water gas (the Bosch process)

Water gas, a mixture of carbon monoxide and hydrogen, is obtained by passing steam through white-hot coke:

$$C(s) + H_2O(g) \rightarrow \underbrace{CO(g) + H_2(g)}_{\text{water gas}}$$

It is then passed, with excess steam, over a heated iron (III) oxide catalyst. The carbon monoxide then reacts with the steam to form more hydrogen:

$$H_2O(g) + CO(g) + H_2(g) \rightarrow CO_2(g) + 2H_2(g)$$

The carbon dioxide is dissolved in water and under pressure, leaving hydrogen.

14.3.2 From natural gas

Much hydrogen is now obtained from natural gas (mainly methane) and other hydrocarbons found with petroleum. For example, natural gas and steam are passed over a nickel catalyst at 700–870 °C.

$$CH_4(g) + H_2O(g) \rightarrow CO(g) + 3H_2(g)$$

There are various methods for removing the carbon monoxide. It can be converted into carbon dioxide as in the Bosch process. Another method consists of converting the carbon monoxide and some of the hydrogen directly to methanol, using a high pressure and a zinc oxide catalyst at 300 °C (see section 25.6.1).

$$CO(g) + 2H_2(g) \rightarrow CH_3OH(g)$$

Other reactions yielding hydrogen from petroleum products are given in chapter 25.

14.3.3 By electrolysis

In the manufacture of sodium hydroxide from common salt solution, much hydrogen is obtained as a by-product (section 21.2.6(*c*)).

14.4 Properties of hydrogen

14.4.1 Hydrogen is a diatomic gas

It is colourless, odourless,, almost insoluble in water and is liquefied only with great difficulty.

14.4.2 Hydrogen is less dense than air

(*a*) A simple method of showing this fact is to blow some bubbles with hydrogen using soap or detergent solution. The bubbles immediately rise towards the ceiling. Alternatively, a rubber balloon can be filled with hydrogen from a cylinder, and then released. If no cylinder is available you can probably devise an apparatus for blowing up the balloon.

(*b*) Another method is to pour hydrogen upwards from one gas jar into another. The jars are held as shown (figure 14.4), and after a few seconds both are tested with a lighted taper. Only the upper jar shows a reaction for hydrogen.

14.4.3 Hydrogen burns in air, but does not support burning

If a lighted taper is plunged into an inverted (why inverted?) gas jar full of hydrogen the hydrogen burns at the mouth of the jar where it is in contact with air. The taper, however, once surrounded by hydrogen, is itself extinguished.

When a lighted taper is applied to a mixture of air and hydrogen there is a sharp pop. There is little risk in doing this experiment in an open gas jar, but

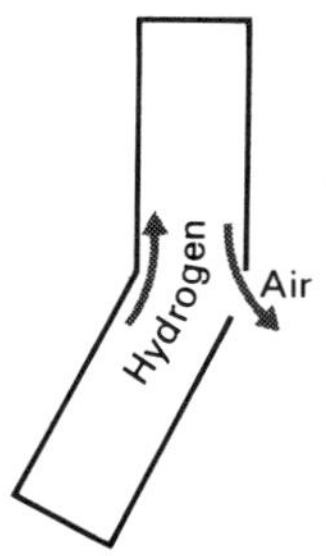

Figure 14.4 Pouring hydrogen upwards.

nasty accidents are possible if the mixture is ignited in a closed or almost closed apparatus so that the expanding gases have no room to escape. The French scientist Rozier actually tried inhaling hydrogen and then igniting his breath. This would have formed the basis of an amusing party trick if the hydrogen had not been—inevitably—mixed with air.

Mixtures of hydrogen and oxygen are even more dangerous. If a small poly(ethene) bottle (e.g. a liquid detergent container) is two-thirds filled with hydrogen and one-third filled with oxygen, and a lighted taper is placed in the mixture, there is a noisy explosion. Larger containers may be used *outside* by lighting with a taper on the end of a long stick, but *on no account* should ordinary glass vessels be used.

14.4.4 Hydrogen forms water when it burns

Hydrogen is passed through a U-tube containing anhydrous calcium chloride to dry it (figure 14.5). After a few moments a test-tube full of the gas is collected and is opened near a flame to ensure that all air has been displaced from the apparatus. The gas issuing from the U-tube is then ignited and the flame is placed under a glass funnel that leads into a condenser.

Drops of a colourless liquid collect in the flask. This liquid turns anhydrous copper(II) sulphate to a blue colour, and boils at 100 °C. It is therefore water:

$$2H_2(g) + O_2(g) \rightarrow 2H_2O(l)$$

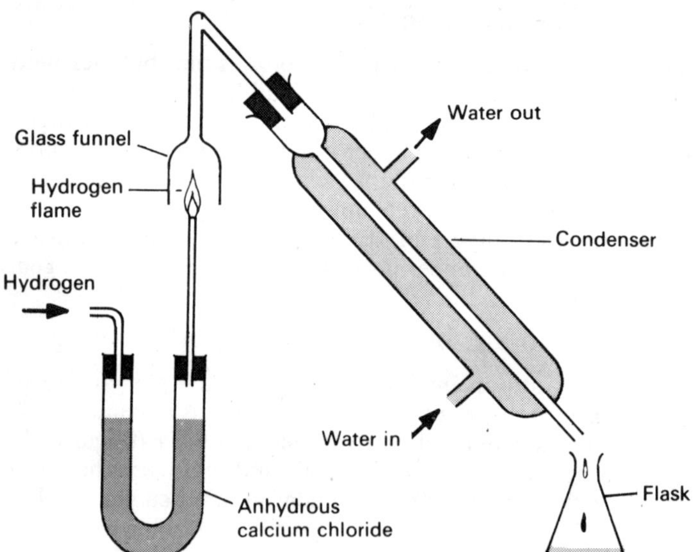

Figure 14.5 Formation of water by burning dry hydrogen in air.

14.4.5 Hydrogen is a reducing agent

A compound is said to be *reduced* when oxygen is removed from it. The process of removing oxygen is termed *reduction*, and the substance responsible for the removal is a reducing agent. (For a fuller treatment of reduction see section 15.8.2).

When hydrogen is passed over the heated oxides of the less-reactive metals they are reduced to the metals, e.g.:

$CuO(s)$ $+ H_2(g) \rightarrow Cu(s) + H_2O(g)$
copper(II) oxide

$PbO(s)$ $+ H_2(g) \rightarrow Pb(l) + H_2O(g)$
lead(II) oxide

Other reductions occur in the presence of mixtures which are producing hydrogen. Thus, if zinc and hydrochloric acid are added to iron(III) chloride solution, iron(II) chloride is formed. In acid solution zinc provides electrons for reducing the iron(III) ions to iron(II) ions:

$$2Fe^{3+}(aq) + Zn(s) \rightarrow 2Fe^{2+}(aq) + Zn^{2+}(aq)$$

For a more detailed treatment, see section 15.8.3(*c*).

Another hydrogen-producing mixture which can be used for reductions is one of zinc dust (or Devarda's alloy) and sodium hydroxide solution. This will reduce nitrates to ammonia. With this type of reduction the reagents for producing hydrogen *must be actually mixed* with the substance to be reduced.

14.4.6 Hydrogen combines directly with the halogens

A gas jar containing hydrogen is placed on top of a gas jar full of chlorine mouth to mouth. The covers are removed and the gases mixed by inverting the jars several times. When a lighted taper is placed at the mouth of one of these jars there is an explosion, and a colourless gas, which fumes in damp air, is formed. The gas is hydrogen chloride:

$$H_2(g) + Cl_2(g) \rightarrow 2HCl(g)$$

Note: This experiment is dangerous and should be performed behind a safety screen by a teacher.

In bright sunlight this reaction takes place spontaneously, while in diffused light the hydrogen and chlorine combine only slowly.

The combination of hydrogen with the other halogens, bromine and iodine, is much less vigorous.

14.4.7 Hydrogen diffuses more rapidly than air

The apparatus is assembled as shown, with the porous pot containing air (figure 14.6). The pot contains many small holes through which the gas molecules present in air are moving in both directions. When a jar of hydrogen is inverted over the pot a jet of liquid shoots out at *A*. This is because hydrogen molecules diffuse into the pot more quickly than air molecules can diffuse out. Therefore a pressure is set up inside the pot, and this causes the jet of liquid.

When the jar of hydrogen is removed the reverse process occurs, and air bubbles are drawn in at *B*. (For a fuller treatment of diffusion see section 7.2).

14.5 Uses of hydrogen

Balloons and airships were filled with this gas, but its inflammability proved a source of danger (figure 14.7). Accordingly, it has to some extent been replaced

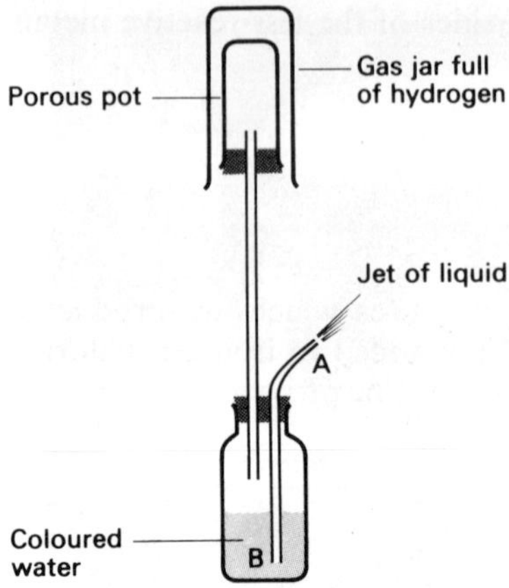

Figure 14.6 Diffusion of hydrogen.

with helium which, although twice as dense as hydrogen, does not burn. It is still used in meteorological balloons, and balloons used in cosmic-ray research (figure 14.8).

Much more important nowadays is the *hydrogenation of oils*, i.e. the combination of oils with hydrogen to give solid fats. These fats are used as margarine or lard substitute, and in the manufacture of soap. Hydrogenation is effected by adding powdered nickel catalyst to a hot oil, such as whale oil or cotton-seed oil. Hydrogen is then passed through under pressure. When the reaction is complete the mixture is filtered while still hot to remove the nickel. The filtrate then solidifies on cooling. (See also section 25.4.2 and figure 25.11).

The *hydrogenation of petroleum* or a mixture of oil and powdered coal at high pressures and a temperature of 450 °C yields a mixture of hydrocarbons from which a synthetic petrol can be extracted by distillation.

Large quantities of hydrogen are used in the *syntheses of ammonia and hydrogen chloride* from their elements and in the *manufacture of methanol.*

As a reducing agent hydrogen is used in the extraction of some of the less-common metals. Thus tungsten, which is used in electric-light filaments, is obtained by the reduction of its oxide at red heat:

$$WO_3(s) + 3H_2(g) \rightarrow W(s) + 3H_2O(g)$$
tungsten(VI) oxide

The oxy-hydrogen blow-torch used in welding produces temperatures of 2000 °C. In this blow-torch hydrogen is burnt at a jet in an atmosphere of oxygen. The *plasma flame spray gun* produces much higher temperatures. In the one illustrated in figure 14.9 temperatures from 5000 to 15 000 °C can be obtained. Here hydrogen, mixed with nitrogen, is passed through an electric arc when the molecules absorb energy and split up into a mixture of atoms, nuclei and electrons (the mixture is known as a *plasma*). When free of the arc, these recombine with the evolution of much heat.

Hydrogen is used as a fuel mixed with other gases in water gas and coal gas. It is also used as a fuel for space rockets. The hydrogen is carried by the rocket as a liquid at very low temperature (the boiling point of hydrogen is −253 °C) and is burnt in the rocket motor with oxygen, (see also section 15.6).

An interesting future for hydrogen would be for it to replace electricity

Figure 14.7 The German airship Hindenburg. This giant 245 m airship successfully crossed the Atlantic; it is shown arriving over New York (top picture). Whilst being moored at Lakehurst, New Jersey, for some reason the inflammable hydrogen gas burst into flames, completely destroying the airship in 2 minutes (bottom picture). The pictures are actual news-photographs taken on May 6th, 1937. (Photos: Fox Photos, Ltd.).

Figure 14.8 A hydrogen filled balloon. The equipment that it carries measures and then sends by radio the weather conditions in the upper atmosphere. (Photo: The Metereological Office).

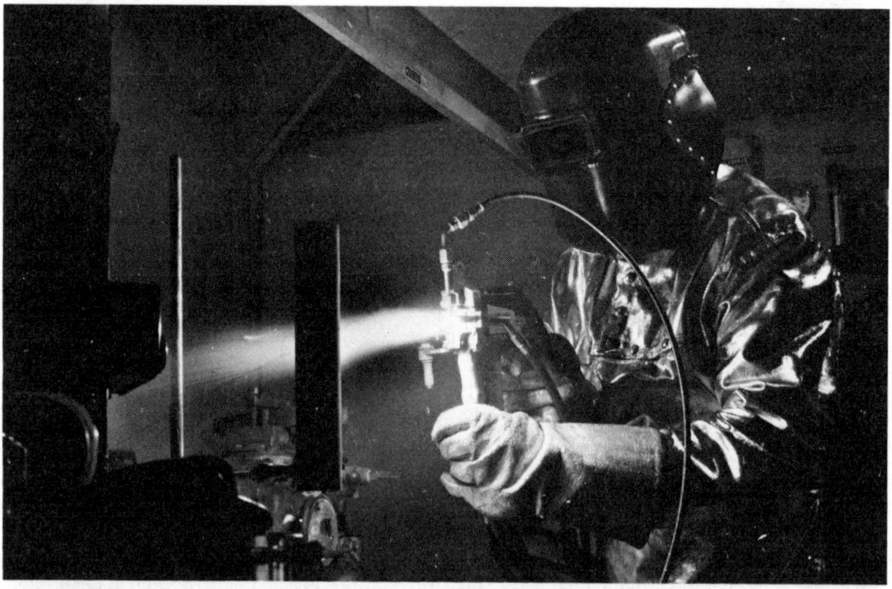

Figure 14.9 A plasma flame spray gun being used for spraying ceramics. Ceramic powder (here alumina) is melted at the very high temperatures produced, and the gas stream is spraying it as a coating on to a graphite tube. (Photo: Ceramics Centre, A.E.R.E. Harwell, U.K.A.E.A.).

as a fuel. It has many attractions; in particular when it is burnt it forms water and, unlike carbon based fuels, does not cause atmospheric pollution (see chapter 28). It could be moved from one place to another in underground pipe-lines and can be stored. This is in contrast to electricity which requires transmission lines above ground and must be generated as required. If and when cheap energy becomes available (see Chapter 28) such a *hydrogen economy* as it is called might become possible.

Questions on chapter 14

1. What volume of hydrogen at s.t.p. could be obtained:
(*a*) by the action of 1 g of zinc on excess sulphuric acid;
(*b*) by the action of 1 g of zinc on excess hydrochloric acid;
(*c*) by the action of 1 g of zinc on excess sodium hydroxide solution;
(*d*) by the action of 1 g of sodium on water?

2. State the conditions necessary, name the products formed, and give the equations for the reactions which take place with each of the following pairs of substances: (*a*) steam and iron; (*b*) calcium and water; (*c*) magnesium and sulphuric acid.
How would you show conclusively that the same gas is liberated in each reaction? (*C.W.B.*)

3. Draw labelled diagrams, one in each case, to show the preparation and collection of a small quantity of hydrogen using (*a*) cold water, (*b*) steam. Describe any changes that may be observed. (Electrolytic methods will not be accepted.)
State *two* large-scale uses of hydrogen.
How can hydrogen be made to react in the laboratory with (*a*) lead(II) oxide, (*b*) chlorine? In each reaction name and describe the appearance of the product. (*C.W.B.*)

4. (*a*) (i) Name the two reagents normally used in the laboratory preparation of hydrogen from a metal and acid.
(ii) Write an ionic equation for the reaction.
(iii) The ionic equation indicates that the reaction is an electron transfer between the two reagents. Which of the reagents is oxidised and which reduced?
(*b*) Hydrogen is also obtained rapidly when steam is passed over heated magnesium. Describe what you would see in this reaction and give a reason why there is such an increase in the rate of reaction under these conditions compared to that between magnesium and cold water.
(*c*) How would you obtain hydrogen by electrolysis in the laboratory? Name the electrolyte used, and state the name and polarity of each electrode and the material of which it is made. Give the name of the product at each electrode and write equations for the reactions occurring during their discharge.
(*d*) What is there in common between the reaction at the cathode in (*c*) and the reaction in (*a*)? (*J.M.B.*)

5. Describe carefully how you would prepare a few gas jars of hydrogen from sulphuric acid and zinc. Draw a diagram of the apparatus you would use.
Explain fully how you would isolate a dry crystalline specimen of the second product of this reaction. (*C.*)

6. (*a*) A small piece of sodium, when placed on water, darts about on the surface and changes its appearance. Explain these observations as far as you can, and state what chemical reaction takes place.
(*b*) Describe (with a diagram of the apparatus you would use) how, starting from hydrochloric acid, you would prepare and collect a specimen of hydrogen. (*C.*)

7. How is hydrogen produced on a large scale? Why is its manufacture important?
How, and under what conditions, does hydrogen react with (*a*) copper(II) oxide, (*b*) sulphur, (*c*) chlorine? (*O. and C.*)

8. Explain how you would show that when dry hydrogen burns in air water is formed. Illustrate your answer by a diagram and state how you would prove the liquid to be pure water. (*J.M.B.*)

9. Hydrogen gas may be prepared in the laboratory by the action of a metal on an acid. Draw a diagram of an apparatus suitable for this preparation and answer the following questions:

(*a*) Which of the following metals would be most suitable? Copper, zinc, magnesium or sodium. Give a reason for rejecting each of the metals you do not select.

(*b*) Which of the following acids would you choose? Dilute sulphuric, dilute nitric, concentrated nitric. Say why you would *not* use each of the acids you reject.

(*c*) It is necessary to run the apparatus for some time before attempting to collect the gas. Why? How do you test when to start collecting?

(*d*) How would you modify your apparatus in order to collect *dry* hydrogen? (*L.*)

10. Briefly outline the apparatus and conditions you would use to prepare and collect hydrogen by the action of water on (i) calcium, (ii) iron. Name the other products of the reactions.

State which of the following metals would give hydrogen when added to dilute hydrochloric acid, and give equations for the reactions: (iii) calcium, (iv) copper, (v) iron, (vi) aluminium.

Describe how aluminium and iron(III) oxide react together.

On the basis of the evidence from these reactions, arrange the four metals in order of increasing chemical reactivity. (*J.M.B.*)

15

Oxygen

15.1 Introduction

Oxygen is more plentiful than any other element. Approximately 20 per cent by volume of the air is uncombined oxygen, nearly 90 per cent of water is combined oxygen, and, combined with other elements, it comprises nearly 50 per cent by mass of the earth's crust.

15.2 Preparation of oxygen

15.2.1 From hydrogen peroxide

Hydrogen peroxide (20 volume) is dropped slowly from a tap funnel on to manganese(IV) oxide in a flask (figure 15.1). The hydrogen peroxide decomposes and a steady flow of oxygen is obtained.

$$2H_2O_2(aq) \rightarrow 2H_2O(l) + O_2(g)$$

hydrogen
peroxide

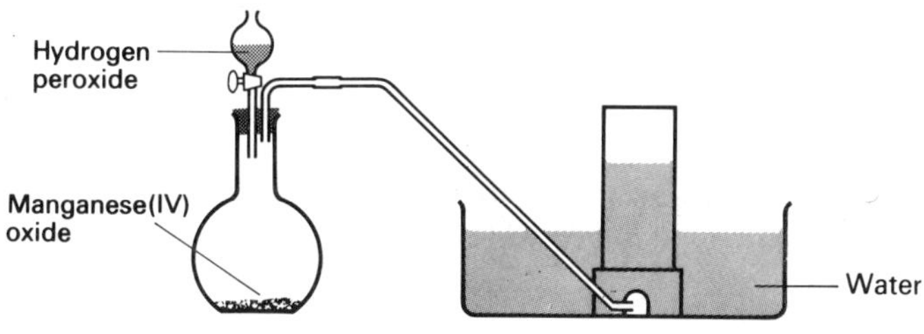

Figure 15.1 The preparation of oxygen.

The manganese(IV) oxide is present as a *catalyst*. On its own, hydrogen peroxide decomposes very slowly indeed but with the catalyst present there is an immediate effervescence.

The catalyst is not used up during the course of the experiment, but it is present in the same quantity at the end of the experiment as at the beginning, i.e. *A catalyst is a substance which alters the rate of a chemical reaction but is chemically unchanged at the end of the reaction.* It is a common error to say that a catalyst takes no part in a chemical reaction. This is certainly not true, or there would be little point in adding the catalyst; it does not, however, affect the products of a chemical reaction. (See section 10.4 for a further discussion of catalysts).

15.2.2 Other reactions which yield oxygen

(a) Action of heat on mercury(II) oxide

When mercury(II) oxide is heated, it decomposes into its elements:

$$2HgO(s) \rightarrow 2Hg(g) + O_2(g)$$

This is the reaction by which Joseph Priestley first obtained oxygen. Mercury(II) oxide is, however, expensive and the yield is low.

(b) Action of heat on the higher oxides of some metals

i.e. oxides where the metal is in its higher valency state: e.g.

$2PbO_2(s)$ lead(IV) oxide	$\rightarrow$	$2PbO(s)$ lead(II) oxide	$+O_2(g)$
$2Pb_3O_4(s)$ dilead(II) lead(IV) oxide	$\rightarrow$	$6PbO(s)$	$+O_2(g)$

(c) Action of heat on nitrates.

On heating, sodium and potassium nitrates yield the nitrites and oxygen:

$$2NaNO_3(s) \rightarrow 2NaNO_2(s) + O_2(g)$$

Most nitrates on heating yield the oxide of the metal and a mixture of the two gases, oxygen and nitrogen dioxide: e.g.

$$2Pb(NO_3)_2(s) \rightarrow 2PbO(s) + 4NO_2(g) + O_2(g)$$

lead(II) nitrate

(d) Action of heat on potassium chlorate(V)

When potassium chlorate(V) is strongly heated, oxygen is evolved:

$$2KClO_3(s) \rightarrow 2KCl(s) + 3O_2(g)$$

potassium
chlorate(V)

If the potassium chlorate(V) is mixed with one-twelfth of its mass of magnesium oxide, the oxygen is much more readily evolved. The magnesium oxide behaves as a catalyst.

(e) Action of water on sodium peroxide

If a little sodium peroxide is added to water, oxygen is formed immediately:

$$2Na_2O_2(s) + 2H_2O(l) \rightarrow 4NaOH(aq) + O_2(g)$$

sodium
peroxide

15.3 Manufacture of oxygen

Most oxygen is manufactured by the fractional distillation of liquid air (see section 13.4). Some is also obtained as a by-product of certain electrolytic processes.

15.4 Properties of oxygen

15.4.1 Introduction

Oxygen is a diatomic gas. It is colourless, odourless, almost insoluble in water and is liquefied only with difficulty. It is only very slightly denser than air and cannot therefore be collected by displacement of air. *If a glowing wooden splint is plunged into oxygen it immediately relights. This is often used as a test for oxygen,* although it is also given by dinitrogen oxide.

Oxygen is a very active element and combines with many other elements, both metallic and non-metallic.

15.4.2 Combustion of non-metals in oxygen

(a) Sulphur

A small piece of sulphur is placed in a deflagrating spoon and then heated until it starts burning with a pale blue flame. The spoon is then plunged into a gas jar full of oxygen (figure 15.2). The sulphur then burns much more brightly to give a gas with a choking smell. The main product is sulphur dioxide, which is colourless, but some white fumes of sulphur(VI) oxide are also formed.

$$S(s) + O_2(g) \rightarrow SO_2(g)$$

If a little water is added to the jar the sulphur dioxide combines with it to give a solution which turns blue litmus red. This is because sulphurous acid is formed:

$$SO_2(g) + H_2O(l) \rightarrow \underset{\text{sulphurous acid}}{H_2SO_3(aq)}$$

(b) Phosphorus

Some phosphorus is ignited in a deflagrating spoon and is then placed into a gas jar full of oxygen. It burns with an exceptionally brilliant flame and produces a dense white smoke. This smoke, which is a mixture of phosphorus(III) oxide and phosphorus(V) oxide, settles after a few minutes:

$$P_4(s) + 3O_2(g) \rightarrow \underset{\text{phosphorus(III) oxide}}{P_4O_6(s)}$$

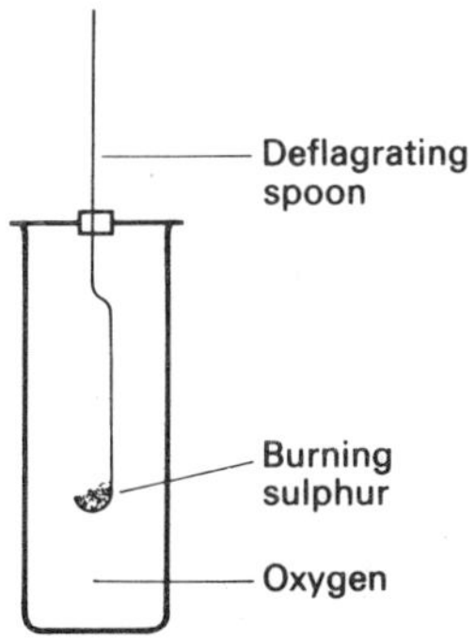

Figure 15.2 Burning sulphur in oxygen.

$$P_4(s) + 5O_2(g) \rightarrow P_4O_{10}(s)$$
phosphorus(V) oxide

These oxides also combine with water to give acids which turn blue litmus red:

$$P_4O_{10}(s) + 6H_2O(l) \rightarrow 4H_3PO_4(aq)$$
phosphoric acid

(c) Carbon
A small lump of carbon in a deflagrating spoon is heated to red heat. When placed in oxygen it then burns with a small orange flame. The colourless gas carbon dioxide is formed:

$$C(s) + O_2(g) \rightarrow CO_2(g)$$

This combines with water to form the weak carbonic acid, which turns blue litmus to a dull red colour:

$$CO_2(g) + H_2O(l) \rightleftharpoons H_2CO_3(aq)$$
carbonic acid

15.4.3 Combustion of metals in oxygen

(a) Magnesium
Some magnesium ribbon, held in a pair of tongs, is ignited and then lowered into a gas jar full of oxygen. It burns even more brilliantly than in air, forming a white powder, magnesium oxide:

$$2Mg(s) + O_2(g) \rightarrow 2MgO(s)$$

With water, magnesium hydroxide is formed which, although only slightly soluble, dissolves to give an alkaline solution:

$$MgO(s) + H_2O(l) \rightarrow Mg(OH)_2(s)$$

(b) Sodium
When clean sodium is heated in a deflagrating spoon it first melts and then burns with a yellow flame. When placed in oxygen it burns much more brightly, occasionally at so high a temperature that the spoon melts. Solid sodium peroxide is left in the spoon:

$$2Na(l) + O_2(g) \rightarrow Na_2O_2(s)$$

If the spoon is dipped *cautiously* into water the sodium peroxide reacts to give a solution of the alkali, sodium hydroxide:

$$2Na_2O_2(s) + 2H_2O(l) \rightarrow 4NaOH(aq) + O_2(g)$$

(c) Iron
A length of thin iron wire is cleaned with sandpaper and made into a small bundle. After heating to red heat in a pair of tongs it is *immediately* plunged into oxygen. It burns, giving a shower of sparks and forming molten beads of iron(II) diiron(III) oxide:

$$3Fe(s) + 2O_2(g) \rightarrow Fe_3O_4(s)$$

(It is advisable, for this experiment, to have a little water in the bottom of the jar to prevent the hot beads from cracking it.)

Iron(II) diiron(III) oxide is insoluble in water, and therefore has no action on litmus.

15.4.4 Combustion of compounds in oxygen

Many compounds when they are burnt or roasted in oxygen form the oxides of their constituent elements, e.g.

(*a*) Most hydrocarbons burn to give a mixture of carbon dioxide and steam: e.g.

$$CH_4(g) + 2O_2(g) \rightarrow CO_2(g) + 2H_2O(g)$$

(*b*) Most compounds of carbon, hydrogen and oxygen also give carbon dioxide and steam:

$$C_2H_5OH(l) + 3O_2(g) \rightarrow 2CO_2(g) + 3H_2O(g)$$
ethanol

$$C_{12}H_{22}O_{11} + 12O_2(g) \rightarrow 12CO_2(g) + 11H_2O(g)$$
sugar

(*c*) Hydrogen sulphide, when burnt in a plentiful supply of oxygen, yields sulphur dioxide and steam:

$$2H_2S(g) + 3O_2(g) \rightarrow 2H_2O(g) + 2SO_2(g)$$

(*d*) Most metallic sulphides when roasted in air give sulphur dioxide and the oxide of the metal:

$$2ZnS(s) + 3O_2(g) \rightarrow 2ZnO(s) + 2SO_2(g)$$
zinc
sulphide

This type of reaction is very important in the extraction of some metals. Zinc, for example, is manufactured by reducing the zinc oxide obtained by roasting naturally occurring zinc sulphide.

(*e*) When ammonia is burnt in oxygen, in the absence of a catalyst, nitrogen and steam (*not* an oxide of nitrogen) are formed:

$$4NH_3(g) + 3O_2(g) \rightarrow 2N_2(g) + 6H_2O(g)$$

15.5 Burning and respiration

When a foodstuff is burnt in oxygen heat is evolved, and carbon dioxide and water vapour are always present in the products. Thus the burning of sugar can be represented by the word equation:

Sugar + Oxygen → Carbon dioxide + Water + Heat energy

(How by a simple experiment would you show that these substances are formed?)

A similar process takes place, although much more slowly, in the body of an animal when food has been eaten, the heat being used to maintain the body temperature. In the lungs haemoglobin in the blood forms a loose compound with oxygen:

Oxygen + Haemoglobin → Oxyhaemoglobin

100 cm^3 of blood absorbs 18 cm^3 of oxygen in this way, but without this chemical reaction it could absorb less than 1 cm^3. The oxyhaemoglobin travels round the body and readily gives up its oxygen to "burn up" some of the food which has been digested, forming carbon dioxide and water. Catalysts in the body enable the reaction to proceed at a low temperature. Exactly the same amount of energy is produced by this oxidation reaction within the body as with the burning in air of the same substance. The blood dissolves the carbon dioxide formed, and when it reaches the lungs once more the carbon dioxide is liberated. Air breathed out, therefore, contains a higher percentage of carbon dioxide (approximately 3 per cent) than is present in ordinary air.

15.6 Uses of oxygen

Oxygen is often sold for industrial purposes in strong steel cylinders in which it is highly compressed or in large containers as liquid oxygen (figure 15.3). Much of this is used in oxy-ethyne (oxy-acetylene) and oxy-hydrogen blow-pipes. The oxy-ethyne flame is intensely hot and is used in welding.

Dinitrogen oxide and other anaesthetics are administered with oxygen, and oxygen is also used medicinally for illnesses in which the lungs are not working properly, the patient often being kept in an atmosphere of oxygen in an enclosed tent.

Oxygen is used in high-altitude flying and climbing; otherwise the atmosphere is so rarefied that ordinary breathing is not possible (figure 15.4).

Liquid oxygen is used to burn the hydrogen or hydrocarbon fuel in rockets (figure 15.5).

Now, however, by far the most important use of manufactured oxygen is in the steel industry (see chapter 24) because it removes carbon and other impurities more rapidly than air. So much oxygen is needed that it is becoming usual to obtain the gas by liquefying air with a special plant constructed in the steel-works. Because of the scale on which it is produced it is called "tonnage oxygen"

15.7 Classification of oxides

When various elements were burnt in oxygen it was noticed that the oxides of non-metals combined with water to form acids, while those oxides of metals which combined with water formed alkalies. Because of differences such as these between oxides of various elements, it is possible to classify them into groups. The most important types of oxide are as follows:

Figure 15.3 A train made up of liquid oxygen waggons going from the liquefying plant to the consumer. (Photo: British Oxygen Co. Ltd.).

Figure 15.4 Sir Edmund Hillary climbing the ridge to the final camp on Everest. The cylinders on his back, made of low-density aluminium alloy, supply the oxygen required at high altitudes. (Photo: The Mount Everest Foundation).

Figure 15.5 A Saturn rocket at lift-off. The first stage rocket burns paraffin in liquid oxygen. (Photo: N.A.S.A.).

15.7.1 Acidic oxides

Oxides which combine with water to form acids are called acidic oxides or acid anhydrides. They also combine with alkalies to form a salt and water *only*. All the common acidic oxides are oxides of non-metals.

e.g. carbon dioxide:

$CO_2(g) + H_2O(l) \rightarrow H_2CO_3(aq)$
carbonic acid

$CO_2(g) + 2NaOH(aq) \rightarrow Na_2CO_3(aq) + H_2O(l)$
sodium carbonate

sulphur dioxide:

$SO_2(g) + H_2O(l) \rightarrow H_2SO_3(aq)$
sulphurous acid

$SO_2(g) + 2NaOH(aq) \rightarrow Na_2SO_3(aq) + H_2O(l)$
sodium sulphite

sulphur(VI) oxide:

$SO_3(s) + H_2O(l) \rightarrow H_2SO_4(aq)$
sulphuric acid

$SO_3(s) + 2NaOH(aq) \rightarrow Na_2SO_4(aq) + H_2O(l)$
sodium sulphate

nitrogen dioxide:

$2NO_2(g) + H_2O(l) \rightarrow HNO_2(aq) + HNO_3(aq)$
nitrous acid, nitric acid

$2NO_2(g) + 2NaOH(aq) \rightarrow NaNO_2(aq) + NaNO_3(aq) + H_2O(l)$
sodium nitrite, sodium nitrate

Nitrogen dioxide, which is unusual in giving a mixture of two acids when it combines with water, is called a mixed anhydride.

15.7.2 Basic oxides

Oxides which combine with acids to form a salt and water *only* are called basic oxides. If they combine with water they form alkalis. All basic oxides are the oxides of metals:

e.g. sodium oxide:

$Na_2O(s) + H_2O(l) \rightarrow 2NaOH(aq)$
sodium hydroxide

$Na_2O(s) + 2HCl(aq) \rightarrow 2NaCl(aq) + H_2O(l)$
sodium chloride

calcium oxide:

$CaO(s) + H_2O(l) \rightarrow Ca(OH)_2(s)$
calcium hydroxide

$CaO(s) + 2HNO_3(aq) \rightarrow Ca(NO_3)_2(aq) + H_2O(l)$
calcium nitrate

copper(II) oxide:

$$CuO(s) + H_2SO_4(aq) \rightarrow \underset{\text{copper(II) sulphate}}{CuSO_4(aq)} + H_2O(l)$$

Like most basic oxides, copper(II) oxide does not combine with water.

15.7.3 Amphoteric oxides

Amphoteric oxides combine some of the properties of acidic oxides with some of the properties of basic oxides, i.e. they combine with acids and with alkalies, in each case yielding a salt and water only. (What other words do you know beginning with amph-, and what do they mean?)

e.g. zinc oxide:

$$ZnO(s) + H_2SO_4(aq) \rightarrow \underset{\text{zinc sulphate}}{ZnSO_4(aq)} + H_2O(l)$$

$$ZnO(s) + 2NaOH(aq) \rightarrow \underset{\text{sodium zincate}}{Na_2ZnO_2(aq)} + H_2O(l)$$

aluminium oxide:

$$Al_2O_3(s) + 3H_2SO_4(aq) \rightarrow \underset{\text{aluminium sulphate}}{Al_2(SO_4)_3(aq)} + 3H_2O(l)$$

$$Al_2O_3(s) + 2NaOH(aq) \rightarrow \underset{\text{sodium aluminate}}{2NaAlO_2(aq)} + H_2O(l)$$

15.7.4 Peroxides

These react with cold dilute acids to yield hydrogen peroxide, e.g.:

sodium peroxide: $Na_2O_2(s) + H_2SO_4(aq) \rightarrow Na_2SO_4(aq) + H_2O_2(aq)$

barium peroxide: $BaO_2(s) + H_2SO_4(aq) \rightarrow BaSO_4(s) + H_2O_2(aq)$

15.7.5 Neutral oxides

Oxides which react with neither acids nor alkalies, and which cannot be included in any of the groups listed above, are called neutral oxides. Common examples include water (H_2O), carbon monoxide (CO), dinitrogen oxide (N_2O) and nitrogen oxide (NO).

15.8 Oxidation and reduction

These terms have already been used on several occasions, and as both have wide meanings nowadays, it is worth considering exactly what constitutes oxidation and reduction.

15.8.1 Oxidation

As the name implies, the term oxidation was originally used just to describe *the addition of oxygen to a substance*. Thus all examples of burning are oxidation:

$C(s) \quad + O_2(g) \rightarrow CO_2(g)$

$2H_2S(g) \quad + 3O_2(g) \rightarrow 2H_2O(g) + 2SO_2(g)$

hydrogen sulphide

In the first example carbon is *oxidized* to carbon dioxide, while in the second hydrogen sulphide is oxidized to water and sulphur dioxide. Oxygen is called an *oxidizing agent*.

Oxidation by addition of oxygen is also effected by certain compounds which contain oxygen. For example, when carbon is heated with concentrated sulphuric acid it is oxidized to carbon dioxide:

$$C(s) + 2H_2SO_4(l) \rightarrow CO_2(g) + 2SO_2(g) + 2H_2O(l)$$

Here sulphuric acid supplies the oxygen, and is therefore the oxidizing agent.

Since hydrogen was in many ways regarded as a chemical opposite to oxygen, the definition of oxidation was later extended to include *the removal of hydrogen from a compound*. Therefore the following equations also illustrate oxidation:

(*a*) $Cl_2(g) + H_2S(g) \rightarrow 2HCl(g) + S(s)$

Here chlorine, the oxidizing agent, has removed hydrogen from hydrogen sulphide, thereby oxidizing it to sulphur.

(*b*) $MnO_2(s) \quad + 4HCl(aq) \rightarrow MnCl_2(aq) + 2H_2O(l) + Cl_2(g)$

manganese(IV) oxide

In this reaction, used in the preparation of chlorine, manganese(IV) oxide oxidizes hydrochloric acid to chlorine.

15.8.2 Reduction.

Reduction is in all respects the opposite to oxidation. Thus *reduction is the removal of oxygen from, or the addition of hydrogen to, a substance*:

e.g. (*a*) $CuO(s) + H_2(g) \rightarrow Cu(s) + H_2O(g)$

Here the copper(II) oxide is *reduced* to copper, hydrogen being the *reducing agent*.

(*b*) $H_2(g) + Cl_2(g) \rightarrow 2HCl(g)$

In this reaction chlorine is reduced to hydrogen chloride.

15.8.3 Electron transfer, oxidation and reduction

Oxygen is an element which gives negative ions. In this respect it is similar to other non-metals and non-metallic radicals. Hydrogen, in giving positive ions, is similar to metals and other basic radicals. Because of these similarities, *an increase in the proportion of negative ions in a substance or a decrease in the proportion of positive ions in a substance is now called oxidation*. It follows that *a decrease in the proportion of negative ions in a substance or an increase in the proportion of positive ions in a substance is reduction.*

This means that the conversion of an iron(II) salt to an iron(III) salt, e.g. by bubbling chlorine through iron(II) chloride solution, is oxidation:

$$\underset{\text{iron(II) chloride}}{2FeCl_2(aq)} + Cl_2(g) \rightarrow \underset{\text{iron(III) chloride}}{2FeCl_3(aq)}$$

The iron(III) chloride contains a higher proportion of negative ions than does iron(II) chloride. Notice that such an oxidation involves an increase in the valency state of the metal concerned. In this case the valency of iron in iron(II) chloride is two and in iron(III) chloride three.

The valency state of an element or ion in an electrovalent compound depends upon the number of electrons it contains, and therefore it is often convenient to consider oxidation and reduction in terms of electron transfer from one atom or ion to another. Thus if we write the ionic equation for the oxidation of iron(II) chloride we get

$$2Fe^{2+}(aq) + 4Cl^-(aq) + Cl_2(g) \rightarrow 2Fe^{3+}(aq) + 6Cl^-(aq)$$

or $2Fe^{2+}(aq) + Cl_2(g) \rightarrow 2Fe^{3+}(aq) + 2Cl^-(aq)$

In being oxidized each iron(II) ion has lost one electron, i.e. *oxidation is the loss of electrons by an atom or ion.* The electrons have been taken by the chlorine, each atom of which has gained one electron and therefore been reduced, i.e. *reduction is the gain of electrons by an atom or ion.*

In reactions of this type electrons are transferred from one species—the species which is said to be oxidized—to another species—the species which is reduced. The whole reaction consists of two parts:

(i) $2Fe^{2+} - 2e \rightarrow 2Fe^{3+}$ This is oxidation.
(ii) $Cl_2 + 2e \rightarrow 2Cl^-$ This is reduction.

and is sometimes called a *redox* (*red*uction-*ox*idation) reaction. Many reactions may be regarded in this light, e.g.

(a) The oxidation of magnesium by burning

$2Mg(s) + O_2(g) \rightarrow 2MgO(s)$

The magnesium is oxidized by electron loss:

$Mg - 2e \rightarrow Mg^{2+}$

The oxygen is reduced by electron gain:

$O_2 + 4e \rightarrow 2O^{2-}$

An ionic equation for the whole reaction is:

$2Mg + O_2 \rightarrow 2Mg^{2+} + 2O^{2-}$

oxidation

reduction

(b) The action of chlorine on potassium iodide solution
In this reaction iodine is liberated:

$2KI(aq) + Cl_2(g) \rightarrow 2KCl(aq) + I_2(aq)$

Iodide ions are oxidized by electron loss:

$2I^- - 2e \rightarrow I_2$

Chlorine is reduced by electron gain:

$Cl_2 + 2e \rightarrow 2Cl^-$

The ionic equation for the whole reaction is:

$$Cl_2 + 2I^- \rightarrow 2Cl^- + I_2$$

reduction

oxidation

(c) The reduction of an iron(III) salt to an iron(II) salt

This reduction occurs when zinc and hydrochloric acid are added to a solution of an iron(III) salt. The iron(III) ions are reduced by electron gain:

$Fe^{3+} + e \rightarrow Fe^{2+}$

The zinc atoms are oxidized by electron loss:

$Zn - 2e \rightarrow Zn^{2+}$

The ionic equation for the whole reaction is:

$$Zn + 2Fe^{3+} \rightarrow Zn^{2+} + 2Fe^{2+}$$

oxidation

reduction

(d) The oxidation of many metals by hydrogen ions

This occurs when a dilute acid is added to metals above hydrogen in the electrochemical series, e.g. when dilute hydrochloric acid is added to zinc. The metal is oxidized by electron loss:

$Zn - 2e \rightarrow Zn^{2+}$

Hydrogen ions are reduced by electron gain:

$2H^+ + 2e \rightarrow H_2$

The ionic equation for the whole reaction is:

$$Zn + 2H^+ \rightarrow Zn^{2+} + H_2$$

oxidation

reduction

(e) The oxidation of an iron(II) salt by acidified potassium manganate(VII) solution

This reaction is rather more complex. Iron(II) ions are oxidized by electron loss:

$$Fe^{2+} - e \rightarrow Fe^{3+}$$

Manganate(VII) ions in acid solution are reduced thus:

$$MnO_4^- + 8H^+ + 5e \rightarrow Mn^{2+} + 4H_2O$$

The ionic equation for the whole reaction is:

$$5Fe^{2+} + MnO_4^- + 8H^+ \rightarrow 5Fe^{3+} + Mn^{2+} + 4H_2O$$

oxidation

reduction

15.8.4 Oxidation numbers

(a) Definition

Any atom which is not combined with an atom of another element is said to have an *oxidation number* (or *oxidation state*) of zero (0).

A simple (i.e. containing one element only) positive ion which is singly charged (i.e. an atom with one electron removed, e.g. Na^+, K^+) has a valency of 1 and is said to have an *oxidation number of* $+1$. Similarly a doubly charged, simple, positive ion (e.g. Ca^{2+}, Mg^{2+}) has a valency of 2 and an *oxidation number of* $+2$. *All simple positive ions have positive oxidation numbers equal to the charge on the ion.* In naming the ion this number is placed directly after the name, e.g. calcium(II), iron(II), iron(III) etc. . .

Simple negative ions (e.g. Cl^-, O^{2-}*) have negative oxidation numbers corresponding to the charge on their ions.* Hence Cl^- has an oxidation number of -1 and O^{2-} of -2.

For complex ions, e.g. SO_4^{2-}, each atom is given an oxidation number and the algebraic sum of these numbers gives the charge on the ion.

Thus for SO_4^{2-}: oxidation number of sulphur + (4 × Oxidation number of oxygen) = -2
∴ oxidation number of sulphur + $(4 \times -2) = -2$
oxidation number of sulphur = $-2 + 8 = +6$

This ion is therefore called sulphate(VI).

(*Note*: The oxidation number of oxygen is always -2; the number after the name is the oxidation number of the other element in the ion.)

Consider the SO_3^{2-} ion (commonly called the sulphite ion.) Here the oxidation number of sulphur = $-2-(3 \times -2) = -2+6 = +4$.
This ion is strictly speaking called sulphate(IV).

Other common complex ions are as follows:

(i) NO_3^- Oxidation number of nitrogen = $-1-(-6) = +5$
Name of ion is nitrate(V).

(ii) MnO_4^- Oxidation number of manganese = $-1-(-8) = +7$
Name of ion is manganate(VII).

(iii) NO_2^- (commonly called nitrite) Oxidation number of nitrogen = $-1-(-4) = +3$
Name of ion is nitrate(III).

(iv) $Cr_2O_7^{2-}$ Oxidation number of chromium $= \dfrac{-2-(-14)}{2} = +6$
Name of ion is dichromate(VI).

In this text-book oxidation numbers are omitted when compounds are named except where their omission would cause doubt. Thus we call Na^+Cl^- sodium chloride and $Fe^{3+}(OH^-)_3$ iron(III) hydroxide.

(b) Oxidation, reduction and oxidation numbers

Oxidation is the loss of electrons by an atom or ion.

∴ *An atom or ion that is oxidized increases its oxidation number.*

(e.g. (i) $Fe^{2+} \rightarrow Fe^{3+}$; oxidation number increases from +2 to +3.

(ii) $2I^- \rightarrow I_2$; oxidation number of each iodine atom increases from −1 to 0.)

Similarly reduction is the gain of electrons by an atom or ion.

∴ *An atom or ion that is reduced decreases its oxidation number.*

(e.g. (i) $MnO_4^- \rightarrow Mn^{2+}$; oxidation number of manganese decreases from +7 to +2.

(ii) $2H^+ \rightarrow H_2$; oxidation number of each hydrogen atom decreases from +1 to 0.)

15.9 Hydrogen peroxide

15.9.1 Preparation

Hydrogen peroxide is formed by the action of a cold dilute acid on a peroxide. Low temperatures are essential, because hydrogen peroxide decomposes when it is warmed.

The most usual reagents are dilute sulphuric acid and barium peroxide, since the insoluble barium sulphate which is also formed can be removed by filtration:

$$BaO_2(s) + H_2SO_4(aq) \rightarrow BaSO_4(s) + H_2O_2(aq)$$

Barium peroxide is itself insoluble, and unless it is finely powdered, an insoluble layer of barium sulphate quickly forms around the peroxide particles, preventing further reaction. Therefore, hydrated barium peroxide is finely powdered and mixed to a paste with water. This paste is added, a little at a time, to ice-cold dilute sulphuric acid. The mixture is well stirred, and addition is continued until the solution is only just acid to litmus. Hydrogen peroxide solution is then obtained by filtration.

15.9.2 Properties of hydrogen peroxide

Pure hydrogen peroxide is a syrupy liquid, which is liable to decompose explosively. A solution of hydrogen peroxide in water is used for all the following reactions.

15.9.3 Decomposition of hydrogen peroxide

(*a*) A solution of hydrogen peroxide is warmed in a test-tube. A gas is evolved which re-ignites a glowing splint, and is therefore oxygen:

$$2H_2O_2(aq) \rightarrow 2H_2O(l) + O_2(g)$$

This decomposition takes place slowly at room temperature, and the gas above hydrogen peroxide kept in a bottle usually responds to this test. Why do you think that on bottles of hydrogen peroxide sold for use in the home there is usually a warning to open the bottle cautiously?

(*b*) Some manganese(IV) oxide is added to hydrogen peroxide solution in a test-tube. It effervesces freely, yielding oxygen. Manganese(IV) oxide catalyses the decomposition of the hydrogen peroxide (see section 10.4.1).

15.9.4 Hydrogen peroxide as an oxidising agent

In its oxidizing actions, hydrogen peroxide loses oxygen and is converted to water. In reactions involving electron transfer it normally oxidizes in acidic solutions by gaining electrons (i.e. the hydrogen peroxide is reduced) thus:

(*a*) Hydrogen peroxide is added to a solution of potassium iodide acidified with dilute sulphuric acid. Iodine is slowly liberated:

$$2KI(aq) + H_2SO_4(aq) + H_2O_2(aq) \rightarrow K_2SO_4(aq) + 2H_2O(l) + I_2(aq)$$

i.e. the iodide ions are oxidized by losing electrons to the hydrogen peroxide:

$$2I^- - 2e \rightarrow I_2$$

$$H_2O_2 + 2H^+ + 2e \rightarrow 2H_2O$$

Adding these equations together gives the ionic equation:

$$2I^-(aq) + H_2O_2(aq) + 2H^+(aq) \rightarrow 2H_2O(l) + I_2(aq)$$

(*b*) A filter paper is dipped in lead(II) nitrate solution and then exposed to hydrogen sulphide. The filter paper, which becomes coated with lead(II) sulphide, is then dipped in hydrogen peroxide solution. The lead(II) sulphide is thereby converted to white lead(II) sulphate:

$$\underset{\text{black}}{PbS} + 4H_2O_2 \rightarrow \underset{\text{white}}{PbSO_4} + 4H_2O$$

or $S^{2-}(s) + 4H_2O_2(aq) \rightarrow SO_4{}^{2-}(s) + 4\,H_2O\,(l)$

This reaction has an interesting application. Lead paints, when exposed to atmosphere containing hydrogen sulphide, blacken as the lead(II) compounds present are converted into lead(II) sulphide. Treatment with hydrogen peroxide oxidizes the lead(II) sulphide to lead(II) sulphate, and the paint is therefore renovated.

15.9.5 Hydrogen peroxide as a reducing agent

Hydrogen peroxide acts as a reducing agent, (i.e. the hydrogen peroxide is oxidized) when it reacts with substances that are strong oxidizing agents. In these reactions oxygen gas is evolved.

(*a*) When hydrogen peroxide is added to a solution of potassium manganate(VII) which has been acidified with dilute sulphuric acid, the purple potassium manganate(VII) is decolourised and oxygen gas is evolved. The manganate(VII) ion reacts as in section 15.8.3(e):

$$MnO_4{}^- + 8H^+ + 5e \rightarrow Mn^{2+} + 4H_2O$$

The hydrogen peroxide reacts to produce oxygen:

$$H_2O_2 \rightarrow 2H^+ + O_2 + 2e$$

The ionic equation for the whole reaction is:

$$2MnO_4{}^-(aq) + 6H^+(aq) + 5H_2O_2(aq) \rightarrow 2Mn^{2+}(aq) + 8H_2O(l) + 5O_2(g).$$

(*b*) A similar reaction occurs when acidified potassium dichromate(VI) solution is treated with hydrogen peroxide solution. The orange potassium dichromate(VI) is reduced to green chromium(III) sulphate.

15.9.6 Uses of hydrogen peroxide

Hydrogen peroxide is usually sold as "10-volume" or "20-volume" solutions. The "volume strength" of hydrogen peroxide is the number of volumes of oxygen, measured at s.t.p., given by one volume of hydrogen peroxide on heating. Thus 1 dm^3 of "10-volume" hydrogen peroxide will give 10 dm^3 of oxygen at s.t.p.

From the equation $2H_2O_2 \rightarrow 2H_2O + O_2$ it can be seen that 2 moles of hydrogen peroxide give 1 mole of oxygen, i.e. 68 g of hydrogen peroxide give 22·4 dm^3 of oxygen at s.t.p. It is therefore possible to calculate the volume strength of hydrogen peroxide if its molarity is known, and vice versa.

A dilute solution of hydrogen peroxide is an excellent antiseptic, and is therefore used as a gargle or mouthwash, and in cleaning infected wounds. Note: Concentrated hydrogen peroxide can cause nasty burns. It is also used in bleaching delicate materials which would be harmed by other bleaching agents. Hydrogen peroxide is used as a source of oxygen for burning alcohol, hydrocarbons or hydrazine in rockets.

Questions on chapter 15

1. Give, with an equation, a fully labelled diagram to show how you would prepare and collect gas jars full of oxygen.

How, and under what conditions, does oxygen react with (*a*) sulphur dioxide, (*b*) carbon monoxide, (*c*) lead(II) oxide, (*d*) nitrogen oxide? (*J.M.B.*)

2. In the laboratory, oxygen is obtained from hydrogen peroxide, using manganese(IV) oxide as a *catalyst*.

(i) Define the term *catalyst*. Describe how you would confirm that this substance does behave as a catalyst in this reaction.

(ii) With the aid of a diagram, describe how you would measure the volume of oxygen evolved in a given time.

(iii) Write down the equation for the reaction.

(iv) Calculate the volume of oxygen at s.t.p. which theoretically could be obtained from 50 cm^3 of a solution of hydrogen peroxide which contains 68 g dm^{-3}.

(One mole of a gas occupies 22·4 dm^3 at s.t.p.). (*O. and C.*)

3. Describe the usual laboratory preparation and collection of oxygen. Give and explain a commercial use of this gas. Describe what happens and say what is formed when the following substances are burnt in oxygen: (*a*) iron filings; (*b*) magnesium; (*c*) carbon; (*d*) sulphur.

(*C.*)

4. Describe briefly the preparation of oxygen from liquid air.

How does (*a*) phosphorus, (*b*) magnesium react with oxygen?

What tests would you use to prove that one product was an acidic and the other was a basic oxide? (*C.*)

5. Describe what you would *see* when the following compounds are heated in air: (*a*) mercury(II) oxide; (*b*) sodium nitrate; (*c*) lead(II) nitrate; (*d*) dilead(II) lead(IV) oxide. Give an equation for each reaction, and state how you would test for the evolution of oxygen.

6. Name *five* different types of oxide and give *one* example of each.

Describe briefly how you would convert a metal into its oxide. (No marks will be given for burning or heating a metal in air or oxygen) *(J.M.B.)*

7. Hydrogen peroxide is a weak, dibasic acid, miscible with water.
(*a*) Explain the terms *weak* and *dibasic*.
(*b*) Write down the empirical formula of this compound.
(*c*) Suggest the formula of calcium peroxide, and give the equation for its reaction with hydrochloric acid.
(*d*) Explain why hydrogen peroxide bleaches hair.
(*e*) Describe an experiment for obtaining and collecting a sample of oxygen from hydrogen peroxide. *(O. and C.)*

8. (*a*) Give different examples, *one* in each case of; (i) a reduction involving a loss of oxygen; (ii) an oxidation involving a loss of hydrogen; (iii) an oxidation involving a change in valency of an element; (iv) a reduction involving a loss of a non-metallic element other than oxygen. (An equation is required for each reaction.)
(*b*) A white powder is known to be either an oxidising agent or a reducing agent. Describe *two* tests that would enable you to find out which it is. *(C.)*

9. What is meant by the terms *oxidation* and *reduction*?
For each of the following substances say whether it usually behaves as an oxidising or as a reducing agent, and describe one reaction in which it is acting according to your classification: (*a*) hydrogen, (*b*) manganese(IV) oxide, (*c*) hydrogen sulphide, (*d*) nitric acid. *(L.)*

10. Define *oxidation*.
Which of the following can act as oxidizing agents and which as reducing agents; hydrogen, dilead(II) lead(IV) oxide, chlorine, carbon monoxide, nitric acid? Give *five* different reactions, *one* for each substance (with equations), to illustrate your answer. *(C.)*

11. How would you prepare a dilute solution of hydrogen peroxide?
Describe what you observe when hydrogen peroxide solution is (*a*) boiled; (*b*) mixed with an acidified solution of potassium permanganate; (*c*) treated with solid manganese(IV) oxide. Describe *one* test not already mentioned in the question, which will distinguish hydrogen peroxide from other oxidizing agents. *(O.)*

12. Define "oxidation" and "reduction" in terms of electron transfer.
Write ionic equations for the reactions between the following pairs of substances: (*a*) magnesium and dilute sulphuric acid; (*b*) iron(III) chloride solution and hydrogen sulphide; (*c*) chlorine and sodium iodide solution; (*d*) zinc and copper(II) sulphate solution; (*e*) sodium and chlorine. In each case state which species is oxidized and which is reduced, giving your reasons.

13. State what you would see and explain the reactions which occur when an excess of hydrogen peroxide reacts with:
(i) a solution of potassium iodide in dilute hydrochloric acid;
(ii) a solution of iron(II) sulphate in dilute sulphuric acid;
(iii) lead(II) sulphide.
Describe what you would see, and state whether hydrogen peroxide is acting as an acid, oxidizing agent or reducing agent in each of the following:
(iv) the reaction with silver(I) oxide,

$$Ag_2O + H_2O_2 \rightarrow 2Ag + H_2O + O_2$$

(v) the reaction with barium hydroxide solution,

$$Ba(OH)_2 + H_2O_2 \rightarrow BaO_2\downarrow + 2H_2O$$

(J.M.B.)

16

Nitrogen and Phosphorus

16.1 Introduction

Nitrogen and phosphorus are the first two elements in Group V (nitrogen, phosphorus, arsenic, antimony and bismuth). Both are non-metals with atoms containing five electrons in their outer shells. Usually they combine covalently with other non-metals but they also form compounds with some metals. These compounds react as if they contain the N^{3-} or P^{3-} ions. Although nitrogen and phosphorus are in the same group, there is little obvious similarity between them.

16.2 Nitrogen

Nitrogen is familiar to us all as the most abundant constituent of the atmosphere. It is also present in various minerals, notably *Chile saltpetre*, sodium nitrate, of which there are large deposits on the Pacific coast of South America.

All living material contains nitrogen, and its important role in nature is discussed later in the chapter.

16.2.1 Preparation of nitrogen

The most satisfactory way of obtaining nitrogen for use in the laboratory is from a nitrogen cylinder. It may however be prepared by the following methods:

(a) From air

Air is passed slowly through a solution of sodium hydroxide, and then through a strongly heated iron or silica tube, which is well packed with copper turnings (figure 16.1).

The sodium hydroxide combines with any carbon dioxide present, and the heated copper with oxygen:

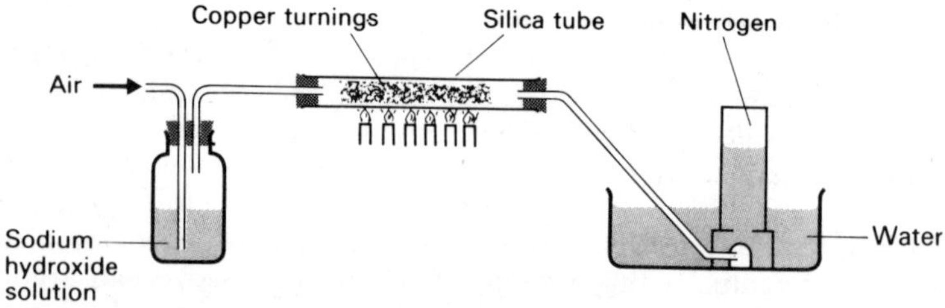

Figure 16.1 Preparation of nitrogen from air.

$$2NaOH(aq) + CO_2(g) \rightarrow Na_2CO_3(aq) + H_2O(l)$$
$$2Cu(s) + O_2(g) \rightarrow 2CuO(s)$$

The resulting gas, which is mainly nitrogen, is collected over water.

(b) From ammonium nitrite

Ammonium nitrite readily decomposes to give nitrogen and water:

$$NH_4NO_2(aq) \rightarrow 2H_2O(l) + N_2(g)$$

As this decomposition takes place slowly at room temperature, ammonium nitrite is not a chemical to be found in a bottle. Normally therefore, a mixture of sodium nitrite and ammonium chloride in dilute solution is gently heated. The ammonium and nitrite ions which are present then decompose to give the required nitrogen:

$$NH_4^+(aq) + NO_2^-(aq) \rightarrow N_2(g) + 2H_2O(l)$$

(c) By other methods

Numerous other reactions yield nitrogen as a product, e.g.

1. The reduction of an oxide of nitrogen by passage over heated copper:

$$2NO(g) + 2Cu(s) \rightarrow 2CuO(s) + N_2(g)$$
nitrogen
oxide

2. The oxidation of ammonia by heated copper(II) oxide:

$$2NH_3(g) + 3CuO(s) \rightarrow 3Cu(s) + 3H_2O(g) + N_2(g)$$

16.2.2 Manufacture of nitrogen

Nearly all nitrogen is manufactured by the fractional distillation of liquid air (section 13.4).

16.2.3 Properties of nitrogen

(a) Nitrogen is a diatomic gas.

It is colourless, odourless, almost insoluble in water and is liquefied only with difficulty.

(b) Nitrogen is slightly less dense than air.

Lord Rayleigh in 1894 performed a series of experiments to find, with great accuracy, the density of nitrogen. He used samples of nitrogen obtained from the air and also samples obtained by chemical methods. The gas was most carefully purified, and he estimated that the experimental error in his results would be not greater than 1 part in 10 000. To begin with, therefore, he was at a loss to explain why "atmospheric nitrogen" was 5 parts in 10 000 denser than nitrogen obtained from its compounds. Eventually this discrepancy led to the discovery of argon and other noble gases. These are present to a small extent in air and are not readily removed.

(c) Nitrogen is noted for its chemical inactivity.
Under normal conditions it does not enter into any chemical reactions and there is no simple positive test for it.
Thus, it does not burn, nor does it support burning.
It does, however, combine with oxygen at very high temperatures, to give a small yield of nitrogen oxide:

$$N_2(g) + O_2(g) \rightleftharpoons 2NO(g)$$

Formerly this reaction was the basis of a method for the manufacture of nitric acid.

Ammonia is formed when nitrogen combines with hydrogen, at a moderate temperature and high pressure, in the presence of a catalyst. (See manufacture of ammonia (section 16.3.2.)).

$$N_2(g) + 3H_2(g) \rightleftharpoons 2NH_3(g)$$

Some red-hot metals (e.g. magnesium or calcium) combine directly with nitrogen to form nitrides:

$$\text{e.g.}\quad 3Mg(s) + N_2(g) \rightarrow \underset{\text{magnesium nitride}}{Mg_3N_2(s)}$$

The formation of nitride can be shown by burning some magnesium or calcium in air. Ammonia, detected by its smell, is evolved when water is added to the ash:

$$Mg_3N_2(s) + 6H_2O(l) \rightarrow 3Mg(OH)_2(s) + 2NH_3(g)$$

16.2.4 Uses of nitrogen

The most important use of nitrogen is in the manufacture of ammonia. The conversion of atmospheric nitrogen into compounds of this sort which can be used as fertilisers or in making fertilisers is a most important development of the present century (figure 16.2). Otherwise the intensive cultivation of soil necessary for a civilised society would have resulted in a gradual depletion of nitrogen compounds in the soil, and a resultant food shortage more serious than any now existing (see chapter 28).

Liquid nitrogen (b. pt. $-196\,^\circ C$) is used as a coolant when low temperatures are required (figure 16.3).

16.3 Ammonia

16.3.1 Preparation of ammonia

Ammonia is prepared by heating an ammonium salt with an alkali:

$$\text{e.g.}\quad \left[\begin{array}{lll}(NH_4)_2SO_4 + 2NaOH & \rightarrow & Na_2SO_4 + 2H_2O + 2NH_3 \\ NH_4^+(aq) + OH^-(aq) & \rightarrow & H_2O(l) + NH_3(g)\end{array}\right.$$

$$\text{or}\quad 2NH_4Cl + Ca(OH)_2 \rightarrow CaCl_2 + 2H_2O + 2NH_3$$

With the caustic alkalies, solutions must be used, but it is more usual in the laboratory to use a mixture of solid calcium hydroxide and solid ammonium chloride.

Ammonia is less dense than air and is collected by upwards delivery (figure 16.4).

Figure 16.2 The inside of a warehouse for storing fertiliser at Billingham, Cleveland county. (Photo: I.C.I. Agricultural Division).

Figure 16.3 Transferring liquid nitrogen from a large storage container (in the van) to smaller portable containers. (Photo: British Oxygen Co. Ltd.).

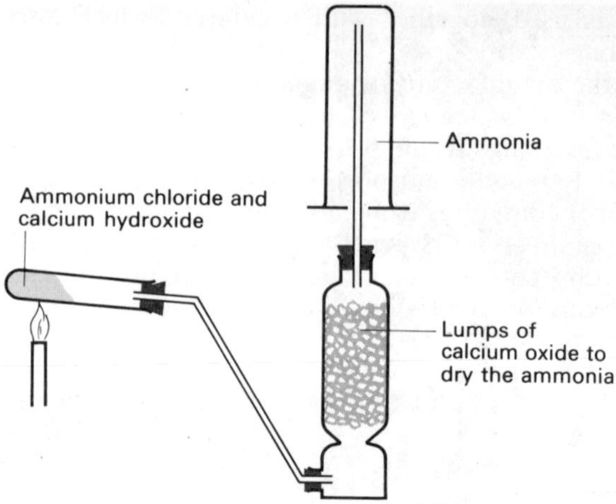

Figure 16.4 Preparation of ammonia.

Notice two things:

(*a*) The tube containing the reagents is tilted downwards. This is to prevent any water which has condensed from trickling back on to the hot glass and cracking it.

(*b*) Calcium oxide is used as the drying agent. Neither concentrated sulphuric acid nor calcium chloride is suitable, as both react with ammonia. The sulphuric acid would combine vigorously to give ammonium sulphate:

$$2NH_3(g) + H_2SO_4(l) \rightarrow (NH_4)_2SO_4(s)$$

Calcium chloride would react to give an addition compound in much the same way as it does with water.

$$CaCl_2(s) + 8NH_3(g) \rightarrow CaCl_2.8NH_3(s)$$

16.3.2 Manufacture of ammonia

Almost all ammonia is now manufactured synthetically by the direct combination of nitrogen and hydrogen using a catalyst to speed up the reaction (see section 10.4) (figure 16.5). This reaction was discovered by a German chemist, Fritz Haber in 1908, and is called the Haber process:

$$N_2 + 3H_2 \rightleftharpoons 2NH_3 \qquad \Delta H = -92\,\mathrm{kJ\,mol^{-1}}$$

The equation shows that 1 molecule of nitrogen combines with 3 molecules of hydrogen to give 2 molecules of ammonia. By Avogadro's Law, therefore, 4 volumes of the reactants give 2 volumes of ammonia. In reversible reactions like this, where there is a decrease in volume, *high pressures give the better yield.*

The equation also indicates that the reaction is *exothermic*, i.e. ammonia is formed with evolution of heat. With reversible reactions which are exothermic, *low temperatures give the better yield.*

At low temperatures, however, chemical reactions proceed slowly, although the yield finally obtained might be high. In practice, therefore, a moderate temperature of about 500 °C is used, together with a catalyst, which also increases the rate of the reaction.

For the Haber process for the manufacture of ammonia, the hydrogen is obtained by the Bosch process from water gas or from natural gas (section 14.3.2), and the nitrogen by the fractional distillation of liquid air. A mixture of the two gases is compressed to 100–300 atmospheres (figure 16.6), and then passed through a catalyst chamber containing iron and molybdenum at 500 °C. The issuing gases, which may contain about 10 per cent of ammonia, are passed through a spiral tube surrounded by cooling water. Under the high pressure the ammonia then liquefies, and can be separated. Unchanged nitrogen and hydrogen are used again.

Figure 16.5 General view of the ammonia works at Billingham, Cleveland County, where ammonia and its compounds are made. (Photo: I.C.I. Agricultural Division).

Figure 16.6 Ammonia compressors operating at a pressure of 250 atmospheres. (Photo: I.C.I. Agricultural Division).

16.3.3 Properties of ammonia

(a) Structure

Ammonia is a gas, and therefore a covalent compound with the structure as illustrated in figure 6.15(*b*).

It is colourless, liquefies at −33·5 °C at normal pressure and is just over half as dense as air. *Its characteristic pungent odour is often sufficient for its identification.*

(b) Ammonia is very soluble in water to give an alkaline solution.

This property is most convincingly shown by the fountain experiment (figure 16.7).

Experiment. A fountain in a flask

A round-bottomed flask is fitted with a rubber bung, through which passes a glass tube drawn out to a jet as shown. This flask is completely filled with ammonia by the normal method, but it is advisable to pass the gas in for some time after the smell of escaping ammonia becomes apparent. The bung and jet are placed in position and the flask is placed with its mouth under water coloured with red litmus. Water rises slowly up the glass tube and eventually trickles over the top of the jet. A large surface area of the liquid is then suddenly exposed to the ammonia, which rapidly dissolves. The resultant decrease of pressure inside the flask is so great that air pressure forces liquid through the jet in the form of a fountain. (If a flat-bottomed flask is used the air pressure is liable to break it.) At the same time the red litmus turns blue in colour, showing that when ammonia dissolves in water an alkali, actually ammonium hydroxide, is formed:

$$NH_3 + H_2O \rightleftharpoons NH_4OH$$

The water takes some time to rise up the glass tubing, but the experiment can be speeded up by using a tube which is wet on the inside, or by cooling the flask when it is in position by dropping a little of a volatile liquid on its outer surface.

(c) Ammonia combines with acids to form salts.
If a few drops of concentrated sulphuric acid are poured down the inside of a gas jar full of ammonia a violent reaction occurs and a white solid is formed. This is ammonium sulphate:

$$2NH_3(g) + H_2SO_4(l) \rightarrow (NH_4)_2SO_4(s)$$

The reactions with nitric and hydrochloric acids are similar:

$$\left[\begin{array}{lcl} NH_3 + HNO_3 & \rightarrow & NH_4NO_3 \\ NH_3 + HCl & \rightarrow & NH_4Cl \\ NH_3(g) + H^+(aq) & \rightarrow & NH_4^+(aq) \end{array}\right.$$

When hydrogen chloride gas, instead of its solution in water (hydrochloric

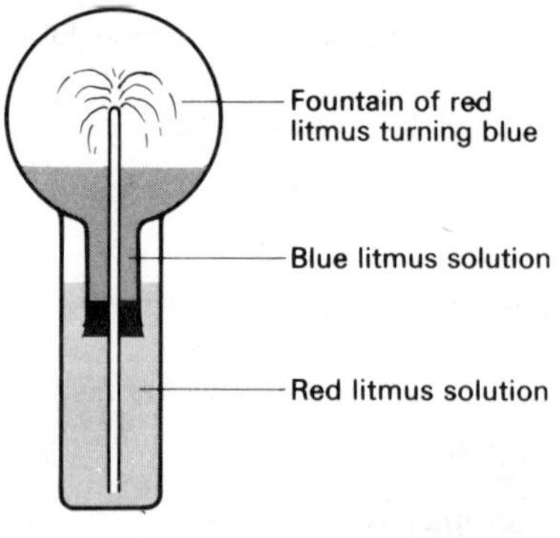

Figure 16.7 The fountain experiment.

acid) is used the ammonium chloride is formed as a dense white smoke:

$$NH_3(g) + HCl(g) \rightarrow NH_4Cl(s)$$

(d) The oxidation of ammonia

(i) Combustion of ammonia
A lighted taper is placed at the mouth of a gas jar full of ammonia. The ammonia burns, but as soon as the taper is removed the flame is extinguished. Ammonia does, however, burn readily in an atmosphere enriched with oxygen (figure 16.8):

$$4NH_3(g) + 3O_2(g) \rightarrow 2N_2(g) + 6H_2O(g)$$

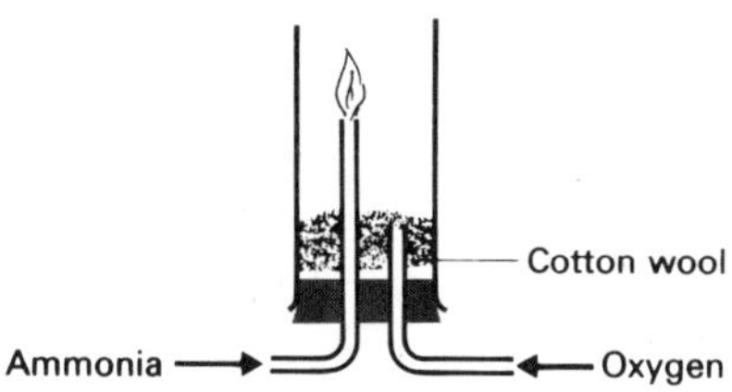

Figure 16.8 Ammonia burning in an atmosphere enriched with oxygen.

(ii) Catalytic oxidation of ammonia
When ammonia mixed with oxygen is passed over a heated platinum catalyst the ammonia is oxidized to nitrogen oxide and steam:

$$4NH_3(g) + 5O_2(g) \rightarrow 4NO(g) + 6H_2O(g)$$

Notice particularly that nitrogen oxide and *not* nitrogen is formed in the presence of a catalyst.

This reaction is used in the manufacture of nitric acid from ammonia (section 16.6.3(*b*)).

(iii) Oxidation of ammonia by copper(II) oxide
Using the apparatus shown in figure 16.9 dry ammonia is passed over heated copper(II) oxide. A colourless liquid condenses in the cooled test-tube. The usual tests prove this to be water. A colourless gas, which does not respond to any of the usual tests for gases, collects in the gas jar. It is therefore nitrogen:

$$3CuO(s) + 2NH_3(g) \rightarrow 3Cu(s) + 3H_2O(g) + N_2(g)$$

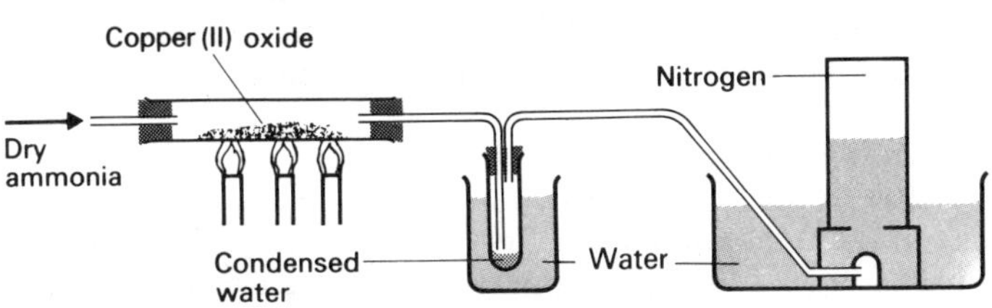

Figure 16.9 Oxidation of ammonia by copper(II) oxide.

This experiment proves that ammonia contains hydrogen and nitrogen. Does it prove that only these two elements are present?

(iv) Oxidation of ammonia by chlorine
A test-tube is fitted with a glass tube as shown in figure 16.10. Concentrated ammonia solution is placed in the test-tube and a few pellets of sodium hydroxide added to displace ammonia. When the end of the glass tube is lowered into a gas jar full of chlorine in a fume cupboard the ammonia ignites spontaneously and is oxidised to nitrogen:

$$2NH_3(g) + 3Cl_2(g) \rightarrow N_2(g) + 6HCl(g)$$

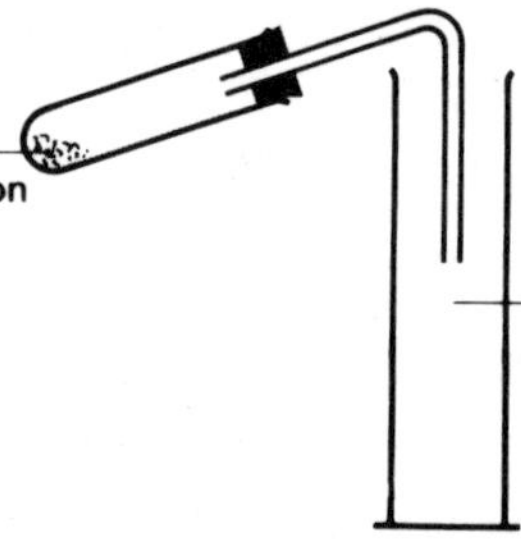

Figure 16.10 Oxidation of ammonia by chlorine.

After the chlorine has been used up the flame goes out, and dense white fumes of ammonium chloride are formed with the excess of ammonia:

$$NH_3(g) + HCl(g) \rightarrow NH_4Cl(s)$$

From the reactions listed above it is seen that the oxidation of ammonia usually yields nitrogen.

16.4 Ammonium hydroxide

16.4.1 Introduction

When ammonia combines with water, ammonium hydroxide is formed:

$$NH_3(g) + H_2O(l) \rightleftharpoons NH_4^+ + OH^-$$

A concentrated solution cannot be prepared just by passing ammonia through an ordinary tube into water, since the high solubility of ammonia would cause the water to suck back into the generating vessel as happened in the fountain experiment.

One method of preventing "sucking back" is to use an inverted funnel (figure 20.6). A better method is to use two bottles connected as in figure 16.11. The liquid is in the bottle on the right and the gas first passes through the left-hand bottle. If "sucking back" occurs the liquid moves in to the left-hand bottle. If pressure builds up in the left-hand bottle the liquid is pushed back into the right-hand bottle.

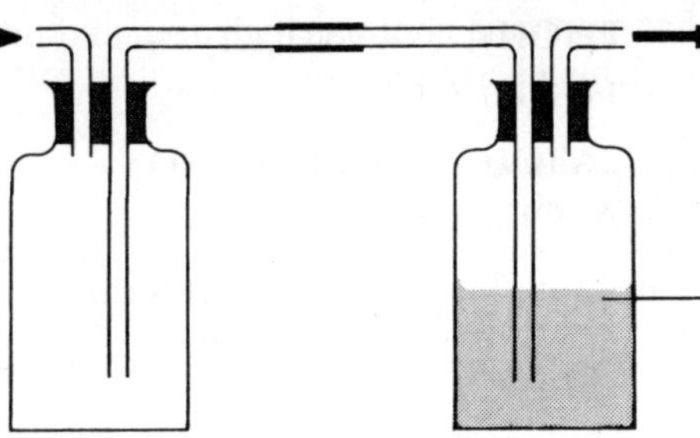

Figure 16.11 Two bottles connected for dissolving very soluble gases.

16.4.2 Properties of ammonium hydroxide

Pure ammonium hydroxide has never been isolated; it is only known in solution. The specific gravity of a saturated solution is 0·880; it is therefore called "eight-eighty ammonia".

16.4.3 Decomposition of ammonium hydroxide

A solution of ammonium hydroxide always smells of ammonia. This is because the reaction of ammonia with water is reversible:

$$NH_4OH(aq) \rightleftharpoons NH_3(g) + H_2O(l)$$

The decomposition is accelerated by heat and also by blowing air through the solution. If large quantities of ammonia are required in the laboratory they are readily obtained by warming concentrated ammonium hydroxide solution in a flask fitted with a delivery tube.

16.4.4 Ammonium hydroxide as an alkali

Ammonium hydroxide is slightly dissociated into ammonium and hydroxide ions, i.e. it is a weak alkali:

$$NH_4OH \rightleftharpoons NH_4^+ + OH^-$$

It therefore has the usual properties of alkalis:

(*a*) It turns red litmus blue;
(*b*) It neutralises acids to form ammonium salts and water only, e.g.:

$$NH_4OH + HCl \rightarrow NH_4Cl + H_2O$$
$$2NH_4OH + H_2SO_4 \rightarrow (NH_4)_2SO_4 + 2H_2O$$

The ionic equation for these reactions is the same as for all neutralizations:

$$H^+(aq) + OH^-(aq) \rightarrow H_2O(l)$$

(*c*) It precipitates the hydroxides of most metals from solutions of their salts:

$$\begin{bmatrix} FeCl_3 + 3NH_4OH \rightarrow Fe(OH)_3 \downarrow + 3NH_4Cl \\ Fe^{3+}(aq) + 3OH^-(aq) \rightarrow Fe(OH)_3(s) \end{bmatrix}$$

$$\begin{bmatrix} Al_2(SO_4)_3 + 6NH_4OH \rightarrow 2Al(OH)_3 \downarrow + 3(NH_4)_2SO_4 \\ Al^{3+}(aq) + 3OH^-(aq) \rightarrow Al(OH)_3(s) \end{bmatrix}$$

With salts of some metals the insoluble hydroxide first formed combines with excess of ammonia to form a soluble compound. Thus, when a solution of ammonium hydroxide is added to a solution of copper(II) sulphate pale blue copper(II) hydroxide is first precipitated, as with the other alkalis. If, however, an excess of ammonium hydroxide is present this precipitate dissolves to give a deep blue solution. This is because ammonia from the excess ammonium hydroxide combines with copper(II) ions to form tetra-ammine copper(II) ions:

$$\begin{bmatrix} CuSO_4 + 2NH_4OH \rightarrow Cu(OH)_2 \downarrow + (NH_4)_2SO_4 \\ Cu^{2+}(aq) + 2OH^-(aq) \rightarrow Cu(OH)_2(s) \end{bmatrix}$$

$Cu^{2+}(aq)$	$+4NH_3(aq)$	$\rightarrow$	$Cu(NH_3)_4{}^{2+}(aq)$
from copper hydroxide	from ammonium hydroxide		tetra-ammine copper(II) ion

16.4.5 Uses of ammonia and ammonium hydroxide

Ammonia is widely used in the manufacture of ammonium salts, nitric acid (see section 16.6.3) and sodium carbonate (see section 21.2.7).

In laundry work a solution of ammonia is often used to remove temporary hardness from water, and it can also be used as a grease remover.

16.4.6 Important ammonium salts

(a) *(Warning: It is easy to confuse the words "ammonia" and "ammonium").*

Ammonia is the name given to the gas of formula NH_3.

Ammonium is the name given to the group of atoms NH_4. This group of atoms has no independent existence. It is present, however, in compounds which ammonia forms with water and acids. Its valency is one, and in its compounds it behaves in many respects like the metals, sodium and potassium. Like them, its ions carry a single positive charge ($NH_4{}^+$) and all its salts are soluble.

(b) Ammonium chloride (NH_4Cl)

The action of heat on ammonium chlorde is of particular interest. It sublimes, i.e. it changes directly from the solid to the vapour state on heating. This can be shown by heating a little ammonium chloride in a test-tube. It disappears from the bottom of the tube, and the solid salt re-forms near the top of the test-tube, where it is cooler. The reason for this sublimation is unusual. When it is heated, ammonium chloride dissociates (i.e. splits up) into two simpler gaseous compounds, ammonia and hydrogen chloride:

$$NH_4Cl(s) \rightleftharpoons NH_3(g) + HCl(g)$$

When cooled, these gases recombine to form solid ammonium chloride.

This sort of decomposition is called *thermal dissociation* to distinguish it from the irreversible *thermal decomposition.*

Thermal dissociation is the splitting up of a compound by heat into simpler substances which recombine on cooling.

Thermal decomposition is the splitting up of a compound by heat into simpler substances, which do not recombine on cooling. (What examples of thermal decomposition do you know?)

It is important to realise that thermal dissociation is not the same as sublimation. It merely provides the explanation for sublimation in this particular case.

Ammonium chloride is used in dry batteries. It is also used as a flux in soldering and galvanising.

(c) Ammonium sulphate ($(NH_4)_2SO_4$)
This compound is of great importance as a fertiliser. In the laboratory it is made by neutralization of sulphuric acid. Industrially, calcium sulphate is used instead of the more expensive sulphuric acid. Ammonia solution is saturated with carbon dioxide, and the mixture is then stirred with powdered calcium sulphate, when calcium carbonate and ammonium sulphate are slowly formed:

$$CaSO_4(s) + CO_2(g) + H_2O(l) + 2NH_3(g) \rightarrow CaCO_3(s) + (NH_4)_2SO_4(aq)$$

The insoluble calcium carbonate is removed by filtration and the ammonium sulphate by crystallization of the filtrate.

On heating, ammonium sulphate decomposes into ammonia and sulphuric acid:

$$(NH_4)_2SO_4(s) \rightarrow 2NH_3(g) + H_2SO_4(l))$$

This reaction is similar to that of ammonium chloride, but no sublimation occurs, as ammonia is much more volatile than sulphuric acid and much of it escapes from the tube before the sulphuric acid volatilises.

(d) Ammonium nitrate (NH_4NO_3)
This is made by the neutralization of nitric acid with ammonia:

$$NH_3(g) + HNO_3(aq) \rightarrow NH_4NO_3(aq)$$

The action of heat on this compound is different from that on other ammonium salts. It decomposes to give dinitrogen oxide and water:

$$NH_4NO_3(s) \rightarrow N_2O(g) + 2H_2O(g)$$

It is used as a source of dinitrogen oxide and also in the manufacture of the explosives, ammonal and amatol.

16.4.7 Test for ammonium salts

All ammonium salts yield ammonia when heated with sodium hydroxide solution:

$$\left[\begin{array}{lllll} NH_4Cl & + NaOH & \rightarrow NaCl & + H_2O & + NH_3 \\ (NH_4)_2SO_4 & + 2NaOH & \rightarrow Na_2SO_4 & + 2H_2O & + 2NH_3 \\ NH_4{}^+(aq) & + OH^-(aq) & \rightarrow H_2O(l) & & + NH_3(g) \end{array}\right.$$

The ammonia evolved can be detected by smell or by its action on moist red litmus paper.

16.5 The oxides of nitrogen

Nitrogen forms three important oxides:

Dinitrogen oxide (nitrous oxide)	N_2O
Nitrogen oxide (nitric oxide)	NO
Nitrogen dioxide	NO_2

16.5.1 Dinitrogen oxide

(a) Preparation of dinitrogen oxide

Dinitrogen oxide is prepared by the action of heat on ammonium nitrate—a method first used by Humphry Davy in 1799:

$$NH_4NO_3(s) \rightarrow N_2O(g) + 2H_2O(g)$$

The solid is gently heated in a boiling tube and the gas evolved collected over water (figure 16.12).

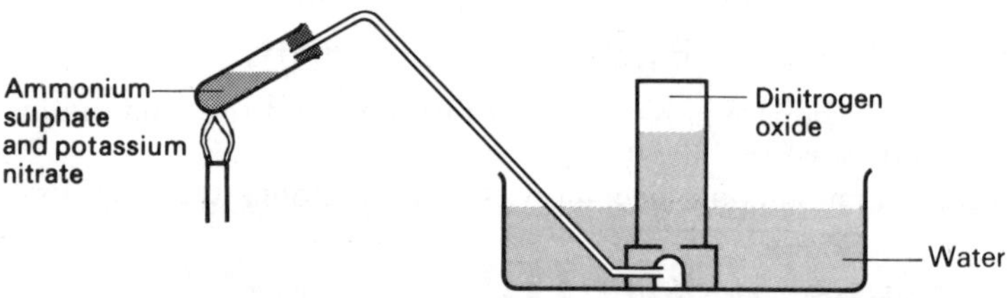

Figure 16.12 Preparation of dinitrogen oxide.

If, in this preparation, the ammonium nitrate is too vigorously heated, or is heated until very little is left in the tube, there is danger of an explosion. This risk is lessened by heating instead a mixture of another ammonium salt and a nitrate, which by double decomposition, gives ammonium nitrate:

$$(NH_4)_2SO_4(s) + 2KNO_3(s) \rightarrow K_2SO_4(s) + 2NH_4NO_3(s)$$
$$NH_4NO_3(s) \rightarrow N_2O(g) + 2H_2O(g)$$

(b) Properties of dinitrogen oxide

(i) Dinitrogen oxide is a colourless gas with a faint, sweet odour. It is denser than air. Cold water dissolves slightly less than its own volume of dinitrogen oxide to give a neutral solution.

Inhalation of dinitrogen oxide soon produces anaesthesia, which lasts for a few minutes. It is therefore used in minor operations, particularly for the extraction of teeth.

The discovery by Davy in 1799 that dinitrogen oxide was an anaesthetic was of major importance at that time. It enabled surgical operations to be performed whilst the patient was unconscious, far preferable to what happened before that time, when the best that could be done before surgery was to get the patient drunk with alcoholic liquid. Davy also discovered that when inhaled in small doses, dinitrogen oxide made people very happy, hence its name "laughing gas". This discovery was of such general interest that it was shown on the music-hall stage to demonstrate this fascinating property of "laughing gas". (figure 16.13).

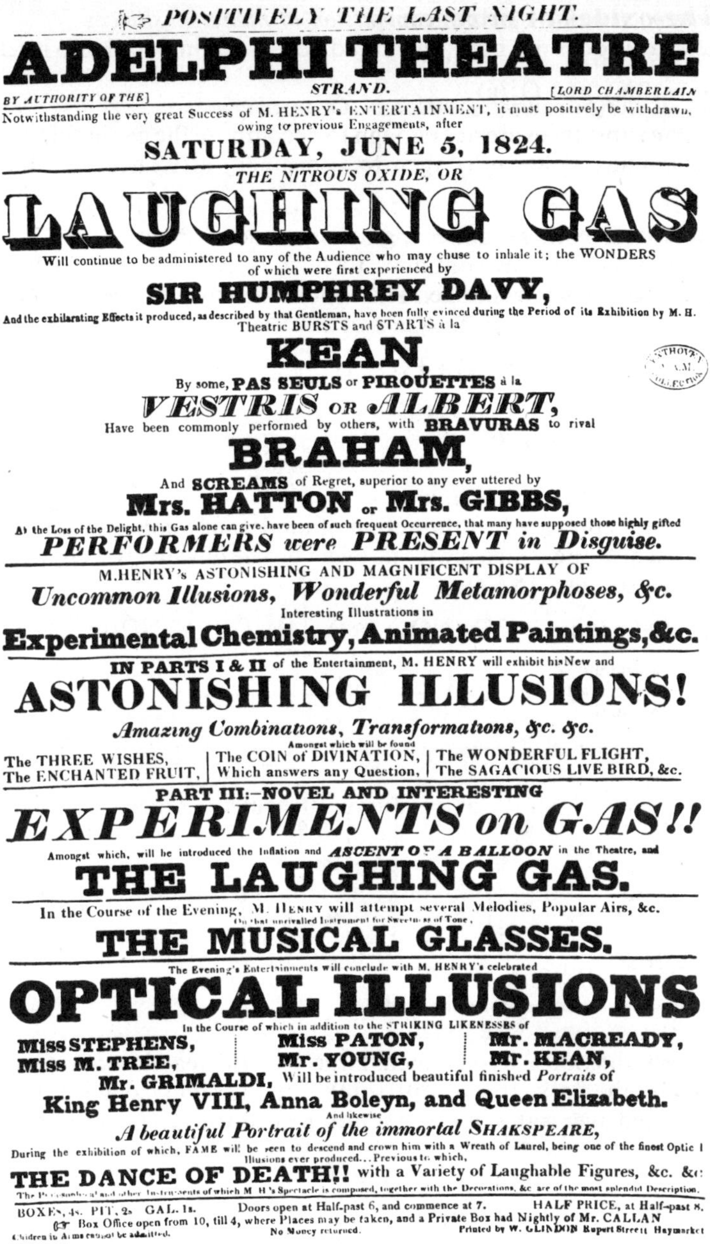

POSITIVELY THE LAST NIGHT.

ADELPHI THEATRE

BY AUTHORITY OF THE] STRAND. [LORD CHAMBERLAIN

Notwithstanding the very great Success of M. HENRY's ENTERTAINMENT, it must positively be withdrawn, owing to previous Engagements, after

SATURDAY, JUNE 5, 1824.

THE NITROUS OXIDE, OR

LAUGHING GAS

Will continue to be administered to any of the Audience who may chuse to inhale it; the WONDERS of which were first experienced by

SIR HUMPHREY DAVY,

And the exhilarating Effects it produced, as described by that Gentleman, have been fully evinced during the Period of its Exhibition by M. H. Theatric BURSTS and STARTS à la

KEAN,

By some, PAS SEULS or PIROUETTES à la

VESTRIS OR ALBERT,

Have been commonly performed by others, with BRAVURAS to rival

BRAHAM,

And SCREAMS of Regret, superior to any ever uttered by

Mrs. HATTON or Mrs. GIBBS,

At the Loss of the Delight, this Gas alone can give, have been of such frequent Occurrence, that many have supposed those highly gifted

PERFORMERS were PRESENT in Disguise.

M. HENRY's ASTONISHING AND MAGNIFICENT DISPLAY OF

Uncommon Illusions, Wonderful Metamorphoses, &c.

Interesting Illustrations in

Experimental Chemistry, Animated Paintings, &c.

IN PARTS I & II of the Entertainment, M. HENRY will exhibit his New and

ASTONISHING ILLUSIONS!

Amazing Combinations, Transformations, &c. &c.

Amongst which will be found

The THREE WISHES, | The COIN of DIVINATION, | The WONDERFUL FLIGHT,
The ENCHANTED FRUIT, | Which answers any Question, | The SAGACIOUS LIVE BIRD, &c.

PART III:—NOVEL AND INTERESTING

EXPERIMENTS on GAS!!

Amongst which, will be introduced the Inflation and ASCENT OF A BALLOON in the Theatre, and

THE LAUGHING GAS.

In the Course of the Evening, M. HENRY will attempt several Melodies, Popular Airs, &c. On that unrivalled Instrument for Sweetness of Tone,

THE MUSICAL GLASSES.

The Evening's Entertainments will conclude with M. HENRY's celebrated

OPTICAL ILLUSIONS

In the Course of which in addition to the STRIKING LIKENESSES of

Miss STEPHENS, | Miss PATON, | Mr. MACREADY,
Miss M. TREE, | Mr. YOUNG, | Mr. KEAN,

Mr. GRIMALDI, Will be introduced beautiful finished Portraits of

King Henry VIII, Anna Boleyn, and Queen Elizabeth.

And likewise

A beautiful Portrait of the immortal SHAKSPEARE,

During the exhibition of which, FAME will be seen to descend and crown him with a Wreath of Laurel, being one of the finest Optic l Illusions ever produced... Previous to which,

THE DANCE OF DEATH!! with a Variety of Laughable Figures, &c. &c.

The [illegible] and other Instruments of which M. H.'s Spectacle is composed, together with the Decorations, &c are of the most splendid Description.

BOXES, 4s. PIT, 2s GAL. 1s. Doors open at Half-past 6, and commence at 7. HALF PRICE, at Half-past 8.

Box Office open from 10, till 4, where Places may be taken, and a Private Box had Nightly of Mr. CALLAN

Children in Arms cannot be admitted. No Money returned. Printed by W. GLINDON Rupert Street Haymarket

Figure 16.13 Playbill of 1824 of a demonstration on the London Music-Hall Stage of the ability of dinitrogen oxide (nitrous oxide) to make people feel happy. What other gas do you think Mr. Henry used during this performance? (Photo: Enthoven theatre collection by permission of the trustees, the Victoria and Albert Museum).

(ii) *Dinitrogen oxide supports combustion of many substances*
Dinitrogen oxide does not burn, but at red heat is decomposed into its elements:

$$2N_2O(g) \rightarrow 2N_2(g) + O_2(g)$$

Most burning substances are hot enough to decompose the gas in this way and so produce a mixture richer in oxygen than ordinary air. They then burn more brightly, e.g.

(*a*) Dinitrogen oxide re-ignites a glowing wooden splint

(*b*) Burning phosphorus continues to burn with a very bright flame, producing clouds of white smoke. This smoke consists of the oxides of phosphorus:

$$P_4(s) + 10N_2O(g) \rightarrow 10N_2(g) + P_4O_{10}(s)$$

16.5.2 Nitrogen oxide

(a) Preparation of nitrogen oxide
Nitrogen oxide is prepared by the action of moderately concentrated nitric acid (concentrated nitric acid mixed with its own volume of water) on copper turnings (figure 16.14); see also 16.6.4(*c*)

$$3Cu(s) + 8HNO_3(aq) \rightarrow 3Cu(NO_3)_2(aq) + 4H_2O(l) + 2NO(g)$$

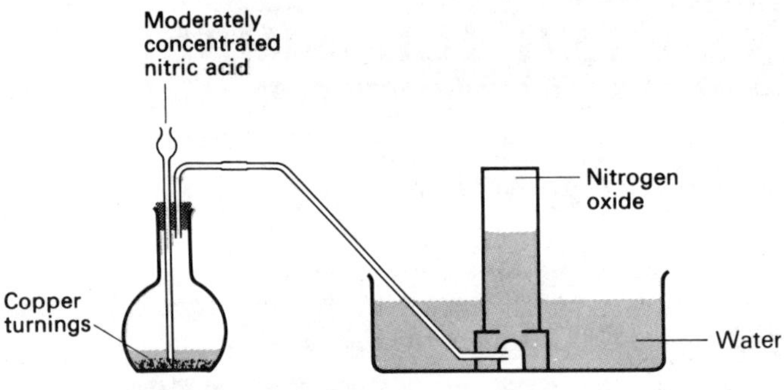

Figure 16.14 Preparation of nitrogen oxide.

Some nitrogen dioxide is also formed, largely by the reaction of nitrogen oxide with oxygen in the flask. It is readily soluble and is removed when the nitrogen oxide is collected over water. A green solution of copper(II) nitrate remains in the flask.

(b) Properties of nitrogen oxide

(i) Nitrogen oxide is a colourless gas which is slightly denser than air. It is almost insoluble in water, the solution being neutral.

(ii) *Nitrogen oxide readily combines with oxygen.* If nitrogen oxide is exposed to air or oxygen at room temperature it immediately combines to form the brown gas nitrogen dioxide:

$$2NO(g) + O_2(g) \rightleftharpoons 2NO_2(g)$$

At high temperatures this combination does not occur.

(iii) *Nitrogen oxide supports the burning of a few strongly burning substances only.* Thus strongly burning phosphorus is hot enough to decompose the gas into its elements and then combine with the oxygen:

$$P_4(s) + 10NO(g) \rightarrow P_4O_{10}(s) + 5N_2(g)$$

(iv) *Nitrogen oxide combines with iron(II) sulphate solution.* If a cold, freshly prepared solution of iron(II) sulphate solution is added to a gas jar full of nitrogen oxide, a dark brown solution is obtained. This is due to the formation of an addition compound:

$$FeSO_4(aq) + NO(g) \rightleftharpoons FeSO_4.NO(aq)$$

This compound is decomposed by heat to give the original substances, and the reaction is sometimes used to prepare pure nitrogen oxide.

16.5.3 Nitrogen dioxide

(a) The formula of this compound depends on the temperature. At low temperatures the gas consists almost entirely of N_2O_4 molecules, but at 140 °C it consists of NO_2 molecules. At intermediate temperatures a mixture of these molecules is present. For convenience the formula NO_2 is used throughout this book.

(b) Preparation of nitrogen dioxide

(i) The action of heat on a heavy metal nitrate.
Dry, powdered lead(II) nitrate is heated in a hard-glass tube. Nitrogen dioxide and oxygen are evolved, and orange lead(II) oxide is left behind:

$$2Pb(NO_3)_2(s) \rightarrow 2PbO(s) + 4NO_2(g) + O_2(g)$$

Nitrogen dioxide readily liquefies, and is separated from the oxygen by passage through a U-tube cooled by a freezing mixture of ice and salt (figure 16.15).

The condensate of nitrogen dioxide should be a pale yellow liquid, but it is often coloured green by impurities caused by the use of moist apparatus.

(ii) The action of concentrated nitric acid on copper.
When *concentrated* nitric acid is added to copper turnings, dense brown fumes of nitrogen dioxide are evolved and a green solution of copper(II) nitrate is formed (see section 16.6.4(c)):

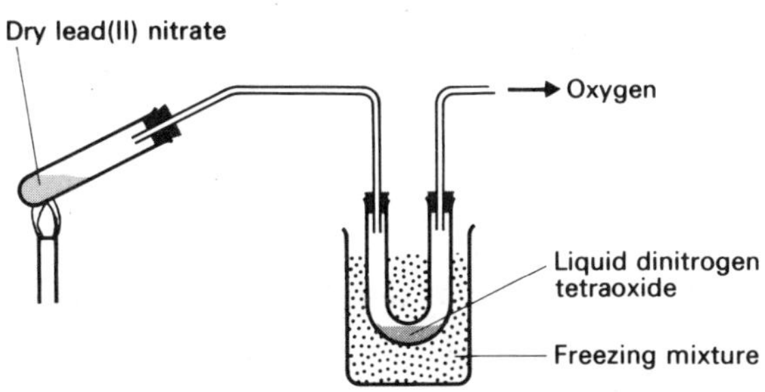

Figure 16.15 Preparation of liquid dinitrogen tetroxide.

$$Cu(s) + 4HNO_3(aq) \rightarrow Cu(NO_3)_2(aq) + 2H_2O(l) + 2NO_2(g)$$

Nitrogen dioxide is very soluble in water, and is therefore collected by upwards displacement of air (figure 16.16).

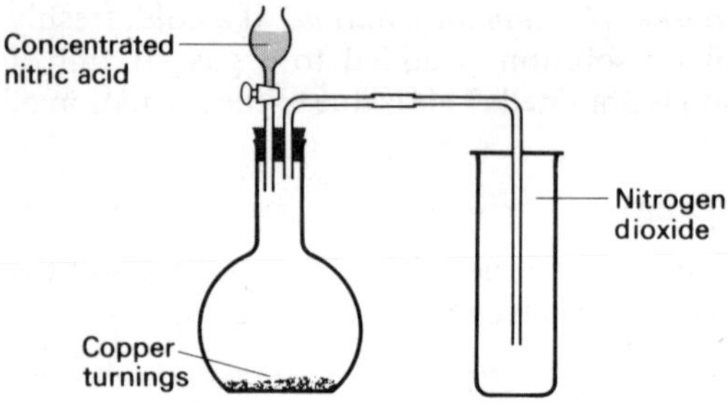

Figure 16.16 Preparation of nitrogen dioxide.

(c) Properties of nitrogen dioxide

Nitrogen dioxide is usually seen as a reddish-brown gas, but it liquefies to a pale yellow liquid at 22 °C, and solidifies at −9 °C.

It is denser than air, has a pungent smell and is *very poisonous.*

(d) Action of heat on nitrogen dioxide

At low temperatures this compound has the formula N_2O_4, and should be called dinitrogen tetraoxide. Dinitrogen tetraoxide is almost colourless. On warming it dissociates into nitrogen dioxide (NO_2), which is dark reddish-brown in colour:

$$\underset{\text{pale yellow}}{N_2O_4} \underset{\text{cool}}{\overset{\text{heat}}{\rightleftharpoons}} \underset{\text{reddish-brown}}{2NO_2}$$

The proportion of nitrogen dioxide present depends on the temperature. At room temperature dinitrogen tetraoxide is under 20 per cent dissociated, but thermal dissociation is complete at 140 °C. A gradual deepening of the colour is therefore observed as the gas is heated from room temperature up to 140 °C.

Above 140 °C the nitrogen dioxide dissociates into nitrogen oxide and oxygen, both colourless gases, this dissociation being complete at 620 °C:

$$\underset{\text{reddish-brown}}{2NO_2} \underset{\text{cool}}{\overset{\text{heat}}{\rightleftharpoons}} \underbrace{2NO + O_2}_{\text{colourless}}$$

(e) Other properties of nitrogen dioxide

(i) Nitrogen dioxide is very soluble in water.

The fountain experiment can be used to illustrate the great solubility of nitrogen dioxide. Nitrogen dioxide forms a mixture of nitrous and nitric acids with water:

$$2NO_2(g) + H_2O(l) \rightarrow \underset{\text{nitrous acid}}{HNO_2(aq)} + \underset{\text{nitric acid}}{HNO_3(aq)}$$

(ii) Nitrogen dioxide does not burn, but supports the combustion of some strongly burning substances.

As a supporter of combustion, nitrogen dioxide is intermediate between dinitrogen oxide and nitrogen oxide.

It readily supports the combustion of strongly burning phosphorus:

$$10NO_2(g) + 2P_4(s) \rightarrow 2P_4O_{10}(s) + 5N_2(g)$$

Strongly burning sulphur or carbon or a vigorously burning taper also continue to burn in nitrogen dioxide.

16.6 Nitric acid

16.6.1 Introduction

Nitric acid has probably been known since the twelfth century. It was Glauber, in 1658, who first prepared it from sulphuric acid and potassium nitrate—the method used in the laboratory nowadays. To the early alchemists it was known as *aqua fortis* (strong water), because of its corrosive action on so many substances.

16.6.2 Preparation of nitric acid

Nitric acid is prepared by the action of concentrated sulphuric acid on a nitrate, usually sodium or potassium nitrate. The hydrogensulphate of the metal is also formed:

$$NaNO_3(s) + H_2SO_4(l) \rightarrow NaHSO_4(s) + HNO_3(l)$$
$$KNO_3(s) + H_2SO_4(l) \rightarrow KHSO_4(s) + HNO_3(l)$$

The nitric acid (b.p. 86 °C) is much more volatile than the other constituents of the mixture obtained, and can therefore be separated by distillation.

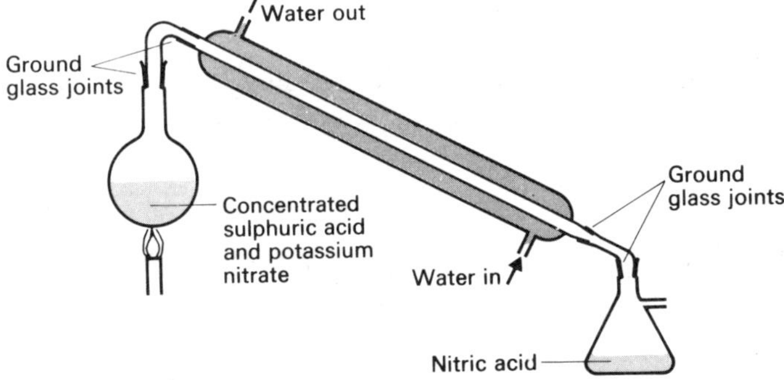

Figure 16.17 Preparation of nitric acid.

An all-glass apparatus must be used, as nitric acid vapour rapidly attacks cork or rubber, and for this reason a distillation apparatus with ground-glass joints is used. Approximately equal masses of the nitrate and concentrated sulphuric acid are placed in the distillation flask. On heating, nitric acid vapour is evolved and condenses as oily drops in the condenser. It is collected in a suitable receiver (figure 16.17).

Some brown fumes are noticed during the heating because of thermal decomposition of the nitric acid to form nitrogen dioxide:

$$4HNO_3(l) \rightarrow 2H_2O(g) + 4NO_2(g) + O_2(g)$$

A little of this nitrogen dioxide dissolves in the acid obtained, giving it a yellowish colour.

16.6.3 Manufacture of nitric acid

(a) From Chile saltpetre

Some nitric acid is manufactured from Chile saltpetre ($NaNO_3$) by a method similar to the laboratory preparation. The retorts are, however, made of iron, which is little attacked by concentrated nitric acid, and the nitric acid is condensed in pipes made of silica. Earthenware jars are used as receivers.

(b) From ammonia

Nearly all nitric acid is now made by the catalytic oxidation of ammonia. There are three stages (figure 16.18):

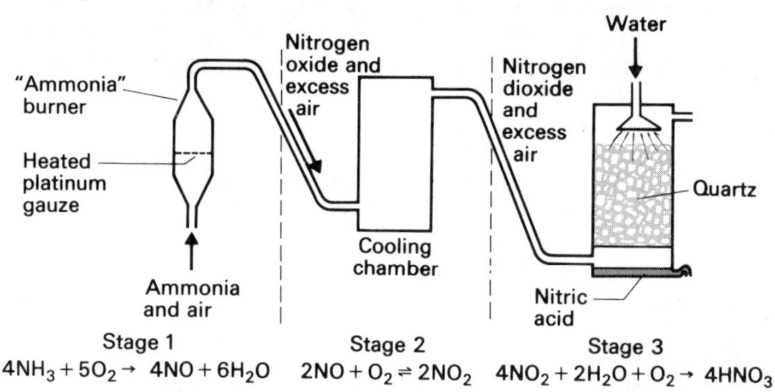

Figure 16.18 Manufacture of nitric acid.

(i) A mixture of purified air and ammonia, containing about 90 per cent air (i.e. an excess of air), is passed through a platinum gauze at 700 °C (figures 16.19 and 16.20). The ammonia is then oxidized to nitrogen oxide:

$$4NH_3(g) + 5O_2(g) \rightarrow 4NO(g) + 6H_2O(g)$$

(The platinum gauze must be heated initially, but, as the oxidation is exothermic, the heat of the reaction then maintains the required temperature.)

(ii) The nitrogen oxide does not combine with excess oxygen in the air until it is cooled in a large empty chamber. Then nitrogen dioxide is formed:

$$2NO(g) + O_2(g) \rightleftharpoons 2NO_2(g)$$

ɡure 16.19 The platinum gauze catalyst for the idation of ammonia to nitrogen oxide, in place in "ammonia burner". (Photo: I.C.I. Agricultural vision).

ɡure 16.20 Ammonia burners for oxidizing monia to nitrogen oxide. (Photo: I.C.I. Agricul- al Division).

(iii) The nitrogen dioxide, mixed with excess air, is then allowed to react with water by passing it up towers packed with quartz, down which water trickles. The nitrogen dioxide combines with the water and excess oxygen to form nitric acid:

$$4NO_2(g) + 2H_2O(l) + O_2(g) \rightarrow 4HNO_3(aq)$$

Nitric acid of up to 50 per cent concentration is obtained (figure 16.21).

In some plants the nitrogen dioxide is absorbed under pressure. Only a single tower is then needed, and more concentrated acid is formed.

16.6.4 Properties of nitric acid

(a) Pure nitric acid is a colourless, fuming, highly corrosive liquid. It rapidly attacks the skin leaving yellow stains. Ordinary concentrated nitric acid, as used in the school laboratory, only contains about 65 per cent nitric acid. Even so, you must take great care when using it.

The chemistry of nitric acid is complicated by the fact that, in addition to being a strong acid, it is also a powerful oxidizing agent.

(b) Nitric acid as an acid

Like other acids, nitric acid:

(i) has a sour taste and turns blue litmus red, and
(ii) with carbonates produces carbon dioxide:

$$Na_2CO_3(s) + 2HNO_3(aq) \rightarrow 2NaNO_3(aq) + H_2O(l) + CO_2(g)$$
$$CuCO_3(s) + 2HNO_3(aq) \rightarrow Cu(NO_3)_2(aq) + H_2O(l) + CO_2(g)$$

Figure 16.21 Towers in which nitrogen dioxide combines with water to form nitric acid. (Photo: I.C.I. Agricultural Division).

The ionic equation for these reactions is:

$$2H^+(aq) + CO_3{}^{2-}(s) \rightarrow CO_2(g) + H_2O(l)$$

(iii) reacts with bases to form salts and water:

$$\begin{bmatrix} NaOH + HNO_3 \rightarrow NaNO_3 + H_2O \\ OH^-(aq) + H^+(aq) \rightarrow H_2O(l) \end{bmatrix}$$

$$\begin{bmatrix} CuO + 2HNO_3 \rightarrow Cu(NO_3)_2 + H_2O \\ CuO(s) + 2H^+(aq) \rightarrow Cu^{2+}(aq) + H_2O(l) \end{bmatrix}$$

Unlike other acids, nitric acid rarely gives hydrogen with metals. An exception is the action of *much diluted* nitric acid on magnesium:

$$\begin{bmatrix} Mg + 2HNO_3 \rightarrow Mg(NO_3)_2 + H_2 \\ Mg(s) + 2H^+(aq) \rightarrow Mg^{2+}(aq) + H_2(g) \end{bmatrix}$$

Usually one or other of the oxides of nitrogen is produced instead.

(c) Nitric acid as an oxidizing agent

Nitric acid behaves as an oxidizing agent whether it is concentrated or dilute. The products are varied, the most usual being nitrogen oxide or nitrogen dioxide. Other reduction products of nitric acid are obtained in its reactions with some of the more reactive metals, e.g. with zinc, dinitrogen oxide or ammonium nitrate can be formed. It often happens that an oxidation reaction of nitric acid is not clear cut, and a variety of products is obtained. Therefore equations which are given in this section must be regarded as only approximate.

(i) The oxidation of copper with concentrated nitric acid.
The copper is oxidized to copper(II) nitrate, and the nitric acid is reduced at the same time to nitrogen dioxide and water:

$$Cu(s) + 4HNO_3(aq) \rightarrow Cu(NO_3)_2(aq) + 2H_2O(l) + 2NO_2(g)$$

i.e. in terms of electron transfer the nitric acid gains electrons and is therefore reduced:

$$4HNO_3 + 2e \rightarrow 2NO_3^- + 2H_2O + 2NO_2$$

The copper loses electrons and is therefore oxidized:

$$Cu - 2e \rightarrow Cu^{2+}$$

(ii) The oxidation of copper with moderately concentrated nitric acid.
The copper is again oxidized to copper(II) nitrate, but the nitric acid is reduced to nitrogen oxide and water:

$$3Cu(s) + 8HNO_3(aq) \rightarrow 3Cu(NO_3)_2(aq) + 4H_2O(l) + 2NO(g)$$

i.e. the nitric acid gains electrons and is reduced thus:

$$8HNO_3 + 6e \rightarrow 6NO_3^- + 4H_2O + 2NO$$

The copper is oxidized as in the previous example.

(iii) The action of nitric acid on other metals.
Gold and platinum are unaffected by nitric acid. Little action occurs between concentrated nitric acid and aluminium or iron, due to the formation of a surface layer of oxide which renders the metal passive.

Tin is oxidized to its oxide, with consequent reduction of the acid to nitrogen dioxide. The oxide formed does not combine with nitric acid to form a nitrate.

The reactions of lead with dilute and concentrated acid correspond with those of copper.

With zinc, as mentioned above, different reduction products are obtained. The concentrated acid yields mainly ammonium nitrate, although some nitrogen dioxide is formed:

$$4Zn(s) + 10HNO_3(aq) \rightarrow 4Zn(NO_3)_2(aq) + NH_4NO_3(aq) + 3H_2O(l)$$

Dinitrogen oxide can be prepared using the dilute acid:

$$4Zn(s) + 10HNO_3(aq) \rightarrow 4Zn(NO_3)_2(aq) + 5H_2O(l) + N_2O(g)$$

(iv) The oxidation of sulphur with concentrated nitric acid
When powdered sulphur is heated with concentrated nitric acid dense brown fumes of nitrogen dioxide are evolved. (If the acid used is that obtained in the laboratory preparation then care must be taken when the acid is added to the sulphur.) If the solution remaining is diluted and decanted from any sulphur it gives a white precipitate of barium sulphate with barium chloride solution. This shows the presence of the sulphate ion, caused by the oxidation of the sulphur to sulphuric acid:

$$S(s) + 6HNO_3(aq) \rightarrow H_2SO_4(aq) + 2H_2O(l) + 6NO_2(g)$$

(v) The oxidation of carbon by concentrated nitric acid
Hot concentrated nitric acid oxidizes carbon to carbon dioxide:

$$C(s) + 4HNO_3(aq) \rightarrow CO_2(g) + 2H_2O(l) + 4NO_2(g)$$

An impressive way of demonstrating this is to drop a *small* lump of red-hot carbon into concentrated nitric acid.

(vi) Oxidation of iron(II) sulphate to iron(III) sulphate.
A little concentrated nitric acid is added to a solution of iron(II) sulphate in dilute sulphuric acid and the mixture heated. The pale green solution turns orange, due to the oxidation of iron(II) sulphate to iron(III) sulphate. The nitric acid is reduced to nitrogen oxide, which on contact with air gives brown fumes of nitrogen dioxide. (Before heating a dark brown colour is observed. Can you suggest a reason for this?).

$$2HNO_3(aq) + 6FeSO_4(aq) + 3H_2SO_4(aq) \rightarrow 3Fe_2(SO_4)_3(aq) + 4H_2O(l) + 2NO(g)$$

i.e. the nitric acid is reduced thus:

$$2HNO_3 + 6H^+ + 6e \rightarrow 4H_2O + 2NO$$

and the iron(II) is oxidized thus:

$$Fe^{2+} - e \rightarrow Fe^{3+}$$

(vii) Other oxidations by nitric acid
Other oxidations are indicated by the following equations, but it should be remembered that the products can be varied by altering the conditions:

$$\underset{\text{red phosphorus}}{P(s)} + 5HNO_3(aq) \rightarrow H_3PO_4(aq) + H_2O(l) + 5NO_2(g)$$

$$\underset{\text{hydrogen sulphide}}{3H_2S(g)} + \underset{\text{dilute}}{2HNO_3(aq)} \rightarrow 4H_2O(l) + 3S(s) + 2NO(g)$$

(d) Action of heat on nitric acid
The apparatus is set up as shown in figure 16.22. The silica tube is strongly heated, and concentrated nitric acid is gradually dropped into it. As the acid

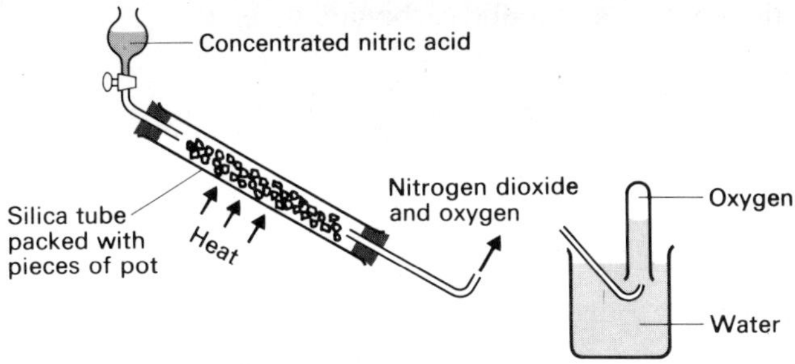

Figure 16.22 Thermal decomposition of nitric acid.

trickles down the tube it is decomposed. Brown fumes are emitted through the opening at the bottom. If the gases are passed into a test-tube inverted over water the tube fills with a colourless gas (what happens to the nitrogen dioxide?), which re-ignites a glowing splint. This gas is oxygen:

$$4HNO_3(aq) \rightarrow 2H_2O(l) + 4NO_2(g) + O_2(g)$$

16.6.5 Uses of nitric acid

Some organic compounds when treated with concentrated nitric acid, usually in the presence of concentrated sulphuric acid, are nitrated. This important reaction is used in the manufacture of methyl-2,4,6-trinitrobenzene (TNT) from methylbenzene (toluene):

$$\underset{\text{methylbenzene}}{C_7H_8} + 3HNO_3 \rightarrow \underset{\text{methyl-2,4,6-trinitrobenzene}}{C_7H_5(NO_2)_3} + 3H_2O$$

T.N.T. is used as a high explosive and is extensively used in mining.

The nitration of cotton can be used to produce guncotton and a compound used in making imitation leather and lacquers.

Propane-1,2,3-triyltrinitrate (nitroglycerin), made from propane-1,2,3-triol (glycerin) and nitric acid, is another explosive which when mixed with a type of sand is known as dynamite.

Much nitric acid is also used in making fertilisers such as ammonium nitrate and in the manufacture of organic compounds required in the synthesis of dyes, drugs and plastics.

16.6.6 Nitrates

(a) Nitric acid forms only one series of salts, known as the nitrates. All these are readily soluble in water. They can therefore be prepared by these standard methods:

(i) By the neutralisation of an acid with an alkali. This method is used in the preparation of sodium, potassium and ammonium nitrates.
(ii) By the action of the acid on a metal. This method cannot be used with the very reactive metals, sodium and potassium, or with those salts which are readily oxidized.

(iii) By the action of the acid on a metallic carbonate or oxide.

(b) Action of heat on nitrates

Sodium and potassium nitrates melt and then decompose on heating to give the nitrites and oxygen:

e.g.	$2KNO_3(l)$	$\rightarrow$	$2KNO_2(l) + O_2(g)$
	potassium nitrate		potassium nitrite

Most nitrates, however, decompose in a different manner:

$$2Pb(NO_3)_2(s) \rightarrow 2PbO(s) + 4NO_2(g) + O_2(g)$$

The nitrogen dioxide is visible as brown fumes, and the oxygen reignites a glowing splint.

Nitrates of metals very low in the electrochemical series decompose similarly, but yield the metal and not its oxide:

$$Hg(NO_3)_2(s) \rightarrow Hg(g) + 2NO_2(g) + O_2(g)$$
mercury(II)
nitrate

Ammonium nitrate falls into a class on its own by decomposing to give dinitrogen oxide and water:

$$NH_4NO_3(s) \rightarrow N_2O(g) + 2H_2O(g)$$

(c) Identification of nitric acid and nitrates

(i) By the action of heat

The action of heat on a nitrate usually results in the evolution of oxygen and nitrogen dioxide (see above).

(ii) By the action of concentrated sulphuric acid

If a nitrate is heated in a test-tube with a small quantity of concentrated sulphuric acid, nitric acid forms:

$$NaNO_3(s) + H_2SO_4(l) \rightarrow NaHSO_4(s) + HNO_3(l)$$

Some of this condenses as oily drops on the side of the test-tube, and some decomposes to give brown fumes of nitrogen dioxide. If some copper is also added to the mixture much nitrogen dioxide is produced by the interaction of the nitric acid with the metal.

(iii) By the "brown ring" test.

The suspected nitrate is dissolved in water in a test-tube, and a freshly prepared solution of iron(II) sulphate in dilute sulphuric acid is added. The test-tube is tilted slightly and concentrated sulphuric acid is poured down the side of the tube. The acid is denser than water and, if the tube is not shaken, it forms a separate layer at the bottom of the tube. A brown ring then forms at the junction of the two liquids (figure 16.23). If the tube is shaken the brown ring disappears. The reason for this brown coloration is as follows:

1. Concentrated sulphuric acid reacts with the nitrate to give nitric acid.
2. The nitric acid formed oxidizes some of the iron(II) sulphate, and is thereby

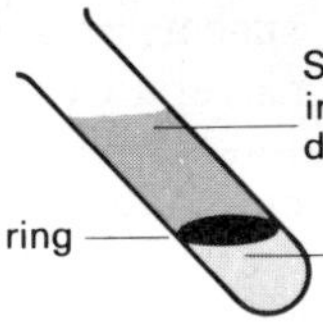

Figure 16.23 The "brown ring" test for a nitrate.

reduced to nitrogen oxide.

3. The nitrogen oxide combines with some of the remaining iron(II) sulphate to give a dark brown addition compound (see section 16.5.2(*b*)).

4. The brown ring disappears if the solution is shaken because the concentrated sulphuric acid and water mix with the evolution of heat which decomposes the addition compound.

(iv) Devarda's alloy test

If a solution of a nitrate in dilute sodium hydroxide solution is heated with Devarda's alloy the nitrate is reduced to ammonia gas. The ammonia gas may be detected by smell or by red litmus turning blue.

16.7 The nitrogen cycle

Compounds of nitrogen are essential to everything living, and any soil in which plants are to grow must contain such compounds. *Plants* obtain nitrogen from the soil by absorbing solutions of nitrates through their roots, and use the nitrogen to build up complicated organic compounds known as plant proteins. *Animals* either obtain nitrogen by eating plants containing plant protein, which is then converted into animal protein, or by eating other animals. Thus nitrates are continually being removed from the soil by plants, and hence by animals. Some combined nitrogen is also lost from the soil by the action of denitrifying bacteria, which aid the oxidation of ammonium compounds to free nitrogen.

Nitrogen compounds are fortunately continually reaching the soil by a variety of methods:

(*a*) During thunderstorms the action of lightning causes combination of some of the nitrogen and oxygen of the air to form nitrogen oxide, which then combines with more oxygen to form nitrogen dioxide. This dissolves in rain water (see section 16.5.3) to yield nitric acid, which forms nitrates in the soil. Various estimates have been made about the quantity of combined nitrogen reaching the soil by this method. A typical one suggests about 5 kg of combined nitrogen per acre per annum.

(*b*) The decay of animal and plant proteins results in the formation of ammonia and ammonium compounds. These compounds are oxidised with the aid of bacteria in the soil first to nitrites and then to nitrates. Very little of the combined nitrogen consumed by man finds its way back to the soil. Much of it is excreted and finds its way through sewers to the sea.

The provident gardener or farmer, however, builds a compost heap of waste vegetable material to promote its decay and uses farmyard manure to help replace that nitrogen which is lost.

(*c*) If you examine the roots of certain leguminous plants, such as peas, beans or clover, you will see many small nodules. These contain bacteria capable of converting nitrogen from the air into soluble nitrogen compounds which can be assimilated by plants. Such plants can be used, therefore, to enrich the soil. The farmer who rotates his crops by growing different plants on a patch of ground in successive years is careful to use clover or another leguminous plant every few years. The plant is then ploughed in so that the nitrogen compounds made by the bacteria can be used by other plants.

(*d*) Because so many nitrogen compounds are wasted in a civilised society by excretion, and because some areas are so intensively cultivated, natural methods of replenishing nitrogen in the soil are no longer adequate. Therefore artificial fertilisers have to be used, (figure 16.2 and section 28.7.2). Until the beginning of this century only two such fertilisers were available: sodium nitrate, which occurs in Chile, and ammonium salts obtained as a by-product in the manufacture of coal gas. Scientists realised the inadequacy of these sources, and great efforts were made to find a method for fixing atmospheric nitrogen, i.e. for converting nitrogen in the atmosphere into its compounds. By far the most important method used for fixing the nitrogen nowadays is the Haber process for the synthesis of ammonia (section 16.3.2). By this means ammonium salts can be made, and also nitrates by the catalytic oxidation of ammonia to nitric acid. This circulation of nitrogen between compounds in the soil, plants, animals and the air is known as the *nitrogen cycle* and is illustrated in figure 16.24.

16.8 Phosphorus

16.8.1 Manufacture of white phosphorus

Originally phosphorus was manufactured from calcium phosphate isolated from bones. It is now obtained from phosphate ores, such as phosphorite, which is impure calcium phosphate, $Ca_3(PO_4)_2$, and fluoroapatite, $CaF_2.3Ca_3(PO_4)_2$. There are immense deposits of these ores in the southern part of the United States, North Africa, on the Nauru and Ocean Islands in the Pacific and in the Kola Peninsula in the U.S.S.R.

The calcium phosphate is strongly heated with a mixture of silicon(IV) oxide chippings and coke or anthracite. To understand what happens, calcium phosphate may be regarded as a compound of the acidic oxide, P_4O_{10}, and the basic oxide, CaO. The silicon(IV) oxide, which is also an acidic oxide, displaces the more volatile phosphorus(V) oxide from this compound to form calcium silicate:

$$2Ca_3(PO_4)_2(s) + 6SiO_2(s) \rightarrow P_4O_{10}(g) + 6CaSiO_3(l)$$

The phosphorus(V) oxide is then reduced by carbon to phosphorus:

$$P_4O_{10}(g) + 10C(s) \rightarrow P_4(g) + 10CO(g)$$

Amalgamating the two equations we get:

$$2Ca_3(PO_4)_2(s) + 6SiO_2(s) + 10C(s) \rightarrow 6CaSiO_3(l) + P_4(g) + 10CO(g)$$

The reactants are fed into furnaces heated electrically by striking an arc between carbon electrodes (figure 16.25). This raises the temperature to over 1000 °C. As the phosphorus is formed it vaporizes. The vapour is led to a series of condensers;

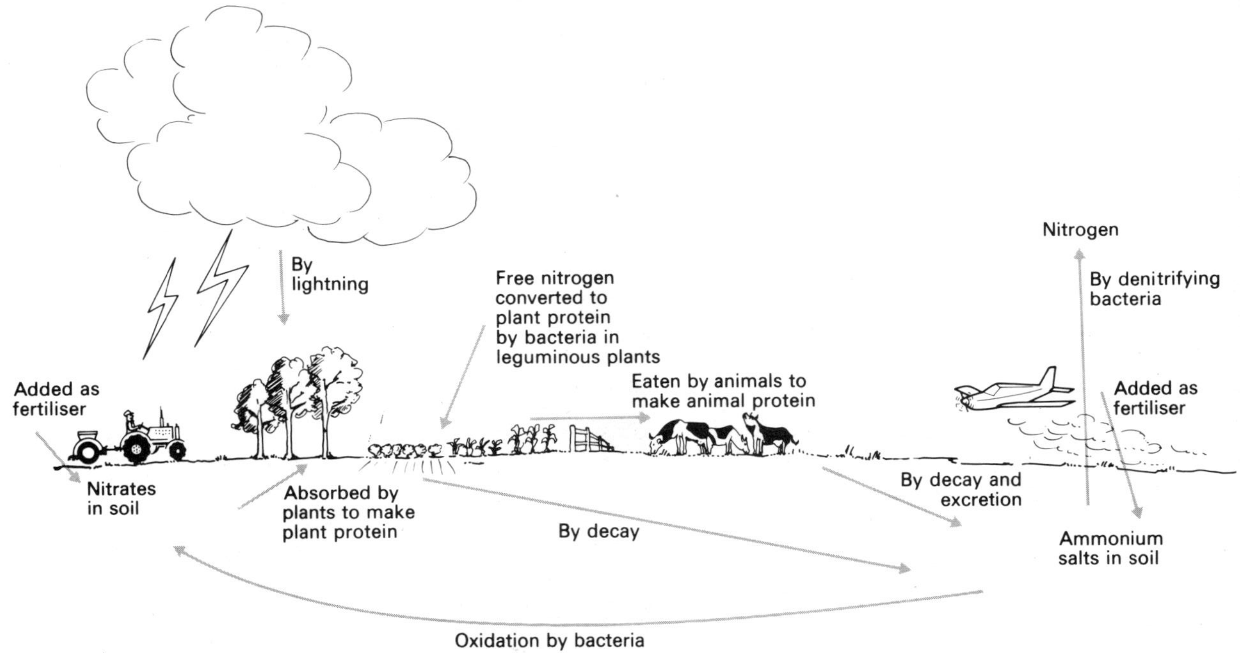

Figure 16.24 The nitrogen cycle. (Drawing: A.M.J. Dixon).

white phosphorus is then collected as a liquid under water. On cooling it solidifies (figure 16.26).

The calcium silicate forms a molten slag, falls to the bottom of the furnace, and runs out through an exit hole. It solidifies on cooling and is used for road construction.

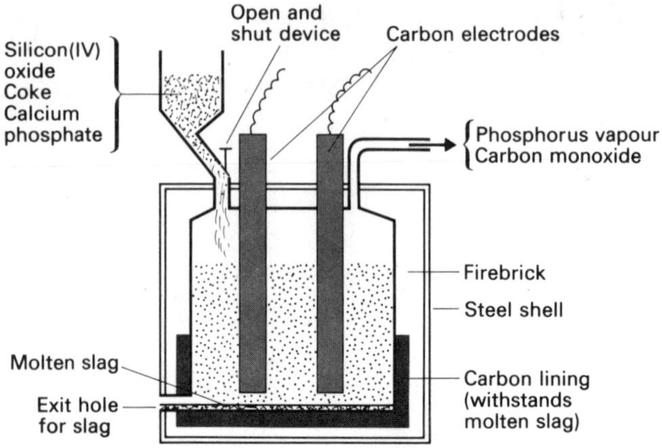

Figure 16.25 The manufacture of phosphorus.

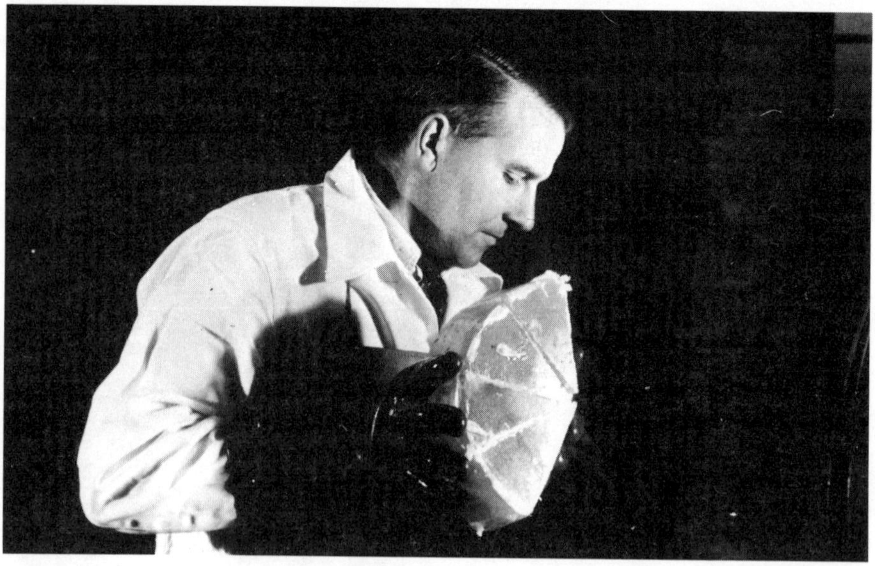

Figure 16.26 A large piece of white phosphorus. (Photo: Albright and Wilson, Ltd.).

16.8.2 Allotropy of phosphorus

Phosphorus exists in two common solid forms, *white* and *red* phosphorus. The form obtained whenever phosphorus vapour is cooled, as in the manufacturing process described above, is white phosphorus. White phosphorus is unstable, and changes very slowly at room temperature to the stable red form. This change is accelerated by heat, and industrially red phosphorus is obtained by heating the white form, *in the absence of air*, for four or five days at 270 °C.

When an element can exist in more than one form without changing its state, it is said to exhibit *allotropy*. The various forms are called *allotropes*. The allotropes of an element differ in their physical properties, and white phosphorus is more chemically reactive than the red form.

The table overleaf shows some of the differences in physical properties between the two forms of phosphorus.

White phosphorus	*Red phosphorus*
White, waxy solid which slowly turns yellow	Dark red powder
Density = 1·8 g cm^{-3}	Density = 2·2 g cm^{-3}
Melting point = 44 °C	Melting point (under pressure) = 500–600 °C
Slightly soluble in water	Insoluble in water
Readily soluble in carbon disulphide and many organic solvents	Insoluble in carbon disulphide and other solvents
Both solid and vapour are very poisonous.	Non-poisonous

Vapour density experiments show the phosphorus molecule to be P_4, the formula used in this section.

White phosphorus ignites at 40–50 °C. For this reason it is always kept under water and must be handled with extreme care. It must always be held in a pair of tongs and cut under water. On burning it forms a dense white smoke consisting of the oxides of phosphorus:

$$P_4(s) + 3O_2(g) \rightarrow P_4O_6(s)$$

$$P_4(s) + 5O_2(g) \rightarrow P_4O_{10}(s)$$

Red phosphorus reacts similarly but at a higher temperature (240 °C).

Both allotropes of phosphorus combine with the halogens (i.e. chlorine, bromine and iodine), but again the white form is much the more reactive. Chlorine gives first phosphorus trichloride, and on further treatment the pentachloride:

$$P_4(s) + 6Cl_2(g) \rightarrow 4PCl_3(l)$$

$$P_4(s) + 10Cl_2(g) \rightarrow 4PCl_5(s)$$

When heated with concentrated sodium hydroxide solution, white phosphorus yields phosphine (a poisonous gas). Red phosphorus does *not* give this reaction.

$$P_4(s) + 3NaOH(aq) + 3H_2O(l) \rightarrow 3NaH_2PO_2(aq) + \underset{\text{phosphine}}{PH_3(g)}$$

16.8.3 Uses of phosphorus

Nearly all phosphorus is converted into phosphoric acid and phosphates. It is also used in making its chlorides and sulphides which are used in the manufacture of insecticides and plasticisers. A little is used in making alloys, e.g. phosphor-bronze, and in rat poisons.

The best known use, however, although it absorbs less than 1 per cent of our total production, is in matches. The type of match that will ignite when rubbed on any rough surface may contain a mixture of a strong oxidizing agent, e.g. potassium chlorate(V) and a sulphide of phosphorus in the match-head. When the match is struck on a rough surface, such as the glass-paper on the side of a match-box, sufficient heat is generated to ignite the mixture. With *safety matches* the kindling mixture, e.g. red phosphorus mixed with antimony sulphide, is on the match-box. The match-head contains the strong oxidizing agent with an abrasive such as ground glass. This must be struck against the side of the match-box for igniting.

The use of white phosphorus in matches is forbidden by law, partly because of a horrible disease known as "phossy-jaw" which afflicted workers in the industry who were continually exposed to phosphorus vapour.

16.8.4 Oxides of phosphorus

Phosphorus(III) oxide is made by burning phosphorus in a limited supply of dry air:

$$P_4(s) + 3O_2(g) \rightarrow P_4O_6(s)$$

Phosphorus(V) oxide is much more important, and is made by burning phosphorus in a plentiful supply of dry air:

$$P_4(s) + 5O_2(s) \rightarrow P_4O_{10}(s)$$

Both oxides are acidic. Thus phosphorus(V) oxide combines with water to give phosphoric acid:

$$P_4O_{10}(s) + 6H_2O(l) \rightarrow 4H_3PO_4(aq)$$

The affinity of this oxide with water is so great that it is a most powerful drying agent.

16.8.5 Phosphorus chlorides

(a) Phosphorus trichloride, PCl_3

Phosphorus trichloride is formed by the direct combination of chlorine and phosphorus (figure 16.27).

White phosphorus is placed in a dry flask fitted with a condenser and receiver. The receiver should be fitted with a drying tube to prevent entry of water vapour from the air. Air is then displaced from the apparatus by a stream of dry carbon dioxide, and the phosphorus is gently heated. A stream of chlorine, dried by passage through concentrated sulphuric acid, is passed over the phosphorus. The phosphorus burns with a pale flame to form the chloride, which collects in the receiver as a colourless liquid.

(b) Phosphorus pentachloride, PCl_5.

This is usually made by the interaction of phosphorus trichloride and chlorine:

$$PCl_3(l) + Cl_2(g) \rightarrow PCl_5(s)$$

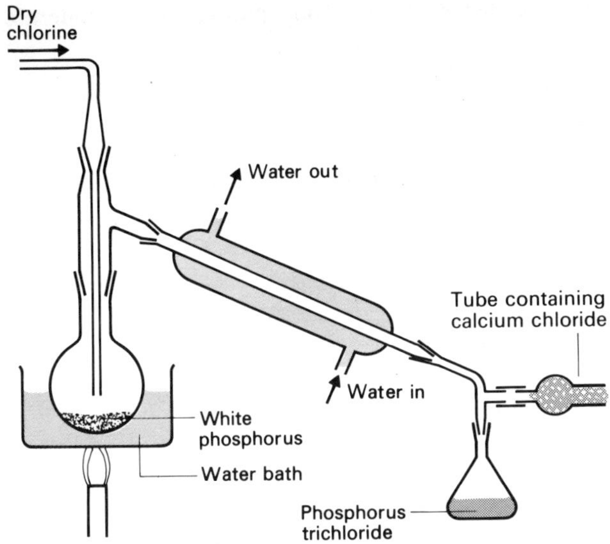

Figure 16.27 Preparation of phosphorus trichloride.

Phosphorus trichloride is dropped slowly into a dry cooled flask, through which a stream of dry chloride is passing (figure 16.28). The pentachloride forms as a white solid on the bottom of the flask.

Like phosphorus trichloride, it reacts vigorously with water. Phosphoric acid and hydrogen chloride are formed:

$$PCl_5(s) + 4H_2O(l) \rightarrow H_3PO_4(aq) + 5HCl(aq)$$

16.8.6 Phosphoric acid, H_3PO_4

Phosphoric acid is made by burning phosphorus in air, and dissolving the

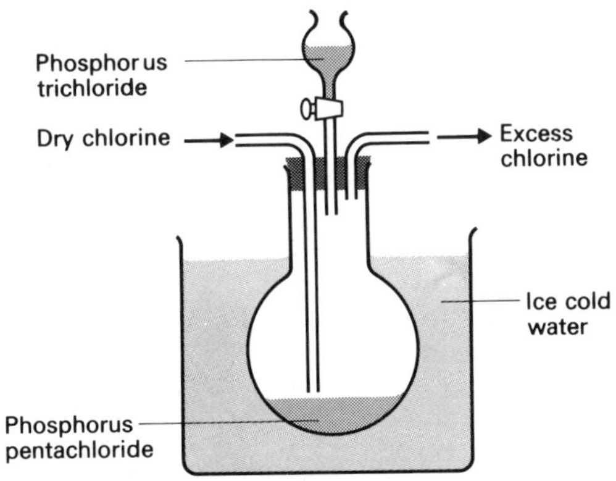

Figure 16.28 Preparation of phosphorus pentachloride

phosphorus(V) oxide formed in water (or in dilute phosphoric acid to moderate the reaction):

$$P_4(s) + 5O_2(g) \rightarrow P_4O_{10}(s)$$
$$P_4O_{10}(s) + 6H_2O(l) \rightarrow 4H_3PO_4(aq)$$

Phosphoric acid is tribasic and therefore forms three series of salts. Thus by treating it with sodium hydroxide three separate sodium salts can be isolated:

$H_3PO_4(aq) + NaOH(aq) \rightarrow NaH_2PO_4(aq) + H_2O(l)$
sodium dihydrogen-phosphate

$H_3PO_4(aq) + 2NaOH(aq) \rightarrow Na_2HPO_4(aq) + 2H_2O(l)$
disodium hydrogen-phosphate

$H_3PO_4(aq) + 3NaOH(aq) \rightarrow Na_3PO_4(aq) + 3H_2O(l)$
trisodium phosphate

16.8.7 Uses of phosphoric acid and phosphates

Like nitrogen, phosphorus is essential to living material. Compounds of phosphorus are continually being removed from the soil by plants and they must be replaced. Calcium phosphate is present in bones, so that bone-meal can be used for this purpose. Unfortunately much phosphorus is wasted in animal waste by sewage disposal, and only a very little of this is regained as fish manure (see section 28.7).

The world therefore relies almost entirely for its phosphate fertilisers on the limited supply of phosphate ores. Calcium phosphate can be used directly, but because of its insolubility, its action is slow. It is more often converted into calcium superphosphate which is moderately soluble. Phosphoric acid is also converted into fertilisers of high phosphate content. A further source of calcium phosphate fertiliser is basic slag, a by-product of the iron and steel industry.

Several complex phosphates made from phosphoric acid are used in water softening, an important one being sold under the trade name of "Calgon". Some of these complex phosphates, e.g. pentasodium triphosphate, $Na_5P_3O_{10}$, are essential constituents of synthetic detergents, as they help the actual detergents to remove dirt and keep it in suspension afterwards.

Phosphates have a variety of other uses in the food industry. For example, acid phosphates, such as calcium hydrogenphosphate, mixed with sodium hydrogencarbonate, are used in baking powders and self-raising flour.

Phosphoric acid itself is used in an important rust-proofing process. Iron articles, dipped into a bath of phosphoric acid, form a crystalline film of iron phosphate on the surface. This film provides a good base for paints and prevents the spread of rust if the paint surface is damaged.

Questions on chapter 16

1. Sketch and label a diagram of an apparatus suitable for obtaining several jars of nitrogen from the air. What impurities would you still expect to be present?

Describe a method used industrially for converting nitrogen of the air into nitrogen compounds.

Why is this fixation of atmospheric nitrogen of great importance to plants and animals? (*D.*)

2. Ammonia is obtained commercially from nitrogen and hydrogen by the Haber process. The equation for this exothermic reaction is

$N_2 + 3H_2 \rightleftharpoons 2NH_3$

(*a*) Under certain conditions the reaction is said to be in a *state of equilibrium*. What do you understand by this term? What is the effect of introducing a catalyst into the mixture of gases when equilibrium has been established?

(*b*) How will the state of equilibrium be affected by increasing (i) the temperature, (ii) the pressure?

(*c*) By what means are the gases heated before they enter the reaction chamber?

(*d*) Name a catalyst used in the process.

(*e*) How is ammonia removed from the nitrogen and hydrogen that remain unconverted? What happens to the unchanged gases?

(*f*) Name *two* commercially important compounds that are obtained from ammonia.

Several reactions are known by which ammonia can be oxidized to water and *either* nitrogen *or* nitrogen oxide, NO. Select *one* of these reactions and state the conditions under which it occurs. Write the equation for the reaction you choose and use this to calculate: (i) the mass of water, (ii) the volume of nitrogen or nitrogen oxide, obtained when 1 dm^3 of ammonia is oxidized. (H = 1; O = 16. Assume that all measurements are made at room temperature and pressure, when 1 mole of gas occupies 24 dm^3.) (*A.E.B.*)

3. In the laboratory, a steady stream of ammonia can be prepared by the reaction between ammonium sulphate and an alkali.

(*a*) Name the alkali you would use.

(*b*) State whether it is necessary to heat the mixture.

(*c*) Write the equation for the reaction.

(*d*) Name the reagent you would use to dry the gas and explain your choice.

(*e*) State how you would collect a sample of dry ammonia. (No diagram is required.)

What change in volume would you expect if 100 cm^3 of ammonia were completely decomposed into its elements? (Assume that all measurements are made at the same temperature and pressure.)

State and explain all that you would see when:

(i) ammonium chloride is gently heated in a test-tube;

(ii) a mixture of ammonia and air is passed over heated platinised asbestos, excess ammonia is removed and the product is allowed to mix with more air. (*A.E.B.*)

4. Describe the preparation of pure, dry ammonia in the laboratory. How, and under what conditions, does ammonia react with (*a*) hydrogen chloride, (*b*) copper(II) oxide, (*c*) copper(II) sulphate?

What volume of hydrogen could be obtained from 10 cm^3 of ammonia at s.t.p.? (*O. and C.*)

5. Ammonia gas dissolves in water to form an alkaline solution.

(*a*) Describe how, starting with ammonium chloride and calcium hydroxide, you would prepare such a solution.

(*b*) Write down the names and formulae of two ions and two molecules that could be detected in the solution.

(*c*) Give two examples of the covalent bond to be found in the substances in the solution.

(*d*) State how each of the ions, quoted in (*b*), has been formed.

(*e*) Suggest a method by which, without applying heat, such a solution could be made to evolve part of the dissolved gas. (*O. and C.*)

6. Give, with an equation, an account of the preparation and collection of nitrogen oxide, stating and explaining what you would expect to *see* during the course of the experiment.

When an iron wire spiral is heated in a given volume of nitrogen oxide the remaining nitrogen occupies half the volume of the original nitrogen oxide (pressure and temperature remaining constant). What conclusions can you draw *from this experiment* about the formula of nitrogen oxide? (*J.M.B.*)

7. How may lead(II) nitrate crystals be prepared from lead(II) oxide? Give practical details.

Describe briefly, with the aid of a diagram, how you would obtain from lead(II) nitrate a sample of an oxide of nitrogen in a liquid condition. If the liquid oxide of nitrogen were sealed in a hard-glass tube, what would be the effect of (*a*) gently heating the liquid, and (*b*) afterwards cooling the liquid to the original temperature? (*B.*)

8. How and under what conditions does nitrogen oxide react with (*a*) copper, (*b*) iron(II) sulphate, (*c*) oxygen?

9. Describe how you would obtain a sample of nitric acid from sodium nitrate. Give a sketch of the apparatus you would use.

State briefly how this acid is obtained on a commercial scale from ammonia.

State very briefly, without sketches, how you would obtain from nitric acid samples of (*a*) oxygen, (*b*) nitrogen oxide, (*c*) dinitrogen tetraoxide, N_2O_4. (*L.*)

10. Describe the preparation of nitric acid in the laboratory.

What substances are formed when nitric acid acts on (*a*) ammonia, (*b*) lead, (*c*) potassium carbonate?

State what you would observe and name the substance produced when each of the three crystalline products from the three reactions is heated. (*B.*)

11. What is a catalyst?

Describe how nitric acid can be made from ammonia. Mention one other example of catalysis. (*O. and C.*)

12. Describe in outline the manufacture of nitric acid from ammonia. (Technical details are not required.)

What is the action of heat on lead(II) nitrate? Given a sample of lead(II) nitrate say briefly how you would prepare metallic lead.

Describe and explain *one* test which can be used to identify the nitrate radical. (*O. and C.*)

13. Here is an extract from a pupil's notebook. Read it and then answer the questions which follow.

"We put about 10 g of potassium nitrate in a retort. We poured in sufficient dilute sulphuric acid to cover the potassium nitrate and noticed that the mixture had become warm. When we gently heated the retort, a pale yellow liquid gradually distilled over and this was collected in a cooled flask.

We dropped some of the pale yellow liquid on to 1·00 g *of copper in a weighed boiling tube. This reaction was carried out in a fume cupboard, and when the vigorous reaction was over and all the copper had dissolved we heated the mixture until no more brown fumes were produced. A black power remained in the boiling tube."*

(*a*) There is one important mistake in the first paragraph of the account. What is this mistake?

(*b*) What was the "pale yellow liquid"?

(*c*) The liquid that the pupil was preparing might have been expected to have no colour. Explain why it was pale yellow.

(*d*) Why was the retort gently heated?

(*e*) Give the equation for the reaction which took place in the retort to produce the liquid which distilled over.

(*f*) If the pupil had used exactly 10 g of potassium nitrate, calculate the maximum mass of "pale yellow liquid" that could have been obtained.

(*g*) Outline the chemistry of the reactions described in the second paragraph and calculate the mass of black powder that would have been produced. (Cu = 64; O = 16; K = 39; N = 14; H = 1) (*L.*)

17

Carbon, Fuels and Flame

17.1 Introduction

There is an unlimited number of carbon compounds, including most of those substances which make up plants and animals. It is said that over a million are known already, and more are being discovered or synthesized the whole time. This amazing fact is due to the unparalleled ability of carbon atoms to join with one another. Compounds are therefore formed which contain long chains of carbon atoms and rings of carbon atoms, e.g.

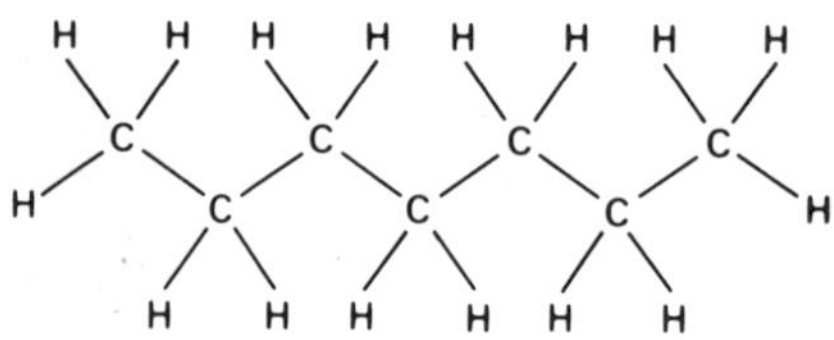

Heptane

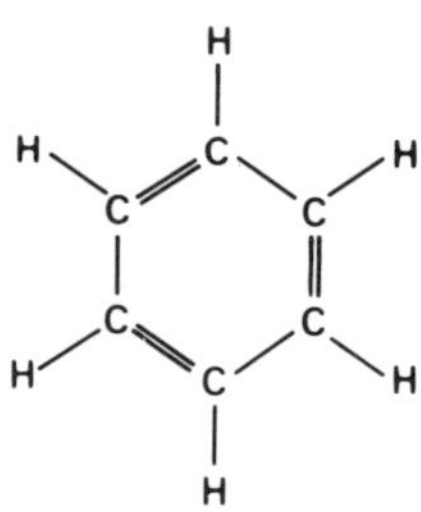

Benzene

The chemistry of carbon atoms is known as *organic chemistry,* and is given a simple treatment in chapters 25 and 26. By convention, carbon dioxide, carbon monoxide, carbonates and a few other carbon compounds are classed as inorganic, and in this chapter and the next one we are mainly concerned with carbon itself and these compounds.

Carbon exists in a variety of solid forms, i.e. it exhibits *allotropy*. The allotropes include such apparently different substances as diamond, graphite, charcoal and soot.

17.2 Diamond

Diamonds are the purest form of carbon occurring naturally, although they are often coloured by slight traces of impurity. Diamonds have a remarkable lustre and hardness. The lustre is caused by the diamond's high refractive index, so that if it is suitably cut there are many internal reflections of the rays of light which enter it. Because of their beautiful, sparkling appearance, together with their hardness and their rarity, diamonds have been a precious gem-stone for over 2000 years (figure 17.1). They have been formed in the earth's crust under very high pressure, and are found only inside extinct volcanos. The main source of diamonds is the African continent, particularly Zaire, but substantial deposits also occur in Russia and other countries. The concentration of diamond, even in rich deposits is very low. On average 1 gram of diamonds are obtained from 20 tonnes of mineral from a diamond mine. The mass of a diamond is measured in a peculiar unit called a carat. 1 carat is 0·2 g. Contrary to general belief, brighter gems than diamonds, consisting of crystals of titanium(IV) oxide, are now known but none are so hard or so rare.

Nowadays, however, the most important property of diamonds is their hardness, for two-thirds of the diamonds mined are used for industrial purposes. They are used in drills for mining, as an abrasive, for cutting metals and in dies for producing wire such as the thin-filament tungsten wire used in electric-light bulbs.

Diamond is so hard because it is a *macromolecule* built up of covalently bonded carbon atoms (figure 6.20; see also section 6.2.7(*b*))

Figure 17.1 The large Cullinan diamond in the head of the Sceptre with cross, a part of the Crown Jewels. This diamond has a mass of 530 carats and is the largest of a pair known as the "Stars of Africa" presented by the Government of Transvaal colony to King Edward VII. The other one, together with many other diamonds and other gem-stones is used elsewhere in the Crown Jewels, and can be seen at the Tower of London. (Photo: H.M.S.O., Crown Copyright reserved).

17.3 Graphite

Graphite, too, occurs naturally, but a superior form is made by heating anthracite with a little sand in an electric furnace.

Graphite consists of greyish-black crystals, which feel soft and greasy. The arrangement of the carbon atoms in graphite is responsible for this, X-ray spectra showing that they are arranged in parallel plates, which therefore readily slide over one another (figure 6.21). Because of this property graphite is used as a lubricant, either alone or suspended in a liquid. Unlike oil, it is non-volatile and is not sticky. It is used, for example, on bicycle chains and for the bearings of some motors.

"Black-lead" is largely graphite, and used to be used as a thin coating over iron articles to prevent them from rusting and to give them a polished appearance. Also the "lead" in lead pencils contains graphite, mixed with clay and baked to make it harder. The percentage of added clay determines the hardness of the pencil.

Another important property of graphite is that it conducts electricity. This fact, coupled with its low chemical reactivity, accounts for its use in electrodes (e.g. see figures 23.1 and 23.2).

17.4 Other forms of carbon

The other forms of carbon are of varying degrees of purity and are not true allotropes. X-ray methods indicate that they consist of minute crystals of graphite bonded together by impurities.

17.4.1 Coke

This is made by heating coal (see section 17.6.3(*a*)) in the absence of air. It is mainly carbon (with impurities that give ash on burning). In addition to its use as a fuel, it is widely used as a reducing agent in the extraction of metals from their ores (e.g. in the iron and steel industry (see chapter 24)).

17.4.2 Wood charcoal

Wood is composed of cellulose and the related carbohydrate called lignin (see section 26.3.1(*d*)). These organic compounds contain the elements carbon, hydrogen and oxygen. When wood is heated in a closed retort so that it cannot burn, it decomposes. Most of the hydrogen and oxygen is lost in the vapours which are a complicated mixture containing methanol, ethanoic acid and propanone, together with an inflammable gas. (An experiment like that described in section 17.6.1 can be used to demonstrate this in the school laboratory.) The product left behind, *wood charcoal*, is mainly carbon (with inorganic impurities).

This process is known as the *destructive distillation* of wood.

Wood charcoal is exceptionally porous and has therefore a low density. It has the remarkable property, connected in some way with its porosity, of being able to absorb large volumes of certain gases and vapours.

To show this, a drop of bromine is placed in a gas jar, which is then closed with a greased plate. The bromine soon vaporizes and fills the jar with a brown vapour. When charcoal is added and the jar is shaken the vapour disappears as it is absorbed by the charcoal.

This property of absorbing gases accounts for the use of a layer of wood charcoal, made from coconut shells, in gas masks to remove poison gases.

17.4.3 Animal charcoal

This is made in a similar manner to wood charcoal by the destructive distillation of bones. It contains about 90 per cent of calcium phosphate and is therefore a very impure form of carbon.

Animal charcoal is effective in removing traces of colouring from solutions. Thus, if some water coloured with litmus is boiled with animal charcoal and the liquid is then filtered, the filtrate is almost colourless.

This method can be used to remove the brown colour from crude sugar and in decolorizing petroleum jelly.

17.4.4 Carbon black and soot

If a smoky flame is burned on a cooled surface a black deposit is obtained. This deposit is now manufactured on a large scale by burning coal-tar residues or, in the U.S.A., methane in a restricted supply of air to prevent complete combustion. It is used to colour rubber tyres black and to make the rubber more resilient. Other uses include the manufacture of black paint, black shoe polish and printers' ink. Soot is formed similarly by the incomplete combustion of coal.

17.5 Properties of carbon

As has been seen, the allotropes of carbon differ considerably in their physical properties. All the allotropes are, however, insoluble in all common solvents and possess similar chemical properties. Diamond and graphite in general show a lower reactivity than the various forms of charcoal.

(a) All forms of carbon burn to form carbon dioxide

Wood charcoal glows when it is heated in a Bunsen flame. If the red-hot charcoal is then plunged into a gas jar full of oxygen it burns with a yellow flame to form carbon dioxide. Much heat is evolved, and because of this carbon and its compounds are widely used as fuels:

$$C(s) + O_2(g) \rightarrow CO_2(g)$$

To make graphite burn, it must be finely powdered and strongly heated in a hard-glass tube, through which oxygen is passing.

This reaction can be used to show that the various forms of carbon are in fact forms of the same element. Known masses of carbon are burned in a stream of oxygen. The carbon dioxide formed is absorbed in sodium hydroxide solution and the gain in mass measured. It is found that equal masses of the allotropes give equal masses of carbon dioxide.

(b) Strong oxidizing agents oxidize carbon to carbon dioxide

This can be shown by heating carbon with either concentrated nitric or sulphuric acids:

$$C(s) + 4HNO_3(l) \rightarrow 2H_2O(l) + 4NO_2(g) + CO_2(g)$$

$$C(s) + 2H_2SO_4(l) \rightarrow 2H_2O(l) + 2SO_2(g) + CO_2(g)$$

(c) Carbon reduces the oxides of the less active metals to the metals

This is conveniently shown by using a block of compressed charcoal. A small hole is scraped in the block and is filled with a mixture of lead(II) oxide and anhydrous sodium carbonate. The mixture is then heated in a blowpipe flame, when the lead(II) oxide is reduced to soft, silvery beads of lead. These are prevented from re-oxidation by a protective coating of sodium carbonate:

$$PbO(s) + C(s) \rightarrow Pb(l) + CO(g)$$

Many other oxides are similarly reduced (see table 8.2), but with a mouth blowpipe the temperature reached is not high enough to reduce zinc oxide. Oxides of the very reactive metals, such as potassium, sodium and calcium, cannot be reduced with carbon.

(d) Carbon combines with sulphur to form carbon disulphide
Carbon disulphide, which has an important use in the manufacture of viscose artificial silk, is manufactured by heating carbon with sulphur vapour in an electric furnace:

$$C(s) + 2S(g) \rightarrow CS_2(g)$$

Figure 17.2 The drilling ship "Offshore Mercury" drilling for natural gas in the Irish Sea. (Photo: British Gas Corporation).

17.6 The use of carbon and its compounds as fuels

A *fuel* is a substance used to *provide energy*. Nearly all fuels are composed of either carbon or its compounds. They may be gaseous, liquid or solid. The *chemical energy* in a fuel is usually obtained by burning it in oxygen (see section

9.5). Energy produced is then used as heat (e.g. to warm a building) or is converted to some other form of energy (e.g. to electricity or the motion of a motor car).

Animals eat carbohydrates and fats. These are also fuels and energy is obtained when they are "burned" in the body. However the processes involved are complicated and will not be discussed here.

The major energy sources at present are coal, oil and natural gas. These have all come from plants and animals as part of the carbon cycle (see figure 18.8) when they have died and become fossilised by being compressed in the earth for millions of years. They are called *fossil fuels*.

17.6.1 Gas

The most widely known "gas" is that which is used at home in your gas-cooker, or at school in your Bunsen burner. It used to be obtained from coal but it now comes from a variety of sources.

(a) Natural gas

Natural gas occurs in many places, often associated with oil, as they are formed by similar processes (see section 25.3.2). In Britain the best known natural gas comes from under the North Sea. It is obtained by drilling (figure 17.2). North Sea gas is mainly methane and vast quantities of it are now used. Some methane is still imported to this country from the Arab states where it is liquefied prior to being transported in large tankers (figure 28.6).

(b) From oil

Several low-boiling fractions are obtained from the distillation of crude oil e.g. naphtha (figure 25.9). Such a mixture is made to react with steam at 700–1000 °C under pressure and in the presence of a catalyst. Essentially the reaction occurs in two stages. Firstly the naphtha is converted into a mixture of carbon monoxide, carbon dioxide, hydrogen and methane. Secondly the hydrogen content of the mixture is increased by these further reactions:

$$CH_4(g) + H_2O(g) \rightleftharpoons CO(g) + 3H_2(g)$$

$$CO(g) + H_2O(g) \rightleftharpoons CO_2(g) + H_2(g)$$

Carbon dioxide is removed by dissolving it in water under pressure. The resulting mixture of gases may have its composition adjusted by addition of more methane before being used.

(c) Coal gas

The manufacture of coal gas may be illustrated by a simple experiment.

Experiment. Manufacture of coal gas

Place powdered coal in a hard glass test-tube as shown in figure 17.3 and heat it strongly. Brownish vapours are evolved. These condense to give an almost black tar, called *coal tar*, and a watery liquid which contains ammonia and is called *ammoniacal liquor*. The condensate collects in the cooled test-tube. A gas which does not condense is also formed. This is impure *coal gas*, and can be ignited at *A*. *Coke* remains in the hard-glass tube.

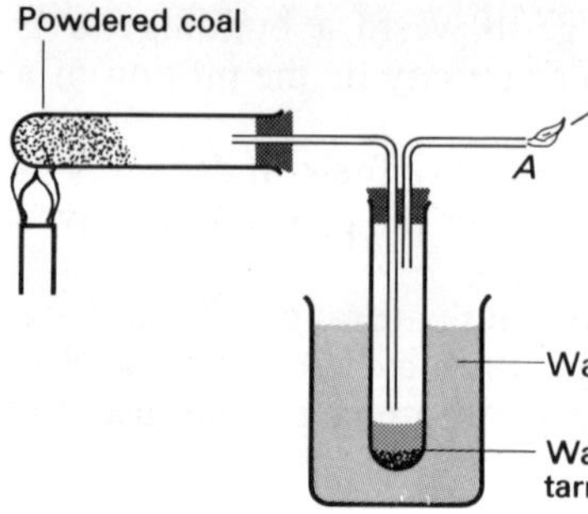

Figure 17.3 Destructive distillation of coal.

This process is known as the *destructive distillation of coal*. The coal does not burn, but is merely heated so that volatile substances are evolved.

Industrially, much higher temperatures are used and the coal gas obtained has to be carefully purified (i) by condensing out all tar and ammoniacal liquor, (ii) by removing traces of sulphur dioxide and ammonia by solution in water and (iii) by removing hydrogen sulphide by combining it with hydrated iron(III) oxide.

The coal tar obtained as a by-product is a most valuable source of organic chemicals. Coke, too, has extremely important uses mentioned earlier. Therefore now that coal gas is of decreasing importance, the main object of much destructive distillation of coal is the production of these substances and not coal gas.

(d) Producer gas and water gas

Producer gas is a mixture of carbon monoxide and nitrogen, made by passing air through red-hot coke. The nitrogen remains unchanged, but the oxygen oxidizes the carbon to carbon monoxide:

$$2C(s) + O_2(g) \rightarrow 2CO(g) + \text{heat}$$

Some carbon dioxide may be formed initially, but this is then reduced by the hot coke:

$$CO_2(g) + C(s) \rightarrow 2CO(g)$$

Producer gas has a low heat value because of its high nitrogen content, but it is inexpensive. It is always made for direct use, such as heating retorts in the manufacture of coal gas and zinc.

Water gas is a mixture of hydrogen and carbon monoxide, made by passing steam through white-hot coke:

$$C(s) + H_2O(g) \rightarrow CO(g) + H_2(g)$$

This reaction is *endothermic*, i.e. it takes in heat, and so the temperature of the coke rapidly decreases and the reaction stops. Therefore after a few minutes air is blown through the coke. Producer gas is then formed by an *exothermic* reaction, i.e. heat is evolved, and a high enough temperature for the production of water gas is once more obtained. Consequently, intermittent blasts of air and steam are always used, but usually only the water gas is collected.

Water gas has a heat value intermediate between those of producer gas and coal gas. It is cheap, and is sometimes added to gas before it is distributed to the consumer.

In addition to its use as a fuel, water gas is used in the manufacture of hydrogen by the Bosch process (see section 14.3.1). Also, by the use of catalysts, it is possible to obtain from water gas such products as methanol, butanol and mixtures of hydrocarbons suitable for use as petrol and lubricating oils.

17.6.2 Liquid fuels

(a) Oil and petrol

Crude oil has been known since very early times, but it was not until 1859 that the first well was drilled in Titusville, Pennsylvania, with the object of obtaining a useful illuminant and fuel oil.

Crude oil is an oily liquid, which varies widely in colour and composition. Essentially, however, it always consists of a mixture of hydrocarbons. It is widely distributed in nature, large deposits being found in Saudi Arabia and the other Arab states, Iran, the U.S.A., South Russia, under the North Sea and in many other places. Usually it is found with salt water, sand and sometimes natural gas, beneath layers of impervious rock which have to be drilled to obtain the oil (figure 25.8). Most scientists believe it to have been formed by the decay of marine animal and vegetable matter under pressure (see also section 25.3.2).

By fractional distillation of crude oil petrol, paraffin, fuel oil and many other substances can be obtained (figures 17.4 and 25.9). Cracking of higher-boiling-point constituents of petroleum gives a further yield of hydrocarbons suitable for petrol.

Figure 17.4 Fractional distillation columns at the Europoort Refinery, Rotterdam, Holland. (Photo: The British Petroleum Co. Ltd.).

(b) Benzene, C_6H_6

Benzene is another important liquid fuel obtained by the fractional distillation of coal tar or by processing petroleum. It is often mixed with petrol for use as a motor fuel. It burns with a smoky flame, forming mainly carbon dioxide and water.

(c) Methanol (methyl alcohol), CH_3OH, and ethanol (ethyl alcohol), C_2H_5OH.

These two alcohols burn readily:

$$2CH_3OH(l) + 3O_2(g) \rightarrow 2CO_2(g) + 4H_2O(g)$$

$$C_2H_5OH(l) + 3O_2(g) \rightarrow 2CO_2(g) + 3H_2O(g).$$

(An experiment to determine how much heat is obtained when an alcohol is burnt is given in section 9.5).

Both methanol and ethanol can be obtained from substances other than oil. They are therefore becoming increasingly attractive as liquid motor fuels either mixed with petrol or used on their own (see section 28.6). To make *methanol* steam is reacted with either methane (natural gas) or coke (see sections 14.3.2 and 17.6.1(d)). These reactions give carbon monoxide and hydrogen which can be made to react with one another to give methanol (see section 25.6.1):

$$CO + 2H_2 \rightarrow CH_3OH.$$

Ethanol can be produced by fermentation (see section 25.6.2) and in countries such as Mexico with an abundance of sugar it is now produced in this way for use as a motor-fuel.

17.6.3 Solid Fuels

The most important solid fuels are coal and coke.

(a) Coal

Coal has been formed by the very slow decomposition of vegetation. This vegetation died forming large layers which decayed to form peat. The chemical changes which took place involved the loss of carbon dioxide, water and methane from the cellulose (see section 26.3.1(d)) in plants with a resultant increase in the percentage of carbon in the material left behind. All this happened millions of years ago and disturbances in the earth's crust then buried the peat.

With the increased pressure and temperature and in the absence of air the process of *carbonisation* continued. More volatile substances were lost and the peat changed to lignite, bitumen, soft coal and finally anthracite. Anthracite may be as much as 95% carbon while the cellulose starting material is about 50% carbon.

Coal is comparatively cheap, but much of the heat from a domestic coal fire escapes up the chimney. Valuable chemicals which could be obtained by the destructive distillation of coal are completely lost. Moreover, in densely populated areas the smoke and gases from coal cause an unhealthy pollution of the atmosphere (see chapter 28). Harder coals, like anthracite, which contain a high percentage of carbon, cause less smoke, but they are hard to ignite and are consequently unsuitable for domestic purposes. Grates and boilers which have been scientifically designed help to ensure that coal is burnt more efficiently.

(b) Coke

Coke is much cleaner in use than coal, and is a valuable fuel, although it is not possible to burn it in the old-fashioned type of grate.

Attempts have been made to produce a smokeless solid fuel, which is easy to ignite. An important method used commercialy consists of heating coal in retorts, as in the manufacture of coal gas, but at a lower temperature. Coal gas of a high heat value, together with tarry liquids, is again produced, but naturally less is obtained than in the normal manufacture of coal gas. The solid product is more expensive than coal or coke, but has a high heat value and is an efficient fuel.

(c) Peat and lignite

Both peat and lignite are used as solid fuels in countries where coal is scarce. They have a low heat value and produce much soot.

(d) Wood

Wood was probably the first fuel used. It is still used today in parts of the world where it is plentiful. Indeed in Eire, for example, in areas from which peat has been removed, quick growing trees such as poplar are grown to produce wood for fuel.

17.6.4 Comparison of fuels

Several factors must be considered when deciding the best fuel for a particular purpose.

(a) Is the fuel available readily and at a reasonable cost?

All fossil fuels exist in limited quantities and one day will be used up (see section 28.6). Estimates of how long present stocks will last vary. They depend on (i) how quickly we use them, (ii) how much gas, oil and coal is, as yet, undiscovered, and (iii) how much we will pay to extract fuel from inaccessible places, e.g. under the ocean or deep in the earth. Perhaps petroleum will last for 20–80 years and coal for 200–300 years.

The cost of a fuel depends on how plentiful it is and how easy it is to extract and deliver it to the customer. As stocks become scarce prices will rise. Some of us during recent oil crises changed from oil heating to gas heating. Was this a wise decision?

(b) Is it a 'clean fuel' and how easy is it to remove harmful impurities from the fuel?

Substances which occur naturally are rarely pure. Coal and oil contain many impurities. These include sulphur, nitrogen compounds and inorganic substances which do not form volatile products on burning and remain as solid ash. The most serious are sulphur and certain of its compounds. Unless these are removed before the fuel is burnt, sulphur dioxide is produced and this will pollute the atmosphere (see section 28.3.2). Because oil is a liquid the sulphur and its compounds can be removed more easily than from solid coal.

When gas is produced from coal, the sulphur dioxide and hydrogen sulphide produced from sulphur containing impurities have to be removed before the gas leaves the gas-works. In contrast, natural gas usually contains few impurities and little treatment is needed before it can be used. (Interestingly it is the sulphur compounds in gas that give it its smell, and very small amounts are actually added to natural gas so that it has a smell and leaks in gas pipes can be detected.)

If the fuel consists only of organic compounds containing only carbon, hydrogen and oxygen the products of complete burning are carbon dioxide and water vapour. Both of these are harmless in the atmosphere and the fuel is said to be a "*clean fuel*". Only gaseous and liquid fuels are clean. Also solid fuels contain large amounts of impurities that will not burn. Therefore they give ash which has to be removed and disposed of.

(c) What is the energy value of the fuel?

The energy value is the amount of energy obtained from a certain mass or volume of fuel. This may be measured using a calorimeter as described in section 9.5. For fuels the value is normally the quantity of heat evolved when a fixed mass or volume (not a mole) is completely burnt (see table 17.1). This is because the mass or volume is an important factor in transport costs which add to the price one pays for a fuel.

Table 17.1 shows the energy values of the various fuels in kilojoules per gram. The fuels with the highest energy values are the liquid fuels (e.g. octane) and the gaseous fuels (e.g. methane). The coals and other solid fuels have much lower energy values. These (in general) increase as the carbon content of the fuel increases.

Table 17.1 CARBON, HYDROGEN AND OXYGEN CONTENTS AND ENERGY VALUES OF FUELS

Fuel	Carbon (%)	Hydrogen (%)	Oxygen (%)	Energy value ($kJ\ g^{-1}$)
Wood	50	6	44	21
Peat	55	6	39	8
Lignite	73	5	22	17
Bituminous coal	84	5	11	33
Anthracite	94	3	3	35
Coke	96	3	1	23
Charcoal	98	1	1	26
Methane	75	25	0	56
Butane	83	17	0	50
Octane	84	16	0	48

Note: For the solid fuels, the other elements present (that give ash on burning) have been ignored in calculating the C, H and O contents.

(d) Is the fuel in a convenient physical state?

Liquid fuels are the most convenient as they can be delivered through pipes and can be supplied to the burners at a controlled rate. They also have a high energy value for a given volume. They are obviously ideal for moving vehicles.

If the fuel burner is static, in a building for example, the fuel does not have to be stored, and a gaseous fuel, which has a much larger volume than a liquid or solid fuel for a certain amount of energy, can become the best choice.

Solid fuels require much more complicated machinery to deliver a steady flow to the burners. Often it needs to be added manually to the fire in which it is being burnt. If it is to be used in a power station to generate electricity, the coal will be ground to a fine powder which can then be moved around the power station in pipes as if it were a liquid. Also some means must be found to remove and dispose of the ash that the solid fuel produces.

It is often desirable not to use any of the carbon based fuels directly but to convert them into electricity which, because it can be transmitted along wires and gives no ash or air pollution when it is used, is often the best fuel of all. However as it must first be made from these other fuels, or from nuclear energy (see section 28.6) it is often the most expensive fuel. Furthermore, electricity is not usually convenient for vehicles which move from place to place. Only if they run along a fixed route, like electric trains, or carry their own electricity supply or a generator, like a milk float or a nuclear submarine, can electricity be their fuel.

Clearly convenience and cost both need to be considered when making a choice.

17.7 Flame

17.7.1 Introduction

When gases combine a flame is often produced. The word flame merely describes the region in which gases combine, with the evolution of light and heat. Usually one of the gases involved is oxygen, but this is not necessarily so. For example, hydrogen will burn in chlorine. Before solids "burn" they must either partly vaporize, e.g. sulphur, or decompose to give vapours, e.g. coal. These vapours then ignite. Non-volatile solids do not give flames when they combine with oxygen or other elements.

17.7.2 The hydrogen flame

Flames are not homogeneous, but contain several defined regions. The hydrogen flame is one of the simplest. Since it is almost invisible in bright light, it is better examined in semi-darkness. It is then seen to consist of a colourless inner zone and an almost colourless outer zone (figure 17.5). If one end of a narrow piece of hard-glass tubing is placed in the colourless portion, and a light is applied at the other end, hydrogen travelling through the tube is ignited. This indicates that the inner zone contains unburnt hydrogen. The presence of this zone is caused by the speed at which the hydrogen leaves the tube, i.e. hydrogen travels out of the tube more quickly than the flame can travel down. When it comes out of the tube the hydrogen slows down as it spreads out, so that a flame is possible. If, however, the jet of hydrogen is powerful enough the flame is blown out, because even with the slowing down of hydrogen on reaching the atmosphere, the flame does not travel downwards quickly enough.

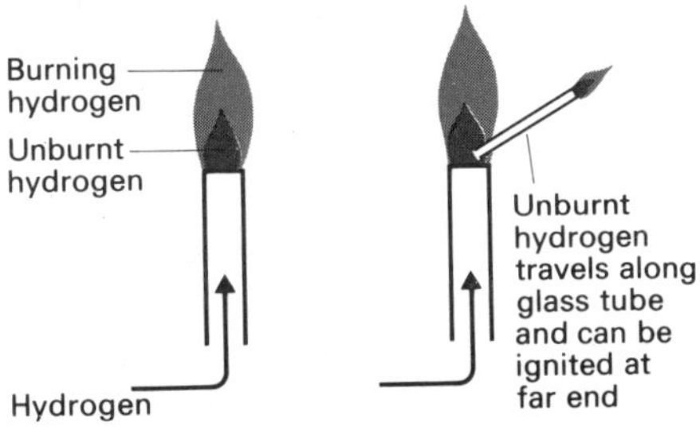

Figure 17.5 The hydrogen flame.

The carbon monoxide flame is identical in structure to the hydrogen flame, but is coloured a characteristic pale-blue.

17.7.3 The Bunsen flame

Before considering the more complex Bunsen flame, your are advised to examine the various parts of a Bunsen burner (figure 17.6). Gas enters the burner tube through a jet at the bottom. When the air-holes are closed a *luminous flame* is produced. Two zones in this flame are easily distinguished (figure 17.7).

(*a*) *A darkish central zone* which contains unburnt gas. This gas can be drawn off with a piece of glass tubing and ignited separately, as with the hydrogen flame. Naturally this zone is quite cool. It is possible to hold a match-head in it for some time before it ignites. Alternatively, if a wooden splint is rapidly brought into position across the central zone and is removed after a few seconds it is found that there is an uncharred central portion.

(*b*) *A bright yellow zone.* If a cooled surface is held in this flame soot is rapidly deposited. This is because carbon particles, caused by incomplete combustion of hydrocarbons in the gas, are present in the flame. e.g.:

$$\underset{\text{methane}}{CH_4(g)} + O_2(g) \rightarrow C(s) + 2H_2O(g)$$

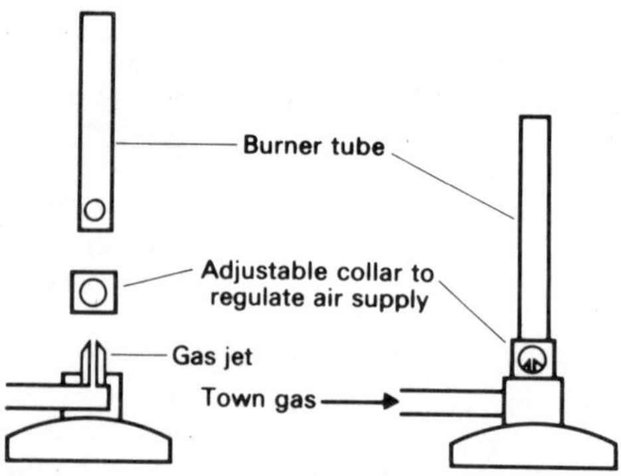

Figure 17.6 The Bunsen burner.

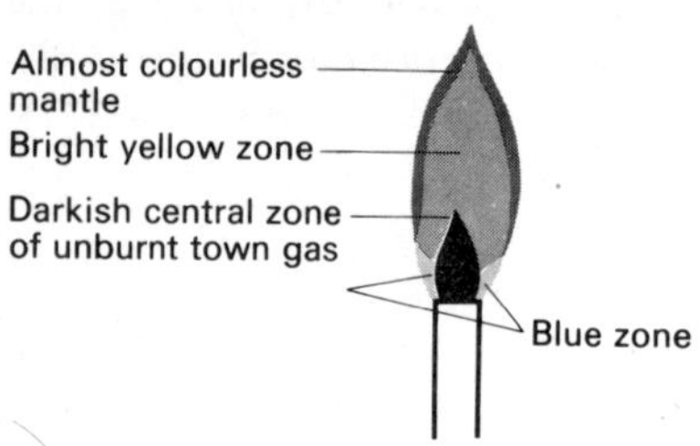

Figure 17.7 The luminous Bunsen flame.

The luminosity of this zone is believed to be due to the heating of these particles to incandescence.

More careful observation will reveal two further zones. One of these is an *almost colourless mantle*, where combustion to carbon dioxide and water is almost complete. The other is a *blue zone* at the bottom of the flame, which receives a plentiful supply of up-rising air, and where combustion is almost complete.

When the air-holes of the Bunsen are open a non-luminous flame is formed. In this case the rush of gas up the burner tube draws in air through the holes, so that a mixture of gas and air is being burned. This ensures plenty of oxygen in the flame, and therefore a much more complete combustion of the gas. The non-luminous flame is much hotter and cleaner than the luminous flame. It contains three main regions (figure 17.8):

(*a*) an inner zone of unburnt gas;
(*b*) a blue cone surrounding this zone—here partial oxidation occurs with oxygen drawn in through the air holes and mixed with the gas;
(*c*) an almost colourless outer zone, where oxidation is completed with oxygen from the air.

The non-luminous Bunsen flame has many applications, other than in laboratories. The burners on a gas cooker, fire, or poker have an air inlet and function similarly to the Bunsen burner. All gas lighting nowadays uses a Bunsen flame, which heats an incandescent mantle, i.e. a mantle which emits light when it is strongly heated. These mantles are usually made from fibre impregnated with thorium and cerium nitrates. On heating, the nitrates decompose to give oxides which are responsible for the incandescence.

When air is mixed with gas in the production of a non-luminous Bunsen flame the speed of burning is considerably greater than when the air-holes are closed. If the gas supply to a Bunsen burner is slowly turned down while the air-holes are fully open a stage is reached when the gas burns downwards more rapidly than it leaves the burner tube. The flame therefore burns down the tube, and ignites at the jet below. The Bunsen burner is then said to have "lit back" or "struck back". When this happens the burner should be turned off immediately and then re-lit with the air-holes open to a suitable extent. Be careful not to touch the burner tube if the gas has been burning at the jet for any length of time as it will be very hot. Wait for it to cool down.

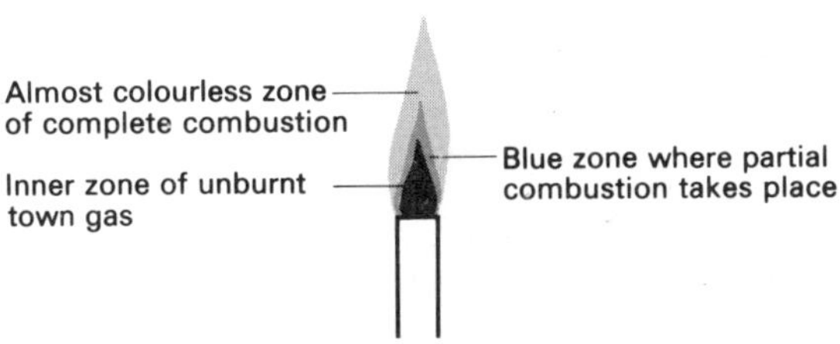

Figure 17.8 The non-luminous Bunsen flame.

17.7.4 The candle flame

A candle flame is similar in structure to the luminous Bunsen burner flame. The candle wax is a mixture of hydrocarbons. When a light is applied these melt and are drawn up the wick by capillary action. They then vaporize, and it is these vapours which ignite. The central zone of the candle flame (figure 17.9) contains unburned vapours of these hydrocarbons, which can be drawn off by a glass tube

Figure 17.9 The candle flame.

and ignited separately. The other zones also correspond very closely to those of the luminous Bunsen flame.

17.8 Explosions

When a mixture of air (or oxygen) and an inflammable gas is ignited the speed of burning depends on the proportions of the gases present. If the quantities are those required by the chemical equation, or are near to these quantities, and the reaction is exothermic, an explosion results. When an explosion occurs the flame travels rapidly through the gas mixture and causes a very rapid increase in pressure. A high temperature is also usually produced, and if the explosion occurs in a confined space so that the gases have no room to expand it can be particularly dangerous.

Explosions can also occur when a mixture of a fine inflammable powder and air are ignited. This unfortunately sometimes happens in a coal mine where the fine dust gets mixed with the air and then ignites. The result of an explosion of this type is shown in figure 17.10; here the exploding mixture of coal-dust and air has been allowed to escape into the open. You can imagine the destruction that can

Figure 17.10 An explosion of a mixture of coal dust and air at the Safety in Mines Research Establishment, Buxton, Derbyshire. (Photo: S.M.R.E. Crown Copyright reserved).

be caused by this happening deep inside a coal mine. Because of the damage that can be caused, many precautions are taken to prevent explosions from happening in mines.

Questions on chapter 17

1. What do you understand by allotropes? Describe *three* physical properties characteristic of wood charcoal and *three* characteristics of graphite.

How would you show experimentally that these substances are allotropes of carbon? (*C.*)

2. The atomic number and relative atomic mass of carbon are 6 and 12 respectively.

(*a*) Give a labelled diagram showing the structure of a carbon atom.

(*b*) Explain what is meant by a covalent bond. Give an electronic diagram for *one* named carbon compound.

Diamond, graphite, and the charcoal obtained from sugar are all pure forms of carbon.

(*c*) How are the atoms arranged in graphite and in diamond? Show how this knowledge enables us to understand the physical characteristics of these substances.

(*d*) Briefly state how it could be shown that graphite and the charcoal obtained from sugar are chemically identical. (Experimental details are *not* required.) (*A.E.B.*)

3. Describe the manufacture of (*a*) producer gas, (*b*) water gas. Explain why it is not possible to use a continuous process for the preparation of water gas. Which of these gases is more efficient as a fuel, and why is this so? How would you demonstrate that both producer gas and water gas contain carbon monoxide? (*C.*)

4. Write an essay, discussing the advantages and disadvantages of important solid, liquid and gaseous fuels.

5. Define *combustion*. How would you show that both carbon dioxide and water vapour are formed when a candle burns in air.

What volume of oxygen is required to burn completely, to carbon dioxide and steam, a mixture of 224 cm^3 methane (CH_4) and 112 cm^3 hydrogen? Assume that all volumes are measured at s.t.p. If the steam formed is condensed to water, what mass of this would be obtained? ($H=1$; $C=12$; $O=16$: one mole of gas occupies 22·4 dm^3 at s.t.p.) (*C.*)

6. Name the products of combustion of a candle in air.

Describe, with the aid of a diagram, an experiment which would demonstrate that the products of combustion of a candle have a greater mass than the candle itself.

Draw a candle flame and state briefly what is taking place in any two of the zones. (*C.W.B.*)

7. Draw the flame produced by a Bunsen burner when its air-hole is open. Account as fully as you can, for what is observed when the air-hole is slowly closed. Under what conditions does a Bunsen burner strike back? Describe an experiment to show that it is possible to burn air in an atmosphere of town gas. (*L.*)

8. Diamond and graphite both consist of giant structures of carbon atoms whereas an iodine crystal is made up of iodine molecules.

Explain the meaning of the above statement and show how the properties of each substance can be related to its structure. (*L.*)

18

Oxides of Carbon, Silica and Glass

Carbon forms two oxides of importance, carbon dioxide, CO_2, and carbon monoxide, CO. Silicon, an element which is similar in some respects to carbon, forms only one oxide—silicon(IV) oxide (silica), SiO_2.

18.1 Preparation of carbon dioxide

18.1.1 A carbonate with acid

When an acid is added to a carbonate or hydrogencarbonate, carbon dioxide is evolved e.g.:

$$\underset{\text{copper carbonate}}{CuCO_3(s)} + 2HNO_3(aq) \rightarrow Cu(NO_3)_2(aq) + H_2O(l) + CO_2(g)$$

$$\underset{\text{sodium hydrogen-carbonate}}{NaHCO_3(s)} + HCl(aq) \rightarrow NaCl(aq) + H_2O(l) + CO_2(g)$$

If the carbonate is powdered, the reaction is vigorous and accompanied by much frothing. Therefore it is customary in the laboratory preparation to use calcium carbonate in the form of marble chips, when the reaction is more readily controlled. Hydrochloric or nitric acid must be used, since the calcium sulphate formed when sulphuric acid is used is insoluble and forms a coating round the marble which prevents further reaction:

$$CaCO_3(s) + 2HCl(aq) \rightarrow CaCl_2(aq) + H_2O(l) + CO_2(g)$$

$$CaCO_3(s) + H_2SO_4(aq) \rightarrow \underset{\text{insoluble}}{CaSO_4(s)} + H_2O(l) + CO_2(g)$$

If the carbon dioxide is required reasonably pure and dry it is first passed through a little water to remove acid spray, and then through concentrated sulphuric acid to dry it. Finally it is collected by downwards delivery (figure 18.1). Otherwise, the carbon dioxide can be collected over water, although a little is then lost by dissolving in water.

Carbon dioxide is a gas that is often required in the laboratory. For this reason it is often prepared using a Kipp's apparatus, or some other similar apparatus. The Kipp's apparatus consists of two parts, as illustrated (figure 18.2), and when assembled is divided into three chambers. Marble chips are placed in the central

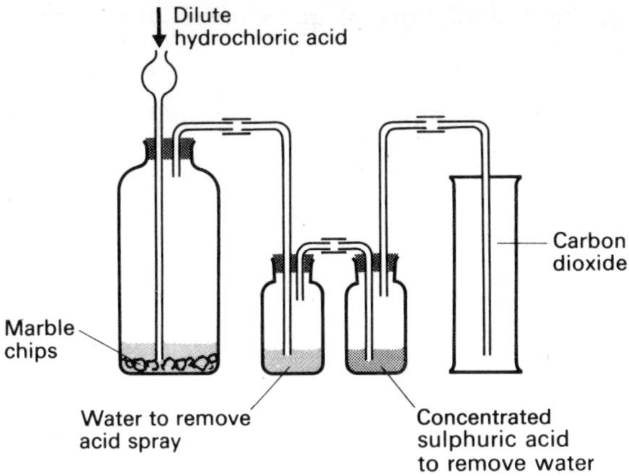

Figure 18.1 Preparation of carbon dioxide.

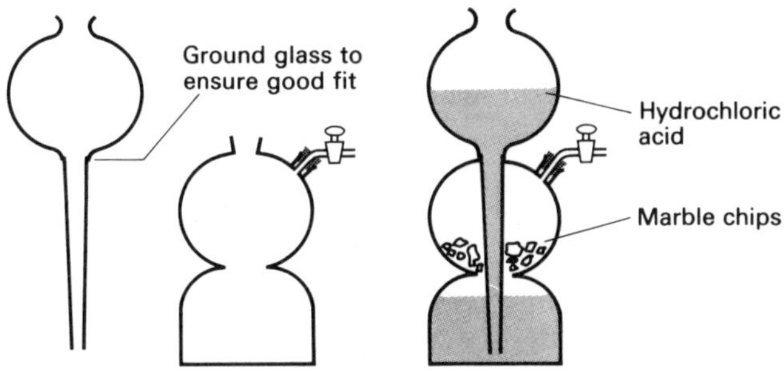

Figure 18.2 Kipp's apparatus.

container and, with the tap open, concentrated hydrochloric acid, diluted with its own volume of water, is poured in until it covers the chips. Carbon dioxide is evolved. If the tap is closed the gas cannot leave the apparatus, and therefore sets up a pressure, which pushes the acid out of contact with the marble and into the top and bottom chambers. Chemical action then stops. If the tap is opened carbon dioxide escapes, and as the pressure is released, acid again contacts the marble to give a steady stream of carbon dioxide. The apparatus can therefore be used many times over until either the acid or the marble is exhausted.

The Kipp's apparatus can be used to prepare any gas formed by the interaction of a cold liquid and lumps of an insoluble solid, e.g. carbon dioxide and hydrogen sulphide. (Why cannot a soluble solid be used?)

18.1.2 Other reactions which yield carbon dioxide

(a) The burning of carbon or carbon containing compounds

(b) The action of heat on carbonates

Except for those of sodium and potassium, most carbonates decompose on heating to give carbon dioxide e.g.:

$$CuCO_3(s) \rightarrow CuO(s) + CO_2(g)$$

$$CaCO_3(s) \rightarrow CaO(s) + CO_2(g)$$

The latter reaction is used in the manufacture of quicklime (CaO), carbon dioxide being a valuable by-product.

(c) The action of heat on sodium or potassium hydrogencarbonates
Both of these decompose on heating to give carbon dioxide, water and the corresponding carbonate:

$$2NaHCO_3(s) \rightarrow Na_2CO_3(s) + H_2O(g) + CO_2(g)$$

(d) The fermentation of sugars by the action of enzymes in yeast

e.g. $C_6H_{12}O_6 \rightarrow 2C_2H_5OH + 2CO_2$

This reaction is mainly used in the manufacture of alcohol and alcoholic drinks, but carbon dioxide is an important by-product. That carbon dioxide is formed is readily shown by warming a 10 per cent solution of glucose to 25–30 °C in a large flask and adding some yeast, preferably brewer's yeast. The flask is connected to a delivery tube which leads to lime water in a flask (figure 18.3), and is stood in a warm place. After some time, the liquid froths, and bubbles of gas pass through the lime water, turning it milky as the white precipitate of calcium carbonate is formed. *This is the test for carbon dioxide.*

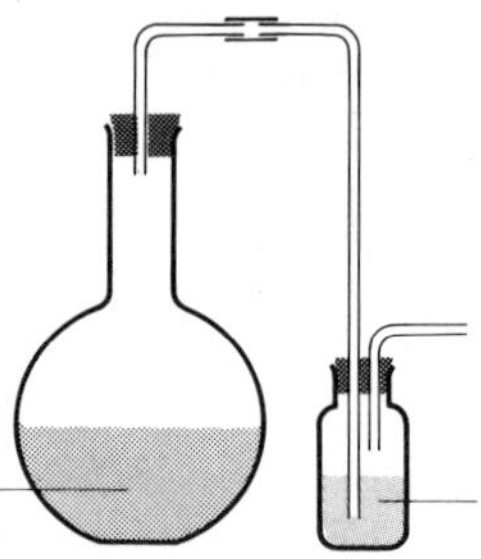

Figure 18.3 Formation of carbon dioxide by fermentation.

18.2 Properties of carbon dioxide

18.2.1 Introduction

Carbon dioxide is a gas and is, therefore, a covalent compound.

It is colourless, almost odourless and has a slightly acidic taste. When compressed and cooled, it readily forms a colourless liquid which is sold in iron cylinders or small sparklet bulbs. If this liquid is allowed to evaporate rapidly, the cooling effect is sufficient to solidify some of the carbon dioxide, forming "dry ice" (see section 18.3).

18.2.2 Carbon dioxide is denser than air.

One of the most interesting ways of showing this is to fill a large tank with carbon dioxide from a Kipp's apparatus. Soap bubbles full of air, blown into the tank, then float on top of the carbon dioxide.

18.2.3 Carbon dioxide will not burn: neither will it support the combustion of most substances.

A burning taper, when plunged into a gas jar full of carbon dioxide, is immediately extinguished. Most other burning substances behave similarly, and for this reason carbon dioxide is used in combating fires.

An exception is burning magnesium, which continues to burn in the gas with a spluttering flame. The magnesium is hot enough to decompose the carbon dioxide into its elements. A mixture of a white powder, magnesium oxide and a black solid, carbon, is formed:

$$CO_2(g) + 2Mg(s) \rightarrow 2MgO(s) + C(s)$$

Dilute acid added to the mixture combines with the magnesium oxide to form a soluble salt, and makes the carbon particles more readily visible.

18.2.4 Carbon dioxide is fairly soluble in cold water

At room temperature water dissolves approximately its own volume of the gas. To show the solubility, a flask, filled with carbon dioxide, is fitted with a cork and a little water is admitted. If, after vigorous shaking, the cork is removed under water some of the water jumps into the flask, showing that some of the gas must have dissolved. A dilute solution of the weak acid, carbonic acid, is formed.

$$CO_2(g) + H_2O(l) \rightleftharpoons H_2CO_3(aq)$$

18.2.5 Carbon dioxide combines readily with solutions of alkalis

If a test-tube full of carbon dioxide is opened under sodium hydroxide solution and is then gently shaken, the solution rises rapidly to the top of the tube. Sodium carbonate is formed:

$$\left[\begin{array}{lll} CO_2 + 2NaOH & \rightarrow & Na_2CO_3 + H_2O \\ CO_2(g) + 2OH^-(aq) & \rightarrow & CO_3^{2-}(aq) + H_2O(l) \end{array}\right.$$

When carbon dioxide is bubbled through the solution, sodium carbonate is again formed at first, but continued passage of the gas produces the hydrogen-carbonate:

$$\left[\begin{array}{lll} Na_2CO_3 + H_2O + CO_2 & \rightarrow & 2NaHCO_3 \\ CO_3^{2-}(aq) + H_2O(l) + CO_2(g) & \rightarrow & 2HCO_3^-(aq) \end{array}\right.$$

With lime water, a solution of calcium hydroxide in water, similar reactions occur. The calcium carbonate first formed is insoluble, and a milky suspension of it is obtained:

$$Ca(OH)_2(aq) + CO_2(g) \rightarrow \underset{\text{calcium carbonate (insoluble)}}{CaCO_3(s)} + H_2O(l)$$

More carbon dioxide converts the carbonate into the hydrogencarbonate which is soluble, and the liquid becomes clear again:

$$CaCO_3(s) + H_2O(l) + CO_2(g) \rightarrow \underset{\substack{\text{calcium}\\\text{hydrogencarbonate}\\\text{(soluble)}}}{Ca(HCO_3)_2(aq)}$$

18.3 Uses of carbon dioxide

(a) Effervescent drinks

Much carbon dioxide is used commercially in the manufacture of mineral waters. In such drinks the carbon dioxide is dissolved under pressure. When the pressure is released, the carbon dioxide comes out of solution, producing the characteristic effervescence.

Carbon dioxide also causes the effervescence in beer, cider and champagne. In these drinks it is not usually dissolved artificially, but is obtained by sealing the containing vessel before fermentation is complete (see section 25.6.2(*b*)).

Various health salts also produce carbon dioxide when they are dissolved in water. These contain an acid, e.g. 2,3-dihydroxybutanedioic (tartaric) acid, and a carbonate, e.g. sodium hydrogencarbonate, which only interact in solution. Therefore no reaction occurs until water is added to the salts.

(b) Fire extinguishers

Many types of fire extinguishers depend on the production of carbon dioxide.

A typical fire extinguisher is illustrated in figure 18.4. When the cylinder is inverted the solutions of sulphuric acid and sodium hydrogencarbonate mix. A large pressure is set up which forces a froth of carbon dioxide through the jet. This is directed onto the fire so that a blanket of carbon dioxide and water smothers the flames.

Foam fire extinguishers use a solution of the acidic aluminium sulphate instead of sulphuric acid and have a substance such as saponin added to render the froth more stable. The carbon dioxide is then less readily dispersed.

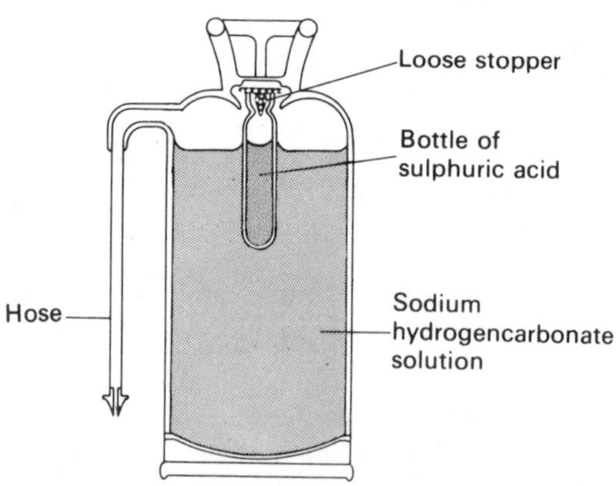

Figure 18.4 A fire extinguisher.

Another type of extinguisher consists of a cylinder containing liquid carbon dioxide which is released as a vapour when a valve is opened. As no water is present such an extinguisher is useful near electrical equipment.

(c) Baking powders

When yeast is used in baking, the enzymes contained in it catalyze the fermentation of sugar in the dough to form carbon dioxide and alcohol. Bubbles of carbon dioxide then cause the dough to rise.

Baking powders, which are more convenient to use than yeast, consist of an acidic substance (e.g. 2,3-dihydroxybutanedioic (tartaric) acid and sodium hydrogencarbonate. These react in the presence of water to produce the required carbon dioxide.

(d) Refrigeration

Solid carbon dioxide or "dry ice" is sold as a refrigerant. It is widely used for keeping ice-cream cold and has replaced freezing mixtures such as ice and salt for this purpose. It provides a low temperature, and on warming changes straight from the solid to the gaseous state so that it is particularly clean in use.

18.4 Carbonates and hydrogencarbonates (bicarbonates)

Carbonic acid is dibasic, and therefore forms two series of salts—carbonates and hydrogencarbonates.

18.4.1 Preparation of carbonates

(a) Of the common carbonates, only those of sodium, potassium, and ammonium are soluble in water. These are prepared in the laboratory from the corresponding alkali:

$$\left[\begin{array}{l} 2NaOH + CO_2 \rightarrow Na_2CO_3 + H_2O \\ 2OH^-(aq) + CO_2(g) \rightarrow CO_3^{2-}(aq) + H_2O(l) \end{array}\right.$$

Full details of this preparation are given in section 21.2.6(*d*).

(b) Other carbonates are prepared by precipitation. Thus calcium carbonate is precipitated when *solutions* of a calcium salt and a carbonate are mixed e.g.:

$$\left[\begin{array}{llllll} CaCl_2 & + Na_2CO_3 & \rightarrow & CaCO_3 & + & 2NaCl \\ \text{calcium} & \text{sodium} & & \text{calcium} & & \text{sodium} \\ \text{chloride} & \text{carbonate} & & \text{carbonate} & & \text{chloride} \end{array}\right.$$

$$Ca^{2+}(aq) + CO_3^{2-}(aq) \rightarrow CaCO_3(s)$$

(N.B. In cases where a metal forms both insoluble hydroxides and carbonates, a mixture of both is often precipitated. For example, if sodium carbonate solution is mixed with lead(II) nitrate solution, a basic carbonate, lead(II)-carbonate hydroxide, $Pb_3(CO_3)_2(OH)_2$, is formed. This is because sodium carbonate solution is alkaline due to hydrolysis and contains enough hydroxide ions to cause the joint precipitation.)

18.4.2 Properties of carbonates

(a) Action of heat

The carbonates of sodium and potassium, metals high in the activity series, *do not decompose on heating.*

All other common carbonates, with the exception of ammonium carbonate, decompose to give carbon dioxide and the oxide of the metal e.g.:

$$PbCO_3(s) \rightarrow PbO(s) + CO_2(g)$$

Ammonium carbonate decomposes thus:

$$(NH_4)_2CO_3(s) \rightarrow 2NH_3(g) + H_2O(g) + CO\ (g)$$

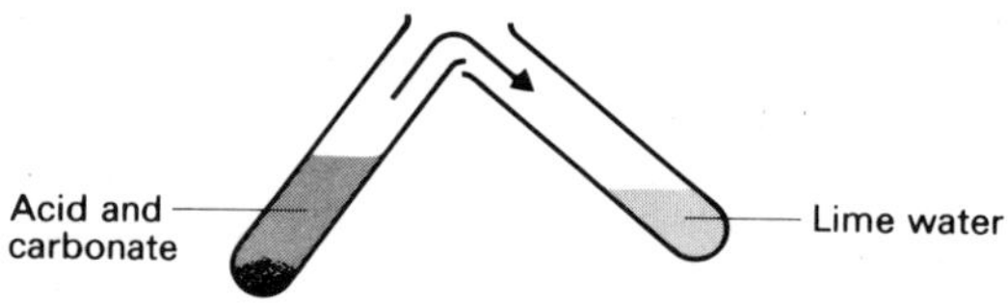

Figure 18.5 Test for a carbonate.

(b) Action of acids

When an acid is added to a carbonate, carbon dioxide, water and a salt are formed e.g.:

$$\left[\begin{array}{l} PbCO_3 + 2HNO_3 \rightarrow Pb(NO_3)_2 + H_2O + CO_2 \\ PbCO_3(s) + 2H^+(aq) \rightarrow Pb^{2+}(aq) + H_2O(l) + CO_2(g) \end{array}\right.$$

To test for a carbonate, calcium hydroxide solution (lime water) is placed in a clean test-tube and acid is added to the suspected carbonate in another. The test-tubes are held as shown in figure 18.5. With a carbonate, there is a ready effervescence, even in the cold, and the dense carbon dioxide falls into the tube containing lime water, which then forms a milky suspension of calcium carbonate on shaking.

18.4.3 Preparation of hydrogencarbonates (bicarbonates)

All hydrogencarbonates are soluble. The only common solid hydrogen-carbonates are those of sodium, potassium and ammonium. Calcium hydrogencarbonate, which is important as a cause of hardness in water, is known only in solution.

The hydrogencarbonates are prepared by passing carbon dioxide through a *cold* solution of the corresponding alkali or carbonate. If the alkali is used the carbonate is first formed e.g.:

$$\left[\begin{array}{l} 2NaOH + CO_2 \rightarrow Na_2CO_3 + H_2O \\ Na_2CO_3 + H_2O + CO_2 \rightarrow 2NaHCO_3 \\ 2OH^-(aq) + CO_2(g) \rightarrow CO_3^{2-}(aq) + H_2O(l) \\ CO_3^{2-}(aq) + H_2O(l) + CO_2(g) \rightarrow 2HCO_3^{2-}(aq) \end{array}\right.$$

Full details of this preparation are given in section 21.2.6(*c*)

18.4.4 Properties of hydrogencarbonates (bicarbonates)

(a) Action of acids

Like carbonates, all hydrogencarbonates yield carbon dioxide with acids e.g.

$$\underset{\text{potassium hydrogen-carbonate}}{KHCO_3} + HNO_3 \rightarrow KNO_3 + H_2O + CO_2$$

$$HCO_3^-(aq) + H^+(aq) \rightarrow H_2O(l) + CO_2(g)$$

(b) Action of heat

All hydrogencarbonates, whether in the solid state or in solution, decompose on heating to give the normal carbonate, carbon dioxide and water e.g.:

$$2NaHCO_3(s) \rightarrow Na_2CO_3(s) + H_2O(g) + CO_2(g)$$

18.5 Carbon monoxide

18.5.1 Preparation of carbon monoxide

(a) By the dehydration of methanoic (formic) acid or methanoates

This is the most convenient method for use in the laboratory. Concentrated sulphuric acid is dropped on to a concentrated solution of methanoic acid or on to solid sodium methanoate (figure 18.6). An effervescence is observed, and the carbon monoxide is then collected over water:

$$\underset{\text{methanoic acid}}{H.COOH} - \underset{\text{by dehydrating action of sulphuric acid}}{H_2O} \rightarrow CO$$

If sodium methanoate is used, methanoic acid is initially formed and is then dehydrated:

$$H.COONa + H_2SO_4 \rightarrow H.COOH + NaHSO_4$$

$$H.COOH - H_2O \rightarrow CO$$

$$H.COONa + H_2SO_4 \rightarrow NaHSO_4 + H_2O + CO$$

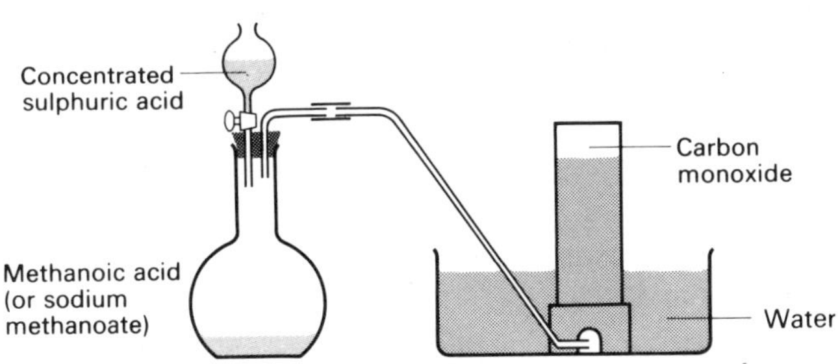

Figure 18.6 Preparation of carbon monoxide.

Ethanedioic acid is sometimes used instead of methanoic acid. When heated, this is readily dehydrated to give a mixture of equal volumes of carbon monoxide and carbon dioxide.

$(COOH)_2 - H_2O \rightarrow CO + CO_2$
ethanedioic
acid

How can pure carbon monoxide be obtained from this mixture?

(b) By reduction of carbon dioxide
This is done by passing a slow stream of carbon dioxide from a Kipp's apparatus through an iron or silica tube, well packed with lumps of charcoal (or coarse iron filings) and heated to red heat (figure 18.7).

$CO_2(g) + C(s) \rightarrow 2CO(g)$
$4CO_2(g) + 3Fe(s) \rightarrow 4CO(g) + Fe_3O_4(s)$

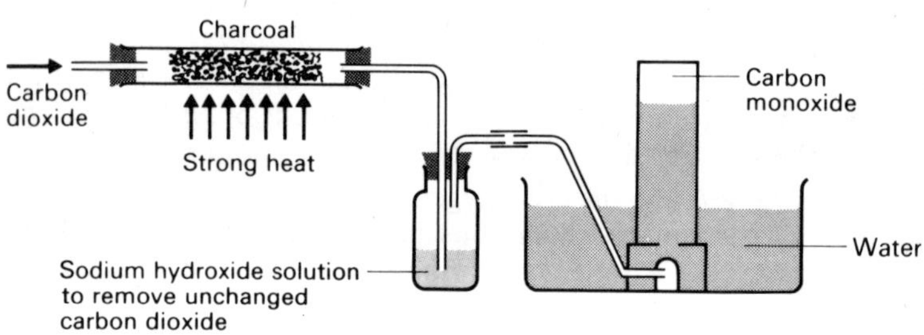

Figure 18.7 Preparation of carbon monoxide.

As the reduction is not complete, unchanged carbon dioxide must be removed by passage through sodium hydroxide solution.

18.5.2 Properties of carbon monoxide

Carbon monoxide is a colourless, odourless and tasteless gas, with a density very slightly less than air. It is almost insoluble in water.
It burns with a pale blue flame to form carbon dioxide:

$2CO(g) + O_2(g) \rightarrow 2CO_2(g)$

(a) Carbon monoxide is a reducing agent
When carbon monoxide is passed over heated copper(II) oxide, the metal is formed together with carbon dioxide:

$CuO(s) + CO(g) \rightarrow Cu(s) + CO_2(g)$

This is readily shown using the apparatus described for the reaction of hydrogen with copper(II) oxide in section 3.3. Other oxides, such as the oxides of lead or iron(III) oxide, are similarly reduced to the metal e.g.:

$$Pb_3O_4(s) + 4CO(g) \rightarrow 3Pb(l) + 4CO_2(g)$$

$$Fe_2O_3(s) + 3CO(g) \rightarrow 2Fe(l) + 3CO_2(g)$$

The last reaction takes place in the blast furnace used in the extraction of iron from its ores (see section 24.2.2).

(b) Carbon monoxide is an active poison

Blood contains a red pigment known as haemoglobin. During breathing this pigment combines with oxygen in the lungs to form a compound, oxyhaemoglobin:

Haemoglobin + Oxygen → Oxyhaemoglobin

The oxyhaemoglobin then travels round the body and decomposes yielding oxygen to tissues which require it:

Oxyhaemoglobin → Haemoglobin + Oxygen

The haemoglobin then returns to the lungs, where it combines with more oxygen. This role of blood as "oxygen carrier" is most important.

If carbon monoxide is inhaled, this also combines with blood to form carboxyhaemoglobin. This compound is much more stable than oxyhaemoglobin, and is unable to take up oxygen. Therefore continued inhalation produces death by starving the tissues of oxygen.

Carbon monoxide is particularly dangerous because of its lack of colour, odour and taste. Little is noticed if it is being inhaled, until the victim suddenly becomes weak, and is unable to escape without aid. First-aid consists of removing the victim to fresh air and applying artificial respiration. Carbon dioxide stimulates breathing, and in the laboratory a tube leading from a Kipp's apparatus producing the gas can be held an inch or two from the nose for this purpose. In severe cases of poisoning, a blood transfusion is essential.

There are several dangerous sources of carbon monoxide encountered in everyday life:

(i) Geysers and gas fires

When town gas burns against a cooled surface or in a confined space, combustion is incomplete and some carbon monoxide escapes. With such appliances, flues for the waste gases or efficient ventilation are essential.

(ii) Coke fires

In coal or coke fires the following reactions occur:

1. At bottom of fire, where there is a good air supply:

$$C(s) + O_2(g) \rightarrow CO_2(g)$$

2. At centre of fire, where there is a limited air supply, carbon dioxide from the above reaction is reduced by red-hot carbon:

$$CO_2(g) + C(s) \rightarrow 2CO(g)$$

3. At top of fire, where there is a good air supply. If the fire is hot enough, the

carbon monoxide from the second reaction burns to form carbon dioxide:

$$2CO(g) + O_2(g) \rightarrow 2CO_2(g)$$

With coke fires some of the carbon monoxide is particularly liable to escape unburned. If the ventilation is good this does not matter, as the gas goes up the chimney. Otherwise the situation is dangerous. Coke braziers of the sort used by night-watchmen must always be used well in the open. A number of deaths has been known because watchmen have used these when drawn too near their huts or in enclosed spaces.

(iii) Car exhausts

The exhaust fumes from motor cars contain carbon monoxide, because the petrol vapour has not been mixed with sufficient oxygen to ensure complete combustion to carbon dioxide and water. It is therefore extremely dangerous to run a car engine in a garage with the doors closed.

18.5.3 Uses of carbon monoxide

Carbon monoxide is becoming increasingly important as a raw material for the chemical industry. It is a constituent of water gas and producer gas. These can be obtained from coke (section 17.6.1(*d*)), from oil or from natural gas (section 14.3.2). It can be converted to other chemicals, e.g. methanol (section 14.3.2) and ethanoic acid (section 25.2.1). It is also used in the refining of nickel.

18.6 The carbon cycle in nature

In the burning of fuels, in respiration and by numerous other methods, large quantities of carbon dioxide are continuously being released to the atmosphere. Yet air contains a small and more or less constant proportion of the gas (0·03 per cent by volume). Therefore there must be means by which carbon dioxide is removed from the atmosphere, i.e. there is a carbon cycle in nature.

18.6.1 Important methods by which carbon dioxide reaches the atmosphere include:

(a) Combustion i.e. burning of coke and other fuels. Any fuel containing carbon, either free or in the combined state, forms carbon dioxide when it is burned.

(b) Respiration by plants and animals (see section 15.5).

(c) By decay of animal and vegetable matter.

(d) From volcanoes.

18.6.2 Important methods by which carbon dioxide is removed from the atmosphere include:

(a) Photo-synthesis (i.e. building up by means of light)

Carbon is an essential constituent of all living matter. Plants obtain their carbon dioxide from the atmosphere. In sunlight they absorb the gas from the air, and in the presence of the green colouring matter, chlorophyll, make it combine with water from the soil to form carbohydrates. A carbohydrate is a compound of

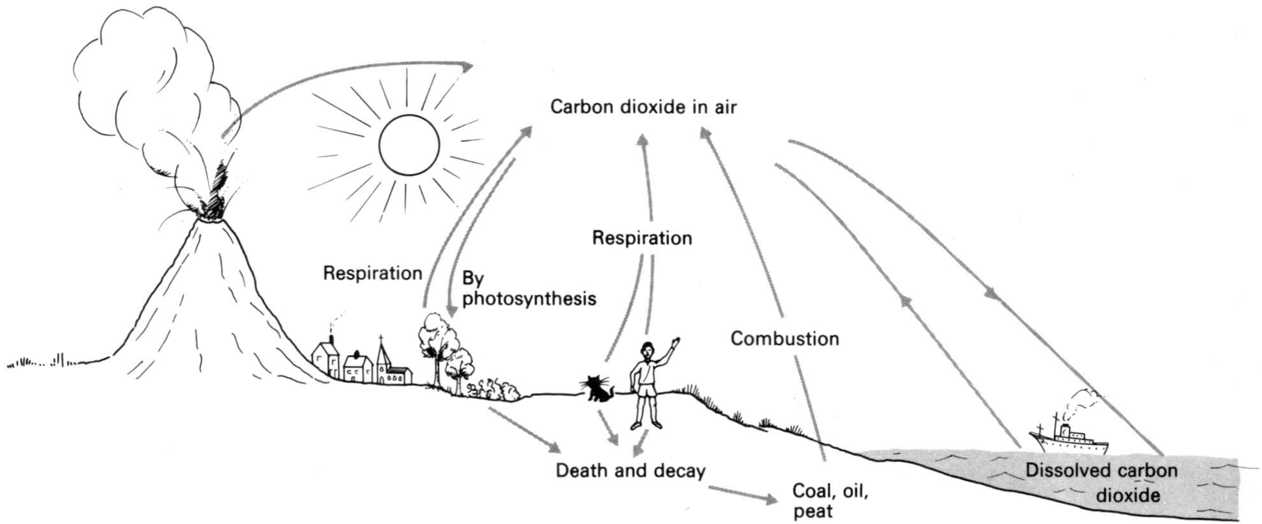

Figure 18.8 The carbon cycle. (Drawing: A.M.J. Dixon).

carbon, hydrogen and oxygen, containing two atoms of hydrogen for each atom of oxygen (see also section 26.3.1).) Sugars are probably formed first, and then starch and cellulose. The synthesis can be represented by a general equation:

$$xCO_2 + yH_2O \rightarrow C_xH_{2y}O_y + xO_2$$
a carbohydrate

e.g. $$6CO_2 + 6H_2O \rightarrow C_6H_{12}O_6 + 6O_2$$
glucose

or $$6nCO_2 + 5nH_2O \rightarrow (C_6H_{10}O_5)_n + 6nO_2$$
starch

These syntheses cannot occur in the absence of light or of chlorophyll. During night-time therefore there is no carbon dioxide being absorbed by plants. On the contrary, carbon dioxide is given out to the atmosphere by respiration. Although respiration almost certainly occurs during day-time, the whole process is masked by photosynthesis, which causes the absorption of carbon dioxide.

It should be noted that photosynthesis is, in effect, the opposite to combustion, i.e. when a carbohydrate burns, it combines with oxygen to form carbon dioxide and water.

(b) By solution in rain and sea water
Carbon dioxide dissolves in natural waters to give carbonic acid. Some of this dissolved carbon dioxide combines with calcium and magnesium carbonates to form the hydrogencarbonates:

$$CaCO_3(s) + H_2O(l) + CO_2(g) \rightarrow Ca(HCO_3)_2(aq)$$
$$MgCO_3(s) + H_2O(l) + CO_2(g) \rightarrow Mg(HCO_3)_2(aq)$$

This happens particularly in sea water, and as all the reactions concerned are reversible, any temporary decrease in the percentage of carbon dioxide in the air results in a partial decomposition of the hydrogencarbonates. This tends to maintain the constancy of the proportion of carbon dioxide in the air.

Figure 18.8 is an illustration of this carbon cycle.

18.7 Silicon

18.7.1 Introduction

The element silicon is the most plentiful in the earth's crust and is widely distributed as silicates (the main constituent of clay) and as its oxide, silica. It is now immensely important as the element on which *solid state electronics* is based. This started with the transistor and has developed with the *silicon chip* in which many thousands of transistors and other electronic components are all present on a minute piece of silicon (see section 28.9.4).

18.7.2 Silicon(IV) oxide, silica, SiO_2

Sand, quartz (figure 1.22) and flint are all forms of silica containing varying amounts of impurity. Certain gems, e.g. amethyst, also consist of silica coloured by impurities.

Silicon(IV) oxide is much less reactive than carbon dioxide, but like that compound, it is an acidic oxide. Thus, if silicon(IV) oxide is heated with concentrated sodium hydroxide solution or molten sodium hydroxide, sodium silicate is formed:

$$SiO_2(s) + 2NaOH(aq) \rightarrow \underset{\text{sodium silicate}}{Na_2SiO_3(aq)} + H_2O(l)$$

$$\text{(cf. } CO_2(g) + 2NaOH(aq) \rightarrow \underset{\text{sodium carbonate}}{Na_2CO_3(aq)} + H_2O(l)\text{)}$$

Silicon(IV) oxide melts at about 1500 °C, and on cooling *fused silica* is obtained. In this respect it is very different from carbon dioxide which, at normal temperatures, is a gas. This difference is explained by the very different way in which the atoms are combined in the two molecules. As explained in section 6.23, carbon dioxide consists of tri-atomic molecules ($O=C=O$); very little attraction exists between these molecules. Silicon(IV) oxide, however, forms giant molecules and should be given the formula $(SiO_2)_n$. The silicon and oxygen atoms are joined covalently by single bonds to form a three dimensional pattern containing millions of atoms. In the silica (quartz) crystal the silicon atoms are arranged in the same way as the carbon atoms in diamond (figure 6.20) with an oxygen atom in between every pair of silicon atoms.

Fused silica has an extremely low coefficient of expansion, and this fact, coupled with a low chemical reactivity, accounts for its use in many vessels used in the laboratory and in industry. Silica test-tubes and crucibles, for example, can be heated to red heat and then plunged into cold water without cracking. (Can you explain why ordinary glass shatters under these conditions?)

18.7.3 Silicic acid and silicates

Clays, and many rocks and minerals are silicates (figure 18.9). The silicates are

Figure 18.9 An example of a silicate rock—the Giant's Causeway in Northern Ireland. It is made of basalt (a volcanic rock) and the peculiar column structure occurred during the shrinking of the molten rock as it cooled. (Photo: Institute of Geological Sciences. Crown Copyright reserved).

usually giant molecules rather like silica, but which contain many other metallic elements in addition to the silicon and oxygen.

A simpler silicate is that of sodium. As already mentioned, silicon(IV) oxide, (silica), being an acidic oxide, combines with alkali to form the salts known as silicates.

Sodium silicate is manufactured by heating sodium carbonate with silicon(IV) oxide, when the more volatile carbon dioxide is displaced:

$$Na_2CO_3(s) + SiO_2(s) \rightarrow \underset{\text{sodium silicate}}{Na_2SiO_3(s)} + CO_2(g)$$

A concentrated solution of sodium silicate in water is sold commercially as "water glass". This was used as a preservative for eggs. Its main uses these days are as an adhesive, and for fireproofing fabrics. Addition of sodium silicate to solutions of other metal salts will give the coloured insoluble precipitate of the metal salt. This can be illustrated beautifully by means of a simple experiment known as a chemical garden.

Experiment. A chemical garden

Make 450 cm^3 of a solution (10–20 %) of sodium silicate in hot water in a 500 cm^3 beaker and allow to cool. When the solution is cold drop into the beaker a few small crystals of copper(II) sulphate, iron(II) sulphate, aluminium iron(III) sulphate, cobalt(II) chloride, nickel(II) sulphate and iron(III) chloride. Leave the beaker undisturbed, when beautiful coloured tree like growths of silicate precipitates will grow from the crystals. See if you can work out how they grow.

Addition of hydrochloric acid to a solution of sodium silicate causes a gelatinous precipitate of silicic acid:

$$Na_2SiO_3(aq) + 2HCl(aq) \rightarrow 2NaCl(aq) + \underset{\text{silicic acid}}{H_2SiO_3(s)}$$

This reaction is similar to the action of an acid on a carbonate, except that when carbonic acid is formed it decomposes in the cold to give carbon dioxide and water. When heated, the silicic acid decomposes in a similar way to give silicon(IV) oxide and water:

$$H_2SiO_3(s) \rightarrow H_2O(l) + SiO_2(s)$$

If silicic acid is partially dehydrated by heating under reduced pressure, it forms a light porous solid known as "silica gel". Silica gel will adsorb large quantities of vapour and is used as a drying agent and for the recovery of vapours from organic solvents. It is also used as a catalyst.

18.8 Glass

18.8.1 Introduction

Glass is a mixture of silicates formed by heating together a mixture of silica with certain oxides or carbonates. Most glass is made by heating together silica, calcium carbonate and sodium carbonate at 1300–1400 °C. Carbon dioxide is evolved, leaving a mixture of sodium and calcium silicates with excess silica:

$$CaCO_3(s) + SiO_2(s) \rightarrow CaSiO_3 + CO_2(g)$$

$$Na_2CO_3(s) + SiO_2(s) \rightarrow Na_2SiO_3(s) + CO_2(g)$$

Such glass has a composition approximating to the formula $CaSiO_3 . Na_2SiO_3 . 4SiO_2$ (or $CaO.Na_2O.6SiO_2$). Small quantities of other constituents, e.g. aluminium and iron(III) oxides, are usually added to impart various desirable properties to the glass.

The molten glass obtained by the above process is a viscous liquid which can be shaped or worked in many ways. As it cools it becomes gradually more viscous, and eventually hard and brittle. There is no sudden change to the solid state, as there is when most liquids are cooled. Although in the everyday sense of the word cold glass is a solid, to the scientist it is a supercooled liquid. The atoms vibrate about fixed positions as in the solid state, but they are arranged randomly, as in a liquid.

18.8.2 Types of glass

(a) Soda glass, made by the method indicated above from silica, sodium carbonate and calcium carbonate with very little else added, comprises most of the glass made today.

If ordinary glass is heated rapidly, it readily cracks. This is because glass is a poor conductor of heat, and therefore unequal expansion sets up a strain which breaks the glass. Rapid cooling of glass also causes it to break.

(b) Heat-resistant glass, such as Pyrex, has a very low coefficient of expansion, and thus is able to withstand reasonable changes of temperature. It has a high silica content, a medium boron(III) oxide content and only small quantities of sodium and aluminium oxides (e.g. 80 per cent silica, 12 per cent boron(III) oxide). These glasses are not only heat resistant but also withstand the attack of chemicals much better than soda glass. They are used both in the laboratory and for cooking utensils.

(c) Coloured glasses are made by adding small quantities of suitable chemicals to molten glass. Cobalt oxide, for example, imparts a deep blue colour; manganese oxide, purple; chromium(III) oxide, green; and iron(III) oxide, yellow. Uranium compounds impart a greenish fluorescence which is effective for fancy glass-ware.

18.8.3 The working of glass

Many articles of glass, particularly hollow-ware, are made by blowing. The glass-blower takes up a certain amount of molten glass on the end of a long iron tube. By suitable manipulation of this tube and by blowing down it, vessels of a variety of shapes can be made. In the final stages of the blowing the glass may be inserted in a mould to obtain the right shape. For ordinary bottles and electric-light bulbs, which have to be made in vast numbers, automatic blowing machines are used. Complex laboratory glassware is made almost entirely by hand (figure 18.10).

Articles such as fruit dishes, motor headlamp lenses, etc., are made by pressing the glass in a mould. Flat glass is made by the float process. Here the molten glass

Figure 18.10 A glassblower making a piece of scientific glassware. (Photo: A. Rendell).

is floated on molten tin, where it cools to a perfect flat finish on the smooth molten tin surface (figures 18.11 and 18.12).

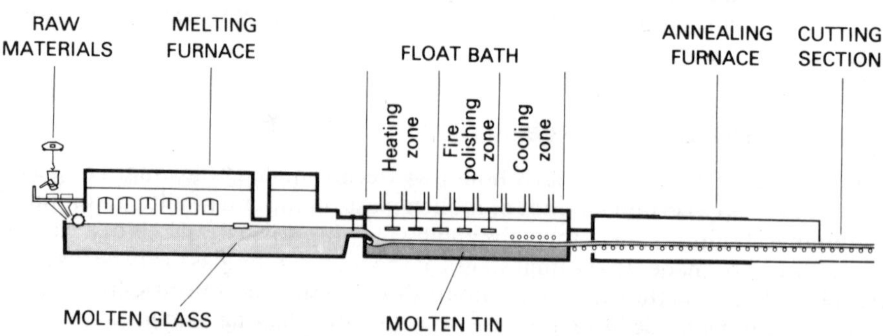

Figure 18.11 The manufacture of float glass.

Figure 18.12 A continuous sheet of float glass coming out of the annealing oven. (Photo: Pilkington Brothers, Ltd.).

18.8.4 Glass fibres

An interesting modern application of glass is the manufacture of glass wool. Glass, in the form of marbles, is dropped on to an electrically heated platinum sheet containing fine holes. The glass melts and drops through these holes. As the globules of glass form on the lower side of the sheet, they are drawn out to fibres which are wound on to a rotating spindle. Long glass filaments are formed. These are becoming increasingly important for "glass fibre optics".

Glass fibres are non-inflammable, non-hygroscopic, do not deteriorate under normal conditions and resist chemical attack. Textiles made from glass are therefore used for electrical insulation, fireproof fabrics and for filtration of corrosive chemicals. Loose glass wool is widely used for heat and sound insulation.

18.9 Earthenware and Porcelain

Clay is a term used to describe certain forms of earth which are plastic when wet. Clays are not usually pure substances, but all contain hydrated aluminium silicate, which binds together particles of silica, quartz and other impurities. One of the purest clays is *kaolin* or *china clay*. One of the main sources of china clay are the quarries in Devon and Cornwall. Here the china clay is washed from the earth with powerful water jets (figure 18.13), and is then purified and dried. China clay is white, and from it are made *porcelain* and good china, and other *ceramic* products. The china clay is mixed with felspar and quartz, and is then moulded to the required shape. When the mixture is heated to 1300 °C in a kiln a non-porous

Figure 18.13 A powerful remotely controlled water-hose washing china clay from the clay-pit face. (Photo: English China Clays Group).

porcelain is formed. Much china clay is also used in paper making as a filler mixed with the cellulose fibres and as a coating (see section 26.3.1(*d*)).

If impure clay and a lower kiln temperature are used a porous solid known as earthenware is formed. The appearance of earthenware articles may be improved by glazing. This consists of coating the ware with some fusible silicate and reheating, when a glaze is formed on the surface.

Questions on Chapter 18

1. Describe, with a diagram, how you would prepare and collect several gas jars of carbon dioxide. How and under what conditions does carbon dioxide react with (*a*) sodium hydroxide, (*b*) magnesium and (*c*) carbon? (*J.M.B.*)

2. Draw a fully labelled diagram for the laboratory preparation and collection of carbon dioxide. Give the equation.

(*a*) State, giving reagents and reaction conditions, how carbon dioxide can be reduced to carbon monoxide. Give an equation for the reaction.

(*b*) Describe carefully how, by a chemical test, you would distinguish between solid samples of hydrated sodium carbonate and sodium hydrogencarbonate. Give the result of the test on each substance.

(*c*) When carbon monoxide and chlorine are passed over a heated catalyst, gaseous carbonyl chloride is formed.

$$CO(g) + Cl_2(g) \rightarrow COCl_2(g)$$

(i) If 20 cm^3 of carbon monoxide were reacted with 20 cm^3 of chlorine, what would be the maximum volume of carbonyl chloride formed if all measurements were made at the same temperature and pressure?

(ii) Using lines to represent covalent bonds, give the probable structural formula of carbonyl chloride. (*J.M.B.*)

3. Draw a labelled digram of a Kipp's apparatus suitably charged to prepare carbon dioxide. Give the equation for the reaction.

How does carbon dioxide react with (*a*) hot sodium hydroxide solution, (*b*) red hot carbon?

What would be the volume at 17 °C and 720 mm of a quantity of carbon dioxide the volume of which was 560 cm^3 at s.t.p.? (*L.*)

4. Name the materials you would use to prepare carbon dioxide. Give the equation for the reaction.

A blue flame is often seen above a clear fire. Explain fully how air and carbon produce this phenomenon.

When a mixture of 30 cm^3 of oxygen and 30 cm^3 of carbon monoxide is sparked, what volume of gas will remain (*a*) after explosion and (*b*) after treating the product with sodium hydroxide solution? (*L.*)

5. By means of a labelled diagram and an equation, show how a sample of carbon dioxide can be made and collected in the laboratory.

Without giving any details of the apparatus used, describe how you would convert carbon dioxide into pure carbon monoxide.

A journalist in a motoring magazine wrote, "On a busy roadway, the proportion of carbon monoxide has varied from 6 parts per million to 180 parts per million".

(*a*) At what time of day would you expect the concentration of carbon monoxide to be high?

(*b*) By what reaction is the carbon monoxide formed?

(*c*) What is the effect of carbon monoxide on blood, and why does this make the gas so poisonous? (*J.M.B.*)

6. How would you demonstrate (*a*) that carbon dioxide is slightly soluble in water, (*b*) that carbon dioxide contains carbon and oxygen, (*c*) that carbon dioxide is produced when a candle burns? (*C.W.B.*)

7. When crystalline ethanedioic acid (oxalic acid) is heated with concentrated sulphuric acid a reaction represented by the following equation takes place:

$$\begin{array}{l} CO_2H \\ | \\ CO_2H \end{array} \longrightarrow CO_2\uparrow + CO\uparrow + H_2O$$

How does the sulphuric acid act in this reaction? How, starting from ethanedioic acid, could your prepare: (*a*) carbon monoxide; (*b*) carbon dioxide. (*J.M.B.*)

8. Describe, with the aid of a labelled diagram, how you would prepare and collect a jar of pure carbon monoxide, starting from carbon dioxide.

Why is it dangerous to run a motor-car engine in a closed garage?

A graduated tube contains a mixture of carbon dioxide and carbon monoxide. How would you determine the percentage composition by volume of the mixture?

20 cm^3 of carbon monoxide were mixed with 20 cm^3 of oxygen and exploded. What is the composition of the remaining mixture? (*C.W.B.*)

9. Describe how you would prepare and collect carbon monoxide free from carbon dioxide.

Illustrate the difference between an exothermic and an endothermic reaction by reference to the industrial preparation of producer gas and water gas.

Give the approximate composition of each of these gases. (*O.*)

10. (*a*) What would be the effect of heating a mixture of sand and sodium carbonate to bright redness, and what would be the properties of the substance produced?

(*b*) What would be the result if some calcium carbonate was added to the mixture before heating?

(*c*) Give reasons for regarding silica as an acid-forming oxide.

(*d*) How would you separate salt from a mixture of sand and salt and how would you prove that the salt was completely removed? (*D.*)

11. Carbon dioxide is obtained by the action of dilute hydrochloric acid on calcium

carbonate. Describe carefully how you would investigate the effects of change in temperature on the reaction rate, and the possible effect of the addition of a copper(II) sulphate crystal to the mixture.

Show by an equation how carbon dioxide may be converted directly into carbon monoxide.

A mixture of carbon dioxide and carbon monoxide is passed into water under pressure. The gas which bubbles through the liquid and escapes now burns with a blue flame. What has happened?

What would you expect to happen if the flow of gas were stopped and the pressure of gas above the liquid were lowered to atmospheric pressure? *(L.)*

19

Sulphur

19.1 Introduction

The element sulphur has been known since prehistoric times. Free sulphur is widely distributed, particularly in volcanic regions. By far the most important deposits are situated in Louisiana and Texas in the United States of America. Appreciable quantities are also obtained from Italy and Japan.

Many important sulphides and sulphates also occur naturally. These include iron pyrites, FeS_2; zinc blende, ZnS; galena, PbS; copper pyrites, $CuFeS_2$; anhydrite, $CaSO_4$ and gypsum, $CaSO_4.2H_2O$.

19.2 Extraction of sulphur

Sulphur deposits in America, in Louisiana and Texas, are usually found more than 500 feet underground, beneath rock and quicksand which render the normal methods of mining impossible. Frasch, in 1890, devised a simple yet ingenious method for obtaining the sulphur; the method is now used for extracting over 5 million tonnes of the element each year (figure 19.1). Three concentric pipes, the outer one with a diameter of only 6 inches, are sunk to the sulphur deposits. Water, superheated under pressure to 170–180 °C, is forced down the space between the two outer pipes. This melts the sulphur underground. Compressed air, which is then pumped down the inner tube, forms an emulsion with the water

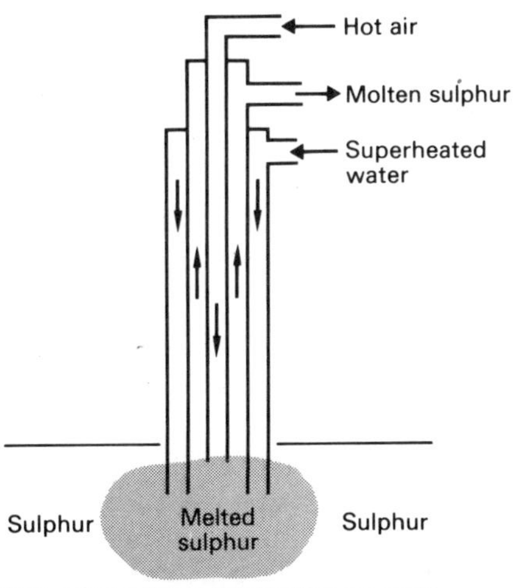

Figure 19.1 Extraction of sulphur.

and molten sulphur. This emulsion is forced up the space between the inner and second tubes and is led to vats, where sulphur of over 99·8 per cent purity solidifies.

Some modern wells are producing as much as 6000 tonnes a day by this method.

19.3 Allotropy of sulphur

Sulphur is another element which exhibits allotropy, i.e. it exists in several different forms, all in the same state. There is little chemical difference between these forms.

19.3.1 α (rhombic) sulphur

This is the only form of sulphur which is stable below 96 °C. Below this temperature all other allotropes of sulphur slowly revert to the rhombic form.

To prepare the characteristic rhombic-shaped crystals (figure 19.2) some powdered sulphur is shaken in a dry test-tube with carbon disulphide. The solution is filtered into a shallow dish or clock glass to remove any undissolved sulphur. It is then placed in a fume cupboard and, as the volatile carbon disulphide evaporates, crystals of rhombic sulphur are formed. (N.B. Carbon disulphide vapour is poisonous and highly inflammable. No lighted Bunsen burners should be in the vicinity when this experiment is being performed, and on no account should the evaporation be speeded up by any form of artificial heat.)

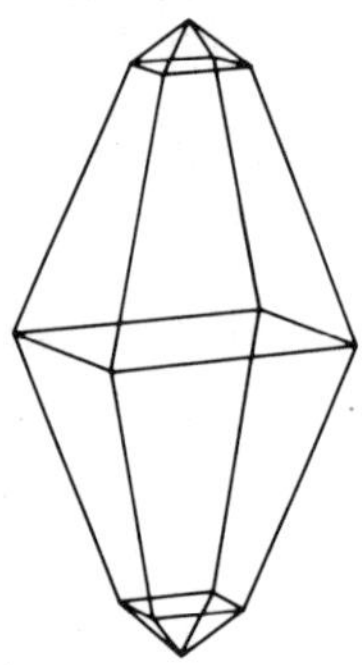

Figure 19.2 Rhombic sulphur.

Crystals of rhombic sulphur are bright yellow in colour. If heated rapidly they melt at 115 °C. Otherwise they change at 96 °C into β-sulphur, which has a higher melting point. Like all other forms of sulphur the rhombic allotrope is insoluble in water.

19.3.2 β (monoclinic) sulphur

This is the only stable form of sulphur between 96 °C and 120 °C. Its needle-like crystals are prepared by gently heating powdered sulphur in a test-tube until it has just melted. It is then poured into a dry filter paper fitted in a funnel and allowed to cool until a thin crust forms on the surface. Two holes (why two holes?) are then pierced in the crust, and any remaining molten sulphur is poured out into a beaker of cold water. The filter paper is then opened to disclose a mass of transparent, needle-like crystals of monoclinic sulphur adhering to it (figure 19.3).

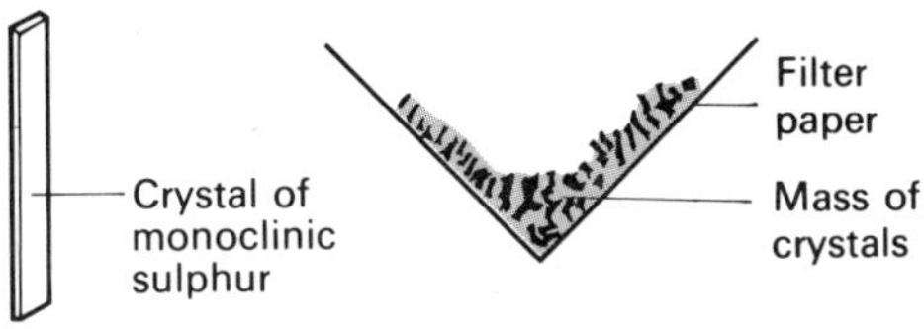

Figure 19.3 Monoclinic sulphur.

Monoclinic sulphur is also yellow in colour, and at room temperature slowly changes to rhombic sulphur. The crystals obtained by the method described above will be found to retain their shape when they are left for several days, but they become opaque, and in fact each "monoclinic" crystal then consists of many minute crystals of rhombic sulphur.

19.3.3 Molten sulphur

Both α-sulphur and β-sulphur consist of S_8 molecules with different ring structures. If sulphur is heated in a test-tube to just above its melting point it forms an amber-coloured mobile liquid in which the S_8 molecules move with ease. On further heating the liquid becomes dark reddish-brown and extremely viscous. This is when the rings break to form chains of atoms which tangle with one another. Near the boiling point of sulphur (444·5° C) the liquid becomes mobile again; this is when the chains of atoms break down to form smaller molecules.

19.3.4 Plastic sulphur

If sulphur near its boiling point is rapidly cooled by pouring it in a thin stream into cold water, plastic sulphur is formed. This is a sticky, elastic substance consisting, presumably, of a tangled mass of chains of sulphur atoms. It slowly hardens and changes to β-sulphur.

19.3.5 Amorphous sulphur

An allotrope of sulphur that is non-crystalline and insoluble in carbon disulphide is formed as an almost white powder if a saturated solution of hydrogen sulphide is oxidized by leaving it to stand in air:

$$2H_2S(aq) + O_2(g) \rightarrow 2S(s) + 2H_2O$$

19.4 Further properties of sulphur

Sulphur is reactive and combines directly with most other elements. Thus on heating in air or oxygen it burns readily with a blue flame to form sulphur dioxide, together with traces of sulphur(VI) oxide:

$$S(s) + O_2(g) \rightarrow SO_2(g)$$

A red glow passes through a mixture of iron filings and sulphur when it is heated, and iron(II) sulphide is formed:

$$Fe(s) + S(s) \rightarrow FeS(s)$$

Many other metals combine similarly.

19.5 Uses of sulphur

19.5.1 Manufacture of sulphur dioxide

About six million tonnes of sulphur are now produced annually. Much of this is burned to give sulphur dioxide, which is then used in the manufacture of sulphuric acid or of hydrogensulphites such as calcium hydrogensulphite, $Ca(HSO_3)_2$.

19.5.2 Vulcanisation of rubber

Large amounts of sulphur are used in vulcanising rubber. Rubber when warm, is soft and sticky and quite unsuitable for most purposes. When heated to 140 °C an intimate mixture of sulphur and rubber combines to give a much stronger and more durable product than ordinary rubber. Addition of up to 50 per cent sulphur produces the very hard vulcanite or ebonite. The properties of rubber can be modified still further by addition of other substances besides sulphur, e.g. carbon black (see section 17.4.4).

19.5.3 Use as an insecticide and fungicide

When sulphur dust is used on vines and hops it prevents a mould fungus which attacks these plants.

19.5.4 Manufacture of explosives and matches

Gunpowder is a mixture of sulphur, carbon and potassium nitrate. When heated, the sulphur and carbon burn explosively with oxygen provided by the potassium nitrate. (N.B. On no account should the budding chemist attempt to prepare gunpowder.) Sulphur is also used, combined with phosphorus, in matches (see section 16.8.3).

19.5.5 Manufacture of other chemicals, including dyes and drugs

Sulphur is used in making many important chemicals. These include carbon disulphide, the so-called sulphur dyes and many drugs.

19.6 Hydrogen sulphide

The only important compound of sulphur with hydrogen is hydrogen sulphide. This compound occurs naturally in some volcanic gases and in certain waters, e.g. at Harrogate, which are said to have medicinal properties. Its smell is well known, particularly to those more malicious boys and girls who have purchased and used stink-bombs. These are solutions of hydrogen sulphide in a volatile solvent.

19.6.1 Preparation of hydrogen sulphide

Hydrogen sulphide is usually prepared by the action of dilute hydrochloric or dilute sulphuric acid on iron(II) sulphide (figure 19.4).

$$FeS(s) + 2HCl(aq) \rightarrow FeCl_2(aq) + H_2S(g)$$

$$FeS(s) + H_2SO_4(aq) \rightarrow FeSO_4(aq) + H_2S(g)$$

It may be collected by downwards delivery.

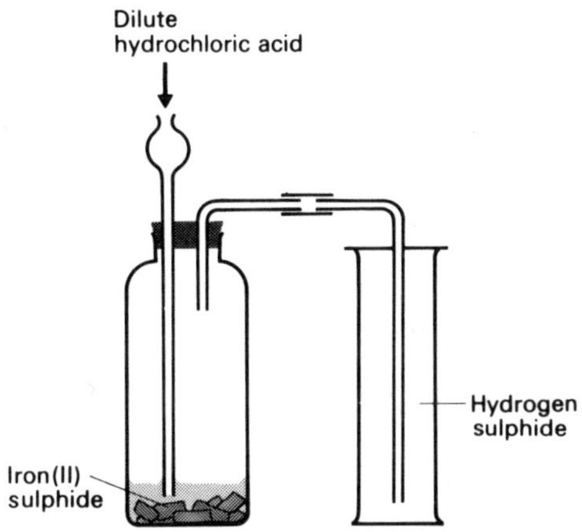

Figure 19.4 Preparation of hydrogen sulphide.

Hydrogen sulphide prepared by this method contains the following impurities:

(*a*) Acid spray carried over by the hydrogen sulphide. This is removed by collecting it over water.

(*b*) Hydrogen. This is formed by the action of the acid on free iron present in the iron(II) sulphide. For most purposes for which hydrogen sulphide is required in the laboratory, the hydrogen impurity is harmless, and it is not usual to remove it.

Like carbon dioxide, hydrogen sulphide is often required in the laboratory. Therefore the gas is usually prepared in a Kipp's apparatus (section 18.1.1). Sticks of iron(II) sulphide are placed in the central compartment, and concentrated hydrochloric acid diluted with its own volume of water is the liquid reagent.

19.6.2 Properties of hydrogen sulphide

(a) Introduction

Hydrogen sulphide is a gas and therefore a covalent compound. It is colourless and very poisonous, has a density slightly greater than air and a characteristic repulsive odour like that of bad eggs which, like other sorts of decaying animal matter, evolve the gas.

The outer shell structure of hydrogen sulphide is identical to that of water (figure 6.7). The sulphur atom uses the next shell of electrons to the oxygen atom.

(b) It is fairly soluble in water to give a weakly acidic solution
One cm^3 of cold water dissolves approximately 5 cm^3 of hydrogen sulphide. The acidity of the solution is caused by the slight ionization of the dissolved gas to give hydrogen ions:

$$H_2S \rightleftharpoons HS^- + H^+ \rightleftharpoons S^{2-} + 2H^+$$

(c) Hydrogen sulphide burns with a pale blue flame
With a good supply of oxygen, steam and sulphur dioxide are the main products:

$$2H_2S(g) + 3O_2(g) \rightarrow 2H_2O(g) + 2SO_2(g)$$

In a limited supply of oxygen, steam and sulphur are formed:

$$2H_2S(g) + O_2(g) \rightarrow 2H_2O(g) + 2S(s)$$

If a light is applied to a gas jar full of hydrogen sulphide, a mixture of these reactions takes place. Sulphur dioxide is readily detected and a deposit of sulphur is seen on the sides of the gas jar.

(d) Hydrogen sulphide is a reducing agent
In many of its reactions hydrogen sulphide behaves as a reducing agent by giving up its hydrogen to elements or compounds, and being itself oxidized to sulphur. Some common examples are as follows:

(i) If a gas jar full of hydrogen sulphide is mixed with one of sulphur dioxide *in the presence of water*, a yellowish deposit of sulphur is formed:

$$2H_2S(g) + SO_2(g) \rightarrow 2H_2O(l) + 3S(s)$$

The same reaction occurs if solutions of the two gases are mixed. Then an almost white suspension of sulphur is obtained.

(ii) With the halogens, the hydrogen halides are formed. Thus, when hydrogen sulphide is mixed with chlorine, in addition to sulphur the colourless, fuming hydrogen chloride is obtained:

$$H_2S(g) + Cl_2(g) \rightarrow 2HCl(g) + S(s)$$

(iii) If hydrogen sulphide is bubbled through a solution of iron(III) chloride, which is yellow in colour, a pale green solution of iron(II) chloride, containing a pale yellow precipitate of sulphur, is produced:

$$2FeCl_3(aq) + H_2S(g) \rightarrow 2FeCl_2(aq) + 2HCl(aq) + S(s)$$

(iv) Hydrogen sulphide decolorises a purple solution of potassium manganate(VII) acidified with dilute sulphuric acid. Again a deposit of sulphur is formed:

$$2KMnO_4(aq) + 3H_2SO_4(aq) + 5H_2S(g) \rightarrow K_2SO_4(aq) + 2MnSO_4(aq) + 8H_2O(l) + 5S(s)$$

By a similar reaction an orange solution of potassium dichromate(VI) is turned green by the formation of chromium(III) sulphate.

19.6.3 Test for hydrogen sulphide

A filter paper, soaked in lead(II) nitrate or lead(II) ethanoate solution, turns black when placed in hydrogen sulphide. Insoluble lead(II) sulphide is formed: e.g.,

$$Pb(NO_3)_2(aq) + H_2S(g) \rightarrow PbS(s) + \underset{\text{nitric acid}}{2HNO_3(aq)}$$

19.6.4 Sulphides

Hydrogen sulphide, when dissolved in water, behaves as a dibasic acid. It therefore forms two series of salts—hydrogensulphides and sulphides.

Sodium, potassium and ammonium sulphides are soluble in water. Most other sulphides are insoluble.

(a) Preparation of sulphides

(i) The *soluble sulphides* are prepared in the laboratory by passing hydrogen sulphide through a solution of the corresponding alkali:

$$2NaOH + H_2S \rightarrow \underset{\text{sodium sulphide}}{Na_2S} + 2H_2O$$

$$NaOH + H_2S \rightarrow \underset{\text{sodium hydrogen-sulphide}}{NaHS} + H_2O$$

Details of the preparation of sodium sulphide and sodium hydrogensulphide are given in section 21.2.7(*j*).

(ii) Several methods are available for preparing the *insoluble sulphides:*

1. Precipitation

If hydrogen sulphide is passed through solutions of the salts of many metals the sulphide of the metal is precipitated, e.g. with copper(II) sulphate solution a black precipitate of copper(II) sulphide in a colourless solution of sulphuric acid is obtained:

$$CuSO_4(aq) + H_2S(g) \rightarrow CuS(s) + H_2SO_4(aq)$$

With lead(II) nitrate solution a black precipitate of lead(II) sulphide is formed:

$$Pb(NO_3)_2(aq) + H_2S(g) \rightarrow PbS(s) + 2HNO_3(aq)$$

Certain sulphides, e.g. iron(II) sulphide, are soluble in acid, so that if hydrogen sulphide is passed through solutions of the corresponding salts the reaction is reversible and incomplete:

$$FeSO_4(aq) + H_2S(g) \rightleftharpoons FeS(s) + H_2SO_4(aq)$$

In such cases it is better to use a solution of ammonium sulphide to effect precipitation:

$$FeSO_4(aq) + (NH_4)_2S(aq) \rightarrow FeS(s) + (NH_4)_2SO_4(aq)$$

2. Direct combination

Many metals combine directly with sulphur to give the sulphide, e.g. iron(II) sulphide is manufactured by heating a mixture of iron and sulphur:

$$Fe(s) + S(s) \rightarrow FeS(s)$$

With some metals, such as zinc, the combination may occur with explosive violence.

(b) Properties of sulphides

(i) When roasted in air or oxygen most sulphides yield sulphur dioxide and the oxide of the metal:

$$2ZnS(s) + 3O_2(g) \rightarrow 2ZnO(s) + 2SO_2(g)$$
$$2PbS(s) + 3O_2(g) \rightarrow 2PbO(s) + 2SO_2(g)$$

This type of reaction is of great importance in the extraction of metals from naturally occurring sulphides.

(ii) Most sulphides yield hydrogen sulphide when heated with either dilute or concentrated hydrochloric acid:

$$FeS(s) + 2HCl(aq) \rightarrow FeCl_2(aq) + H_2S(g)$$
$$Na_2S(s) + 2HCl(aq) \rightarrow 2NaCl(aq) + H_2S(g)$$

This fact is used in identifying sulphides, as the hydrogen sulphide is readily detected by smell or with lead(II) nitrate or lead(II) ethanoate paper.

19.7 The oxides of sulphur

There are two important oxides of sulphur, sulphur dioxide and sulphur(VI) oxide. Both of these combine with water, to give sulphurous and sulphuric acids, respectively.

19.7.1 Sulphur dioxide

(a) Preparation

The burning of sulphur yields sulphur dioxide:

$$S(s) + O_2(g) \rightarrow SO_2(g)$$

In the laboratory, however, sulphur dioxide is usually obtained from a siphon or by *the action of an acid on a sulphite or hydrogensulphite*. Using the apparatus shown (figure 19.5), dilute sulphuric acid is added to sodium sulphite crystals. Sulphur dioxide is readily evolved when the mixture is warmed. It is collected by upward displacement of air:

$$\left[\begin{array}{l} \underset{\text{sodium sulphite}}{Na_2SO_3} + H_2SO_4 \rightarrow \underset{\text{sodium sulphate}}{Na_2SO_4} + H_2O + SO_2 \\ SO_3^{2-}(aq) + 2H^+(aq) \rightarrow H_2O(l) + SO_2(g) \end{array}\right.$$

This reaction is very similar to the action of an acid on a carbonate (section 12.1.1(*d*)). Heating is necessary here, because sulphur dioxide is a very soluble gas, and it would otherwise remain dissolved in the water from the dilute acid.

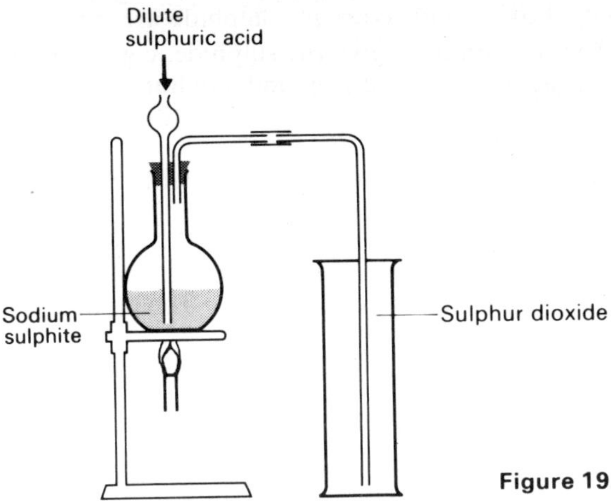

Figure 19.5 Preparation of sulphur dioxide.

(b) Properties of sulphur dioxide

Sulphur dioxide is a colourless gas, with a characteristic choking smell. It is poisonous, and care should be taken to prevent much of the gas escaping into the laboratory.

It is more than twice as dense as air and is readily liquefied. The liquefaction can be shown quite easily by passing sulphur dioxide through a tube immersed in a freezing mixture of ice and salt (figure 19.6). The liquid sulphur dioxide boils at $-9\,^{\circ}C$. Sulphur dioxide is also liquefied by compression at room temperatures, and most laboratories have a supply of liquid sulphur dioxide kept under pressure in disposable aluminium containers.

Sulphur dioxide is very soluble in water, giving a solution of sulphurous acid:

$$H_2O(l) + SO_2(g) \rightleftharpoons H_2SO_3(aq)$$

The fountain experiment can be used to illustrate this high solubility (section 16.3.3(*b*)).

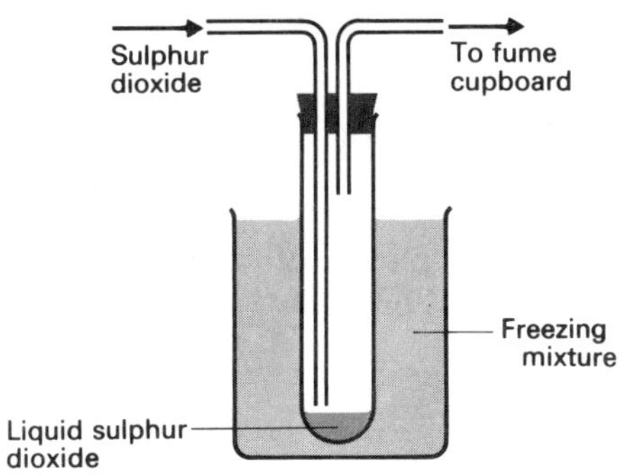

Figure 19.6 Liquefaction of sulphur dioxide.

When gas jars full of sulphur dioxide and hydrogen sulphide are mixed, a yellow deposit of sulphur forms:

$$SO_2(g) + 2H_2S(g) \rightarrow 2H_2O(l) + 3S(s)$$

This reaction only occurs in the presence of moisture.

(c) The reducing action of sulphur dioxide

(i) Sulphur dioxide reduces purple potassium manganate(VII) solution to manganese(II) sulphate. This is shown by placing a filter paper, dipped into potassium manganate(VII) solution, in contact with the gas. The solution is decolorised:

$$\underset{\text{potassium manganate(VII)}}{2KMnO_4(aq)} + 5SO_2(g) + 2H_2O(l) \rightarrow \underset{\text{potassium sulphate}}{K_2SO_4(aq)} + \underset{\text{manganese(II) sulphate}}{2MnSO_4(aq)} + 2H_2SO_4\ (aq)$$

It may be noted that the sulphur dioxide has, in effect, taken some of the oxygen from the potassium manganate(VII). Sulphur(VI) oxide is then formed, which, with water, gives sulphuric acid.

With potassium dichromate(VI) a similar reaction occurs. In this case the orange dichromate(VI) solution turns green as chromium(III) sulphate is formed.

Both these reactions can be used to help confirm the presence of sulphur dioxide or of sulphurous acid. Other reducing agents will, however, behave similarly. Hydrogen sulphide, for example, produces the same colour reactions, although with that gas a pale yellow deposit of sulphur is formed simultaneously.

(ii) Sulphur dioxide will bleach various dyes. This can be shown by dipping certain moist flowers, e.g. a red rose or a blue aster, into the gas. The dye is reduced; possibly by this sort of reaction:

$$\text{Dye} + SO_2 + 2H_2O \rightarrow \underset{\text{colourless}}{(\text{Dye} + \text{Hydrogen})} + H_2SO_4$$

This is essentially a different reaction from that of most bleaching agents, which oxidize the dye to a colourless compound. This reducing bleach is often reversed by the action of oxygen in the air on the reduced dye.

(iii) When sulphur dioxide is passed through a solution of an iron(III) salt it reduces it to an iron(II) salt. This is indicated by a colour change from orange to pale green:

$$\underset{\text{iron(III) sulphate}}{Fe_2(SO_4)_3(aq)} + SO_2(g) + 2H_2O(l) \rightarrow \underset{\text{iron(II) sulphate}}{2FeSO_4(aq)} + 2H_2SO_4(l)$$

As the reducing actions listed above all require the presence of water, it is more correct to consider them as the reactions of sulphurous acid.

(d) Sulphurous acid

A solution of sulphurous acid is prepared by passing sulphur dioxide into water.

$$SO_2(g) + H_2O(l) \rightleftharpoons H_2SO_3(aq)$$

If the sulphur dioxide is being prepared by the action of concentrated sulphuric acid on copper, it is advisable to use a device to prevent sucking back of the water into the generating vessel (see figure 16.11).

Pure sulphurous acid has never been isolated. Its solution decomposes into sulphur dioxide and water, even at room temperature. Attempts to concentrate the solution result in an acceleration of this decomposition.

It is a weak dibasic acid. Accordingly, it forms two series of salts, hydrogensulphites and sulphites, e.g. with sodium hydroxide solution these reactions occur:

$$NaOH + H_2SO_3 \rightarrow \underset{\text{sodium hydrogensulphite}}{NaHSO_3} + H_2O$$

$$OH^-(aq) + H_2SO_3(aq) \rightarrow HSO_3^-(aq) + H_2O(l)$$

$$2NaOH + H_2SO_3 \rightarrow \underset{\text{sodium sulphite}}{Na_2SO_3} + 2H_2O$$

$$2OH^-(aq) + H_2SO_3(aq) \rightarrow SO_3^{2-}(aq) + 2H_2O(l)$$

The laboratory preparation of sodium sulphite and sodium hydrogensulphite is given in section 21.2.7(i).

Sulphurous acid, on exposure to air, is slowly converted into sulphuric acid:

$$2H_2SO_3(aq) + O_2(g) \rightarrow 2H_2SO_4(aq)$$

All the reducing actions earlier ascribed to sulphur dioxide are given by sulphurous acid.

(e) Uses of sulphur dioxide

The fumes from burning sulphur, i.e. sulphur dioxide, have been used for fumigation and bleaching for thousands of years. Sulphur dioxide still has a limited use for both these purposes. Its use as a bleaching agent is confined to those materials, such as wool and silk, which would be damaged by the more usual bleach, chlorine.

Sodium hydrogensulphite ($NaHSO_3$) and calcium hydrogensulphite ($Ca(HSO_3)_2$) both need sulphur dioxide in their preparation. The former is needed in the manufacture of sodium sulphinate, $Na_2S_2O_4$, an important reducing agent used in dyeing with vat dyes. The calcium hydrogensulphite is needed in preparing sulphite pulp from wood. This pulp is used in the manufacture of paper and artificial silk.

By far the most important use of sulphur dioxide is in the large-scale production of sulphuric acid.

19.7.2 Sulphur(VI) oxide

(a) Introduction

The fact that when sulphur burns sulphur dioxide is the main product indicates that sulphur dioxide can have no great affinity with oxygen. On heating, sulphur dioxide and oxygen combine slowly, but a catalyst is necessary to obtain a high rate of reaction.

To prepare sulphur(VI) oxide in the laboratory, a mixture of *dry* oxygen and *dry* sulphur dioxide is passed over a heated catalyst of platinised asbestos (asbestos coated with a fine deposit of platinum), when the following reaction occurs:

$$2SO_2(g) + O_2(g) \rightleftharpoons 2SO_3(g)$$

Figure 19.7 indicates the apparatus needed. If the whole apparatus is completely dry at the beginning of the experiment a white, solid deposit of sulphur(VI) oxide collects in the cooled tube.

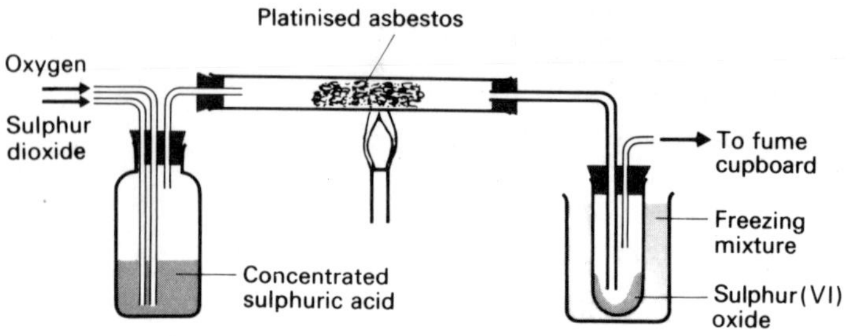

Figure 19.7 Preparation of sulphur(VI) oxide.

(b) Properties of sulphur (VI) oxide

Sulphur(VI) oxide is a white fuming solid, often obtained as silky needles. It chars corks and attacks rubber and in the experiment above it is preferable to use all glass apparatus.

It has a great affinity with water. This can be shown by carefully adding a few drops of water to the specimen prepared in the above experiment. There is a hissing noise and clouds of white fumes are formed. Sulphuric acid is the product:

$$SO_3(g) + H_2O(l) \rightarrow H_2SO_4(aq)$$

Sulphur(VI) oxide combines with concentrated sulphuric acid, forming fuming sulphuric acid, or oleum:

$$H_2SO_4(l) + SO_3(g) \rightarrow \underset{\text{oleum}}{H_2S_2O_7(l)}$$

19.7.3 Manufacture of sulphuric acid

(a) Sulphur dioxide must first be manufactured

There are three methods that can be used industrially. The third is, by far, the most important.

(i) By roasting sulphides in air.

The most important sulphide for this purpose is iron pyrites, FeS_2. It contains a high percentage of sulphur, which on roasting gives the dioxide. Iron(III) oxide is the other product:

$$4FeS_2(s) + 11O_2(g) \rightarrow 2Fe_2O_3(s) + 8SO_2(g)$$

Zinc blende, ZnS, is another source. In the manufacture of zinc the blende is first roasted to obtain zinc oxide:

$$2ZnS(s) + 3O_2(g) \rightarrow 2ZnO(s) + 2SO_2(g)$$

Sulphur dioxide is also formed, and it is therefore usual to combine the production of zinc with a sulphuric acid plant.

(ii) From anhydrite (naturally occurring calcium sulphate).
This method was used for many years by I.C.I., the anhydrite coming from the large deposits in the North-East of England. It is currently an uneconomic route to sulphur dioxide.

Anhydrite is heated with carbon when it is reduced to calcium sulphide. This reacts with more calcium sulphate to give sulphur dioxide. The overall reaction is:

$$4CaSO_4(s) + 2C(s) \rightarrow 4CaO(s) + 2CO_2(g) + 4SO_2(g)$$

Acidic oxides, silica (SiO_2) and alumina (Al_2O_3), are added to the reaction mixture. They combine with the calcium oxide to give calcium silicate ($CaSiO_3$) and calcium aluminate ($Ca(AlO_2)_2$). If iron(III) oxide is also added the solid product is a clinker that on grinding is Portland cement.

(iii) By burning sulphur in air:

$$S(s) + O_2(g) \rightarrow SO_2(g)$$

Elemental sulphur is mined and, for convenience, transported as a liquid in insulated tanks with steam heaters. It is stored as a liquid and then pumped to the sulphur burners in a modern sulphuric acid plant (figure 19.8).

Figure 19.8 A modern sulphuric acid plant. Here liquid sulphur is burnt to give sulphur dioxide, which is then converted to sulphur(VI) oxide (figure 19.9) for sulphuric acid manufacture. (Photo: I.C.I. Agricultural Division.)

(b) Sulphuric acid is then manufactured by the contact process

The chemistry of the contact process is very simple. Sulphur dioxide is oxidized to sulphur(VI) oxide, using platinum or vanadium(V) oxide as catalyst:

$$2SO_2(g) + O_2(g) \rightleftharpoons 2SO_3(g)$$

The sulphur(VI) oxide is then allowed to react with water to give sulphuric acid:

$$SO_3(g) + H_2O(l) \rightarrow H_2SO_4(aq)$$

In practice, there are several difficulties to be overcome. Some of these are indicated in the following three main stages of the manufacture.

(i) Purification of the gases.
A careful purification and drying of the sulphur dioxide is essential. It contains impurities, such as arsenic(III) oxide, particularly if it is obtained from iron pyrites. These impurities "poison" the platinum catalyst, i.e., render it ineffective for catalysis. Vanadium(V) oxide as a catalyst is less susceptible to poisoning and costs less.

(ii) Conversion of sulphur dioxide to sulphur(VI) oxide.
The gases are passed through converters containing layers of hot catalyst deposited in finely divided form on silica. A temperature of 400–500 °C is normally used:

$$2SO_2 + O_2 \rightleftharpoons 2SO_3 + \text{Heat}$$

The reaction is exothermic, and the heat evolved maintains the temperature of the catalyst. By using several converters in series and a slight excess of oxygen, 95–98 per cent conversion of sulphur dioxide into sulphur(VI) oxide is achieved (figure 19.9).

Figure 19.9 Sulphur dioxide to sulphur(VI) oxide converters at Billingham, Cleveland county. (Photo: I.C.I. Agricultural Division).

(iii) Absorption of the sulphur(VI) oxide.
Direct absorption of sulphur(VI) oxide in water results in an extremely vigorous reaction and the production of a mist, consisting of droplets of sulphuric acid. The oxide is therefore absorbed in concentrated sulphuric acid, in which it dissolves with ease. The oxide is passed up towers, down which the acid is trickling:

$$H_2SO_4(l) + SO_3(g) \rightarrow H_2S_2O_7(l)$$

The oleum so obtained is carefully diluted with the correct quantity of water to give concentrated sulphuric acid:

$$H_2S_2O_7(l) + H_2O(l) \rightarrow 2H_2SO_4(l)$$

Sulphuric acid is a very important chemical. The world production, excluding that made in the U.S.S.R., China and countries of Eastern Europe, is approximately 30 million tonnes per year.

19.7.4 Properties of sulphuric acid

Concentrated sulphuric acid (Oil of Vitriol) is a colourless oily liquid of specific gravity 1.84.

It has a great affinity with water, and much heat is evolved when it is diluted with water. *The only safe way to dilute the acid is to add it slowly to water, stirring the whole time.* If the stirring is omitted the denser acid forms a layer at the bottom of the vessel.

Because of the ease with which sulphuric acid absorbs water, it is often used as a drying agent for those gases which do not combine with it, e.g., chlorine, sulphur dioxide, hydrogen chloride and hydrogen. This drying action can be illustrated by leaving a beaker half full of concentrated sulphuric acid in the laboratory. The beaker slowly fills up as moisture in the air is absorbed and the acid becomes more dilute.

Most of the other properties of sulphuric acid can be considered under four main headings.

(a) Sulphuric acid as an acid

The reactions of concentrated sulphuric acid are often complicated by its oxidizing action, but dilute sulphuric acid behaves as a typical acid.

(i) With metals above hydrogen in the activity series, e.g., iron and zinc, hydrogen is evolved, and a solution of the sulphate of the metal is formed:

$$Zn + H_2SO_4 \rightarrow ZnSO_4 + H_2$$
$$Zn(s) + 2H^+(aq) \rightarrow Zn^{2+}(aq) + H_2(g)$$

Little action occurs with lead.

(ii) With hydrogencarbonates and carbonates, carbon dioxide is evolved and a salt and water are also formed:

$$2NaHCO_3(aq) + H_2SO_4(aq) \rightarrow Na_2SO_4(aq) + 2H_2O(l) + 2CO_2(g)$$
$$Na_2CO_3(aq) + H_2SO_4(aq) \rightarrow Na_2SO_4(aq) + H_2O(l) + CO_2(g)$$

Calcium carbonate reacts in this way, but the calcium sulphate initially produced is insoluble. It therefore forms a coating round any remaining calcium carbonate, preventing further reaction. You may remember this as a reason why carbon dioxide is not prepared by the action of sulphuric acid on marble.

(iii) With a base, a salt and water are formed. As the acid is dibasic, it gives rise to two series of salts. Thus, with sodium hydroxide, sodium sulphate or sodium hydrogensulphate are produced, depending on the quantities of reagents used:

$$H_2SO_4(aq) + NaOH(aq) \rightarrow \underset{\text{sodium hydrogensulphate}}{NaHSO_4(aq)} + H_2O(l)$$

$H_2SO_4(aq) + 2NaOH(aq) \rightarrow Na_2SO_4(aq) + 2H_2O(l)$
sodium
sulphate

(b) Sulphuric acid as an oxidizing agent
Hot, concentrated sulphuric acid behaves as an oxidizing agent. In its action with non-metals each molecule of sulphuric acid provides one atom of oxygen for this oxidation, and is thereby reduced to sulphur dioxide. With metals the reaction is sometimes complex and the conventional equations are only approximate. Sulphur dioxide is again formed.

(i) If concentrated sulphuric acid is heated with powdered charcoal the carbon is oxidized to carbon dioxide and the sulphuric acid reduced to sulphur dioxide and water.

$C(s) + 2H_2SO_4(l) \rightarrow CO_2(g) + 2H_2O(l) + 2SO_2(g)$

(ii) Similarly, concentrated sulphuric acid oxidizes sulphur to sulphur dioxide while the acid itself is reduced to sulphur dioxide and water.

$S(s) + 2H_2SO_4(l) \rightarrow 2H_2O(l) + 3SO_2(g)$

(iii) One method for the preparation of sulphur dioxide in the laboratory involves the oxidation of copper to its sulphate by hot concentrated sulphuric acid:

$Cu(s) + 2H_2SO_4(l) \rightarrow CuSO_4(aq) + 2H_2O(l) + SO_2(g)$

Another product of this reaction is copper(I) sulphide, Cu_2S.

(iv) If hydrogen sulphide is bubbled through concentrated sulphuric acid it is oxidized to sulphur:

$H_2S(g) + H_2SO_4(l) \rightarrow S(s) + 2H_2O(l) + SO_2(g)$

For this reason concentrated sulphuric acid is an unsuitable drying agent for hydrogen sulphide.

(c) Sulphuric acid as a dehydrating agent
In addition to being a drying agent, concentrated sulphuric acid is able to remove the elements of water from certain compounds. Thus, if the acid is added to some sugar, or to a concentrated solution of sugar, heat is evolved and a voluminous mass of carbon is obtained:

$C_{12}H_{22}O_{11} - 11H_2O \rightarrow 12C$
sugar

This process is known as *dehydration* and must not be confused with drying. The sugar may be completely dry before the reaction; the sulphuric acid extracts hydrogen and oxygen atoms from the compound, in the ratio in which they are present in water.

Other compounds dehydrated by sulphuric acid include methanoic (formic) acid and ethanol.

$HCOOH - H_2O \rightarrow CO$
methanoic
acid

$C_2H_6O - H_2O \rightarrow C_2H_4$
ethanol ethene

Another type of dehydration occurs if blue copper(II) sulphate crystals ($CuSO_4.5H_2O$) are warmed with concentrated sulphuric acid. The crystals change to white anhydrous copper(II) sulphate as the water is removed:

$$\underset{\text{blue}}{CuSO_4.5H_2O} - 5H_2O \rightarrow \underset{\text{white}}{CuSO_4}$$

(d) Sulphuric acid in the preparation of other acids
If concentrated sulphuric acid is warmed with potassium nitrate, nitric acid is expelled:

$$KNO_3(s) + H_2SO_4(l) \rightarrow HNO_3(g) + \underset{\substack{\text{potassium}\\\text{hydrogensulphate}}}{KHSO_4(s)}$$

This is because sulphuric acid is less volatile than nitric acid, and therefore displaces it from its salts. The preparation of hydrogen chloride from sodium chloride is a similar reaction:

$$NaCl(s) + H_2SO_4(l) \rightarrow HCl(g) + NaHSO_4(s)$$

In some cases the volatile acid is formed initially but reacts with the sulphuric acid. With sodium methanoate, for example, the methanoic acid produced is dehydrated:

$$\underset{\substack{\text{sodium}\\\text{methanoate}}}{HCOONa(s)} + H_2SO_4(l) \rightarrow \underset{\substack{\text{methanoic}\\\text{acid}}}{HCOOH(l)} + NaHSO_4(s)$$

$$HCOOH(l) \rightarrow H_2O(l) + CO(g)$$

19.7.5 Uses of sulphuric acid

Over 30 per cent of the sulphuric acid manufactured in the United Kingdom is used in making the two fertilizers (see figure 16.2), calcium hydrogenphosphate (or "superphosphate") and ammonium sulphate.

Calcium hydrogenphosphate is made by the action of concentrated sulphuric acid on calcium phosphate, which occurs naturally:

$$\underset{\substack{\text{calcium}\\\text{phosphate}}}{Ca_3(PO_4)_2(s)} + 2H_2SO_4(l) \rightarrow \underset{\substack{\text{calcium}\\\text{hydrogenphosphate}}}{Ca(H_2PO_4)_2(s)} + 2CaSO_4(s)$$

Ammonium sulphate is obtained by the neutralization of sulphuric acid with ammonia:

$$2NH_3(g) + H_2SO_4(aq) \rightarrow \underset{\substack{\text{ammonium}\\\text{sulphate}}}{(NH_4)_2SO_4(s)}$$

Viscose, a form of cellulose used as an artificial silk or in making one type of transparent paper, needs sulphuric acid in the final stage of its manufacture. Much sulphuric acid is used in the treatment of metals to remove rust, prior to such processes as tinning and galvanizing. Further large quantities are needed in accumulators, in the manufacture of titanium(IV) oxide and synthetic detergents, in making dyestuffs and in the dyeing process itself.

The acid is still used on a large scale in the production of hydrochloric acid from rock salt. Increasing amounts are being used in the refining of petroleum.

The following table shows the percentages of our total production of the acid used in the more important industries:

Superphosphate	24%	Detergents	11%
Chemicals	16%	Viscose fibre	9%
Pigments for paints		Ammonium sulphate	8%
(e.g. TiO_2, $BaSO_4$)	15%	Others	17%

19.7.6 Sulphates

(a) Preparation

Sulphuric acid forms two series of salts, sulphates and hydrogensulphates. Most of these are freely soluble in water, with the important exceptions of lead, barium and calcium sulphates. The standard methods, details of which are given in chapter 12, are suitable for their preparation.

(i) By neutralization of an acid with an alkali. The preparation of sodium sulphate and sodium hydrogensulphate by this method is given in section 12.4.5.

(ii) By the action of a dilute acid on a metal. This method cannot be used with:

1. the very reactive metals, such as sodium and potassium;
2. metals occurring below hydrogen in the activity series;
3. metals, such as lead, forming insoluble sulphates.

(iii) By the action of dilute sulphuric acid on a metallic carbonate or oxide. This method cannot be used for preparing insoluble sulphates.

(iv) By double decomposition. This kind of reaction is used for preparing the insoluble sulphates.

(b) Identification of sulphates

The presence of the sulphate ion in solutions of sulphuric acid or sulphates is shown by addition of dilute hydrochloric acid and barium chloride solution. A white precipitate of barium sulphate is formed:

$$\underset{}{BaCl_2} + \underset{\text{sodium sulphate}}{Na_2SO_4} \rightarrow \underset{\text{barium sulphate}}{BaSO_4\downarrow} + 2NaCl$$

$$Ba^{2+}(aq) + SO_4^{2-}(aq) \rightarrow BaSO_4(s)$$

The dilute hydrochloric acid prevents the precipitation of other insoluble barium compounds. Thus, if barium chloride solution is added to sodium sulphite solution, a white precipitate of barium sulphite is obtained:

$$BaCl_2 + Na_2SO_3 \rightarrow BaSO_3\downarrow + 2NaCl$$

$$Ba^{2+}(aq) + SO_3^{2-}(aq) \rightarrow BaSO_3(s)$$

This is not possible in the presence of hydrochloric acid which reacts with the sulphite, with evolution of sulphur dioxide:

$$Na_2SO_3 + 2HCl \rightarrow 2NaCl + H_2O + SO_2$$

$$SO_3^{2-}(aq) + 2H^+(aq) \rightarrow H_2O(l) + SO_2(g)$$

Questions on chapter 19

1. Give, with an equation, a fully labelled diagram to show how you would prepare and collect gas jars of sulphur dioxide.

How, and under what conditions, does sulphur dioxide react with hydrogen sulphide?

Starting with sulphur dioxide how would you obtain a sample of normal sodium sulphite? (*J.M.B.*)

2. Describe how you would prepare and collect some hydrogen sulphide.

State what you would see, name the products formed and write the equations for the reactions which take place when hydrogen sulphide is

(*a*) bubbled through copper(II) sulphate solution,
(*b*) mixed with moist sulphur dioxide,
(*c*) bubbled through iron(III) chloride solution,
(*d*) burned in a plentiful supply of air. (*C.*)

3. Sulphuric acid can act as

(*a*) an acid,
(*b*) an oxidizing agent,
(*c*) a dehydrating agent.

Illustrate this statement by describing *two* reactions of each type, giving the conditions for each reaction and equations wherever possible. In each pair of reactions, choose as reagents with the sulphuric acid substances which are different in character. (*J.M.B.*)

4. A gaseous oxide of sulphur contains half its weight of oxygen, and its density is double that of oxygen. Explain clearly how these facts enable you to calculate the molecular formula of this oxide.

Describe how you would prepare a reasonably pure specimen of it. ($O = 16$; $S = 32$.) (*C.*)

5. How may sulphur dioxide be prepared and collected in the laboratory?

Why is this gas important commercially and how is it usually prepared for such purposes?

Explain briefly the bleaching action of sulphur dioxide and compare it with that of chlorine.

What reaction takes place when sulphur dioxide is passed into (*a*) moist hydrogen sulphide, (*b*) sodium hydroxide solution?

Give one test to distinguish between a solution of sodium sulphite and a solution of sodium sulphate. (*C.W.B.*)

6. Starting with concentrated sulphuric acid in each case, how would you prepare (*a*) sodium sulphate, (*b*) lead(II) sulphate, (*c*) sulphur dioxide? (*C.*)

7. Give labelled diagrams to show how you would prepare the two common oxides of sulphur.

Describe briefly *three* reactions which indicate the affinity of sulphuric acid for water, or for the elements of water. (*L.*)

8. How may (*a*) sulphur dioxide, (*b*) sulphur(VI) oxide, (*c*) hydrogen sulphide be obtained from sulphur?

What (i) mass, (ii) volume at s.t.p. of sulphur dioxide could be obtained from 16 g of sulphur?

How can you convert rhombic sulphur into monoclinic sulphur? (*O. and C.*)

9. What do you understand by *allotropy*?

Starting from roll sulphur, describe how you would prepare *two* different forms of sulphur so that the crystals would be large enough for their shape to be observed?

How, and under what conditions, does sulphuric acid react with (*a*) copper, (*b*) sodium nitrate, (*c*) copper(II) sulphate crystals? What property of sulphuric acid does each of these three reactions illustrate? (*O.*)

10. Outline *one* method for the manufacture of sulphuric acid.

Give *one* example to illustrate *each* of the following uses of sulphuric acid (*a*) as a source of hydrogen, (*b*) as an oxidizing agent, and (*c*) in the preparation of another acid. (*D.*)

11. "Rhombic and monoclinic sulphur are allotropes". Briefly explain this statement. (Details of the preparation of these forms of sulphur are not required.)

Describe the contact process for the preparation of sulphur(VI) oxide, paying particular attention to the effects of the catalyst and changes of temperature and pressure on 1. the equilibrium concentration of sulphur(VI) oxide, and 2. the rate of the reaction. (The formation of sulphur(VI) oxide is exothermic.) *(J.M.B.)*

12. Describe the action of sulphuric acid on the following substances, stating carefully, in each case, whether concentrated or dilute acid is required and *the conditions under which the reaction takes place:* (*a*) ethanedioic acid (oxalic acid); (*b*) ethanol; (*c*) metallic copper; (*d*) calcium carbonate; (*e*) copper(II) nitrate.

Give equations in each case to illustrate the reactions. *(C.)*

13. Describe the Frasch method of extracting sulphur from the underground deposits in Texas and Louisiana.

Describe and explain what happens when hydrogen sulphide is passed (*a*) into bromine water, (*b*) into a solution of sulphur dioxide, (*c*) over paper moistened with lead(II) ethanoate solution. *(L.)*

14. How would you prepare and collect a jar of hydrogen sulphide starting with iron filings and sulphur? What impurity would you expect to be present in the gas?

How and under what conditions does hydrogen sulphide react with (*a*) oxygen, (*b*) sulphur dioxide, (*c*) a metal such as tin?

What would be the relative volume of the gas remaining after the reaction with tin, if its volume was measured at the same temperature and pressure as the hydrogen sulphide? *(C.W.B.)*

20

The Halogens

20.1 Introduction

The elements of Group VII of the Periodic Table are fluorine, chlorine, bromine, iodine and astatine. They are all non-metals whose reactivity decreases as the group is descended. Fluorine is exceptionally reactive and this fact makes its properties and those of some of its compounds very different from those of the other halogens. The name, halogen, means "salt producer"; you will see why in this chapter.

20.2 Extraction of the halogens

20.2.1 Fluorine

The principal ores of fluorine are calcium fluoride (fluorspar), CaF_2, and sodium hexafluoroaluminate (cryolite), Na_3AlF_6. The only known method of obtaining the element is by electrolysis. For this process a solution of potassium fluoride in anhydrous hydrogen fluoride is used. (No water must be present as fluorine reacts with water.) The anode is of carbon and the cathode, together with most of the cell components, is made of steel. These materials are fairly resistant to attack by fluorine (figure 20.1).

Ions present in the solution are potassium and fluoride ions from the potassium fluoride and a small number of hydrogen and fluoride ions from the little ionized hydrogen fluoride.

At the anode, fluoride ions are discharged:

$$F^- - e \rightarrow F$$
$$2F \rightarrow F_2$$

At the cathode, hydrogen ions are discharged:

$$H^+ + e \rightarrow H$$
$$2H \rightarrow H_2$$

20.2.2 Chlorine

Most chlorine is manufactured by the electrolysis of sodium chloride solution (see section 21.2.7(*c*) and figures 21.4 and 21.6).

In the laboratory, there are two main methods of preparation.

(a) By the oxidation of hydrochloric acid

Hydrochloric acid may be oxidized using manganese(IV) oxide or other substances rich in oxygen, such as lead(IV) oxide or potassium manganate(VII) (permanganate).

Figure 20.1 Maintenance work being carried out on fluorine cells. (Photo: I.C.I. Mond Division).

If manganese(IV) oxide is used, it is mixed with concentrated hydrochloric acid, and the mixture is gently heated. Chlorine is formed:

$$\left[\begin{array}{l} MnO_2 + 4HCl \rightarrow MnCl_2 + 2H_2O + Cl_2 \\ MnO_2(s) + 4H^+(aq) + 2Cl^-(aq) \rightarrow Mn^{2+}(aq) + 2H_2O(l) + Cl_2(g) \end{array}\right.$$

Traces of hydrogen chloride from the hydrochloric acid are removed by passage through water; the chlorine gas is then dried with concentrated sulphuric acid and collected by upwards displacement of air (figure 20.2).

Chlorine is very poisonous and a fume cupboard should be used for this preparation. If the gas is not required dry or if an efficient fume cupboard is not available, it can be collected over water.

Another good method is to add concentrated sulphuric acid to a mixture of sodium chloride and manganese(IV) oxide. The manganese(IV) oxide oxidizes chloride ions:

$$\left[\begin{array}{l} 2NaCl(s) + 3H_2SO_4(l) + MnO_2(s) \rightarrow 2NaHSO_4(s) + MnSO_4(s) + 2H_2O(l) + Cl_2(g) \\ 2Cl^- \rightarrow Cl_2 + 2e \\ MnO_2 + 4H^+ + 2e \rightarrow Mn^{2+} + 2H_2O \end{array}\right.$$

(b) By the action of acid on bleaching powder

Chlorine is liberated by the action of any dilute acid on bleaching powder. No heating is necessary.

$$\left[\begin{array}{l} CaOCl_2 + 2HNO_3 \rightarrow Ca(NO_3)_2 + H_2O + Cl_2 \\ CaOCl_2(s) + 2H^+(aq) \rightarrow Ca^{2+}(aq) + H_2O(l) + Cl_2(g) \end{array}\right.$$

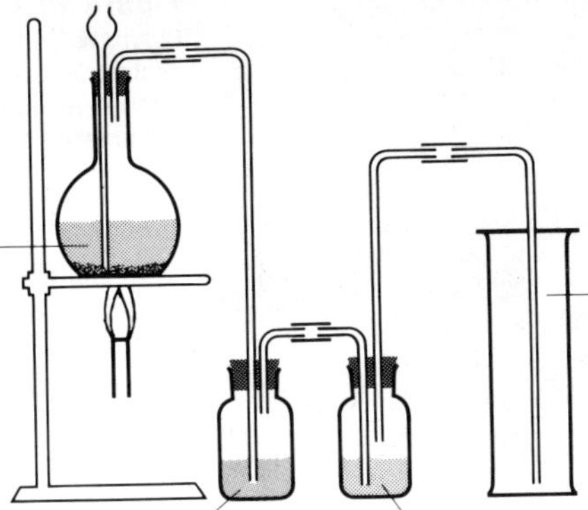

Figure 20.2 Preparation of chlorine.

20.2.3 Bromine

Most bromine is obtained from sea-water which contains small quantities of bromides (figure 20.3). Chlorine is used to displace the bromine:

$$2Br^-(aq) + Cl_2(g) \rightarrow Br_2(aq) + 2Cl^-(aq)$$

In the laboratory it is obtained in a similar way to chlorine but because hydrobromic acid is not a normal laboratory reagent it is usual to warm a mixture of a bromide and manganese(IV) oxide with concentrated sulphuric acid:

$$2KBr(s) + 3H_2SO_4(l) + MnO_2(s) \rightarrow 2KHSO_4(s) + MnSO_4(s) + 2H_2O(l) + Br_2(g)$$

An all-glass apparatus is used as in the preparation of nitric acid. Originally retorts were used (figure 1.9), but apparatus with ground glass joints is more convenient. A reddish-brown vapour is formed, which condenses to a reddish-brown liquid in the cooled receiver.

20.2.4 Iodine

The ash from burnt seaweed contains small quantities of iodides. The most important source, however, is Chile saltpetre which contains about 0·1 % of sodium iodate(V).

In the laboratory the preparation of iodine may be demonstrated by warming a mixture of manganese(IV) oxide and potassium iodide with concentrated sulphuric acid. If this is done in a test-tube, beautiful violet fumes of iodine are formed which condense to the solid on the cooler part of the tube:

$$2KI(s) + 3H_2SO_4(l) + MnO_2(s) \rightarrow 2KHSO_4(s) + MnSO_4(s) + 2H_2O(l) + I_2(g)$$

Figure 20.3 A plant for extrac bromine from sea-water at Amlwc Wales. The small picture shows large sea-water pumps. (Photos: Associated Octel Co. Ltd.).

20.3 Properties of the halogens

20.3.1 Physical properties

In physical properties the halogens show the gradation expected from a group of elements in the periodic table. The atoms increase in size and the density increases from fluorine to iodine:

(*a*) *Fluorine* is a pale yellow exceptionally poisonous gas which liquefies at $-187\,^\circ$C. It has a pungent odour.

(*b*) *Chlorine* is a pale yellow-green poisonous gas which liquefies at $-101\,^\circ$C. It has a pungent odour.

(*c*) *Bromine* is a reddish-brown liquid, very poisonous, which boils at 59 °C. It has a pungent odour.

(*d*) *Iodine* is a black solid. When gently heated it evaporates without melting to give a violet vapour. This vapour, when cooled, condenses directly to the solid, this being an example of sublimation. Its odour is similar to that of chlorine.

20.3.2 Reactions with hydrogen

All of the halogens combine directly with hydrogen. With *fluorine* this reaction can occur with explosive violence and evolution of much heat.

$$H_2(g) + F_2(g) \rightarrow 2HF(g) \quad \Delta H = -542 \text{ kJ mol}^{-1}$$

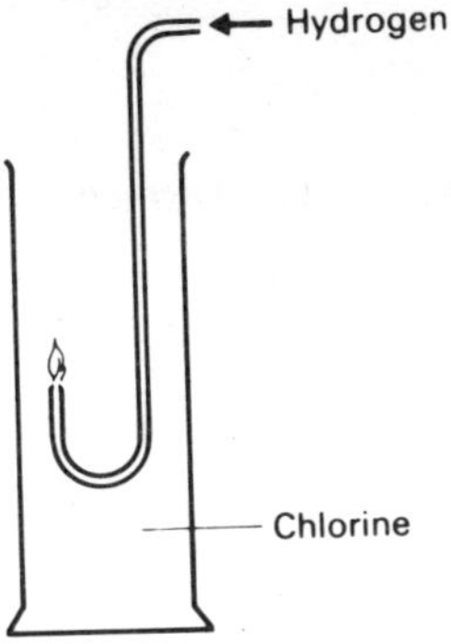

Figure 20.4 Preparation of hydrogen chloride by burning hydrogen in chlorine.

With *chlorine* combination is slow in diffused daylight but in bright sunlight or if, as discussed in section 14.4.6, a light is applied, reaction occurs with explosive violence

$$H_2(g) + Cl_2(g) \rightarrow 2HCl(g) \quad \Delta H = -184 \text{ kJ mol}^{-1}$$

This reaction can be safely demonstrated by introducing a jet of burning hydrogen into a gas jar full of chlorine (figure 20.4). The formation of hydrogen chloride is soon apparent, as misty fumes form at the mouth of the jar.

With *bromine*, combination occurs only on heating or by using a platinum catalyst and with iodine there is very little reaction.

$$H_2(g) + Br_2(g) \rightarrow 2HBr(g) \quad \Delta H = -73 \text{ kJ mol}^{-1}$$
$$H_2(g) + I_2(g) \rightarrow 2HI(g) \quad \Delta H = +53 \text{ kJ mol}^{-1}$$

20.3.3 Direct union with other elements

Fluorine combines vigorously with most other elements except nitrogen, oxygen and the noble gases.

Chlorine, whilst very reactive, combines less readily than fluorine, e.g. if dry white phosphorus is lowered into a gas jar full of chlorine it ignites spontaneously to give white fumes of phosphorus trichloride and phosphorus pentachloride:

$$P_4(s) + 6Cl_2(g) \rightarrow 4PCl_3(l)$$
$$P_4(s) + 10Cl_2(g) \rightarrow 4PCl_5(s)$$

Most metals combine with chlorine e.g. if copper foil, iron wire or sodium are heated in chlorine they readily form chlorides. Anhydrous metallic chlorides such as iron(III), aluminium and tin(IV) chlorides are often prepared by this method of direct combination. Remember that where a metal has more than one valency the chloride formed is of the higher one, e.g. iron(III) chloride is formed and *not* iron(II) chloride. This is because chlorine is an oxidizing agent.

The reactivity of bromine with other elements is less than chlorine although, being a liquid and therefore more concentrated, it is capable of combining explosively with potassium. Iodine is the least reactive of the elements.

20.3.4 Reactions with water and alkalis

(*a*) *Fluorine* reacts vigorously with water:

$$2H_2O(l) + 2F_2(g) \rightarrow 4HF(g) + O_2(g) \quad \text{(some } F_2O \text{ and } O_3 \text{ are also formed)}$$

(*b*) *Chlorine* is moderately soluble in water and reacts with it to give a mixture of hydrochloric and chloric(I) acids

$$H_2O(l) + Cl_2(g) \rightleftharpoons HCl(aq) + HClO(aq)$$

The solution obtained is called "chlorine water" and has two interesting properties:

(*a*) in sunlight, it decomposes to give oxygen:

$$2H_2O(l) + Cl_2(g) \rightarrow 4HCl(aq) + O_2(g)$$

(*b*) it bleaches many colouring matters, e.g. if *moist* litmus paper, grass or flowers are dipped into a gas jar containing chlorine, they are bleached. Ink writing is also bleached, provided it is not printers' ink, which is mainly carbon. Blue litmus paper is turned red by the acids present, prior to bleaching. The bleaching is effected by the chloric(I) acid which readily gives its oxygen atom to dyes, thereby converting them into colourless compounds.

If chlorine is passed into a cold dilute solution of an alkali, e.g. sodium or potassium hydroxides, or a suspension of calcium hydroxide in water, then the salts of hydrochloric and chloric(I) acids are formed:

$$\left[\begin{array}{l} Cl_2 + 2NaOH \rightarrow NaCl + NaClO + H_2O \\ Cl_2(g) + 2OH^-(aq) \rightarrow Cl^-(aq) + ClO^-(aq) + H_2O(l) \end{array}\right.$$

If the solution is hot and concentrated the chlorate(I) is probably formed first but is then converted into a chlorate(V) and a chloride:

$$\left[\begin{array}{l} 3(Cl_2 + 2NaOH \rightarrow NaCl + NaClO + H_2O) \\ \quad 3NaClO \rightarrow 2NaCl + NaClO_3 \\ \hline 3Cl_2 + 6NaOH \rightarrow 5NaCl + NaClO_3 + 3H_2O \end{array}\right.$$

$$\text{or } 3Cl_2(g) + 6OH^-(aq) \rightarrow 5Cl^-(aq) + ClO_3^-(aq) + 3H_2O(l)$$

When chlorine is passed over dry calcium hydroxide, bleaching powder is formed:

$$Ca(OH)_2(s) + Cl_2(g) \rightarrow CaOCl_2(s) + H_2O(l)$$

The reaction is complex, but the equation given is an adequate one for our purpose.

(*c*) *Bromine* is moderately soluble in water to give bromine water, which is more stable than chlorine water. Bromine water will bleach, but far less readily than chlorine.

With alkalis, bromine behaves similarly to chlorine.

$$\left[\begin{array}{l} Br_2 + 2NaOH \rightarrow NaBr + NaBrO + H_2O \\ Br_2(aq) + 2OH^-(aq) \rightarrow Br^-(aq) + BrO^-(aq) + H_2O(l) \end{array}\right.$$

$$\begin{bmatrix} 3Br_2 + 6NaOH \rightarrow 5NaBr + NaBrO_3 + 3H_2O \\ 3Br_2(aq) + 6OH^-(aq) \rightarrow 5Br^-(aq) + BrO_3^-(aq) + 3H_2O(l) \end{bmatrix}$$

(*d*) *Iodine* is slightly soluble in water but is readily soluble in alcohol and in potassium iodide solution. With potassium iodide the solubility is due to the formation of potassium tri-iodide:

$$KI + I_2 \rightleftharpoons KI_3$$
$$I^-(aq) + I_2(s) \rightleftharpoons I_3^-(aq)$$

Iodine reacts with alkalis similarly to chlorine and bromine.

20.3.5 The affinity of halogens for hydrogen

The vigorous reactions of fluorine and chlorine with hydrogen have already been discussed. The affinity of these elements for hydrogen is shown by their ability to remove hydrogen from many compounds. With fluorine the reactions are extraordinarily vigorous. Those with chlorine are more important and are easily demonstrated:

(*a*) When a burning taper is lowered into a gas jar full of chlorine it continues to burn with a reddish flame, producing clouds of soot and a gas which fumes in damp air. The fumes are intensified if some ammonia is blown over the mouth of the gas jar.

The taper is composed of hydrocarbons (compounds consisting of carbon and hydrogen only). Hydrogen chloride is the fuming gas produced by combination of chlorine with the hydrogen, thus leaving the carbon in the form of soot. With ammonia, the hydrogen chloride forms dense white fumes of ammonium chloride.

This type of reaction can be even more strikingly illustrated by using some glass wool, soaked in *hot* turpentine ($C_{10}H_{16}$), instead of the burning taper:

$$C_{10}H_{16}(l) + 8Cl_2(g) \rightarrow 10C(s) + 16HCl(g)$$

(*b*) When gas jars full of hydrogen sulphide and moist chlorine are mixed a pale yellow deposit of sulphur is formed and hydrogen chloride is again produced. Alternatively, chlorine may be bubbled through hydrogen sulphide solution:

$$H_2S(aq) + Cl_2(g) \rightarrow 2HCl(aq) + S(s)$$

(*c*) If a few drops of concentrated ammonia solution are dropped from a pipette into a gas jar full of chlorine, ammonia gas from the solution burns and white clouds of ammonium chloride are formed. This reaction occurs in two stages:

$$2NH_3 + 3Cl_2 \rightarrow N_2 + 6HCl$$
$$6NH_3 + 6HCl \rightarrow 6NH_4Cl$$

$$\begin{bmatrix} 8NH_3 + 3Cl_2 \rightarrow 6NH_4Cl + N_2 \\ 8NH_3(g) + 3Cl_2(g) \rightarrow 6NH_4Cl(s) + N_2(g) \end{bmatrix}$$

Bromine and iodine show a much lower tendency to enter into reactions of this sort, but both will oxidize hydrogen sulphide to sulphur:

e.g., $H_2S(aq) + Br_2(l) \rightarrow 2HBr(aq) + S(s)$

20.3.6. Other oxidizing actions of the halogens

All of the reactions in the previous section are examples of oxidation by removal of hydrogen. Chlorine, bromine and iodine enter into a wide range of other oxidation reactions. Here are two examples:

(*a*) The passage of chlorine through a solution of sulphurous acid converts it to sulphuric acid:

$$H_2SO_3 + H_2O + Cl_2 \rightarrow H_2SO_4 + 2HCl$$

sulphurous acid

$$SO_3^{2-}(aq) + H_2O(l) + Cl_2(g) \rightarrow SO_4^{2-}(aq) + 2H^+(aq) + 2Cl^-(aq)$$

Similarly sulphites are converted to sulphates:

(*b*) Pale green iron(II) chloride solution turns yellow when chlorine is bubbled through it. This is because the iron(III) salt is formed.

$$2FeCl_2 + Cl_2 \rightarrow 2FeCl_3$$

$$2Fe^{2+}(aq) + Cl_2(g) \rightarrow 2Fe^{3+}(aq) + 2Cl^-(aq)$$

Bromine and iodine behave similarly to chlorine but react with less vigour.

20.3.7 The displacement of halogens from their salts

Chlorine displaces both bromine and iodine from solutions of their salts and bromine displaces iodine from solutions of its salts. Passing a few bubbles of chlorine through the colourless solutions of potassium bromide or iodide produces a reddish, or brownish colouration due to the formation of free bromine or iodine, e.g.:

$$Cl_2 + 2KI \rightarrow 2KCl + I_2$$

$$Cl_2(g) + 2I^-(aq) \rightarrow 2Cl^-(aq) + I_2(aq)$$

These reactions show again the gradual decrease of oxidizing power as the group is descended, i.e. of these three halogens chlorine acquires extra electrons the most readily and iodine the least. Fluorine, by far the most powerful oxidizing agent of the halogens, will displace all the others from their salts but if water is present it reacts with this as well.

20.4 Uses of the halogens

(*a*) *Fluorine* is used in making uranium hexafluoride(UF_6). This fluoride is volatile and can therefore be used for separating the isotopes of uranium by a diffusion method. The fluoride of uranium-235 diffuses more rapidly than that of uranium-238.

(*b*) *Chlorine* is widely used for bleaching cotton, linen and rayon and in the paper and pulp industry.

The germicidal action of chlorine accounts for its use in the treatment of sewage and the purification of water. Very small quantities of chlorine suffice to destroy bacteria responsible for certain epidemic diseases such as typhoid. The smell often associated with swimming pools is that of chlorine.

Chlorine is used to prepare many inorganic chlorides and numerous organic compounds e.g. tetrachloromethane (CCl_4) and trichlorethene, which are both used as solvents; trichloromethane (chloroform), an anaesthetic; the insecticide DDT and the plastic poly(chloroethene) (polyvinyl chloride (P.V.C.)).

Much hydrogen chloride is made by burning hydrogen in chlorine.

(*c*) *Bromine* is used in making 1,2-dibromoethane, which in conjunction with tetraethyllead(IV), is used to prevent "knocking" in internal combustion engines.

(*d*) *Iodine* has many uses in medicine. "Tincture of iodine" (a solution of iodine in ethanol) is well known.

20.5 Hydrogen Halides

20.5.1 Preparation

As explained earlier in the chapter, all of the hydrogen halides may be obtained by direct synthesis and hydrogen chloride is manufactured by this method. In the laboratory the following methods are suitable.

20.5.2 Hydrogen fluoride

Calcium fluoride or another fluoride is heated with concentrated sulphuric acid:

$$CaF_2(s) + H_2SO_4(l) \rightarrow CaSO_4(s) + 2HF(g)$$

Note. This reaction should *never* be carried out in the laboratory, as hydrogen fluoride is a most unpleasant gas. Inhalation of quite small quantities have caused permanent loss of voice and even death.

20.5.3 Hydrogen chloride

A similar method is used, i.e. a metallic chloride is treated with concentrated sulphuric acid (Figure 20.5). In practice, the cheapest chloride, common salt, is always used. When the salt is powdered, however, a rather vigorous reaction occurs, because of the large surface area of salt in contact with the acid. It can be moderated by using lumps of rock salt or fused salt (salt which has been melted, and then allowed to solidify), or by slightly diluting the acid.

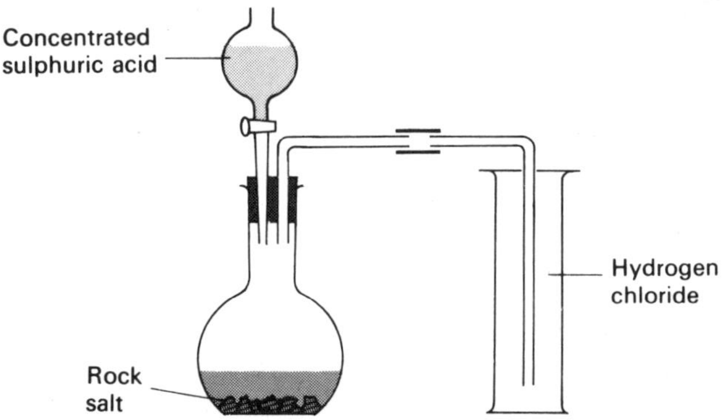

Figure 20.5 Preparation of hydrogen chloride.

The hydrogen chloride formed is denser than air and is collected by downwards delivery; steamy fumes of the gas are seen at the mouth of the jar when it is full. Sodium hydrogensulphate, i.e., the acid sulphate, is the other product:

$$NaCl(s) + H_2SO_4(l) \rightarrow NaHSO_4(s) + HCl(g)$$

Only at higher temperatures than those normally used in the laboratory is the normal sulphate formed:

$$2NaCl(s) + H_2SO_4(l) \rightarrow Na_2SO_4(s) + 2HCl(g)$$

20.5.4 Hydrogen bromide

You might expect, in view of the similarities existing between the halogens, that hydrogen bromide would be prepared by the action of concentrated sulphuric acid on a bromide. Indeed some hydrogen bromide *is* formed by this method but because bromine has a lower affinity with hydrogen than either fluorine or chlorine the hydrogen bromide formed is easily oxidized. It is in part oxidized by the concentrated sulphuric acid to bromine, and the sulphuric acid is simultaneously reduced to sulphur dioxide:

$$NaBr(s) + H_2SO_4(l) \rightarrow NaHSO_4(s) + HBr(g)$$
$$2HBr(g) + H_2SO_4(l) \rightarrow 2H_2O(l) + Br_2(g) + SO_2(g)$$

It is therefore necessary to use an acid which is not an oxidizing agent such as phosphoric acid. Alternatively bromine may be dropped onto a mixture of red phosphorus and water. Phosphorus tribromide is formed which then reacts with the water.

$$P_4(s) + 6Br_2(l) \rightarrow 4PBr_3(aq)$$
$$PBr_3(aq) + 3H_2O(l) \rightarrow H_3PO_3(aq) + 3HBr(g)$$

20.5.5 Hydrogen iodide

Hydrogen iodide is even more readily oxidized than hydrogen bromide. It may be prepared by dropping water on to a mixture of finely ground iodine and red phosphorus:

$$P_4(s) + 6I_2(s) \rightarrow 4PI_3(aq)$$
$$PI_3(aq) + 3H_2O(l) \rightarrow H_3PO_3(aq) + 3HI(g)$$

20.5.6 Properties of the hydrogen halides

The trend in properties from hydrogen chloride to hydrogen iodide is what would be expected from the positions of chlorine, bromine and iodine in the Periodic Table. Hydrogen fluoride is unusual both in its physical and chemical properties. At room temperature all the hydrogen halides, (including hydrogen fluoride) are colourless gases with irritating odours, but hydrogen fluoride departs from the general trend so far as boiling points and melting points are concerned.

	b.p.	m.p.
Hydrogen fluoride	19° C	− 83° C
Hydrogen chloride	−85° C	−114° C
Hydrogen bromide	−67° C	− 87° C
Hydrogen iodide	−36° C	− 51° C

Hydrogen fluoride is very stable indeed to heat. Hydrogen chloride shows little tendency to decompose on heating. Hydrogen bromide decomposes on strong heating whereas hydrogen iodide decomposes readily to give violet iodine vapour if a red hot glass rod is plunged into a jar full of the gas:

$$2HI(g) \rightarrow H_2(g) + I_2(g)$$

All of the hydrogen halides are very soluble in water and all give strong acids except hydrogen fluoride (which gives the weak hydrofluoric acid); one volume of water at 0 °C dissolves 535 volumes of hydrogen chloride. This property is convincingly shown by the fountain experiment, described in section 16.3.3. The details are exactly the same as given there, except, of course, that blue litmus is used.

The dry hydrogen halides do *not* show acidic properties. Thus hydrogen chloride has no effect on dry litmus paper. Moreover, if a solution is made by dissolving dry hydrogen chloride in methylbenzene, acidic properties associated with the presence of hydrogen ions are absent. Such a solution will not conduct electricity and, if completely dry, will not react with litmus, metals or carbonates. With ammonia, ammonium chloride, being ionic and therefore insoluble in methylbenzene, is formed as a white precipitate. Only when the hydrogen chloride dissolves in water does it give acidic reactions because of a reaction involving the formation of hydrogen (hydroxonium) ions and chloride ions:

$$HCl(g) + H_2O(l) \rightarrow H_3O^+(aq) + Cl^-(aq)$$

The other hydrogen halides behave in the same way. Typically acidic properties, which are due to hydrogen ions, develop only in the presence of water, when ionization is able to occur.

All of the hydrogen halides react with ammonia. Thus when a gas jar full of hydrogen chloride is inverted over a gas jar full of ammonia, dense white clouds of ammonium chloride are formed:

$$NH_3(g) + HCl(g) \rightarrow NH_4Cl(s)$$

20.5.7 The halogen acids

(a) Preparation

As stated above the hydrogen halides dissolve in water to give acids, the most important being hydrochloric acid. Concentrated hydrochloric acid contains approximately 36 % by weight of hydrogen chloride. This solution cannot be prepared just by passing hydrogen chloride through a tube into water. The high solubility of the gas would probably result in an unwanted repetition of the fountain experiment; if the hydrogen chloride is being prepared using concentrated sulphuric acid the result could be disastrous, as water would flow back onto the sulphuric acid.

Some device to prevent "sucking back" is essential; an inverted funnel, used as shown in figure 20.6, is the simplest one. The opening of the funnel must be just beneath the surface of the water. When water begins to suck back, the level of the water in the beaker becomes lower so that the rim of the funnel is soon exposed. Air then enters the funnel and all the liquid falls back into the beaker. The process begins again and continues doing so until a saturated solution is obtained. Then the hydrogen chloride will bubble through the concentrated hydrochloric acid. (*Note:* Figure 16.11 shows an improved method.)

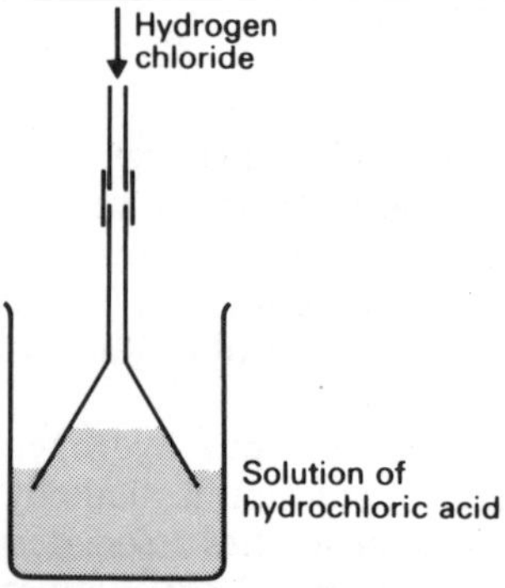

Figure 20.6 Device to prevent "sucking-back".

(b) Properties

All of the concentrated acids are colourless, fuming liquids, the fumes being due to escape of the hydrogen halide from the solution into the atmosphere. The dilute acids are odourless.

The most important acid is hydrochloric, which is typical of them all, and behaves similarly whether dilute or concentrated. Thus hydrochloric acid has the following reactions:

(i) With metals above hydrogen in the activity series, e.g. zinc and iron, hydrogen is evolved:

$$Zn + 2HCl \rightarrow ZnCl_2 + H_2$$
$$Zn(s) + 2H^+(aq) \rightarrow Zn^{2+}(aq) + H_2(g)$$

(ii) With carbonates and hydrogencarbonate, carbon dioxide is evolved:

$$Na_2CO_3 + 2HCl \rightarrow 2NaCl + H_2O + CO_2$$
$$CO_3{}^{2-}(s) + 2H^+(aq) \rightarrow H_2O(l) + CO_2(g)$$

(iii) With bases, a salt and water are formed:

$$MgO + 2HCl \rightarrow MgCl_2 + H_2O$$
$$MgO(s) + 2H^+(aq) \rightarrow Mg^{2+}(aq) + H_2O(l)$$

(c) Uses of the acids

Hydrochloric acid is far the most important. It is widely used in the removal of rust from iron which is to be galvanized, and in cleaning metals prior to electroplating. Hydrogen chloride is used in the synthesis of chloroethene which is important in the manufacture of some plastics (see section 26.1).

(d) The halides

The salts of hydrochloric acid are the most important. They are all freely soluble, with the exceptions of lead(II), silver(I) and mercury(I) chlorides.

They can be prepared by the standard methods:

(i) By neutralization of an acid with an alkali, e.g. sodium and potassium chlorides.

(ii) By the action of an acid on a metal. The method cannot be used with the very reactive metals e.g. sodium and potassium, or with those occurring below hydrogen in the activity series.

(iii) By the action of the acid on a metallic carbonate or oxide.

(iv) By precipitation.

(v) By direct combination of the metal with chlorine.

Other halides are prepared by similar methods.

(e) Identification of chlorides, bromides and iodides (see also section 27.3.7 (*a*))

(i) When a *chloride* is treated with concentrated sulphuric acid steamy fumes of hydrogen chloride are formed:

$$KCl(s) + H_2SO_4(l) \rightarrow KHSO_4(s) + HCl(g)$$

These fumes are intensified if ammonia is blown over the mouth of the test-tube (see section 20.5.6).

Similarly, with a *bromide*, hydrogen bromide is formed but because this compound is easily oxidized it is mixed with reddish-brown bromine vapour and with sulphur dioxide.

With an *iodide*, concentrated sulphuric acid gives a vigorous reaction forming iodine (violet fumes), sulphur dioxide, sulphur and hydrogen sulphide.

(ii) If a mixture of a *chloride* with manganese(IV) oxide is treated with concentrated sulphuric acid, pale yellowish-green chlorine is formed. This is detected by its smell or by its bleaching action on damp litmus paper.

Similarly a *bromide* yields brownish red bromine vapour and an *iodide* violet iodine vapour.

(iii) If dilute nitric acid, followed by silver nitrate solution, is added to a solution of a *chloride*, a white precipitate of silver chloride is immediately formed:

$$Ag^+(aq) + Cl^-(aq) \rightarrow AgCl(s)$$

The dilute nitric acid prevents precipitation of various other silver salts. The white precipitate is readily soluble in ammonia solution to form a complex ion:

$$AgCl(s) + 2NH_3(aq) \rightarrow Ag(NH_3)_2^+(aq) + Cl^-(aq)$$

With a *bromide*, a pale yellow precipitate of silver bromide is formed. This precipitate is only slightly soluble in excess of ammonia solution.

With an *iodide*, a yellow precipitate of silver iodide, insoluble in ammonia solution is formed.

Questions on chapter 20

1. Give, with an equation, a fully labelled diagram to show how in the laboratory you would prepare and collect gas jars of dry chlorine free from hydrogen chloride.

What is the action of chlorine on (*a*) a cold dilute solution of sodium hydroxide, (*b*) heated iron filings, (*c*) hydrogen sulphide and (*d*) turpentine ($C_{10}H_{16}$)? Give equations. (*J.M.B.*)

2. Give, with the equations, *two* fully labelled diagrams to show how, starting from sodium chloride, you would prepare (*a*) a solution of hydrogen chloride, (*b*) a gas jar of pure chlorine.

Describe briefly how you would prepare anhydrous specimens of (i) iron(II) chloride, (ii) iron(III) chloride. (*J.M.B.*)

3. The element fluorine has an atomic number of 9 and a mass number of 19. At room temperature, fluorine is a gas and consists of diatomic molecules, F_2. Fluorine reacts vigorously with cold water, forming hydrogen fluoride and oxygen.

(*a*) Draw a simple diagram to show the structure of the fluorine atom and the arrangement of its electrons.

(*b*) Give the formulae of the compounds you would expect fluorine to form with hydrogen; carbon and calcium and say whether you would expect the bonding in each compound to be covalent or ionic.

(*c*) Explain in terms of electrons why fluorine exists as F_2 molecules whereas the noble gas neon (of atomic number 10) exists as single atoms.

(*d*) Fluorine is normally prepared by an electrolytic method. Which of the following would be the most suitable liquid to use as the electrolyte for this preparation:

(i) pure liquid hydrogen fluoride,
(ii) molten potassium fluoride,
(iii) a solution of potassium fluoride in water?

Give reasons for your choice. (*C.*)

4. Describe the preparation and collection of chlorine in the laboratory.

Give examples of its action on (*a*) one metal element, (*b*) one non-metal element.

How and under what conditions does chlorine react with (*c*) water, (*d*) slaked lime (calcium hydroxide)? (*D.*)

5. Group VII of the periodic table includes the elements chlorine, bromine and iodine; the compounds of these elements with potassium are crystalline solids whose aqueous solutions are electrolytes. State in terms of electrons how the particles in these solids are held together.

What are the results of bubbling chlorine into:

(*a*) cold dilute sodium hydroxide solution,
(*b*) hot concentrated potassium hydroxide solution,
(*c*) potassium iodide solution?

How would you release (i) hydrogen, (ii) chlorine from an aqueous solution of hydrogen chloride? A different method should be used for each element. (*O.*)

6. Describe how you would prepare pure dry chlorine. Using chlorine how would you distinguish between sodium bromide and sodium iodide?

Outline an experiment by which you would demonstrate the rate of diffusion of chlorine compared with that of air.

7. Bromine (atomic mass = 80) is a dark reddish-brown, corrosive liquid boiling at 59 °C. It is a member of the halogen family of elements and closely resembles chlorine in its chemical properties. 1 g of bromine vapour occupies a volume of 140 cm^3, the volume being corrected to s.t.p.

(*a*) Name one member of the halogen family other than chlorine or bromine.

(*b*) Calculate the molecular mass and the atomicity of bromine vapour.

(*c*) Bromine can be prepared by heating a mixture of sodium bromide and manganese(IV) oxide with concentrated sulphuric acid. Draw a diagram of the apparatus you would use for this preparation and explain why manganese(IV) oxide

and concentrated sulphuric acid are used to obtain bromine from sodium bromide.
(*d*) Describe one other reaction in which bromine is formed from sodium bromide.
(*e*) What particles are present in a crystal of sodium bromide?

8. The halogens have the following boiling points: fluorine (– 188 °C), chlorine (– 34 °C), bromine (59 °C) and iodine (184 °C). The heats of dissociation of the molecules are:

$$F_2 \rightarrow 2F \qquad \Delta H = 318 \text{ kJ mol}^{-1}$$
$$Cl_2 \rightarrow 2Cl \qquad \Delta H = 234 \text{ kJ mol}^{-1}$$
$$Br_2 \rightarrow 2Br \qquad \Delta H = 188 \text{ kJ mol}^{-1}$$
$$I_2 \rightarrow 2I \qquad \Delta H = 146 \text{ kJ mol}^{-1}$$

(*a*) State how the stabilities of the halogen molecules vary.
(*b*) At 1700 °C one halogen only is completely dissociated into atoms. Name this element.
Both the boiling points and the heats of dissociation show a progressive change through the family of halogens. Show that this pattern of change is also followed in their reactivity with (i) hydrogen, (ii) metals and (iii) in the displacement of halogens from halides.
How does chlorine react with sodium hydroxide solution? (*J.M.B.*)

9. The halogens have the following atomic numbers: fluorine 9, chlorine 17, bromine 35, iodine 53.
(*a*) Give the electronic configuration for
(i) an atom of fluorine,
(ii) an atom of chlorine.
(*b*) How many valency electrons has an iodine atom?
(*c*) Write down the names of the four halogens and after each state whether it is a solid, a liquid or a gas under ordinary laboratory conditions.
(*d*) How, and under what conditions, does chlorine react with (i) hydrogen, (ii) iron, (iii) potassium bromide?
(*e*) State whether iodine is more or less reactive than chlorine. Give an example to illustrate your answer. (*J.M.B.*)

10. Chlorine is often prepared in the laboratory by the reaction between manganese(IV) oxide and hydrochloric acid.
(*a*) State the conditions necessary to obtain a reasonable quantity of chlorine.
(*b*) What is oxidized in this reaction?
(*c*) Name the other products of the reaction, and give the equation.
(*d*) What is the main impurity in the gas, other than moisture? State how you would remove it.
(*e*) Give a diagram of the apparatus you would use for preparing, purifying, and collecting the chlorine.
The rest of this question refers to an experiment performed to find the formula of hydrogen chloride. 40 cm^3 of hydrogen and 30 cm^3 of chlorine were mixed in subdued light. When the mixture was passed over hot platinum it lost its colour. On cooling to the original temperature the total volume was found to be unchanged. The remaining gas was pumped into water, when all but 10 cm^3 dissolved.
(i) Why was the hydrogen mixed with chlorine in subdued light?
(ii) What was the purpose of the platinum?
(iii) Show clearly how the results of this experiment enable you to calculate the formula of hydrogen chloride. (You may assume that hydrogen and chlorine are diatomic gases.)
(iv) State the chemical law you use in (iii). (*A.E.B.*)

21

Metals and Non-Metals; the Alkali Metals

21.1 Metals and non-metals

The terms "metal" and "non-metal" have been used frequently throughout this book. Before studying the detailed properties of some metals and their compounds, we should discuss thoroughly what particular characteristics determine whether an element is one or the other. At an elementary level our decision is based mainly on differences in physical properties. Consider those metals encountered in everyday life, such as iron, copper, tin, lead and mercury. Thus metals are lustrous and can be highly polished. Except for copper and gold, they are all silver to grey in colour. They have comparatively high densities and, with the exception of mercury, are solids usually with high melting and boiling points. They are good conductors of heat and electricity, are malleable and ductile, and have great tensile strength. All of these properties are explained by the way in which the atoms in a metal are arranged. It appears that they form a rigid lattice with the atoms packed as closely together as possible, this giving the high densities. The outer electrons of the atoms are not closely associated with individual atoms but move around amongst the atoms holding them together—there is an "electron glue". The atoms can slip over one another but remain firmly held together by the electrons. This is why metals can be hammered into thin sheets, i.e., are malleable, or drawn into thin wires, i.e., are ductile. When an electrical potential difference is applied, the electrons move readily from atom to atom, i.e., a current flows.

Non-metals, on the other hand, are usually non-lustrous and have low densities. Many are gases; the others normally have low boiling points and melting points. They are poor conductors of heat and electricity, are neither malleable nor ductile and have little tensile strength.

Differences in chemical properties are, however, more important. Thus:

(*a*) *The normal oxides of metals are usually basic*, e.g. copper(II) oxide, sodium oxide.

Oxides of non-metals are usually acidic, e.g. carbon dioxide, nitrogen dioxide and sulphur(VI) oxide. A few non-metallic oxides are neutral.

(*b*) *Chlorides of metals are salts*, i.e., they are electrovalent and give solutions which conduct electricity. In general, they are not decomposed by water.
Chlorides of non-metals are not salts. They are covalent gases or liquids and usually react chemically with water.

(*c*) *A metal is capable of replacing the hydrogen of an acid either directly or indirectly.*

A non-metal will not replace the hydrogen of an acid.

In terms of electrons, *the characteristic behaviour of a metal is that in chemical reactions it loses electrons to form positively charged ions*, e.g.

$Na - e \rightarrow Na^+$

$Cu - 2e \rightarrow Cu^{2+}$

This means that metals are reducing agents, these electrons being supplied to the substance that the metal reacts with. *Non-metals*, on the other hand, *combine with metals by gaining electrons to form negatively charged ions*, e.g.

$Cl_2 + 2e \rightarrow 2Cl^-$

$S \ + 2e \rightarrow S^{2-}$

This means that non-metals are oxidizing agents. When non-metals combine with one another they usually share electrons to form covalent compounds.

The classification of elements into metals and non-metals is not entirely satisfactory. As we have seen in section 5.3.4, the metals and non-metals are the groups of elements on the left hand side and right hand side of the periodic table. Where these two groups meet there will be elements such as germanium, arsenic, selenium and iodine that cannot be properly called metals or non-metals.

21.2 The alkali metals

Lithium, sodium, potassium, rubidium, caesium and francium are known as the *alkali metals*. Francium has not been found in nature, but has been prepared artificially in recent years.

The alkali metals and their compounds show a remarkable similarity to one another. Sodium and potassium are by far the most important and the discussion in this book is restricted mainly to these two of the alkali metals.

21.2.1 Occurrence of the alkali metals

The alkali metals are far too reactive to occur in the uncombined state. Compounds of sodium and potassium are, however, widely distributed, the most important and most abundant being *sodium chloride*, *common salt*. Vast quantities of sodium chloride are present in the sea and in a number of lakes (such as the Dead Sea and the Great Salt Lake). It has been leached from the land by rain and carried to the sea and lakes by rivers. Solid sodium chloride deposits, known as *rock salt*, are found in many countries (figure 21.1). Other compounds of sodium found in nature include *Chile saltpetre*, $NaNO_3$, *borax*, $Na_2B_4O_7.10H_2O$, and silicates such as *felspar*, $NaAlSi_3O_8$. Potassium compounds are also present in sea water and in most clays and silicate rocks. The most important sources of potassium compounds are deposits formed by the evaporation of inland seas, e.g. at Stassfurt in Germany where large deposits of *carnallite*, $KMgCl_3.6H_2O$ are found.

21.2.2 Extraction of the alkali metals

Sodium was first extracted by Humphry Davy by the electrolysis of molten sodium hydroxide in a platinum crucible, using the crucible itself as anode and a platinum wire as cathode, as shown in figure 21.2.

Figure 21.1 Sources of sodium chloride—rock salt and the sea. (*a*) Inside the Winsford salt mine in Cheshire. here a layer of rock salt 2.5 m thick lies 160 m below the surface. (Photo: I.C.I. Mond Division). (*b*) The Dry Creek saltfields, near Adelaide in Australia. Here sea-water flows slowly through shallow ponds and the water is evaporated by the sun. Eventually solid salt remains and this is collected by the mechanical harvesters. (Photo: I.C.I. Australia Ltd.).

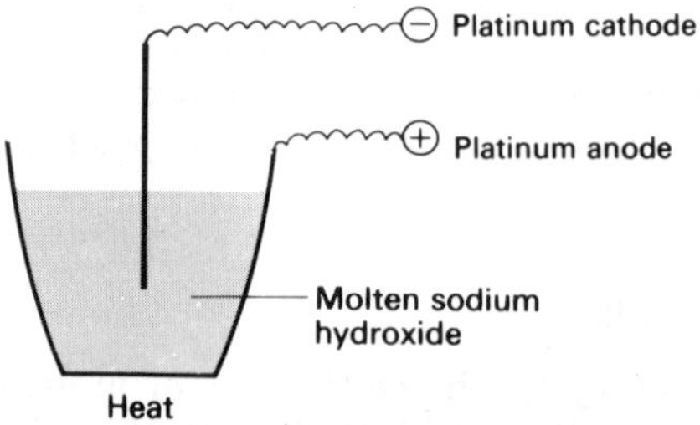

Figure 21.2 Electrolysis of molten sodium hydroxide.

Sodium and hydroxide ions are present. The former move to the cathode, where they gain electrons: $Na^+ + e \rightarrow Na$. Molten sodium is formed and ignites in the air.

The hydroxide ions are discharged at the anode, forming water and oxygen:

$$OH^- - e \rightarrow OH$$

$$4OH \rightarrow 2H_2O + O_2.$$

(N.B. It is a common error of students to state that sodium hydroxide *solution* is electrolyzed. This is quite unsuitable, as hydrogen ions from water would then be discharged in preference to sodium ions (see section 8.3.4(*b*)). Even assuming that it could be formed, the sodium would immediately react with water.)

Industrially, all of the alkali metals are obtained nowadays by electrolysis of the molten chloride, e.g. the Downs Process for the extraction of sodium.

21.2.3 The Downs process

This method uses the cheap sodium chloride and has chlorine as an important by-product. The melting point of pure sodium chloride (800 °C) is lowered to approximately 600 °C by addition of another salt, e.g. calcium chloride.

The vessel is made of steel lined with fire-brick and has a central anode of carbon (which, unlike anodes made of metals, will resist attack by chlorine). The anode is surrounded by a circular steel cathode. To separate the products of electrolysis, the cathode is in turn surrounded by a wire gauze, as shown in figure 21.3.

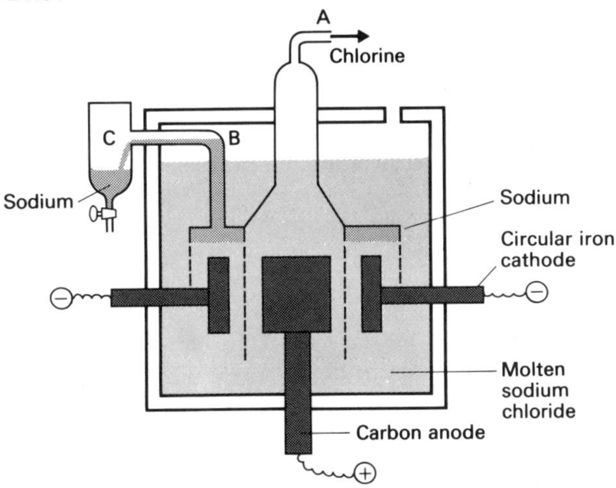

Figure 21.3 The Downs' process for the manufacture of sodium.

During electrolysis chloride ions from the molten sodium chloride are discharged at the anode:

$$Cl^- - e \rightarrow Cl$$

$$2Cl \rightarrow Cl_2.$$

The liberated chlorine is directed through the earthenware pipe A.

Sodium ions gain electrons at the cathode: $Na^+ + e \rightarrow Na$. Sodium metal then collects above the cathode and is forced through the pipe *B* into the collector *C*.

21.2.4 Properties of the alkali metals

The physical properties of the alkali metals are given in table 21.1. They are all soft metals with low densities. Their melting points and boiling points decrease from lithium to caesium. (Caesium is only just a solid at room temperature; caesium in a glass tube melts if held in the hand.) They are good conductors of heat and electricity. When cut they have a silvery lustre but quickly tarnish by oxidation and by subsequent reactions with water vapour and carbon dioxide in the air to form the carbonate. For this reason they are kept in an inert liquid, e.g.

TABLE 21.1 PHYSICAL PROPERTIES OF THE ALKALI METALS

	Lithium	Sodium	Potassium	Rubidium	Caesium
Melting point (° C)	180	98	64	39	28·5
Boiling point (° C)	1317	892	774	688	690
Density (g cm^{-3})	0·53	0·97	0·86	1·53	1·87
Molar volume (cm^3 mol^{-1})	13·1	23·7	95·5	55·8	71

sodium is kept under methylbenzene.

The alkali metals combine readily with non-metals, for example:

(*a*) When heated, lithium burns to form the oxide:

$$4Li(s) + O_2(g) \rightarrow 2Li_2O(s)$$

Sodium, in a limited supply of air, also gives the oxide but more usually it forms the peroxide:

$$2Na(s) + O_2(g) \rightarrow Na_2O_2(s)$$

The other alkali metals form superoxides, e.g.

$$K(s) + O_2(g) \rightarrow KO_2(s)$$

(*b*) They combine with chlorine to form chlorides, e.g.

$$2Na(s) + Cl_2(g) \rightarrow 2NaCl(s)$$

(*c*) When hydrogen is passed over a heated alkali metal, a hydride is formed, e.g.

$$2Na(s) + H_2(g) \rightarrow 2NaH(s)$$

All of the alkali metals react vigorously with cold water to give a solution of the hydroxide and hydrogen, e.g.

$$\left[\begin{array}{l} 2Na \quad + 2H_2O \rightarrow 2NaOH \quad + H_2 \\ 2Na(s) + 2H_2O(l) \rightarrow 2Na^+(aq) + 2OH^-(aq) + H_2(g) \end{array}\right.$$

By pressing small pieces of sodium under mercury, a solution, or alloy, or sodium in mercury is obtained. It is known as sodium *amalgam*. An alloy of sodium can also be formed with lead, and sodium in either of these alloys reacts more safely with water than does the pure metal.

21.2.5 Relative reactivity of the alkali metals

In all of their chemical reactions atoms of the alkali metals lose electrons to form positive ions:

$$Na - e \rightarrow Na^+$$

The higher the atomic number of the metal, the more easily are the electrons lost because (*a*) the outer electrons are shielded from the attraction of the nucleus by increasing numbers of electron shells, and (*b*) the outer electrons are further away from, and less attracted to, the positive nucleus. This difference in reactivity is readily shown by placing very small pieces of lithium, sodium and potassium in water (see section 11.3.1).

21.2.6 Uses of the alkali metals

Lithium is used to harden lead and aluminium alloys. Sodium is used in making certain of its compounds e.g. sodium peroxide, sodium hydride and sodium cyanide, NaCN. This last compound is used in electroplating and in the extraction of gold. Sodium is needed in the manufacture of tetraethyl-lead(IV)—an additive to petrol which prevents "knocking". Liquid sodium is used as a coolant in nuclear reactors. Sodium is also an important reagent in the organic chemical industry.

In the extraction of the increasingly important metal titanium, molten sodium is used to reduce titanium(IV) chloride:

$$TiCl_4(g) + 4Na(l) \rightarrow Ti(s) + 4NaCl(l)$$

21.2.7 Compounds of the alkali metals

(a) Introduction

Because the alkali metals combine so readily with other elements their compounds are stable. The stability increases as the group is descended, e.g., (*a*) lithium carbonate and hydroxide decompose to the oxide on strong heating, but the corresponding compounds of the other alkali metals cannot be decomposed using normal laboratory methods of heating; (*b*) lithium nitrate gives the oxide on heating but nitrates of the metals lower in the group decompose only to the nitrite.

(b) Sodium peroxide

The only oxide of importance is *sodium peroxide*, Na_2O_2. It is manufactured by heating sodium in a stream of dry, carbon dioxide free air:

$$2Na(s) + O_2(s) \rightarrow Na_2O_2(s)$$

Sodium peroxide is a pale yellow solid, which reacts vigorously with cold water to give oxygen:

$$\left[\begin{array}{l} 2Na_2O_2 + 2H_2O \rightarrow 4NaOH + O_2 \\ 2Na_2O_2(s) + 2H_2O(l) \rightarrow 4Na^+(aq) + 4OH^-(aq) + O_2(g) \end{array}\right.$$

Like other peroxides, it forms hydrogen peroxide with ice-cold dilute acids:

$$\left[\begin{array}{l} Na_2O_2 \quad +2HCl \rightarrow 2NaCl+H_2O_2 \\ Na_2O_2(s)+2H^+(aq) \rightarrow 2Na^+(aq)+H_2O_2(aq) \end{array}\right.$$

(c) Sodium hydroxide

(i) Introduction

Sodium hydroxide, NaOH, is the most important alkali used in industry and in the laboratory. It is manufactured either by the electrolysis of a solution of sodium chloride or from sodium carbonate and calcium hydroxide (slaked lime).

(ii) Electrolytic manufacture

Two types of cell are used for the electrolysis of sodium chloride solution (brine) to manufacture sodium hydroxide. These are the mercury cathode cell and the diaphragm cell.

1. The mercury cathode cell

In this process (figures 21.4 and 21.5) brine is circulated through a large rectangular steel tank lined with hard rubber to prevent corrosion. The cathode consists of mercury which flows along the steel floor of the tank. The anodes, which until recently were made of carbon, are made of titanium coated with a precious metal oxide. These titanium electrodes have 30 times the electrical conductivity of the carbon ones that they have replaced.

Ions present in the solution are Na^+ and Cl^- from the electrovalent sodium chloride, and H^+ and OH^- from the slight ionization of the water. During electrolysis Cl^- and OH^- ions move towards the anode where, because of their high concentration, Cl^- ions are preferentially discharged to form gaseous chlorine: $2Cl^- - 2e \rightarrow Cl_2$.

Na^+ and H^+ ions move towards the mercury cathode. *At a mercury cathode it is the Na^+ ions which are preferentially discharged:*

$Na^+ + e \rightarrow Na$

The sodium forms a dilute amalgam with the mercury, which is then led into iron tanks containing water. The sodium combines with this water to form sodium hydroxide and hydrogen:

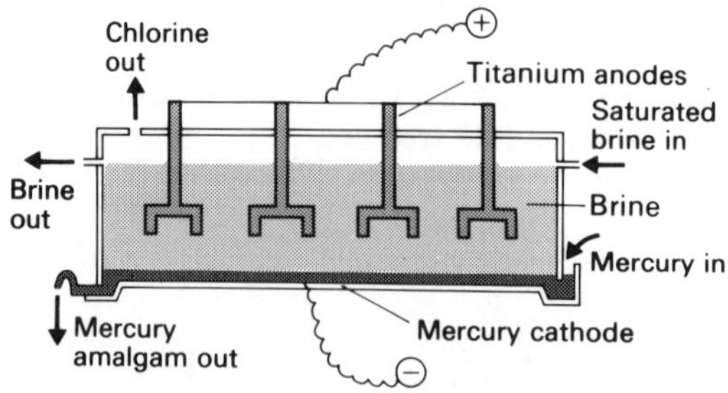

Figure 21.4 Mercury cathode cells for the manufacture of sodium hydroxide.

Figure 21.5 Mercury cathode cells (Photo: I.C.I. Mond Division).

$$\left[\begin{array}{llll} 2Na & +2H_2O & \rightarrow 2NaOH & +H_2 \\ 2Na(s) & +2H_2O(l) & \rightarrow 2Na^+(aq) & +2OH^-(aq)+H_2(g) \end{array}\right.$$

Solid sodium hydroxide is obtained by evaporation of the solution to dryness.

Because mercury vapour is toxic strict safety precautions are taken with mercury cathode cells. Inevitably some mercury is discharged and can become a pollutant which is toxic to marine life (see chapter 28). For this reason the mercury cell is being replaced by the diaphragm cell. In Japan this change has been enforced by law.

2. The diaphragm cell

This too consists of a rubber-lined steel tank with titanium anodes. Long steel-mesh tubes coated with an asbestos fibre or poly(tetrafluroethene) diaphragm are suspended between the anodes, the steel mesh being the cathode of the cell. Brine enters the anode compartment and, the liquid level being above the cathode tube, passes under pressure through the diaphragm into the cathode compartment (figure 21.6).

As in the mercury cell Cl^- and OH^- ions move to the anode where the Cl^- ions are preferentially discharged as gaseous chlorine:

$$2Cl^- - 2e \rightarrow Cl_2$$

Na^+ and H^+ ions move towards the steel cathode where H^+ ions are discharged as hydrogen gas:

$$2H^+ + 2e \rightarrow H_2$$

The solution is continually flowing from the anode to cathode compartments and the excess Na^+ and OH^- ions that remain after chlorine and hydrogen gases have been discharged, flow out of the cathode compartment with excess sodium chloride.

In practice less than half of the sodium chloride entering the cell is converted to sodium hydroxide.

The solution leaving the cell is concentrated in evaporators when most of the sodium chloride crystallises out leaving a concentrated solution of sodium hydroxide.

A new type of diaphragm is in small scale use and will soon become more

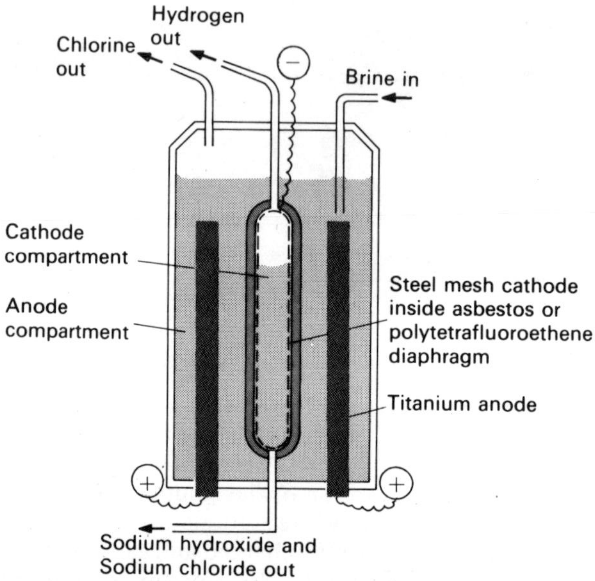

Figure 21.6 Diaphragm cell for the manufacture of sodium hydroxide. (Note that the level of the brine is higher in the anode compartment to provide the pressure to drive it through the pores in the diaphragm.)

important. This uses a polymer membrane for the diaphragm that will not allow chloride ions to pass through it. Therefore sodium hydroxide solution free of sodium chloride is produced.

(iii) Manufacture from sodium carbonate

This process consists of heating sodium carbonate solution with calcium hydroxide (slaked lime):

$$\left[\begin{array}{llll} Na_2CO_3 & +Ca(OH)_2 & \rightarrow CaCO_3 & +2NaOH \\ CO_3{}^{2-}(aq) & +Ca(OH)_2(s) & \rightarrow CaCO_3(s) & +2OH^-(aq) \end{array}\right.$$

After removal of the insoluble calcium carbonate by filtration, evaporation of the filtrate gives solid sodium hydroxide.

(iv) Properties of sodium hydroxide

Sodium hydroxide is a white crystalline solid, usually sold as flakes, pellets or sticks. It is very deliquescent, i.e. it absorbs moisture from the air to form a solution. When pellets of sodium hydroxide are exposed to the air, the solution formed becomes coated with a white crust. This crust is caused by combination of the sodium hydroxide with carbon dioxide from the atmosphere to give sodium carbonate:

$$\left[\begin{array}{llll} 2NaOH & +CO_2 & \rightarrow Na_2CO_3 & + H_2O \\ 2OH^-(aq) & +CO_2(g) & \rightarrow CO_3{}^{2-}(aq) & +H_2O(l) \end{array}\right.$$

Like other deliquescent solids, sodium hydroxide is very soluble in water. Solutions should be made in beakers or thin glass vessels, since so much heat is evolved that thick glass is cracked.

In its chemical reactions, sodium hydroxide behaves as a typical alkali:

1. It will neutralize an acid to give a salt and water only:

e.g. $$\left[\begin{array}{llll} NaOH & + HNO_3 & \rightarrow & NaNO_3 + H_2O \\ OH^-(aq) & + H^+(aq) & \rightarrow & H_2O(l) \end{array}\right.$$

2. With acidic oxides, salts are also formed:

e.g. $$\left[\begin{array}{llll} 2NaOH & + SO_2 & \rightarrow & Na_2SO_3 + H_2O \\ 2OH^-(aq) & + SO_2(g) & \rightarrow & SO_3{}^{2-}(aq) + H_2O(l) \end{array}\right.$$

3. Solutions of sodium hydroxide precipitate insoluble metallic hydroxides from solutions of the appropriate salts:

e.g. $$\left[\begin{array}{llll} CuSO_4 & + 2NaOH & \rightarrow & Cu(OH)_2\downarrow \quad + Na_2SO_4 \\ Cu^{2+}(aq) & + 2OH^-(aq) & \rightarrow & Cu(OH)_2(s) \end{array}\right.$$

or $$\left[\begin{array}{llll} FeCl_3 & + 3NaOH & \rightarrow & Fe(OH)_3\downarrow \quad + 3NaCl \\ Fe^{3+}(aq) & + 3OH^-(aq) & \rightarrow & Fe(OH)_3(s) \end{array}\right.$$

4. When heated with ammonium salts, sodium hydroxide yields ammonia:

e.g. $$\left[\begin{array}{llll} NH_4Cl & + NaOH & \rightarrow & NaCl \quad + H_2O + NH_3 \\ NH_4{}^+(aq) & + OH^-(aq) & \rightarrow & H_2O(l) + NH_3(g) \end{array}\right.$$

The hydroxides of the other alkali metals are very similar to those of sodium hydroxide.

(v) Uses of sodium hydroxide

Sodium hydroxide is a major raw material for the manufacture of other inorganic and organic chemicals. Much is used in paper manufacture and in the purification of bauxite for aluminium extraction. Sodium hydroxide is also used in the manufacture of viscose rayon (section 26.3.1(*d*)) and of soap.

Soap is made by heating oils and fats, such as tallow, coconut oil or palm oil, with sodium hydroxide. This reaction occurs:

$$\text{Fat} + \text{Alkali} \rightarrow \text{Glycerol} + \text{Soap}$$

Sodium chloride (common salt) added to the mixture reduces the solubility of the soap which separates as an upper layer. The soap is then purified and has various colours and perfumes added.

The chemistry of the reaction is:

$$\begin{array}{lcccc} CH_2OOC.C_{17}H_{35} & & & CH_2OH & \\ | & & & | & \\ CHOOC.C_{17}H_{35} & + & 3NaOH \rightarrow & CHOH & + \; 3C_{17}H_{35}COONa \\ | & & & | & \\ CH_2OOC.C_{17}H_{35} & & & CH_2OH & \end{array}$$

propane-1,2,3, triyl trioctadecanoate (glyceryl stearate)	glycerol	sodium octadecanoate (sodium stearate)

The following list shows the percentages of sodium hydroxide used by different industries:

Inorganic chemicals	21 %	Aluminium manufacture	7 %
Organic chemicals	17 %	Soap manufacture	4 %
Pulp and paper	14 %	Rayon manufacture	4 %
Neutralisation	12 %	Others	21 %

(d) Sodium carbonate

By far the most important carbonate is sodium carbonate which is sold commercially either anhydrous or as its decahydrate, washing soda, $Na_2CO_3.10H_2O$.

(i) Manufacture of sodium carbonate.

It is manufactured by the ammonia-soda or Solvay Process. The main stages are:

1. *The preparation of "ammonical brine".* A saturated solution of sodium chloride is passed down a tower fitted with baffle plates. Ammonia is passed up the tower, dissolves in the water and saturates it.

2. *The preparation of sodium hydrogencarbonate.* The ammoniacal brine is then run down towers, also fitted with baffle-plates, up which carbon dioxide is blown. In these towers, known as Solvay towers, the fundamental reactions of the whole process take place. Ammonium hydroxide and carbon dioxide combine to give a solution of ammonium hydrogencarbonate:

$$\left[\begin{array}{lll} NH_4OH + CO_2 & \rightarrow & NH_4HCO_3 \\ OH^-(aq) + CO_2(g) & \rightarrow & HCO_3^-(aq) \end{array}\right.$$

The hydrogencarbonate ions in the solution and sodium ions from the sodium chloride interact to precipitate sodium hydrogencarbonate, which is only slightly soluble in salt solution:

$$\left[\begin{array}{lll} NH_4HCO_3 + NaCl & \rightarrow & NaHCO_3 + NH_4Cl \\ HCO_3^-(aq) + Na^+(aq) & \rightarrow & NaHCO_3(s) \end{array}\right.$$

3. *Formation of sodium carbonate.* The sodium hydrogencarbonate, formed as a white sludge at the bottom of the tower, is filtered off and then heated in furnaces. It decomposes to give anhydrous sodium carbonate, soda ash:

$$\left[\begin{array}{lll} 2NaHCO_3 & \rightarrow & Na_2CO_3 + H_2O + CO_2 \\ 2HCO_3^-(s) & \rightarrow & CO_3^{2-}(s) + H_2O(g) + CO_2(g) \end{array}\right.$$

If sodium carbonate-10-water is required, the anhydrous salt is dissolved in a small quantity of hot water, so that the crystals form on cooling:

$$Na_2CO_3(aq) + 10H_2O(l) \rightarrow Na_2CO_3.10H_2O(s)$$

Potassium carbonate cannot be manufactured by this process because potassium hydrogencarbonate is too soluble to be precipitated in the Solvay tower.

Both the history and economy of this process are of interest. It was discovered by a Belgian, Ernest Solvay, in 1869. Considering all the stages together there are only two starting materials–sodium chloride and calcium carbonate–and two products–sodium carbonate and calcium chloride.

The carbon dioxide for stage 2 is obtained from stage 3 and also by heating limestone (calcium carbonate):

$$CaCO_3 \rightarrow CaO + CO_2$$

Ammonia for stage 1 is obtained by heating calcium oxide from the above reaction with ammonium chloride:

$$CaO + 2NH_4Cl \rightarrow CaCl_2 + H_2O + 2NH_3$$

Ammonium chloride for the above reaction is a product of stage 2; theoretically no ammonium chloride is used up.

These statements are made clearer by figure 21.7 which shows that the *sum* of *all* the reactions involved is:

$$2NaCl + CaCO_3 \rightarrow CaCl_2 + Na_2CO_3$$

It now needs another Ernest Solvay to discover a large scale use for calcium chloride!

(ii) Naturally occurring sodium carbonate

Sodium carbonate has been obtained since prehistoric times from a naturally occurring compound of sodium carbonate and sodium hydrogencarbonate ($Na_2CO_3.NaHCO_3.2H_2O$). This compound, sodium sesquicarbonate, is known as *trona* (from the Arabic word, natrum), and is found in Egypt, Kenya and particularly the United States of America where a huge deposit exists at about 500 metres under Wyoming. The cost of mining is much cheaper than the Solvay process and over 70% of the U.S. production is from trona. Already most ammonia-soda plants in the U.S. have closed, and soon, it is likely, all will be closed.

The ore contains over 80% $Na_2CO_3.NaHCO_3.2H_2O$. It is heated to give sodium carbonate:

$$2Na_2CO_3.NaHCO_3.2H_2O \rightarrow 3Na_2CO_3 + 5H_2O + CO_2$$

The impure product is purified by recrystallisation.

(iii) Laboratory preparation of sodium carbonate

1. Carbon dioxide is bubbled through a moderately concentrated solution of sodium hydroxide. Eventually a white precipitate of sodium hydrogencarbonate forms:

$$2NaOH + CO_2 \rightarrow Na_2CO_3 + H_2O$$
$$2OH^-(aq) + CO_2(g) \rightarrow CO_3^{2-}(aq) + H_2O(l)$$
$$Na_2CO_3 + H_2O + CO_2 \rightarrow 2NaHCO_3 \downarrow$$
$$2Na^+(aq) + CO_3^{2-}(aq) + H_2O(l) + CO_2(g) \rightarrow 2NaHCO_3(s)$$

This precipitate is filtered off, washed with a little cold water and heated to

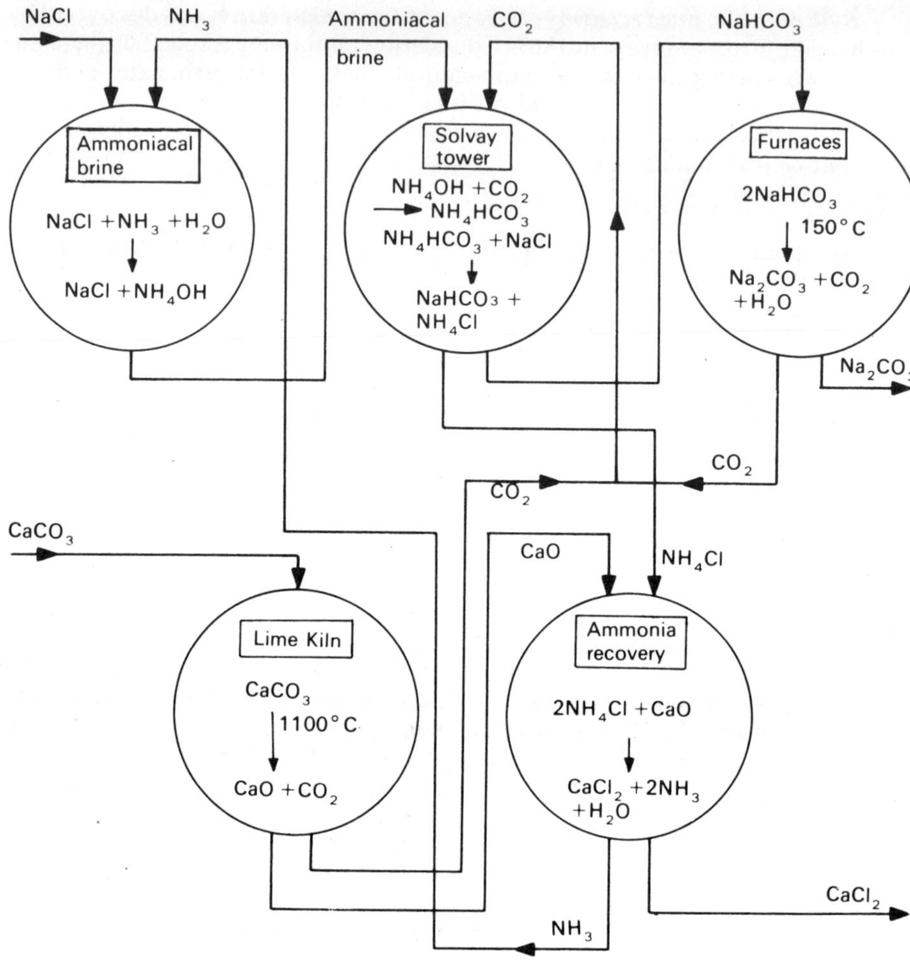

Figure 21.7 The Solvay process.

constant weight in an evaporating dish. The sodium hydrogencarbonate then decomposes to give the carbonate:

$$2NaHCO_3(s) \rightarrow Na_2CO_3(s) + H_2O(g) + CO_2(g)$$

2. Alternatively, a *hot*, concentrated solution of sodium hydroxide is saturated with carbon dioxide. Under these conditions the hydrogencarbonate is unstable. On cooling, crystals of sodium carbonate-10-water, $Na_2CO_3.10H_2O$, separate. These can be removed by filtration and dried.

(iv) Properties of the carbonates

Apart from lithium carbonate, the alkali metal carbonates are readily soluble in water.

Sodium carbonate-10-water crystals, washing soda, are efflorescent, i.e., they give up water of crystallization to the atmosphere. A monohydrate is formed as a white powder on the surface of the crystals:

$$Na_2CO_3.10H_2O(s) \rightarrow Na_2CO_3.H_2O(s) + 9H_2O(g)$$

When heated, washing soda loses all its water of crystallization, but no further chemical decomposition occurs:

$$Na_2CO_3.10H_2O(s) \rightarrow Na_2CO_3(s) + 10H_2O(g)$$

(v) Uses of sodium carbonate

Glass manufacturers (section 18.8.1) are by far the biggest users of sodium carbonate (45 %). Substantial amounts are used in the manufacture of sodium phosphate (12 %), silicates (5 %), the paper industry (4 %), in alkaline cleaners (5 %) and for many other purposes (29 %).

(e) Sodium hydrogencarbonate, baking soda, $NaHCO_3$

This, the only important alkali metal hydrogencarbonate, is manufactured by the Solvay process.

In the laboratory it is prepared by passing carbon dioxide through a *cold*, concentrated solution of either sodium hydroxide or sodium carbonate:

$$NaOH(aq) + CO_2(g) \rightarrow NaHCO_3(s)$$

$$Na_2CO_3(aq) + H_2O(l) + CO_2(g) \rightarrow 2NaHCO_3(s)$$

The hydrogencarbonate is formed as a white precipitate, which must be filtered off, washed with a little water and dried *without heating*.

Like carbonates, hydrogencarbonates give carbon dioxide with acids. One important reaction, peculiar to hydrogencarbonates, is their decomposition to the carbonates on heating:

$$2NaHCO_3(s) \rightarrow Na_2CO_3(s) + H_2O(g) + CO_2(g)$$

The use of sodium hydrogencarbonate as "baking soda" depends on this reaction. Mixed with dough, the hydrogencarbonate decomposes on heating to give bubbles of carbon dioxide which make the dough rise. More usually baking powders, which are a mixture of baking soda and an acidic substance, are used. The production of carbon dioxide is then caused by the action of the acidic substance on the hydrogencarbonate in the presence of moisture.

(f) Chlorides

Both sodium and potassium chlorides are very important. Sodium chloride is the raw material for manufacturing sodium itself and nearly all its compounds. Deposits of sodium chloride, known as *rock salt*, are either mined by conventional methods (figure 21.1) or else salt solution, obtained by dissolving salt underground, is pumped to the surface and evaporated to dryness. Some idea of the importance of this chemical is obtained from figure 21.8.

Potassium chloride may be extracted from carnallite, $KMgCl_3$, by fractional crystallization, as potassium chloride is less soluble at low temperatures than magnesium chloride. Large quantities of *sylvine*, which is mainly potassium chloride, are now obtained from Cleveland (the mines in the North Yorkshire moors National Park have been the subject of much controversy). Potassium is

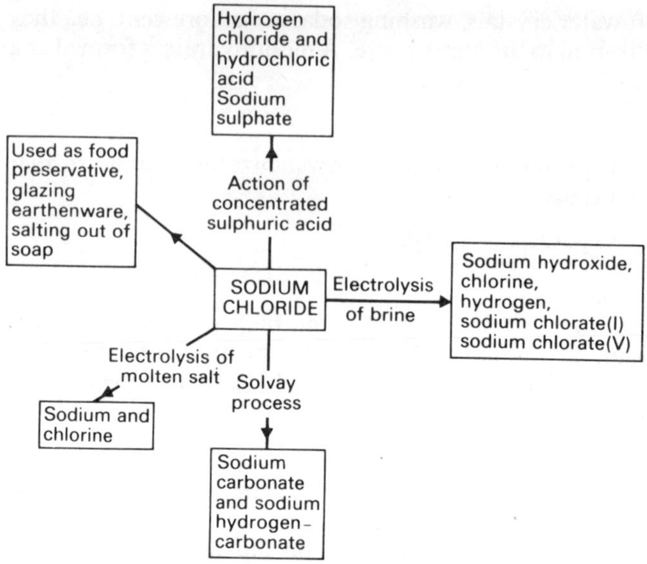

Figure 21.8 Uses of common salt (sodium chloride).

an essential element for plants and potassium chloride is needed as a constituent of many common fertilizers.

(g) Salts of chloric(I) and chloric(V) acid

Both sodium chlorate(I) (sodium hypochlorite), NaClO, and sodium chlorate(V)(sodium chlorate), $NaClO_3$, are made by the action of chlorine on solutions of sodium hydroxide. Sodium chlorate(I) is formed with a cold dilute solution of the alkali, and sodium chlorate(V) with a hot concentrated solution:

Cold, dilute NaOH

$$\left[\begin{array}{l} 2NaOH + Cl_2 \rightarrow NaCl + NaClO + H_2O \\ 2OH^-(aq) + Cl_2(g) \rightarrow Cl^-(aq) + ClO^-(aq) + H_2O(l) \end{array}\right.$$

Hot, concentrated NaOH

$$\left[\begin{array}{l} 6NaOH + 3Cl_2 \rightarrow 5NaCl + NaClO_3 + 3H_2O \\ 6OH^-(aq) + 3Cl_2(g) \rightarrow 5Cl^-(aq) + ClO_3^-(aq) + 3H_2O(l) \end{array}\right.$$

As both sodium hydroxide and chlorine are formed by the electrolysis of brine, the manufacture is carried out in a special electrolytic cell. In this cell chlorine liberated at the anode is allowed to mix with sodium hydroxide solution produced at the cathode.

Solutions of sodium chlorate(I) are widely used for bleaching purposes. Sodium chlorate(V) is used as a weed killer.

Potassium chlorate(V) which is made similarly, is used for making matches and fireworks. The cheaper sodium chlorate(V) cannot be used because it absorbs moisture from the air.

(h) Sulphates and hydrogensulphates

These are readily prepared by neutralization. The details for making sodium sulphate-10-water and sodium hydrogensulphate are given in section 12.4.5.

(i) Sulphites and hydrogensulphites

These are obtained from sulphur dioxide and the appropriate alkali, e.g., when sulphur dioxide is passed through a solution of sodium hydroxide, sodium sulphite is the first product. With more sulphur dioxide, the hydrogensulphite is formed:

$$\begin{bmatrix} 2NaOH & +SO_2 & \rightarrow Na_2SO_3 & +H_2O \\ 2OH^-(aq) & +SO_2(g) & \rightarrow SO_3^{2-}(aq) & +H_2O(l) \end{bmatrix}$$

$$\begin{bmatrix} NaOH & +SO_2 & \rightarrow NaHSO_3 \\ OH^-(aq) & +SO_2(g) & \rightarrow HSO_3^-(aq) \end{bmatrix}$$

It should be noted from the equations that, for a given weight of sodium hydroxide, twice as much sulphur dioxide is required to prepare the hydrogensulphite as to prepare the sulphite.

To obtain the sulphite, therefore, a solution of sodium hydroxide is divided into two equal portions. One portion is saturated with sulphur dioxide to give a solution of the hydrogensulphite. The other portion of sodium hydroxide solution is then added. Crystals of sodium sulphite are then obtained by evaporating the solution to the point of crystallization and cooling.

(j) Sulphides and hydrogensulphides

These are made by the action of hydrogen sulphide on a solution of the alkali:

$$\begin{bmatrix} 2NaOH & +H_2S & \rightarrow Na_2S & +2H_2O \\ 2OH^-(aq) & +H_2S(g) & \rightarrow S^{2-}(aq) & +2H_2O(l) \end{bmatrix}$$

$$\begin{bmatrix} NaOH & +H_2S & \rightarrow NaHS & +H_2O \\ OH^-(aq) & +H_2S(g) & \rightarrow HS^-(aq) & +H_2O(l) \end{bmatrix}$$

The details of the preparations are similar to those given above for the sulphites.

(k) Nitrates

Large natural deposits of sodium nitrate, Chile saltpetre, are found in Chile. It is separated from insoluble impurities by solution in water. The solution is evaporated by the heat of the sun until it reaches a certain density, and the nitrate is then allowed to crystallize out.

Potassium nitrate, commonly called saltpetre, occurs naturally but it is usually manufactured by boiling concentrated solutions of sodium nitrate and potassium chloride:

$$NaNO_3(aq) + KCl(aq) \rightarrow KNO_3(aq) + NaCl(s)$$

Sodium chloride, which is less soluble than potassium chloride, is deposited leaving potassium nitrate in solution. The salt crystallizes out on cooling.

The alkali metal nitrates, in common with all nitrates, are soluble in water. With the exception of lithium nitrate, they decompose on heating to give the nitrite and oxygen:

$2NaNO_3(s) \rightarrow 2NaNO_2(s) + O_2(g)$

Lithium nitrate decomposes further to give the oxide.

Both sodium and potassium nitrates are used as fertilisers. Potassium nitrate is also used in making gunpowder and fireworks. The less expensive sodium nitrate is unsuitable for this purpose, since it absorbs moisture from the air.

21.2.8 Tests for alkali metals

The presence of a particular ion in a compound is often detected by the formation of an insoluble compound. Various examples can be found throughout this book.

Sodium and potassium form no simple insoluble compounds, and in elementary work must be detected by the "flame test". For this test a small piece of platinum wire sealed in a glass rod is used. (Nichrome wire forms a cheap substitute for platinum.) The end of the wire is dipped in a mixture obtained by adding a few drops of concentrated hydrochloric acid to a little of the compound being tested. When the wire is placed in a non-luminous flame, the flame is coloured according to the ion present. The hydrochloric acid produces the chloride of the metal, which is comparatively volatile.

Metals which can be detected in this way include:

Sodium	Intense yellow	Calcium	Brick red
Potassium	Lilac	Strontium	Red
Lithium	Crimson	Copper	Blue-green
Barium	Light green		

A coloured picture of these flames is given on the back cover of this book.

Questions on Chapter 21

1. Tabulate *four* physical differences and *two* chemical differences between metals and non-metals.

State *briefly two* general methods in each case by which metals can be converted into (*a*) oxides, (*b*) chlorides.

In which class of oxides would you include (i) the two oxides of carbon, (ii) zinc oxide? (*C.W.B.*)

2. Define an *electrolyte*.

How can sodium be obtained by electrolysis? Answer by means of a labelled diagram which should show the materials of the containing vessel and electrodes, the names and signs of the electrodes, the products and the place where each product is obtained.

State *three* reasons (of which at least one must be chemical) for believing sodium is a metal. (*J.M.B.*)

3. When fused sodium chloride is electrolyzed, how do the products differ from those obtained by the electrolysis of sodium chloride solution?

Describe how each of these processes is used in industry and for what purposes. (*O.*)

4. Given a piece of sodium about the size of a pea, state clearly how you would safely prepare from it first a dilute solution of sodium hydroxide, and then, from this, a sample of pure common salt.

Outline the most interesting properties, both chemical and physical, of metallic sodium. (*L.*)

5. Describe *one* modern method for the manufacture of sodium hydroxide from sodium chloride solution.

What is the action of chlorine on (*a*) a cold dilute solution of sodium hydroxide, (*b*) a hot concentrated solution of sodium hydroxide? Give the equations. (*J.M.B.*)

6. How may a dilute solution of sodium hydroxide be prepared in the laboratory from washing soda (sodium carbonate-10-water)? How and under what conditions does sodium hydroxide react with (*a*) carbon dioxide, (*b*) chlorine, (*c*) silica (silicon dioxide), (*d*) mutton fat, (*e*) sodium hydrogensulphate? (*D.*)

7. Describe how you would prepare (*a*) sodium hydroxide from sodium carbonate-10-water, (*b*) sodium carbonate-10-water from sodium hydroxide?

Name *three* gases which react with sodium hydroxide, and in each case give the equation for the reaction and name the products. (*C.*)

8. Describe in detail how you would prepare from sodium hydroxide (*a*) sodium carbonate-10-water crystals, (*b*) sodium hydrogencarbonate.

Give a simple test which would help you to distinguish between the two compounds named in (*a*) and (*b*).

State one everyday use for each of these two compounds and explain why they are used for the purpose. (*C.W.B.*)

9. Explain the modern commercial process for the production of anhydrous sodium carbonate.

Write equations for the essential reactions on which this process is based.

Starting from sodium carbonate, explain in full detail how you would prepare a pure, dry specimen of sodium sulphate. (*J.M.B.*)

10. State what you would *see*, and in each case give an equation, when sodium hydroxide solution is mixed with (*a*) a cold solution of copper(II) sulphate, (*b*) a freshly prepared air-free solution of iron(II) sulphate, (*c*) a solution of iron(III) chloride, (*d*) a solution of calcium chloride. (*J.M.B.*)

11. Describe carefully how sodium carbonate is manufactured by the ammonia-soda process. No details of the plant are required.

Give *two* reasons why this is a very economical process.

Explain briefly how glass is manufactured. (*J.M.B.*)

12. Identify *A*, *B*, *C*, and *D*, which are all compounds of sodium, and explain the reactions described below.

When *A* is heated, it melts and then gives off oxygen. When *A* is warmed with copper and concentrated sulphuric acid, a brown gas is produced.

B reacts with dilute hydrochloric acid to give a strong-smelling gas that turns moist blue litmus red but does not give a dark precipitate with lead(II) nitrate solution.

When *C* is heated, water vapour is produced but no carbon dioxide and a white solid residue remains that gives carbon dioxide when treated with dilute sulphuric acid.

A solution of *D* gives a white precipitate when added to lead(II) nitrate solution. This precipitate disappears when the liquid is heated. (*C.*)

22

The Alkaline Earth Metals, Zinc and Mercury

22.1 The alkaline earth metals

22.1.1 Introduction

Beryllium, magnesium, calcium, strontium and barium are known as the alkaline earth metals. They are so called because their compounds are widespread in nature and their oxides are alkaline compounds. Like the alkali metals they show a great similarity to one another (although beryllium shows quite a few differences from the rest). Magnesium and calcium are the most important and, as their behaviour is typical of the others, our study is mainly devoted to these two and their salts.

22.1.2 Occurrence of the alkaline earth metals

Magnesium, calcium, strontium and barium all occur as carbonates, e.g. marble, limestone, chalk, $CaCO_3$, dolomite, $CaCO_3.MgCO_3$ and as sulphates, e.g. Epsom salt, $MgSO_4.7H_2O$, gypsum, $CaSO_4.2H_2O$, anhydrite $CaSO_4$. Other very important ores include carnallite, $KCl.MgCl_2.6H_2O$ and fluorspar, CaF_2.

22.1.3 Extraction of the alkaline earth metals

These are normally obtained by the electrolysis of the molten chloride. Thus magnesium is obtained by the electrolysis of molten anhydrous carnallite. Magnesium ions are discharged at the cathode in preference to potassium ions and float to the top of the melt:

$$Mg^{2+} + 2e \rightarrow Mg$$

Chlorine is formed at the anode.

Much magnesium chloride for this electrolysis is now obtained from sea water which is an increasingly important source of raw materials. To do this magnesium hydroxide is precipitated from the water in huge settling tanks by the addition of calcium hydroxide (slaked lime), $Ca(OH)_2$: (figure 22.1).

$$Mg^{2+}(aq) + 2OH^-(aq) \rightarrow Mg(OH)_2(s)$$

The separated magnesium hydroxide is converted to the chloride by treatment with hydrogen chloride:

$$Mg(OH)_2(s) + 2HCl(g) \xrightarrow{heat} MgCl_2(s) + 2H_2O(g)$$

Figure 22.1 Magnesium hydroxide settling tanks at Hartlepool. Here 250 000 tonnes of magnesium hydroxide are produced each year from sea-water (Photo: The Steetly Co. Ltd.).

22.1.4 Properties of the alkaline earth metals

Magnesium and calcium are silvery grey metals of low density. They are very good conductors of heat and electricity.

Atoms of all the metals have two electrons in their outer shell which are readily lost to form positive ions:

$$Be - 2e \rightarrow Be^{2+}$$

As with the alkali metals, the tendency for these electrons to be lost increases with atomic mass and for similar reasons (see section 21.2.5). This means that beryllium is the least reactive and barium the most reactive of the alkaline earth metals. None of them is as reactive as the alkali metals.

All of the metals will burn, e.g. when heated in air both magnesium and calcium burn vigorously to form the oxide together with some nitride:

$$2Mg(s) + O_2(g) \rightarrow 2MgO(s)$$
$$3Mg(s) + N_2(g) \rightarrow Mg_3N_2(s)$$

The presence of the nitride can be shown by adding a little water to the residue, when ammonia is formed:

$$Mg_3N_2(s) + 6H_2O(l) \rightarrow 3Mg(OH)_2(s) + 2NH_3(g)$$

Magnesium reacts little or not at all with cold water but will burn in steam, carbon dioxide and many other gaseous oxides:

$$Mg(s) + H_2O(g) \rightarrow MgO(s) + H_2(g)$$
$$2Mg(s) + CO_2(g) \rightarrow 2MgO(s) + C(s)$$

The more reactive calcium reacts readily with cold water, but less readily than the alkali metals. Calcium hydroxide and hydrogen are formed:

$$Ca(s) + 2H_2O(l) \rightarrow Ca(OH)_2(s) + H_2(g)$$

All of the alkaline earth metals react readily with dilute acids to give hydrogen:

$$\left[\begin{array}{l} Mg + 2HCl \rightarrow MgCl_2 + H_2 \\ Mg(s) + 2H^+(aq) \rightarrow Mg^{2+}(aq) + H_2(g) \end{array}\right.$$

22.1.5 Uses of the alkaline earth metals

Magnesium is used in many alloys which combine low density with high strength, e.g. electron (95% Mg, 4·5% Zn, 0·5% Cu) and magnalium (30% Mg, 70% Al). It is also used for cathodic protection to underground pipes etc.; its greater reactivity ensures that it corrodes in preference to iron. Some photoflash bulbs and incendiary bombs also contain magnesium. Magnesium is very widespread in nature. The green colouring matter of all plants, called chlorophyll, is a complicated organic molecule in the centre of which is a magnesium atom.

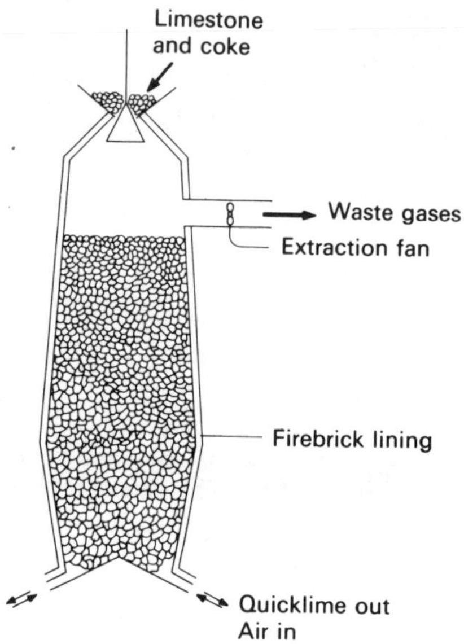

Figure 22.2 A Lime-kiln.

22.1.6 Oxides

(a) Magnesium oxide, MgO

Industrially magnesium oxide is made by heating the carbonate. In the laboratory the carbonate, nitrate or hydroxide may be heated:

$$\left[\begin{array}{lll} MgCO_3(s) & \rightarrow & MgO(s) + CO_2(g) \\ 2Mg(NO_3)_2(s) & \rightarrow & 2MgO(s) + 4NO_2(g) + O_2(g) \\ Mg(OH)_2(s) & \rightarrow & MgO(s) + H_2O(g) \end{array}\right.$$

To ensure complete decomposition, heating to constant mass is necessary. Magnesium oxide is a white powder with an extremely high melting point. For this reason it is used for making bricks for lining furnaces.

It is a basic oxide and slowly combines with water to form magnesium hydroxide, which is sufficiently soluble in water to give an alkaline reaction. Like other basic oxides, magnesium oxide combines with acids to form salts and water, e.g.

$$\left[\begin{array}{lll} MgO + 2HCl & \rightarrow & MgCl_2 + H_2O \\ MgO(s) + 2H^+(aq) & \rightarrow & Mg^{2+}(aq) + H_2O(l) \end{array}\right.$$

(b) Calcium oxide, quicklime, CaO
Industrially all calcium oxide is obtained by heating one of the forms of calcium carbonate, e.g. limestone or, in some parts of America, oyster shells. This is done in a *lime kiln* (figure 22.2). Heat is provided either by coke mixed with the limestone or by producer gas (see section 17.6.1(*d*)). The decomposition is reversible:

$$CaCO_3(s) \rightleftharpoons CaO(s) + CO_2(g)$$

Fans are used to draw the air through the kiln. This removes carbon dioxide and prevents its recombination with calcium oxide.

In the laboratory calcium oxide is best prepared by heating the nitrate or hydroxide.

Calcium oxide is used to manufacture calcium dicarbide and calcium hydroxide.

22.1.7 Hydroxides

Both calcium and magnesium hydroxides may be prepared by precipitation e.g. by adding sodium hydroxide solution to a solution of a chloride:

$$\left[\begin{array}{lll} MgCl_2 \quad + 2NaOH & \rightarrow & Mg(OH)_2 \downarrow + 2NaCl \\ Mg^{2+}(aq) + 2OH^-(aq) & \rightarrow & Mg(OH)_2(s) \end{array}\right.$$

Calcium hydroxide is more usually prepared from the oxide. Thus if water is added to a lump of calcium oxide (quicklime), the lump begins to swell and break up. It gets very hot and some of the water is changed to steam. Calcium hydroxide (slaked lime) is the product:

$$CaO(s) + H_2O(l) \rightarrow Ca(OH)_2(s)$$

Calcium hydroxide is manufactured in this manner.

Calcium hydroxide is slightly soluble in water to give *lime water*. Magnesium and calcium hydroxides are strong bases and react with acids to form salts, e.g.:

$$Mg(OH)_2(aq) + 2HNO_3(aq) \rightarrow Mg(NO_3)_2(aq) + 2H_2O(l)$$

Uses of the hydroxides
Magnesium hydroxide, as indigestion tablets, or its suspension in water is used to neutralize excess acid in the stomach.

The much more important calcium hydroxide has many uses:

(i) Mortar is a mixture of calcium hydroxide and sand, usually with some cement added. The setting of mortar depends to some extent on the reaction of the calcium hydroxide with carbon dioxide in the atmosphere. Interlocking crystals of calcium carbonate are formed:

$$Ca(OH)_2(s) + CO_2(g) \rightarrow CaCO_3(s) + H_2O(l)$$

(ii) Calcium hydroxide (slaked lime) is added to soil to correct acidity and to make heavy soils lighter.

(iii) Calcium hydroxide has many other uses as a cheap alkali, e.g., by heating it with the ammonium chloride obtained in the ammonia-soda process (see section 21.2.7(*d*)) ammonia is recovered.

(iv) Temporary hardness of water may be removed by adding the correct proportion of calcium hydroxide (see section 22.2.4(*c*)).

(v) A suspension of calcium hydroxide in water (milk of lime) is used in the manufacture of sodium hydroxide.

22.1.8 Carbonates

(a) Occurrence
Both magnesium and calcium carbonates occur naturally. Like all the alkaline earth metal carbonates they are insoluble in water; this means that they may be prepared by precipitation. Thus calcium carbonate is formed when solutions of calcium chloride and sodium carbonate are mixed:

$$\left[\begin{array}{lll} CaCl_2 + Na_2CO_3 & \rightarrow & CaCO_3\downarrow + 2NaCl \\ Ca^{2+}(aq) + CO_3{}^{2-}(aq) & \rightarrow & CaCO_3(s) \end{array}\right.$$

Attempts to precipitate magnesium carbonate using sodium carbonate solution give a basic carbonate (see section 18.4.1); sodium hydrogencarbonate solution precipitates the pure carbonate:

$$\left[\begin{array}{lll} MgCl_2 + 2NaHCO_3 & \rightarrow & MgCO_3\downarrow + 2NaCl + H_2O + CO_2 \\ Mg^{2+}(aq) + 2HCO_3{}^{-}(aq) & \rightarrow & MgCO_3(s) + H_2O(l) + CO_2(g) \end{array}\right.$$

If carbon dioxide is passed through a suspension of an alkaline earth carbonate in water the hydrogencarbonate is formed:

$$MgCO_3(s) + H_2O(l) + CO_2(g) \rightleftharpoons Mg(HCO_3)_2(aq)$$

(b) Uses of the carbonates

(i) Magnesium carbonate is used to manufacture the oxide for lining furnaces.

(ii) Much calcium carbonate is used in manufacturing calcium oxide, glass, iron and phosphorus.

(iii) *Cement* is made by strongly heating a mixture of clay and limestone until a partially fused mixture is obtained. After cooling, the product, which contains calcium silicate and calcium aluminate, is finely ground to give Portland cement. *Concrete* is a mixture of cement with sand or gravel.

22.1.9 Sulphates

(a) Magnesium sulphate
Beryllium and magnesium sulphates are prepared by the usual methods for soluble sulphates, e.g. magnesium sulphate is prepared by the action of dilute sulphuric acid on magnesium carbonate:

$$MgCO_3(s) + H_2SO_4(aq) \rightarrow MgSO_4(aq) + H_2O(l) + CO_2(g)$$

From solutions, magnesium sulphate crystallises as the hepta-hydrate, ($MgSO_4.7H_2O$), commonly known as Epsom salt.

(b) Calcium sulphate
Calcium sulphate is found naturally as gypsum, $CaSO_4.2H_2O$, and as anhydrite, $CaSO_4$. As it is only sparing soluble in water, it may be prepared in the laboratory by precipitation, e.g. by mixing solutions of sulphuric acid and calcium chloride

$$\left[\begin{array}{lll} CaCl_2 & +H_2SO_4 & \rightarrow CaSO_4\downarrow + 2HCl \\ Ca^{2+}(aq) + SO_4^{2-}(aq) & & \rightarrow CaSO_4(s) \end{array}\right.$$

Strontium and barium sulphates, which are even less soluble, are prepared similarly.

When gypsum is heated at 110 °C it changes to calcium sulphate-$\frac{1}{2}$-water:

$$CaSO_4.2H_2O \rightarrow CaSO_4.\tfrac{1}{2}H_2O + 1\tfrac{1}{2}H_2O$$

At higher temperatures the anhydrous salt is formed. The hemi-hydrate is known as *Plaster of Paris*. When mixed with water it sets to a hard mass of the dihydrate and expands slightly. Because of these properties Plaster of Paris is used in making casts, particularly for broken limbs and for cheap ornaments, and as a plaster for other purposes.

Calcium sulphate is also important for making sulphuric acid by the anhydrite process (see section 19.7.3(*a*)).

22.1.10 Chlorides

All the chlorides of the alkaline earth metals are soluble in water. Calcium chloride is obtained commercially as a by-product in the manufacture of sodium carbonate by the ammonia-soda process. It forms crystals with the formula $CaCl_2.6H_2O$.

Calcium chloride is extremely soluble in water and is very deliquescent. Accordingly, it is widely used as a drying agent for gases and liquids with which it does not combine. It is *not* suitable for drying ammonia, as a chemical reaction occurs to form the addition compound $CaCl_2.8NH_3$.

22.1.11 Nitrates

All the nitrates of the alkaline earth metals are soluble in water and may be prepared by reacting nitric acid with the metal or its carbonate or hydroxide. On a large scale, calcium nitrate is made by neutralizing nitric acid with chalk and the product is sold as a nitrogenous fertiliser.

The nitrates decompose on heating to give the oxides, e.g.:

$$2Ca(NO_3)_2(s) \rightarrow 2CaO(s) + 4NO_2(g) + O_2(g)$$

22.1.12 Bleaching powder, $CaOCl_2$

Bleaching powder is manufactured by passing chlorine over slaked lime until no more is absorbed:

$$Ca(OH)_2(s) + Cl_2(g) \rightarrow CaOCl_2(s) + H_2O(l)$$

The formula given is an approximation but serves to explain many of the reactions of bleaching powder.

Bleaching powder is a white solid which readily yields chlorine with dilute acids:

$$\left[\begin{array}{lll} CaOCl_2 \quad +2HNO_3 & \rightarrow Ca(NO_3)_2 + H_2O & + Cl_2 \\ CaOCl_2(s) + 2H^+(aq) & \rightarrow Ca^{2+}(aq) & + H_2O(l) + Cl_2(g) \end{array}\right.$$

The powder always smells of chlorine, since carbonic acid from damp air is sufficiently acidic to liberate the gas.

22.2 Hardness of water

22.2.1 Causes

(a) Hard water
Hard water is water which does nto readily lather with soap. This is because it contains dissolved substances which combine with soap to form an insoluble scum.

There are two types of hardness: *temporary hardness* and *permanent hardness.*

(b) Temporary hardness
Temporary hardness is caused by the action of rain water, containing dissolved carbon dioxide, on chalk or limestone in the earth:

$$CaCO_3(s) + H_2O(l) + CO_2(g) \rightleftharpoons Ca^{2+}(aq) + 2HCO_3^-(aq)$$

The calcium carbonate reacts to give a solution containing calcium ions and hydrogencarbonate ions. If magnesium carbonate is present it can give a solution containing magnesium ions and hydrogencarbonate ions. *This temporary hardness can be removed by boiling.* The hydrogencarbonate ion decomposes, calcium carbonate is reformed and precipitates when the water is boiled:

$$Ca^{2+}(aq) + 2HCO_3^-(aq) \rightarrow CaCO_3(s) + H_2O(l) + CO_2(g)$$

The precipitate forms a coating known as "fur" in kettles, boiler and pipes as shown in figure 22.3.

(c) Permanent hardness
Permanent hardness is caused mainly by calcium sulphate and to a lesser extent by calcium chloride and corresponding magnesium salts. These compounds are present in water which has passed over or through soil containing them. *Permanent hardness is not removed by boiling.*

Figure 22.3 Fur in a hot water pipe. (Photo: The Permutit Co. Ltd.).

22.2.2 Action of hard water upon soap

Most soap is made of the soluble salt, sodium octadecanoate (stearate), $NaOOC.C_{17}H_{35}$. Calcium octadecanoate is insoluble, and is therefore precipitated if hard water is mixed with soap:

$$CaSO_4 + 2NaOOC.C_{17}H_{35} \rightarrow \underset{\text{insoluble scum}}{Ca(OOC.C_{17}H_{35})_2\downarrow} + Na_2SO_4$$

$$Ca(HCO_3)_2 + 2NaOOC.C_{17}H_{35} \rightarrow Ca(OOC.C_{17}H_{35})_2\downarrow + 2NaHCO_3$$

$$Ca^{2+}(aq) + 2(OOC.C_{17}H_{35}^-)(aq) \rightarrow Ca^{2+}(OOC.C_{17}H_{35}^-)_2(s)$$

No lather can be obtained with hard water until sufficient soap has been added to combine with all calcium and magnesium salts present. In addition to the waste of soap when hard water is used, calcium octadecanoate is a sticky solid. It absorbs dirt and clings to cloth and to the sides of baths and basins, causing the familiar "tide-mark". "Fur" caused by temporary hardness is a poor conductor of heat. Furthermore, explosions may result in boilers if the "fur" cracks and allows water to contact overheated metal.

It is therefore often very desirable to remove calcium salts from solution in water.

22.2.3 Experiment—comparision of the hardness of various types of water

Hardness may be compared by using a dilute solution of soft soap in a mixture of alcohol and water. (20 g soap in 1 dm^3 of 50 % alcohol gives a solution of suitable concentration. The alcohol prevents the solution from becoming cloudy.)

To compare, for example, the hardness of rain-water and tap-water, place 50 cm^3 of each in separate conical flasks. Fill a burette with soap solution and note its reading. Run soap solution into one of the samples a little at a time. After each addition cork the flask and shake it well. When a permanent lather is obtained (or one lasting for approximately a minute), note the reading of the burette once more.

Perform a similar titration with the second sample. It will be found that tap-water in most places requires far more soap solution to produce a permanent lather than rain-water.

(N.B. For a class experiment a commercial brand of soap flakes may be used. The number of flakes to produce a permanent lather may be counted. The result is fairly accurate.)

Can you suggest a method whereby it would be possible to find out if the water in your district is temporarily hard?

22.2.4 Removal of hardness from water

(a) By boiling

As mentioned earlier, boiling destroys temporary hardness by decomposing calcium hydrogencarbonate into the insoluble calcium carbonate.

As figure 22.3 shows this deposit can block pipes, and it is also a poor conductor of heat. It is usual, therefore, to soften water that is to be used in boilers by other methods and use the softened water in the boilers. This then avoids the need for expensive and troublesome cleaning of calcium carbonate from boilers and pipes.

(b) By distillation

Distillation removes all dissolved solids but it is far too expensive for large-scale use. Distilled water is, however, used in laboratories for making solutions of substances which combine with dissolved impurities. It is also used in accumulators.

(c) By addition of calcium hydroxide

When calcium hydroxide (slaked lime) is added to solutions of calcium hydrogencarbonate, both compounds are converted into the carbonate and precipitated:

$$Ca(HCO_3)_2(aq) + Ca(OH)_2(s) \rightarrow 2CaCO_3(s) + 2H_2O(l)$$

Only temporary hardness is removed. It is important to add the *exact* quantity of calcium hydroxide required since excess would itself cause hardness.

(d) By addition of sodium carbonate-10-water

When sodium carbonate-10-water (washing soda) is added to water containing dissolved calcium salts, calcium carbonate is precipitated, and the water is softened:

$$\left[\begin{array}{llll} CaSO_4 & + Na_2CO_3 & \rightarrow CaCO_3\downarrow & + Na_2SO_4 \\ Ca(HCO_3)_2 & + Na_2CO_3 & \rightarrow CaCO_3\downarrow & + 2NaHCO_3 \\ Ca^{2+}(aq) & + CO_3^{2-}(aq) & \rightarrow CaCO_3(s) & \end{array}\right.$$

Sodium salts remain in solution, but in small quantities these have no effect on the lathering of soap.

(e) By ion exchange

If the hard water is passed through a column filled with a particular type of material, the calcium ions are removed and replaced by sodium ions. One such exchanger is Permutit, an insoluble sodium aluminium silicate which may be represented by the simplified formula Na_2Perm. When used for domestic purposes, a cylindrical container is filled with this material and the hard water is passed through it. The following reaction takes place:

$$\left[\begin{array}{llll} \underset{\text{insoluble}}{Na_2Perm} & + CaSO_4 & \rightarrow \underset{\text{insoluble}}{CaPerm} & + Na_2SO_4 \\ Na_2Perm(s) & + Ca^{2+}(aq) & \rightarrow CaPerm(s) & + 2Na^+(aq) \end{array}\right.$$

i.e. the calcium ions present in water are replaced by sodium ions, which, in small quantity, have no effect on the lathering of soap. The reaction is reversible, and the original Permutit may be regenerated, when it has all been converted, by pouring a concentrated solution of common salt through the container (figure 22.4):

$$\left[\begin{array}{llll} CaPerm & + 2NaCl & \rightarrow Na_2Perm & + CaCl_2 \\ CaPerm(s) & + 2Na^+(aq) & \rightarrow Na_2Perm(s) & + Ca^{2+}(aq) \end{array}\right.$$

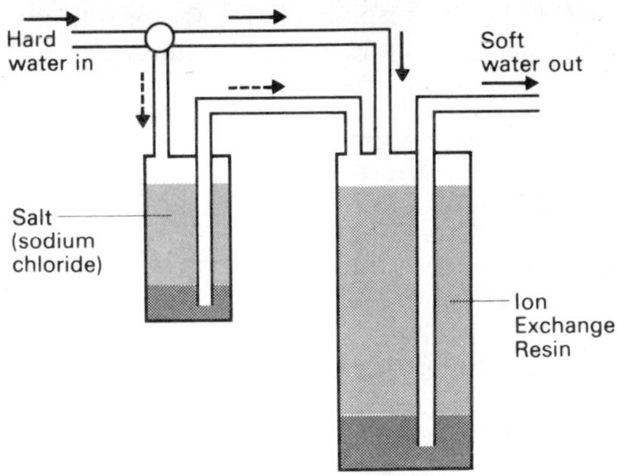

Figure 22.4 A water softener. During use the hard water flows into the softener at the top and soft water flows out from the bottom. When the column is recharged with sodium chloride (salt) solution, the water flows from the salt container and washes the salt solution through the column.

Other compounds, similar in function to Permutit, have been developed, including some which convert all salts present in water to their corresponding acids:

RH	+ NaCl →	RNa	+ HCl
positive ion exchanger (insoluble)		insoluble	

(R = a complex group of atoms)

Such a compound, which exchanges the positive ions, may be used in conjunction with a *negative ion* exchanger which will remove acids from solution:

ROH	+ HCl →	RCl	+ H_2O
negative ion exchanger (insoluble)		insoluble	

Thus "demineralised" water, purer than water which has only been distilled once, may be obtained at a considerably lower cost than distilled water.

22.2.5 Hard water and synthetic detergents

Many synthetic or soapless detergents are now marketed under a variety of trade names (see section 26.5). These detergents do not react with calcium ions and are equally effective in both hard and soft water. They have widespread use for washing in industry and the home. At home synthetic detergents are used for washing clothes, dishes and floors, but most people still prefer the less harsh (usually less alkaline) soap for washing themselves.

22.2.6 Stalactites and stalagmites

The solvent action of rain-water containing dissolved carbon dioxide upon natural calcium carbonate has already been discussed above; this is how water gets its *temporary hardness*. This action is also responsible for the beautifully impressive stalactites and stalagmites which are found in caves in limestone and chalk districts (see figure 22.5).

Figure 22.5 Stalactites and stalagmites in a particularly impressive formation in Nepture's Grotto, Sardinia. (Photo: J. Allan Cash).

The caves themselves were formed over millions of years by limestone being dissolved by the dilute carbonic acid that made up streams and rivers. Rainwater, percolating through the soil and rocks above, forms a dilute solution of calcium hydrogencarbonate and drips into the caves:

$$CaCO_3(s) + H_2O(l) + CO_2(g) \rightleftharpoons Ca(HCO_3)_2(aq)$$

Drops of this solution partially evaporate on the roof of the cave, and by the reverse of the above reaction a minute deposit of calcium carbonate is obtained:

$$Ca(HCO_3)_2(aq) \rightleftharpoons CaCO_3(s) + H_2O(l) + CO_2(g)$$

After hundreds or thousands of years, icicle-like *stalactites* are formed, hanging from the roof. When drops of water hit the floor of the cave, *stalagmites* are built upwards. Some stalactites and stalagmites join together to give pillars, and varied formations are produced.

22.3 Zinc, Cadmium and Mercury

22.3.1 Introduction

These three elements comprise Group IIb of the Periodic Table. Zinc and cadmium are fairly similar to one another but show few resemblances to mercury. Cadmium is of little importance at this stage.

22.3.2 Extraction of zinc

Zinc is most widely distributed as its sulphide, *zinc blende*, ZnS. Large deposits of this ore are found in Australia and in the United States. Another important ore is *calamine*, zinc carbonate, $ZnCO_3$.

To extract zinc the ore is first converted into zinc oxide by roasting in air. Zinc blende is then oxidized to the oxide and sulphur dioxide, while calamine undergoes thermal decomposition to give the oxide:

$$2ZnS(s) + 3O_2(g) \rightarrow 2ZnO(s) + 2SO_2(g)$$
$$ZnCO_3(s) \rightarrow ZnO(s) + CO_2(g)$$

A mixture of the zinc oxide with briquettes made from powdered coke is then fed into the top of a vertical retort strongly heated by burning producer gas (see section 17.6.1(*d*)). The zinc oxide is reduced to the metal, which, having a low boiling point (907 °C), vaporizes.

$$ZnO(s) + C(s) \rightarrow Zn(g) + CO(g)$$

The zinc vapour is cooled to form liquid zinc which is then run into moulds.

Zinc is also manufactured by electrolysis of zinc sulphate solution. This solution is made by reacting zinc oxide with dilute sulphuric acid. During electrolysis very pure zinc is deposited on aluminium cathodes:

$$Zn^{2+}(aq) + 2e \rightarrow Zn(s)$$

At the anodes oxygen is liberated:

$$4OH^-(aq) - 4e \rightarrow 2H_2O(l) + O_2(g)$$

22.3.3 Properties of zinc

Zinc is a bluish-grey metal which melts at 420 °C and boils at 907 °C. Its density is 7 g cm^{-3}. It has a high tensile strength and is a good conductor of heat and electricity.

At ordinary temperatures it undergoes very little corrosion. This is because of a surface film of zinc carbonate hydroxide. When strongly heated in air, however, it burns to form zinc oxide:

$$2Zn(s) + O_2(g) \rightarrow 2ZnO(s)$$

Cold water has no action on zinc, but the red-hot metal decomposes steam with formation of hydrogen:

$$Zn(s) + H_2O(g) \rightarrow ZnO(s) + H_2(g)$$

22.3.4 Action of acids on zinc

Like most other metals, zinc yields hydrogen with non-oxidizing acids, e.g. dilute sulphuric acid or hydrochloric acid, and this is the basis of the normal laboratory preparation of the gas:

$$\left[\begin{array}{l} Zn + H_2SO_4 \rightarrow ZnSO_4 + H_2 \\ Zn(s) + 2H^+(aq) \rightarrow Zn^{2+}(aq) + H_2(g) \end{array}\right.$$

With very pure zinc the action is slight. It may be accelerated by addition of a little copper(II) sulphate, when some copper is deposited on the zinc.

Hot concentrated sulphuric acid gives sulphur dioxide as the main gaseous product:

$$Zn(s) + 2H_2SO_4(l) \rightarrow ZnSO_4(s) + 2H_2O(l) + SO_2(g)$$

Depending on concentration and temperature, a variety of products may be obtained with nitric acid. These include, in addition to zinc nitrate, the oxides of nitrogen and ammonium nitrate (see section 16.6.4(c iii)).

22.3.5 Action of alkalis on zinc

Unlike other metals studied in this chapter, zinc reacts with hot solutions of caustic alkalis. Hydrogen and a zincate are formed:

$$\left[\begin{array}{llll} Zn & + 2NaOH & \rightarrow Na_2ZnO_2 & + H_2 \\ & & \quad \text{sodium zincate} & \\ Zn(s) & + 2OH^-(aq) & \rightarrow ZnO_2^{\ 2-}(aq) & + H_2(g) \end{array}\right.$$

22.3.6 Uses of zinc

Because of its resistance to atmospheric corrosion, zinc is used to protect iron and steel by galvanizing (figure 22.6). This iron, after a thorough cleansing in acid, is dipped in molten zinc and then withdrawn. Some of the metal adheres to the

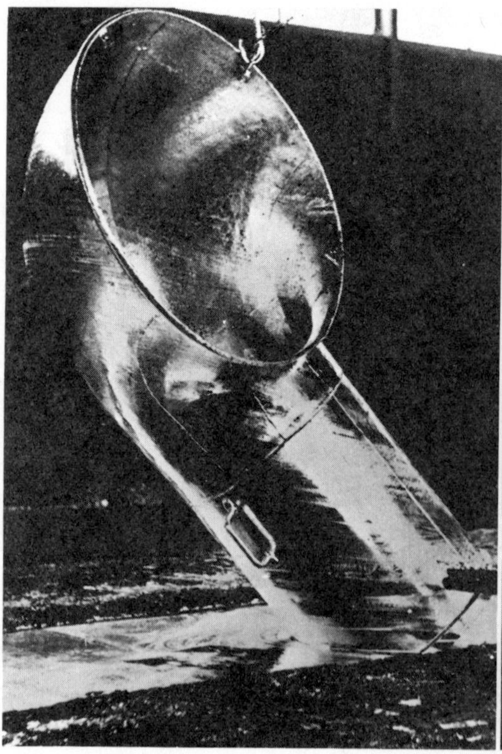

Figure 22.6 Ship's ventilator after galvanizing being removed from the bath of molten zinc (Photo: Zinc Development Association).

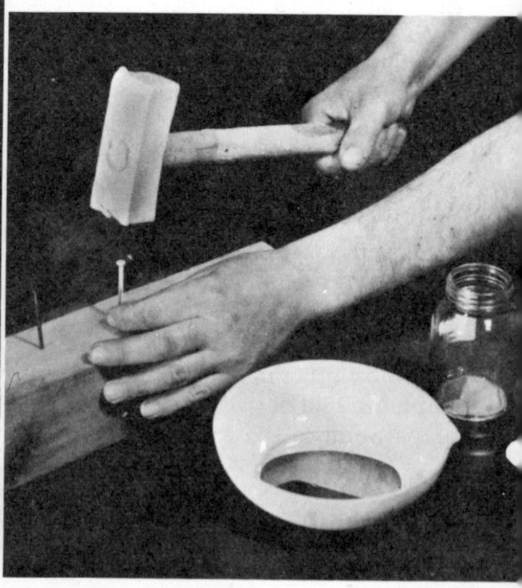

Figure 22.7 A mercury hammer, made by cooling mercury in liquid air. (Photo: A Rendell).

iron and, on cooling, a thin and tenacious layer of crystalline zinc is formed. If the zinc coating is scratched the zinc corrodes in preference to the iron. Water containing dissolved carbon dioxide acts as the electrolyte for a zinc/iron cell. Electrons move from the zinc (i.e. the more electropositive metal) to the iron. Zinc ions go into the solution, i.e. it corrodes. This is an example of "sacrificial" protection.

Numerous alloys contain zinc, e.g. brass, which contains 1 part zinc to 2 parts copper. A large amount of zinc is used also as electrodes in electric batteries.

22.3.7 Zinc oxide

Zinc oxide is made by burning zinc vapour. In the laboratory it is more conveniently prepared by heating the carbonate, nitrate or hydroxide of the metal to constant mass:

$$ZnCO_3(s) \rightarrow ZnO(s) + CO_2(g)$$
$$2Zn(NO_3)_2(s) \rightarrow 2ZnO(s) + 4NO_2(g) + O_2(g)$$
$$Zn(OH)_2(s) \rightarrow ZnO(s) + H_2O(g)$$

Zinc oxide is a white, insoluble powder which turns yellow when hot. It is amphoteric, dissolving in solutions of either acids or alkalis:

e.g.
$$\left[\begin{array}{l} ZnO + 2HCl \rightarrow ZnCl_2 + H_2O \\ ZnO(s) + 2H^+(aq) \rightarrow Zn^{2+}(aq) + H_2O(l) \end{array}\right.$$
$$\left[\begin{array}{l} ZnO + 2NaOH \rightarrow Na_2ZnO_2 + H_2O \\ ZnO(s) + 2OH^-(aq) \rightarrow ZnO_2{}^{2-}(aq) + H_2O(l) \end{array}\right.$$

Zinc oxide is used mainly as a white pigment. It is also used in various ointments and as a filler for rubber.

22.3.8 Zinc hydroxide

Zinc hydroxide is insoluble in water and may therefore be prepared by precipitation, e.g. by adding a solution of sodium hydroxide to zinc sulphate solution:

$$\left[\begin{array}{l} ZnSO_4 + 2NaOH \rightarrow Zn(OH)_2 \downarrow + Na_2SO_4 \\ Zn^{2+}(aq) + 2OH^-(aq) \rightarrow Zn(OH)_2(s) \end{array}\right.$$

However, like zinc oxide it is amphoteric and, if excess of the alkali is added, will combine to give a solution of sodium zincate:

$$\left[\begin{array}{l} Zn(OH)_2 + 2NaOH \rightarrow Na_2ZnO_2 + 2H_2O \\ Zn(OH)_2(s) + 2OH^-(aq) \rightarrow ZnO_2{}^{2-}(aq) + 2H_2O(l) \end{array}\right.$$

In this reaction zinc hydroxide shows acidic properties. Like other hydroxides, it will combine also with acids to give salts and water.

22.3.9 Zinc carbonate

This is a white insoluble powder, which decomposes on heating:

$$ZnCO_3(s) \rightarrow ZnO(s) + CO_2(g)$$

Attempts to prepare it by precipitation using sodium carbonate solution result in the formation of zinc carbonate hydroxide.

22.3.10 Zinc chloride

Zinc chloride can be prepared by the standard methods for soluble salts, i.e., by reacting the appropriate acid (here hydrochloric acid) with the metal, its carbonate or hydroxide. It is a white deliquescent solid, which may be used as a dehydrating agent. Its solution in concentrated hydrochloric acid is used as a flux in soldering.

22.3.11 Zinc sulphate and zinc nitrate

These may be prepared (like the chloride) by the standard methods for soluble salts.

22.4 Mercury

Mercury is found mainly as *cinnabar*, mercury(II) sulphide, HgS. It is readily extracted from this compound by roasting the ore in air. As with zinc sulphide, the oxide is formed, but this dissociates at the temperature used:

$$2HgS(s) + 3O_2(g) \rightarrow 2HgO(s) + 2SO_2(g)$$
$$2HgO(s) \rightarrow 2Hg(l) + O_2(g)$$

Mercury vapour is evolved, which is condensed to give the familiar dense silvery liquid.

Mercury is the only metal to exist as a liquid at normal temperatures. It solidifies at $-39\,^{\circ}C$. Then it is more like a typical metal in its physical properties, e.g., a mercury hammer, made by cooling mercury in liquid air, may be used to hammer a nail into a piece of wood (figure 22.7).

Questions on Chapter 22

1. How would you obtain crystals of Epsom salt from a mixture of Epsom salt and sand?

From this salt ($MgSO_4.7H_2O$) how would you prepare (*a*) magnesium hydroxide, (*b*) magnesium oxide?

How would you show that Epsom salt (*c*) contains water of crystallization, (*d*) is a sulphate? (*L.*)

2. Give an account of the preparation of quicklime from limestone on a large scale and state some of the properties and uses of lime.

How would you prepare, in as pure a state as possible, from calcium carbonate (*a*) crystals of calcium nitrate, (*b*) dry calcium hydroxide, (*c*) dry calcium sulphate? Give equations in each case. (*C.*)

3. What are limestone and gypsum? What important product is made from each? For what purposes are these two products chiefly used and upon what behaviour does the use depend?

Describe how, starting from limestone, you would prepare some temporarily hard water. How would you (*a*) show that this water was hard, (*b*) soften it? (*O.*)

4. Explain how it is that certain natural waters contain calcium hydrogencarbonate.

Starting from lime water, how would you prepare some water possessing temporary hardness?

Give *two* methods, other than distillation, by which temporary hardness may be removed. (*L.*)

5. What is meant by *temporary* and *permanent* hardness in water? Name *one* substance which causes *temporary* and another causing *permanent* hardness.

Write equations for the reactions which occur when (*a*) slaked lime and (*b*) washing soda are used to remove hardness.

In an experiment to determine the relative amounts of *temporary* and *permanent* hardness in a sample of tap-water, 50 cm^3 samples of (*c*) distilled water, (*d*) tap-water and (*e*) boiled tap-water required respectively 1·2, 27·2 and 11·6 cm^3 of soap solution. Calculate the ratio of *temporary* to *permanent* hardness in the tap water. (*D*)

6. Name *two* naturally occurring compounds of calcium and give their formulae.

What calcium compound causes (*a*) the permanent hardness of water, (*b*) the temporary hardness of water.

Given a supply of calcium carbonate, explain with all necessary experimental detail how you would obtain from it a pure sample of calcium oxide. (The action of heat on calcium carbonate will not be accepted in answer to this question.) (*J.M.B.*)

7. Describe very briefly how zinc is obtained from its oxide. Give *two* industrial uses of the metal zinc. Describe the changes which occur when (*a*) zinc is added to dilute sulphuric acid, (*b*) excess of zinc is added to copper(II) sulphate solution, (*c*) zinc carbonate is heated. (*L.*)

8. What is the chief ore of zinc? Outline the manufacture of zinc from this ore. How would you prepare zinc hydroxide from the metal? How does this hydroxide behave with (*a*) hydrochloric acid, (*b*) caustic soda? (*O. and C.*)

9. A divalent metal, *X*, reacted vigorously with cold water with the immediate evolution of a gas, *A*, and after a short time, the formation of a white precipitate, *B*. The precipitate, *B*, was separated from the colourless liquid, *C*, by filtration. After being dried in a desiccator, *B* was heated and gave the solid, *D*, and a vapour, *E*. This vapour condensed to a colourless liquid, and when this cold liquid was added to *D* a rise in temperature was noted. When carbon dioxide was passed through the liquid, *C*, a white precipitate, *F*, was formed, but with the passage of more carbon dioxide, this disappeared leaving a colourless solution, *G*.

Identify *X*, *A*, *B*, *C*, *D*, *E*, *F* and *G* and write equations for each of the reactions.

What happens when the colourless solution, *G*, is boiled? (*J.M.B.*)

10. The element barium is a member of Group II of the periodic table and is just above calcium in the activity series. It is found naturally as the carbonate which is insoluble in water.

(*a*) Describe briefly the steps by which you would attempt to obtain a sample of the element starting from the carbonate.

(*b*) Describe how you would expect the element to react with water.

(*c*) A solution of the hydroxide left standing in the air develops a white crust. Suggest an explanation. Indicate an experiment which you would make to test your suggestion. (*L.*)

11. Calcium belongs to Group II of the periodic table; state whether electrons are gained or lost when a calcium atom becomes a calcium ion and give the number of electrons per atom concerned.

Mention two reactions as a result of which calcium ions are formed from calcium atoms.

Calcium is a very reactive metal. What method can you suggest for converting calcium ions into calcium atoms? (No description of a process or naming of materials is required).

Describe how (*a*) calcium oxide, (*b*) calcium hydroxide, can be obtained from calcium carbonate, and mention two large-scale uses of calcium carbonate. (*O.*)

12. When limestone (calcium carbonate) is strongly heated, it decomposes into calcium oxide and carbon dioxide. The change is reversible.

(*a*) Write the equation for the reaction.

(*b*) Explain what would happen if some limestone were strongly heated (i) in a closed vessel, (ii) in a current of air.

(*c*) Describe the changes that occur when (i) water is slowly added to freshly made calcium oxide, (ii) a mixture of calcium oxide with excess coke is heated in an electric furnace.

(*d*) State *one* commercial use for the principal products obtained in (*c*) (i) and (*c*) (ii).

How do you account for the presence of calcium hydrogencarbonate in some natural waters? Explain *one* reason why the presence of this substance may be disadvantageous and mention *one* reason why it may be beneficial. (*A.E.B.*)

13. Zinc is extracted from zinc blende (impure zinc sulphide, ZnS) by first heating the zinc blende in air to give zinc oxide and sulphur dioxide. The zinc oxide is then mixed with carbon and heated, when zinc vapour and carbon monoxide are formed. The zinc vapour is condensed and the zinc can be separated from its impurities by fractional distillation.

(*a*) Write equations for the *two* chemical reactions referred to above.

(*b*) Calculate the maximum weight of zinc and of sulphur dioxide obtainable from 1000 g of pure zinc sulphide.

(*c*) Zinc vapour burns readily in air. Why do you think this does not happen in the process described above?

(*d*) Give *two* large-scale uses each for zinc and sulphur dioxide.

(*e*) Say what is meant by the terms *condensed* and *fractional distillation*, and name *one* other mixture that can be separated by fractional distillation.

(*f*) Give a test, other than heating in air or oxygen, by which you could show that a given material is a metallic sulphide. (*C.*)

23

Aluminium, Tin and Lead

23.1 Aluminium

23.1.1 Introduction

Much aluminium is found in clays and rocks, e.g. felspar, $KAlSi_3O_8$, and kaolin, $Al_2O_3.2SiO_2.2H_2O$. The metal is extracted from a hydrated oxide, *bauxite*, $Al_2O_3.2H_2O$, its most important ore. Also important is *cryolite*, Na_3AlF_6.

23.1.2 Extraction of aluminium

To obtain aluminium, silica and iron(III) oxide, which are normally present in bauxite, are removed by reacting powdered bauxite with hot concentrated sodium hydroxide solution, when a solution of sodium aluminate is formed:

$$Al_2O_3(s) + 2NaOH(aq) \rightarrow 2NaAlO_2(aq) + H_2O(l)$$

Insoluble impurities, including iron(III) oxide, are filtered off and the solution of sodium aluminate is diluted with water when aluminium hydroxide precipitates :

$$NaAlO_2(aq) + 2H_2O(l) \rightarrow Al(OH)_3(s) + NaOH(aq)$$

The aluminium hydroxide is filtered off, washed and dried. Heating gives pure aluminium oxide:

$$2Al(OH)_3(s) \rightarrow Al_2O_3(s) + 3H_2O(l)$$

Electrolysis of this compound, when melted, yields aluminium. The melting point of the aluminium oxide has to be lowered by mixing it with cryolite (Na_3AlF_6) and fluorspar (CaF_2).

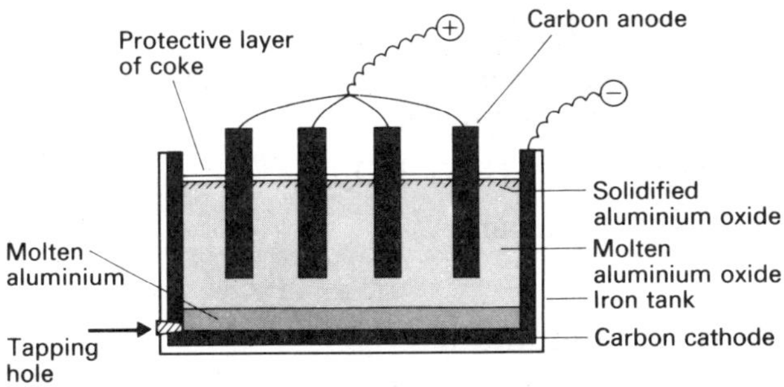

Figure 23.1 Extraction of aluminium.

The three chemicals are placed in an iron tank lined with carbon, which also acts as a cathode (figures 23.1 and 23.2). Anodes, also made of carbon, dip into the tank, and the passage of the electric current serves to maintain the temperature of the molten mixture.

Aluminium ions are discharged at the cathode, forming molten aluminium at the bottom of the tank. This is tapped off at regular intervals and is very pure:

$$Al^{3+} + 3e \rightarrow Al.$$

Oxygen ions are discharged at the anodes and these gradually oxidize the anodes to carbon monoxide. Therefore the anodes need replacing frequently:

$$O^{2-} - 2e \rightarrow O$$

$$O + O \rightarrow O_2$$

Under correct conditions very little of the cryolite and fluorspar is used up. Therefore, to obtain more aluminium only aluminium oxide need be added.

At the surface of the melt solidification occurs. This crust, together with a layer of coke which is added, prevents any atmospheric oxidation.

23.1.3 Properties of aluminium

Aluminium is a silvery-white metal of low density. It is a good conductor of heat and electricity and is reasonably malleable and ductile.

From its position in the activity series (see section 8.5.2), aluminium would be expected to be very reactive and to be readily corroded. In fact, the metal forms a very thin layer of aluminium oxide on its surface when it is exposed to air. This protects it from further atmospheric corrosion and makes it less reactive chemically. On strong heating in air or oxygen aluminium will, however, burn to form a mixture of aluminium oxide and aluminium nitride:

$$4Al(s) + 3O_2(g) \rightarrow 2Al_2O_3(s)$$

$$2Al(s) + N_2(g) \rightarrow 2AlN(s)$$

23.1.4 Action of acids on aluminium

Hydrochloric acid, particularly when hot and concentrated, reacts with aluminium to yield hydrogen:

$$\left[\begin{array}{l} 2Al + 6HCl \rightarrow 2AlCl_3 + 3H_2 \\ 2Al(s) + 6H^+(aq) \rightarrow 2Al^{3+}(aq) + 3H_2(g) \end{array}\right.$$

Dilute sulphuric acid is without appreciable effect on the metal, but with the hot, concentrated acid, sulphur dioxide is formed.

$$2Al(s) + 6H_2SO_4(l) \rightarrow Al_2(SO_4)_3(s) + 6H_2O(l) + 3SO_2(g)$$

Nitric acid has no reaction with aluminium.

23.1.5 Action of alkalis on aluminium

Solutions of caustic alkalis react readily with aluminium. Hydrogen and a soluble aluminate are formed:

Figure 23.2 Aluminium electrolysis cells at Invergordon in Scotland. On the overhead crane is the tapping crucible which collects the molten aluminium from the reduction cells on the left of the picture. On the right are the carbon anode blocks ready for use in the cells. (Photo: The British Aluminium Co. Ltd.).

Figure 23.3 Aluminium strip being produced in a hot rolling mill in South Wales. (Photo: The British Aluminium Co. Ltd.).

$$\left[\begin{array}{llllll} 2Al & +2NaOH & +2H_2O & \rightarrow & \begin{array}{l}2NaAlO_2\\ \text{sodium}\\ \text{aluminate}\end{array} & +3H_2 \\ 2Al(s) & +2OH^-(aq) & +2H_2O(l) & \rightarrow & 2AlO_2^-(aq) & +3H_2(g) \end{array}\right.$$

23.1.6 Uses of aluminium

Becauses of the difficulty of its extraction by chemical methods, aluminium was little more than a chemical curiosity until near the end of the last century. Now it is one of our most important metals.

Since aluminium is a good conductor of heat, resists corrosion and has a low density, it is widely used in cooking utensils. These utensils should not be washed with washing soda or other alkaline solutions. (Why not?)

Weight for weight, aluminium is a better conductor or electricity than copper, its low density making it particularly suitable for overhead cables.

Since aluminium is an excellent reflector of light, it is used for silvering mirrors. When rolled into thin foil, it is also used as a wrapping material and for making milk bottle tops; increasingly uses are being found for it in the kitchen at home (figure 23.3).

Aluminium alloys which combine lightness with strength, such as duralumin (Al, Mg, Cu, Mn) and magnalium (Al, Mg), are widely used in aeroplane construction and in making parts for all other forms of transport. Aluminium and its alloys are much used in the construction of space-craft where the need for high strength and low weight materials is very important.

Another important use of aluminium is in *Thermite Processes.*

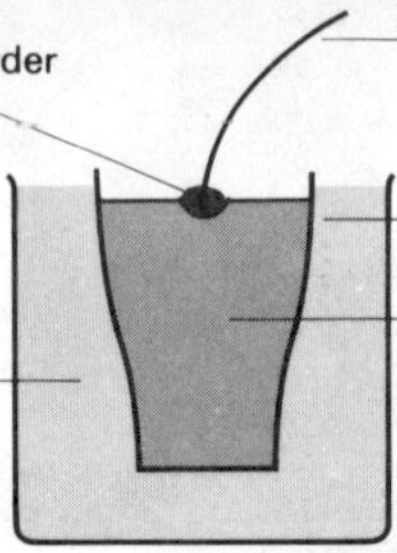

Figure 23.4 The Thermite Process.

Experiment. The Thermite Process

The Thermite Process is the reduction by aluminium of the oxide of a less reactive metal.
In this example chromium(III) oxide is used.
Assemble the apparatus shown in figure 23.4.
The thermite reaction can be very dangerous as molten metal and showers of sparks and smoke are produced. The reaction must be carried out out of doors and you should stand well away from the experiment.

The magnesium fuse is lit, and when the flame reaches the barium peroxide/aluminium powder mixture, sufficient heat is evolved to start a reaction between the aluminium and the chromium(III) oxide:

$$Cr_2O_3(s) + 2Al(s) \rightarrow Al_2O_3(s) + 2Cr(l) \qquad \Delta H = 460 \text{ kJ.mol}^{-1}$$

The reaction is exothermic, and such a high temperature is reached that a bead of molten chromium sinks to the bottom of the crucible. When the apparatus has cooled down, this may be extracted.

Industrially chromium may be made by this method.

A mixture of aluminium powder and iron(III) oxide ignited with magnesium ribbon reacts similarly:

$$Fe_2O_3(s) + 2Al(s) \rightarrow Al_2O_3(s) + 2Fe(l)$$

This mixture may be used in welding broken steel castings, etc.. The molten iron formed is extremely hot and is moulded round the ends of the broken metal.

Much aluminium is *anodized.* This is to say, it is made the anode for the electrolysis of sulphuric or chromic(VI) acid solution. Instead of oxygen just being released at the anode, some is used for oxidizing the surface of the aluminium to give a layer of oxide about a thousand times thicker than normal. This layer of oxide adsorbs organic dyes or coloured inorganic compounds so that in additon to giving a greater resistance to corrosion, aluminium articles can be produced that have a permanent coloured finish.

Experiment. Anodizing of aluminium

Aluminium can be readily anodized in the apparatus shown in figure 23.5. This uses alternating current so that each electrode becomes alternately the

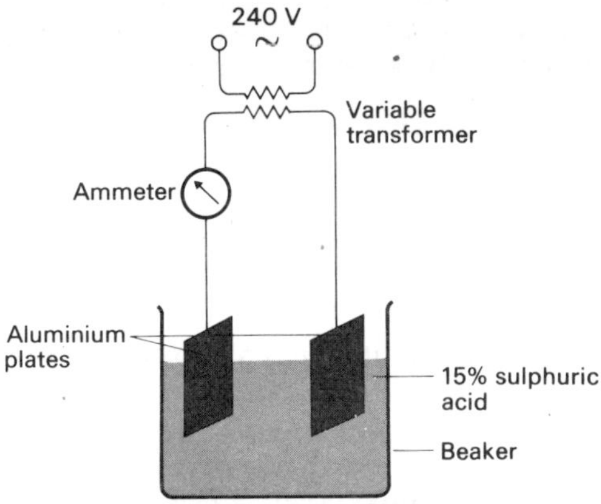

Figure 23.5 Anodizing of aluminium.

anode and the cathode and both the electrodes become anodized.

Clean two pieces of aluminium sheet by dipping for a few seconds in a solution containing about 50 g dm^{-3} of sodium hydroxide in water at about 50 °C. Rinse the plates and put them into the sulphuric acid (150 cm^3 dm^{-3}). Pass the electric current, adjusting the variable transformer to maintain about 1.5 amps during the electrolysis. Electrolyse for about 20 minutes, switch off the current and rinse the plates in water.

The presence of the oxide film can be dramatically demonstrated by connecting a small bulb to a battery through the aluminium that was out of the acid; the bulb will light up. If the bulb is moved onto the anodized part of the aluminium it will go out because of the increased electrical resistance due to the oxide film.

The anodized aluminium film can be coloured by absorbing a chemical in the porous oxide film and then adding a second chemical so that a chemical reaction takes place inside the oxide layer. In the table are given 5 pairs of chemicals that are required to give the five colours shown:

First reagent	Second reagent	Possible chemical compound produced	Colour
potassium hexacyanoferrate(II)	iron(III) nitrate	prussian blue	Blue
copper(II) sulphate	10% v/v yellow ammonium sulphide	copper(II) sulphide	Green
silver nitrate	potassium chromate(VI)	silver chromate(VI)	Bronze
lead(II) ethanoate	potassium chromate(VI)	lead chromate(VI)	Yellow
3,4,5-trihydroxybenzoic acid (gallic acid)	ammonium iron(II) sulphate	—	Violet

To colour the anodized aluminium, dip it into an aqueous solution of the first reagent (at about 35 °C, concentration not critical) for about 30 seconds, rinse it in water and dip it into the second reagent (also at about 35 °C, concentration not critical). Rinse again and repeat until the colour is deep enough. The colour may then be permanently fixed by putting the aluminium into boiling water for about 10 minutes, when the aluminium oxide forms the hydroxide:

$$Al_2O_3 + 3H_2O(l) \rightarrow 2Al(OH)_3(s)$$

This is not porous; the colour is permanently sealed into the surface layer.

23.1.7 Aluminium oxide

This can be prepared in the laboratory by heating aluminium hydroxide, nitrate or sulphate:

$$2Al(OH)_3(s) \rightarrow Al_2O_3(s) + 3H_2O(l)$$

It is a white insoluble powder of high melting point. Chemically it is amphoteric and will normally react with both acids and alkalis. When prepared at high temperatures it is, however, insoluble in acids.

23.1.8 Aluminium hydroxide

Aluminium hydroxide is insoluble, and is therefore usually prepared by precipitation, e.g. by adding ammonium hydroxide solution to aluminium sulphate solution:

$$\left[\begin{array}{l} Al_2(SO_4)_3 + 6NH_4OH \rightarrow 2Al(OH)_3\downarrow + 3(NH_4)_2SO_4 \\ Al^{3+}(aq) + 3OH^-(aq) \rightarrow Al(OH)_3(s) \end{array}\right.$$

It is better not to use caustic alkalis for precipitation, as aluminium hydroxide is amphoteric. With excess of strong alkali, therefore, a soluble aluminate is formed:

$$Al(OH)_3 + NaOH \rightarrow NaAlO_2 + 2H_2O$$
$$Al(OH)_3(s) + OH^-(aq) \rightarrow AlO_2^-(aq) + 2H_2O(l)$$

23.1.9 Aluminium chloride

Anhydrous aluminium chloride may be obtained by heating aluminium in a stream of dry chlorine or hydrogen chloride:

$$2Al(s) + 3Cl_2(g) \rightarrow 2AlCl_3(s)$$

A suitable apparatus is that used for iron(III) chloride (see figure 24.12).

A solution of aluminium chloride may be prepared by dissolving aluminium in hydrochloric acid.

Anhydrous aluminium chloride fumes in damp air because it reacts so readily with water:

$$AlCl_3 + 3H_2O(g) \rightarrow Al(OH)_3(s) + 3HCl(g)$$

23.1.10 The alums

An alum is a compound of the general formula $X^+ Y^{3+}(SO_4{}^{2-})_2.12H_2O$ where X^+ is a monovalent cation (e.g. Na^+, K^+, $NH_4{}^+$) and Y^{3+} is a trivalent cation (e.g. Al^{3+}, Cr^{3+}, Fe^{3+}). All alums have a similar crystalline shape. The most important is aluminium potassium sulphate-12-water (commonly called potash alum or alum).

The alums are *double salts* and in solution potash alum is indistinguishable from a mixture of potassium and aluminium sulphates, i.e. it contains potassium ions, aluminium ions and sulphate ions, just as a mixture of the individual salts would do.

This fact indicates a method of preparation. The formula shows that in potash alum there is one mole of potassium sulphate for each mole of aluminium sulphate. A solution is made in the minimum of hot water containing these salts in this proportion. (What would be convenient masses to use?) On cooling crystals of the double salt separate out. It is possible to grow large crystals of alums. An experiment to do this is given in section 1.14, and some examples of alum crystals are shown in figure 1.20.

23.2 Tin

Tin and lead are the two most important metals in Group IV of the periodic table. Their relationship with carbon, silicon and germanium is discussed in section 5.3.4(*f*).

23.2.1 Extraction of tin

The most common ore of tin is cassiterite or tinstone, SnO_2, much of which is found in the Malay States. As mined, the ore is mixed with a high proportion of impurities, and it must first be concentrated. To obtain tin, the oxide is reduced by heating with coke in a furnace:

$$SnO_2(s) + 2C(s) \rightarrow Sn(l) + 2CO(g)$$

Molten tin, which is cast into ingots, is formed.

23.2.2 Properties and uses of tin

The ordinary form of tin, white tin, is a silvery-white, lustrous metal. It has a high density (7·3 g cm^{-3}), is malleable, but not very ductile.

At room temperature tin is very resistant to corrosion, and it is this property which renders it valuable.

Tin is widely used for plating steel in the manufacture of "tins" for containing foodstuffs. It is much more resistant to attack by acids in foodstuffs than is zinc. However if the tin coating is damaged the iron corrodes more rapidly than if it were not plated at all. In the presence of an electrolytic solution the two metals form a simple cell; iron is more reactive than tin and corrodes more readily than it (see table 8.4).

It is also used in a number of alloys, e.g. bronze (Cu, Sn) pewter (Sn, Pb), solder (Pb, Sn) and type metal (Pb, Sb, Sn).

23.3 Lead

23.3.1 Occurrence

The most important ore of lead is galena, lead(II) sulphide, PbS. It is usually associated with zinc blende, ZnS, and is widely distributed in nature. Important deposits are found in New South Wales and in North America.

23.3.2 Extraction of lead

After concentration of the galena, it is roasted in air, when it is largely converted into lead(II) oxide:

$$2PbS(s) + 3O_2(g) \rightarrow 2PbO(s) + 2SO_2(g)$$

The lead(II) oxide is then reduced by heating with coke and a little iron in a blast furnace:

$$PbO(s) + C(s) \rightarrow Pb(l) + CO(g)$$

The iron reduces any remaining lead(II) sulphide:

$$PbS(s) + Fe(l) \rightarrow Pb(l) + FeS(s)$$

Molten lead forms at the base of the furnace.

23.3.3 Properties of lead

When cut, lead is seen to be a greyish, lustrous metal. It has a high density ($11{\cdot}4\ g\,cm^{-3}$) and is very malleable. It is a good conductor of heat and electricity, but has a lower melting point than most metals (327 °C).

In damp air, lead is rapidly corroded on the surface, forming lead(II) hydroxide and lead(II) carbonate. This protects the metal from further corrosion.

Lead does not react with pure water, but soft water containing dissolved air has a slight reaction to form lead(II) hydroxide. Lead(II) hydroxide has a slight solubility in water, and cases of lead poisoning may result from the continual drinking of soft water which has passed through lead pipes. With hard water a protective layer of corrosion is formed. Because of this it was sometimes necessary to harden water artificially in very soft water districts. Nowadays copper and plastic are used for water pipes, and it is only in old houses that lead piping is still found.

Non-oxidizing acids have little action on lead, but it reacts readily with hot, concentrated sulphuric acid, and with dilute or concentrated nitric acid. With sulphuric acid, sulphur dioxide is formed:

$$Pb(s) + 2H_2SO_4(l) \rightarrow PbSO_4(s) + 2H_2O(l) + SO_2(g)$$

With nitric acid, nitrogen oxide and nitrogen dioxide are formed:

$$\left[\begin{array}{l} 3Pb + \underset{(50\,\%)}{8HNO_3} \rightarrow 3Pb(NO_3)_2 + 4H_2O + 2NO \\ 3Pb(s) + 8H^+(aq) + 2NO_3^-(aq) \rightarrow 3Pb^{2+}(aq) + 4H_2O(l) + 2NO(g) \end{array}\right.$$

$$\left[\begin{array}{l} Pb + \underset{\text{(concentrated)}}{4HNO_3} \rightarrow Pb(NO_3)_2 + 2H_2O + 2NO_2 \\ Pb(s) + 4H^+(aq) + 2NO_3^-(aq) \rightarrow Pb^{2+}(aq) + 2H_2O(l) + 2NO_2(g) \end{array}\right.$$

Alkalis are without reaction on lead.

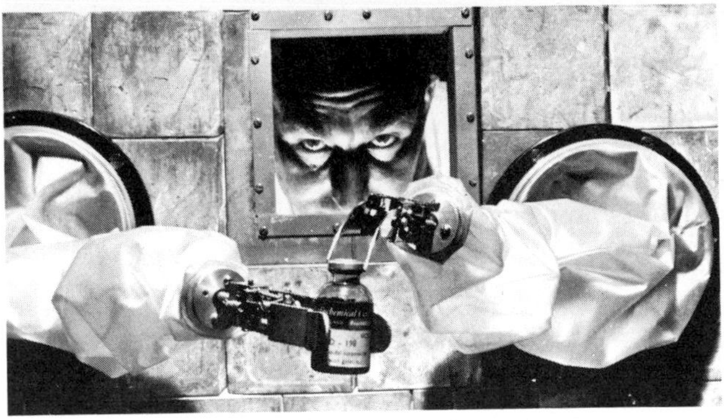

Figure 23.6 Handling a radioactive substance by remote control. The wall is made of lead bricks and the windows of lead glass, both of which absorb harmful radiation. (Photo: U.K.A.E.A.).

23.3.4 Uses of lead

Because of its malleability, softness and resistance to corrosion, lead is used in piping and as a roofing material. It is also used in accumulators and for covering electric cables. It protects workers handling radioactive material, as it absorbs harmful radiation (see figure 23.6), and is a constituent of many alloys, notably type metal (Pb, Sb, Sn) and solder (Pb, Sn).

23.3.5 Compounds of lead

Lead forms two series of compounds:

Lead(II) (plumbous) compounds in which lead has a valency of *two*.

Lead(IV) (plumbic) compounds in which lead has a valency of *four*. Lead(II) compounds are far more commonly used in the laboratory.

(a) Oxides of lead

Lead forms three important oxides:

(i) Lead(II) oxide, litharge, PbO.

(ii) Lead(IV) oxide, PbO_2.

(iii) Dilead(II) lead(IV) oxide, red lead, Pb_3O_4, which behaves chemically as a mixture of lead(II) oxide and lead(IV) oxide with a composition given by the formula $2PbO.PbO_2$.

(i) Lead(II) oxide

This oxide is usually made by heating lead(II) carbonate:

$$PbCO_3(s) \rightarrow PbO(s) + CO_2(g)$$

In the laboratory the action of heat on lead(II) nitrate or lead(II) hydroxide is equally suitable:

$$2Pb(NO_3)_2(s) \rightarrow 2PbO(s) + 4NO_2(g) + O_2(g)$$

$$Pb(OH)_2(s) \rightarrow PbO(s) + H_2O(g)$$

Lead(II) oxide is a yellow or orange solid. It is *amphoteric*. Thus, when warmed with dilute nitric acid, a solution of lead(II) nitrate is formed:

$$\left[\begin{array}{lll} PbO & +2HNO_3 & \rightarrow Pb(NO_3)_2+H_2O \\ PbO(s)+2H^+(aq) & & \rightarrow Pb^{2+}(aq)+H_2O(l) \end{array}\right.$$

The action of hydrochloric and sulphuric acids is less pronounced, as the products, lead(II) chloride and lead(II) sulphate, are both insoluble. Therefore an insoluble coating is formed on lead(II) oxide which is added to either of these two acids and this prevents a complete reaction. With solutions of alkalis, a plumbate(II) is formed:

$$\left[\begin{array}{ll} PbO + 2NaOH & \rightarrow Na_2PbO_2 + H_2O \\ PbO(s)+2OH^-(aq) & \rightarrow PbO_2{}^{2-}(aq)+H_2O(l) \end{array}\right.$$

As lead is comparatively unreactive, lead(II) oxide is readily reduced to the metal by heating in a stream of hydrogen or carbon monoxide, or by heating with carbon:

$$PbO(s)+H_2(g) \rightarrow Pb(l)+H_2O(g)$$

$$PbO(s)+CO(g) \rightarrow Pb(l)+CO_2(g)$$

$$PbO(s)+C(s) \rightarrow Pb(l)+CO(g)$$

(ii) Lead(IV) oxide

This is prepared by warming dilead(II) lead(IV) oxide, red lead, with dilute nitric acid:

$$\left[\begin{array}{l} Pb_3O_4 + 4HNO_3 \rightarrow PbO_2 + 2Pb(NO_3)_2 + 2H_2O \\ Pb_3O_4(s)+4H^+(aq) \rightarrow PbO_2(s)+2Pb^{2+}(aq) + 2H_2O(l) \end{array}\right.$$

The lead(IV) oxide is filtered off, washed with water and dried at a low temperature.

Lead(IV) oxide is a dark brown, insoluble solid. When heated to 300 °C, it decomposes into oxygen and the lower oxide:

$$2PbO_2(s) \rightarrow 2PbO(s)+O_2(g)$$

Hot, concentrated hydrochloric acid is oxidized to chlorine by lead(IV) oxide:

$$PbO_2(s)+4HCl(aq) \rightarrow PbCl_2(s)+2H_2O(l)+Cl_2(g)$$

With hot, concentrated sulphuric acid, oxygen is evolved:

$$2PbO_2(s)+2H_2SO_4(l) \rightarrow 2PbSO_4(s)+2H_2O(l)+O_2(g)$$

Nitric acid does not react with lead(IV) oxide.

Like lead(II) oxide, lead(IV) oxide is readily reduced to the metal:

$$PbO_2(s)+2H_2(g) \rightarrow Pb(s)+2H_2O(g)$$

(iii) Dilead(II) lead(IV) oxide, red lead Pb_3O_4

This is a scarlet powder, made by heating lead(II) oxide in air at 450 °C for a long time:

$$6PbO(s)+O_2(g) \rightarrow 2Pb_3O_4(s)$$

The reaction is reversible, and if the temperature much exceeds 450 °C, the dilead(II) lead(IV) oxide reverts to the lower oxide:

$2Pb_3O_4(s) \rightarrow 6PbO(s) + O_2(g)$

If dilead(II) lead(IV) oxide is regarded as a mixture of lead(II) oxide and lead(IV) oxide, $2PbO.PbO_2$, its chemical reactions are easily understood. Thus with dilute nitric acid, the reaction is just what would be expected for a mixture of the simpler oxides:

$PbO_2(s) \rightarrow PbO_2(s)$ (no reaction)

and $2PbO(s) + 4HNO_3(aq) \rightarrow 2Pb(NO_3)_2(aq) + 2H_2O(l)$

Adding these together gives:

$Pb_3O_4(s) + 4HNO_3(aq) \rightarrow PbO_2(s) + 2Pb(NO_3)_2(aq) + 2H_2O(l)$

The reactions with concentrated hydrochloric acid and with concentrated sulphuric acid may be deduced similarly:

	$PbO_2(s)$	$+4HCl(aq)$	$\rightarrow PbCl_2(s)$	$+2H_2O(l) + Cl_2(g)$
and	$2PbO(s)$	$+4HCl(aq)$	$\rightarrow 2PbCl_2(s)$	$+2H_2O(l)$
gives:	$Pb_3O_4(s)$	$+8HCl(aq)$	$\rightarrow 3PbCl_2(s)$	$+4H_2O(l) + Cl_2(g)$
	$2PbO_2(s)$	$+2H_2SO_4(l)$	$\rightarrow 2PbSO_4(s)$	$+2H_2O(l) + O_2(g)$
and	$4PbO(s)$	$+4H_2SO_4(l)$	$\rightarrow 4PbSO_4(s)$	$+4H_2O(l)$
gives:	$2Pb_3O_4(s)$	$+6H_2SO_4(l)$	$\rightarrow 6PbSO_4(s)$	$+6H_2O(l) + O_2(g)$

(b) Lead(II) hydroxide, $Pb(OH)_2$

Lead(II) hydroxide is insoluble in water and is prepared by precipitation. If solutions of sodium or potassium hydroxides are used for this precipitation, the hydroxide combines with excess of alkali, forming a soluble plumbate(II):

$$\left[\begin{array}{l} Pb(OH)_2 + 2NaOH \rightarrow Na_2PbO_2 + 2H_2O \\ Pb(OH)_2(s) + 2OH^-(aq) \rightarrow PbO_2{}^{2-}(aq) + 2H_2O(l) \end{array}\right.$$

(c) Lead(II) nitrate, $Pb(NO_3)_2$

Lead(II) nitrate may be prepared by the action of dilute nitric acid on lead, lead(II) oxide or lead(II) carbonate.

It is the only common salt of lead to be readily soluble in water, and it is therefore used for preparing insoluble salts. *These insoluble salts cannot be prepared directly from lead; lead(II) nitrate must be prepared first.*

The following equations illustrate the use of lead(II) nitrate in preparing other lead compounds. The first requires heat.

$2Pb(NO_3)_2(s) \rightarrow 2PbO(s) + 4NO_2(g) + O_2(g)$

$$\left[\begin{array}{l} Pb(NO_3)_2 + H_2SO_4 \rightarrow PbSO_4\downarrow + 2HNO_3 \\ Pb^{2+}(aq) + SO_4{}^{2-}(aq) \rightarrow PbSO_4(s) \end{array}\right.$$

$$\left[\begin{array}{l} Pb(NO_3)_2 + 2HCl \rightarrow PbCl_2\downarrow + 2HNO_3 \\ Pb^{2+}(aq) + 2Cl^-(aq) \rightarrow PbCl_2(s) \end{array}\right.$$

$$\left[\begin{array}{l} Pb(NO_3)_2 + H_2S \rightarrow PbS\downarrow + 2HNO_3 \\ Pb^{2+}(aq) + S^{2-}(aq) \rightarrow PbS(s) \end{array}\right.$$

(d) Lead(II) chloride, $PbCl_2$

Lead(II) chloride is prepared, as indicated above, by precipitation. It is one of the few substances to be almost insoluble in cold water, but to have a much greater solubility in hot water.

(e) Lead(II) carbonate, $PbCO_3$

This has to be prepared by mixing sodium hydrogencarbonate and lead(II) nitrate solutions:

$$\left[\begin{array}{l} Pb(NO_3)_2 + 2NaHCO_3 \rightarrow PbCO_3\downarrow + 2NaNO_3 + H_2O + CO_2 \\ Pb^{2+}(aq) + 2HCO_3^-(aq) \rightarrow PbCO_3(s) + H_2O(l) + CO_2(g) \end{array}\right.$$

If sodium carbonate is used, a basic carbonate, lead(II) carbonate hydroxide (white lead), $PbCO_3 . 2Pb(OH)_2$, is formed.

In the past white lead was used in making paints, although it has the disadvantages of being poisonous and of blackening in air containing hydrogen sulphide. It has now been almost completely replaced by another white pigment, titanium(IV) oxide, which does not deteriorate in this way.

Questions for Chapter 23

1. Describe how aluminium is manufactured from pure aluminium oxide by an electrolytic process.

Explain why (*a*) aluminium is often used for overhead power cables, and (*b*) molten iron can be obtained by igniting a mixture of iron(III) oxide and aluminium powder.

The same quantity of electricity was passed through two appropriate cells. If 4·6 g of sodium was liberated in the first cell, what mass of aluminium was liberated in the second cell? (Na = 23, Al = 27) (*A.E.B.*)

2. Describe briefly the manufacture of aluminium.

Mention two uses of this metal.

What is the action, if any, of aluminium hydroxide with (*a*) hydrochloric acid, (*b*) sodium hydroxide, (*c*) ammonium hydroxide? (*O. and C.*)

3. Which of the metals lead, aluminium and magnesium (*a*) burn in air, (*b*) dissolve in dilute hydrochloric acid, (*c*) react with steam when heated? Represent by equation, or otherwise, what happens when reaction does occur.

Mention *one* use of each of these metals. (*O. and C.*)

4. In what form is lead most commonly found in nature?

Draw and label a diagram showing how you would produce carbon monoxide and use it to reduce lead(II) oxide to lead. (*O. and C.*)

5. Describe one physical and one chemical test by which you would identify a given metal as lead.

State *three* important uses of this metal.

Starting from lead(II) carbonate, describe with all necessary practical details how you would prepare dry samples of (*a*) lead(II) oxide, (*b*) lead(II) nitrate crystals, (*c*) lead(II) sulphate. (*C.W.B.*)

6. Briefly compare and contrast the properties of metallic lead with those of iron.

Starting with a piece of lead how would you proceed to prepare (*a*) lead(II) oxide, (*b*) dilead(II) lead(IV) oxide?

How and under what conditions do these oxides react with (*c*) concentrated hydrochloric acid, (*b*) carbon monoxide? (*D.*)

7. Describe how you would prepare a pure, dry sample of lead(II) chloride by precipitation, giving an equation for the reaction used.

A small amount of lead(II) chloride was placed in a boiling tube which was then about half filled with water. The tube and contents were heated until a colourless solution *A* was obtained. Hydrogen sulphide was bubbled through the hot solution and a silvery black precipitate *B* was formed. After filtering and washing, the precipitate was treated with hydrogen peroxide solution which converted the black precipitate into a white solid *C*.

(*a*) Identify *A*, *B* and *C* and write an equation for the formation of *B* from *A*.

(*b*) What type of reaction is the conversion of *B* into *C*? Write an equation for this reaction.

(*c*) How may *C* be prepared in one step from lead(IV) oxide? Write the equation for this exothermic reaction and say what you would see during its course. (*J.M.B.*)

8. Two pieces of lead foil are dipped, side by side but not touching, into a beaker of fairly concentrated sulphuric acid. After some days' immersion the plates are covered in a white deposit. The two pieces of lead are now connected to a 6 V d.c. supply, and a current passes. After some hours, one piece of lead has lost its white deposit and bubbles of hydrogen are streaming off it. The other piece of lead now appears to be coated with a dark brown deposit, replacing the white deposit.

(*a*) Draw a labelled diagram, showing the apparatus when the current is being passed through it.

(*b*) Suggest a name for the white deposit first formed on each plate.

(*c*) Suggest a name for the dark brown deposit. If some of this deposit could be scraped off, describe how you would attempt to identify it.

(*d*) Suggest any change that you would expect to occur in the liquid during the passage of the current. (*O. and C.*)

9. A scarlet powder *A* when warmed with dilute nitric acid produces a brown substance *B* that is removed by filtration. The filtrate on treatment with potassium chromate(VI) (K_2CrO_4) solution gives a yellow precipitate *C*. On heating *B* in a tube, a gas is evolved that ignites a glowing splinter of wood, and a yellow solid *D* remains in the tube.

When *D* is strongly heated in a stream of hydrogen, drops of a colourless liquid condense at the cooler end of the tube through which hydrogen passes and globules of a metal *E* are produced. These globules solidify on cooling and are found to be malleable and mark paper.

When either *A* or *B* is heated with hydrochloric acid and a greenish gas that turns starch iodide paper blue is evolved. The resulting solution deposits white crystals. *F*, on cooling.

Name the substances *A* to *F*. Give equations for the various changes that are described. (*L.*)

10. How is dilead(II) lead(IV) oxide made from metallic lead? Name *one* important use of dilead(II) lead(IV) oxide.

Describe how you would prepare from dilead(II) lead(IV) oxide, specimens of (*a*) crystalline lead(II) chloride, (*b*) lead(IV) oxide, (*c*) lead(II) sulphate. (*L.*)

11. Starting from metallic lead, describe in detail how you would prepare a specimen of lead(II) nitrate. Describe also, in outline only, how you would obtain from this salt specimens of (*a*) lead(II) chloride, (*b*) lead(II) oxide.

How could you show that lead(II) oxide contains lead and oxygen. (*C.*)

24

Two Transition Elements—Iron and Copper

24.1 Introduction

In section 5.3.4(*g*) we have seen that the transition elements are all metals. Those in the fourth period, i.e. the first transition series, include a group of metals of extreme industrial importance. The transition metals have a number of features in common, i.e.

1. variable oxidation states, i.e. variable valency;
2. coloured ions which are usually hydrated;
3. an ability to catalyse reactions;
4. those properties normally associated with metals, i.e. high melting point, excellent conductivity, good malleability and ductility.

In this book our detailed study will be confined to iron and copper.

24.2 Iron

24.2.1 Introduction

Aluminium is the most plentiful metal in the earth's crust, but, despite the fact that aluminium is now produced in large quantities at an economical price, iron remains by far the most important.

The principal ores are:

(*a*) haematite—impure iron(III) oxide, Fe_2O_3,
(*b*) limonite—hydrated iron(III) oxide, Fe_2O_3,
(*c*) magnetite—iron(II) diiron(III) oxide, Fe_3O_4,
(*d*) iron pyrites—iron(II) disulphide, FeS_2,
(*e*) siderite—iron(II) carbonate, $FeCO_3$.

Such ores, in varying degrees of purity, are widely distributed. The extent to which ores are worked in a particular country depends not only on the supply of iron ore but also on the supply of coal and of cheap transport.

24.2.2 Extraction of iron

Most ores are roasted in air. This drives off water, carbon dioxide, etc. and leaves iron(III) oxide as the main product.

The essential stage in the manufacture of iron is, however, the reduction of the iron(III) oxide in a blast furnace.

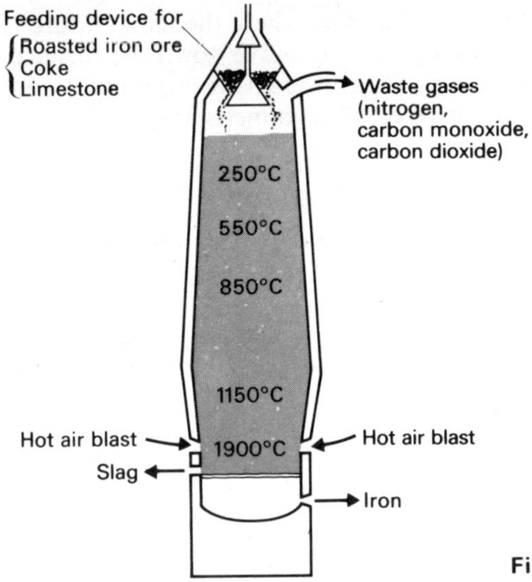

Figure 24.1 The blast furnace.

A blast furnace (figure 24.1) may be more than 70 m high, with a hearth diameter of 11 m. It is made of firebricks in an iron shell. These blast furnaces can produce 7000 to 8000 tonnes of molten iron every day. Four materials are fed into the furnace:

1. Roasted iron ore . . . . (14 000 tonnes)
2. Coke . . . . (5600 tonnes)
3. Limestone . . . . (3500 tonnes)
4. Hot air . . . . (28 000 tonnes)

(The figures given in brackets are the approximate quantities of materials required daily for a furnace producing 7000 tonnes of iron per day.)

The iron ore, coke and limestone are fed intermittently by a charging device into the top of the furnace. A hot-air blast is introduced through tuyeres as the pipes near the base of the furnace are called.

There are many chemical reactions taking place in the blast furnace. The important ones are:

(a) Formation of carbon monoxide

Coke at the base of the furnace burns in the hot-air blast to form carbon dioxide, which is then reduced to carbon monoxide by more carbon:

$$C(s) + O_2(g) \rightarrow CO_2(g)$$

$$CO_2(g) + C(s) \rightarrow 2CO(g)$$

In the neighbourhood of the tuyeres, the temperature may be raised by this reaction to as high as 1900 °C.

(b) Reduction of iron(III) oxide

The carbon monoxide rises up the furnace and is largely responsible for the reduction of the iron oxide:

$$Fe_2O_3(s) + 3CO(g) \rightarrow 2Fe(s) + 3CO_2(g)$$

This reaction takes place near the top of the furnace, where the temperature is comparatively low (550 °C). As the charges gradually move down the furnace, iron which has been formed absorbs carbon. This lowers its melting point and, with the high temperature lower in the furnace, the iron melts and sinks to the bottom of the furnace.

(c) Formation of a fusible slag

The limestone introduced into the furnace decomposes into quicklime and carbon dioxide:

$$CaCO_3(s) \rightarrow CaO(s) + CO_2(g)$$

Carbon dioxide from this reaction, when reduced by red-hot coke, provides more carbon monoxide.

The main purpose of the limestone is, however, to provide quicklime. This oxide is basic and combines with earthy acidic impurities, such as silica, to form a slag:

$$\underset{}{CaO(s)} + \underset{\text{silica}}{SiO_2(s)} \rightarrow \underset{\text{calcium silicate}}{CaSiO_3(l)}$$

The slag melts, and, like the iron, it drips to the bottom of the furnace. Fortunately, the molten slag and the molten iron do not mix, and two layers are formed with the slag floating on top of the iron.

The molten slag and the molten iron are tapped off through separate holes (figure 24.2). Some of the slag is tipped, but uses are being found for it, e.g. for

Figure 24.2 Tapping a blast furnace (Photo: The British Steel Corporation).

Figure 24.3 Pig-iron casting. (Photo: The British Steel Corporation).

road surfacing and in making cement. Iron, that is not to be converted into steel while still molten, is run into sand moulds or into a "pig"-casting machine, where it solidifies into lumps of a convenient size (figure 24.3). Waste gases which leave the top of the furnace are largely nitrogen (60 per cent), carbon monoxide (30 per cent) and carbon dioxide (10 per cent). These gases are burnt and the heat used in heating up the incoming air blast.

24.2.3 Pig-iron and cast iron

The iron obtained directly from the blast furnace is known as pig-iron. It is very impure and contains 3–5 per cent of carbon and smaller quantities of silicon, phosphorus, sulphur and manganese. Some pig-iron is remelted and cast into moulds in the shape of the required article. It is then known as cast iron.

Cast iron is brittle and cannot be welded. Its tensile strength is low, but, because of its comparative cheapness and ease of melting, it is used in making a wide range of products, e.g. lamp-posts, grates, stoves, cookers, radiators, railings, hot-water pipes, chairs for railway tracks, machinery parts which undergo no great strain, grids and man-hole covers.

24.2.4 Wrought iron

Wrought iron is the purest form of iron and contains only 0·1–0·25 per cent of carbon. It is made by removing most of the impurities from pig-iron. This is done by heating pig-iron with iron ore (iron(III) oxide) to a high temperature in a furnace. Oxygen from the iron ore oxidizes carbon to carbon monoxide, which bubbles out of the iron and ignites. Silicon, phosphorus, manganese, etc., are also oxidized and removed as a slag.

As the impurities are removed, the melting point of the iron rises, and accordingly it begins to solidify. The pasty mass of wrought iron is hammered under a steam hammer to squeeze out slag and is then made into bars.

Wrought iron is softer and less brittle than cast iron, and it resists rusting. Unlike cast iron, it can be welded, i.e. pieces of wrought iron can be joined together by heating and hammering. Nowadays wrought iron has been replaced almost completely by mild steel, which is cheaper. It is still used to a limited extent in making chains, in ornamental work and for agricultural implements.

24.2.5 Steel

Steel is an alloy of iron containing between 0.02 per cent and 1.7 per cent of carbon. The properties depend to a considerable extent on the percentage of carbon present, and therefore this value has to be controlled very carefully. Steel containing a low percentage of carbon, up to 0.25 per cent, is known as mild steel. Wrought iron may therefore be considered as a form of mild steel. High carbon steels are much harder, but more brittle.

During the 20th century steelmaking has undergone many changes and new processes have been continually developed to give increasing quantities of better quality steel faster than before. Today there are three major steelmaking processes—the Open Hearth, the Electric arc and the Basic Oxygen processes. During the last 20 years the Basic Oxygen and Electric arc processes have grown in importance and the Open Hearth process has declined. This is because the Open Hearth method is much slower than the other two; a large open-hearth furnace may take 10 hours to produce 350 tonnes of steel.

(a) The Open-Hearth Process

Molten iron from the blast furnace, scrap iron and limestone are fed into the furnace. The mixture is melted by the method described below. Carbon is oxidized to carbon monoxide and the mixture appears to boil. Oxides of silicon, phosphorus, etc., combine with quicklime to form a slag (cf. the formation of slag in the blast furnace).

The carbon content is either reduced to the required percentage or is removed almost completely, and then a calculated quantity of carbon is added.

(b) Heating in the open hearth process

The open-hearth process is economical because of the ingenious method used for

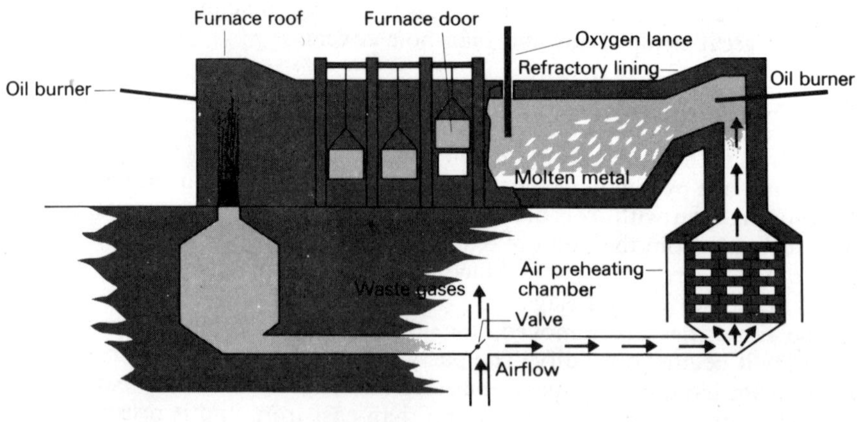

Figure 24.4 The open hearth furnace.

heating the furnace. Heat from waste gases is used to pre-heat the air used for burning the fuel.

Natural gas or oil is burnt in a stream of air that has passed through chambers containing a chequer-work of hot firebricks. The burning gases create a very high temperature (approx. 1700 °C) (figure 24.4). Burnt gases pass out of the opposite end and hence through another chamber containing a chequer-work of bricks. The bricks extract a large proportion of heat from the waste gases.

After 20–30 minutes the direction of flow of the gases is altered, i.e. the air is passed through the chambers previously heated. The direction of flow is alternated in this way for about 10 hours—the time taken for the production of steel from each charge.

Oxygen is also blown into the molten mixture through water-cooled lances. This helps the gases to burn more efficiently and the impurities are oxidized more rapidly.

(c) The electric arc process

Unlike the Open Hearth and Basic Oxygen processes the Electric Arc process uses only cold scrap metal; no molten iron is used. The process was originally developed to make special high quality steels as it gave very precise control over the composition of the steel but today a considerable amount of all steels is made by this method; about 150 tonnes can be produced in one furnace every 4 hours.

The furnace consists of a circular bath with a moveable roof through which three carbon electrodes are raised or lowered (figure 24.5). Steel scrap is charged into the furnace and a powerful electric current passed through the electrodes when the heat from the electric arc that forms melts the metal. Limestone and iron oxide are then added and these combine with the impurities to give a slag. This is tapped off the metal and the steel is then poured into a teeming ladle suspended from an overhead crane. "Teeming" is the name used in the iron and steel industry for casting. The steel is poured from the teeming ladle into moulds where steel ingots of between 5 and 40 tonnes are produced (figure 24.6).

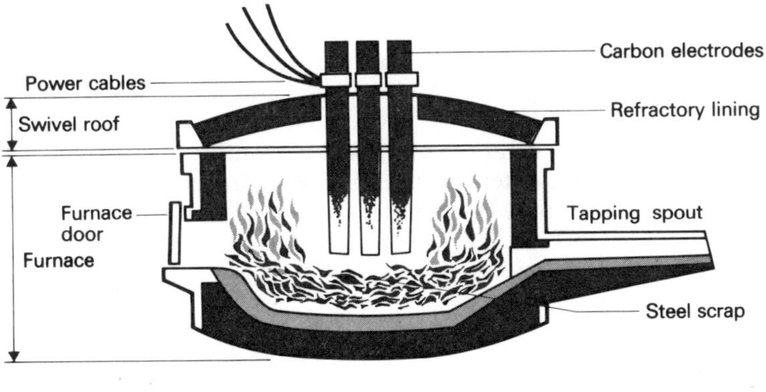

Figure 24.5 The electric arc furnace.

Figure 24.6 Making steel ingots by teeming. (Photo: The British Steel Corporation).
Figure 24.7 Charging a basic oxygen furnace with molten iron. (Photo: The British Steel Corporation).

(d) The basic oxygen process (also called the LD process)

This process, which is fast becoming the major method of making steel, is based on the now obsolete Bessemer converter. Modern Basic Oxygen furnaces can produce 350 tonnes of steel in 40 minutes.

The furnace is tilted from its vertical position and scrap iron followed by molten iron (figure 24.7) is added to it. A water cooled oxygen lance is lowered into the furnace and pure oxygen is blown into the metal at high speed (figures 24.8 and 24.9). The oxygen combines with the carbon and other acidic impurities (oxides of silicon and phosphorus etc.) combine with limestone that is added during this "blow" stage to give a slag. It is because basic limestone is added to combine with the acidic impurites and oxygen gas is blown through the furnace that it is called "Basic Oxygen". After the steel has been produced it is poured into a teeming ladle and cast into ingots or steel bars or sheets.

This process was first used in Linz and Donawitz in Austria and is often called the LD process.

(e) Special steels

Nowadays many special steels are made containing other alloying elements besides carbon. Stainless steels always contain added chromium and sometimes nickel. "Staybrite" steel, for example, contains 18 per cent chromium, 8 per cent nickel, the remainder being iron. The addition of 10–15 per cent of manganese produces a steel that is very resistant to wear. Tool steels often contain tungsten, chromium, vanadium and sometimes cobalt, molybdenum and uranium as well.

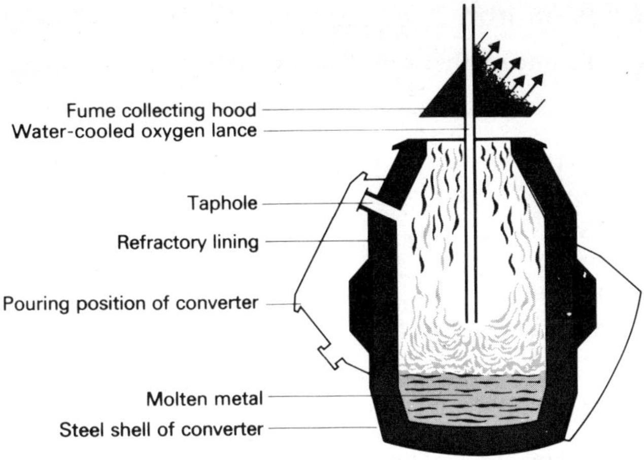

Figure 24.8 The basic oxygen furnace.

Figure 24.9 A basic oxygen furnace during the blow. (Photo: The British Steel Corporation).

24.2.6 Properties of iron

Pure iron is a white lustrous metal and is one of the few magnetic elements. It has a high density (7·86 g cm^{-3}) and a high melting point (about 1500 °C). It is very malleable, has a high tensile strength and is a good conductor of heat and electricity.

Iron does not combine with oxygen at ordinary temperatures, but when strongly heated it burns to form iron(II) diiron(III) oxide:

$$3Fe(s) + 2O_2(g) \rightarrow Fe_3O_4(s)$$

This reaction can be shown by heating some steel wool, held in a pair of tongs, to red heat, and then immediately plunging it into a gas jar full of oxygen.

When steam is passed over red-hot iron, iron(II) diiron(III) oxide is again formed, this time with the liberation of hydrogen (see section 10.6.2):

$$3Fe(s) + 4H_2O(g) \rightarrow Fe_3O_4(s) + 4H_2(g)$$

24.2.7 Action of acids on iron

With hydrochloric acid and dilute sulphuric acid, an iron (II) salt and hydrogen are formed:

$Fe + 2HCl$	$\rightarrow$	$FeCl_2$ iron(II) chloride	$+ H_2$
$Fe + H_2SO_4$	$\rightarrow$	$FeSO_4$ iron(II) sulphate	$+ H_2$
$Fe(s) + 2H^+(aq)$	$\rightarrow$	$Fe^{2+}(aq)$	$+ H_2(g)$

Dilute nitric acid gives a variety of products, including the oxides of nitrogen, depending on the exact concentration and temperature. When concentrated, it renders the iron "passive", i.e., there is no apparent reaction. Afterwards the iron will not take part in its normal reactions. Thus, if the concentrated acid is mixed with iron, and then water is carefully added to dilute the acid, no action takes place. But when the iron is scratched, reaction begins. The passivity is possibly caused by the formation of a protective layer of oxide on the surface, which is partially removed by the scratching.

24.2.8 Rusting of iron

Rust is hydrated iron(III) oxide, $Fe_2O_3.xH_2O$. The causes of rusting are complex and not fully understood. A series of simple laboratory experiments serves to illustrate some of the essential conditions for rusting to occur.

Experiment. Rusting

1. Boil some water in a test-tube for some minutes to remove dissolved air. Immerse some bright iron nails in the water and boil it again for a short while. Melt some Vaseline and pour it on the surface of the water so that air is completely excluded (figure 24.10).

2. Suspend some more bright nails in a test-tube containing lumps of fused calcium chloride. Close the tube with a well-fitting rubber bung. Any moisture in the tube is absorbed by the drying agent.

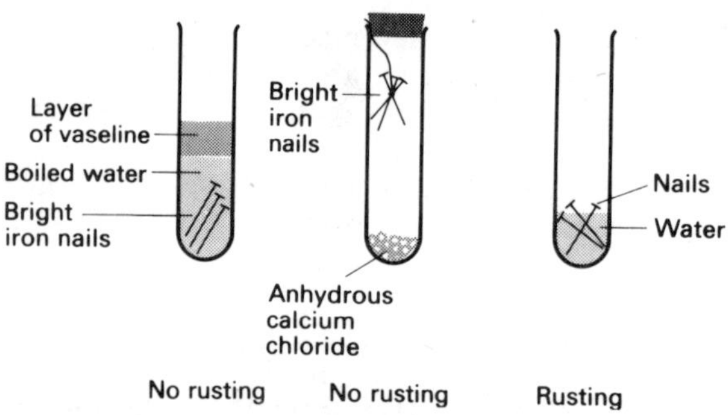

Figure 24.10 The rusting of iron.

3. Place some more nails with water in an open test-tube.

After a few days you will find that rusting has occurred only in the case of the nails in the third tube. This indicates that both air and moisture are necessary for rusting. (N.B. Because of the difficulty of completely excluding air, it is sometimes found that a little rusting occurs in the first test-tube.)

Carbon dioxide is also believed to be essential for rusting. Can you devise an experiment to show this?

24.2.9 Prevention of rusting

The corrosion of iron wastes a huge amount of money, and iron articles are usually protected in some way to prevent them from rusting. The need is to prevent water and air from coming into contact with the iron. Several methods of preventing rust are in common use.

(a) Painting
Iron bridges, cars, machinery, ships and many other iron objects are painted to prevent rusting. Unfortunately if the paint layer gets broken the iron begins to rust and the rust layer can spread underneath the paint. The underbodies of motor cars rust very readily. They are continually getting wet, and in winter when salt is put onto the roads the dilute salt solution accelerates rusting. For this reason cars are "undersealed". They are treated with a tar-like liquid that prevents the water reaching the steel underbody. Modern underseals (such as "Ziebart") also contain chemicals that hold the underseal to the metal. They also remain liquid so that if a stone breaks the underseal surface it flows and closes the crack before rusting can occur.

(b) Galvanizing
The iron or steel is covered with a thin layer of zinc. (See section 22.3.6 and figure 22.6). If the surface is damaged, then unlike a painted surface the iron does not rust. This is because an electrochemical cell is set up with the zinc and iron as the electrodes. As zinc is higher than iron in the electrochemical series it dissolves and rusting does not occur. Also the zinc hydroxide that is formed blocks up the cracks in the zinc and prevents further corrosion.

(c) Tin-Plating
Tin can be used in the same way as zinc, and is used when steel cans are to be used for canning food—hence the name "tin cans". In contrast to a zinc-plated surface, a tin-plated iron surface rusts more rapidly than pure iron. This is because tin is below iron in the electrochemical series and the iron becomes the anode of the tin-iron electrochemical cell and rapidly dissolves i.e. rusts.

Other metals are also used for plating iron. For example, car-bumpers are plated with chromium, as this both prevents rusting and looks attractive.

(d) Greasing and oiling
Iron and steel is prevented from rusting by having a layer of oil on its surface. This is the common method in moving machinery, the oil film being constantly replaced as it is rubbed off.

(e) Sacrificial anodes
Large blocks of a metal higher in the electrochemical series than iron are sometimes fastened to iron surfaces such as a ship's hull; magnesium or zinc are usually used.

The magnesium or zinc dissolves as it is the anode of the electrochemical cell. The iron is thus prevented from rusting by the "sacrifice" of the magnesium or zinc. The blocks can, of course, be readily replaced. Underground pipes and iron piers can be similarly protected.

24.2.10 Compounds of iron

Iron forms two series of compounds:

(*a*) *iron(II)* or ferrous compounds, in which iron has a valency of *two*;
(*b*) *iron (III)* or ferric compounds, in which iron has a valency of *three*.

(a) Oxides of iron
Iron forms three oxides:

(i) *iron(II) oxide*, FeO;
(ii) *iron(III) oxide*, Fe_2O_3;
(iii) *iron(II) diiron(III) oxide*, Fe_3O_4. This oxide may be regarded as a compound of iron(II) oxide and iron(III) oxide.

(i) *Iron(II) oxide*. Iron(II) oxide is a black powder which readily oxidizes in air to form iron(III) oxide. It is of little importance.

(ii) *Iron(III) oxide*. Iron(III) oxide occurs naturally as haematite. In the laboratory it may be prepared by heating either iron(III) hydroxide or iron(II) sulphate to constant mass:

$$2Fe(OH)_3(s) \rightarrow Fe_2O_3(s) + 3H_2O(g)$$
$$2FeSO_4(s) \rightarrow Fe_2O_3(s) + SO_2(g) + SO_3(g)$$

Prepared from iron(II) sulphate, it is a red powder known as *rouge*, and is used in pigments and as a polishing powder.

Iron(III) oxide is a basic oxide but it does not react readily with acids.

(iii) *Iron(II) diiron(III) oxide*. Iron(II) diiron(III) oxide is a strongly magnetic oxide obtained when iron is heated in air or in steam:

$$3Fe(s) + 4H_2O(g) \rightleftharpoons Fe_3O_4(s) + 4H_2(g)$$

(b) Iron(II) sulphate-7-water (green vitriol), $FeSO_4.7H_2O$
Iron(II) sulphate is the most important iron(II) salt. It is usually made by the action of dilute sulphuric acid on iron, and concentrating the solution obtained to the point of crystallization. On cooling, pale green crystals of the heptahydrate are deposited.

The most interesting reaction of iron(II) sulphate crystals takes place on heating. First steam is formed from the water of crystallization and a white solid remains. On further heating, a mixture of sulphur dioxide and sulphur(VI) oxide is evolved (the latter visible as white fumes), leaving *iron(III) oxide* as a red solid:

$$FeSO_4.7H_2O(s) \rightarrow FeSO_4(s) + 7H_2O(g)$$
$$2FeSO_4(s) \rightarrow Fe_2O_3(s) + SO_2(g) + SO_3(g)$$

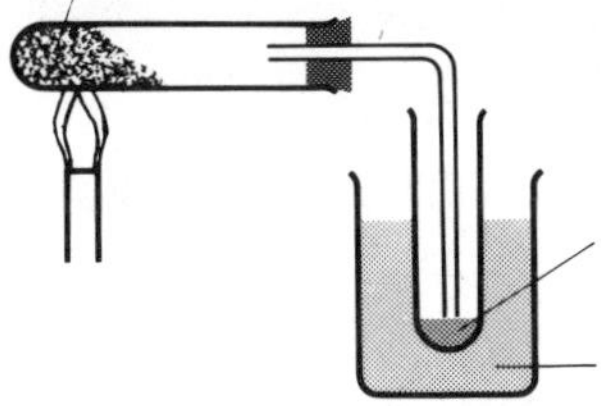

Figure 24.11 The action of heat on iron(II) sulphate crystals.

If the products are passed into a cooled receiver (figure 24.11), sulphur(VI) oxide and water combine to form sulphuric acid:

$$SO_3(g)+H_2O(l) \rightarrow H_2SO_4(aq)$$

The sulphuric acid will be contaminated with sulphurous acid from the combination of sulphur dioxide and water. The chemical reaction has certainly been known since mediaeval times, and was used by the alchemists for preparing sulphuric acid.

Like other iron(II) salts, iron(II) sulphate is readily oxidized to the iron(III) state. Even on exposure to air it becomes coated with a brownish substance—probably a basic iron(III) sulphate. To obtain complete conversion into iron(III) sulphate, oxidizing agents such as nitric acid, potassium manganate(VII) or hydrogen peroxide may be used. For example, if a dilute solution of iron(II) sulphate in dilute sulphuric acid is warmed with a little concentrated nitric acid, an orange solution of iron(III) sulphate is formed:

$$Fe^{2+}(aq)-e \rightarrow Fe^{3+}(aq)$$

$$2HNO_3(aq)+6H^+(aq)+6e \rightarrow 4H_2O(l)+2NO(g)$$

This particular oxidation provides the basis for the brown ring test for nitrates (see section 16.6.6(*c*)).

Iron(II) sulphate is used in manufacturing ink. Most inks contain a mixture of iron(II) sulphate, tannin and a dark blue dye. Fresh writing in ink is blue in colour, the colour of the dye, but iron(II) tannate formed in solution is slowly oxidized to iron(III) tannate. The iron(III) tannate imparts a permanent black colour.

(c) Iron(II) chloride, $FeCl_2$

Pale green crystals of iron(II) chloride, $FeCl_2.4H_2O$, may be prepared by dissolving iron in dilute hydrochloric acid, followed by crystallization in the usual manner.

The anhydrous salt is best prepared by heating iron in a stream of dry hydrogen chloride:

$$Fe(s)+2HCl(g) \rightarrow FeCl_2(s)+H_2(g)$$

Iron(II) chloride is readily oxidized to iron(III) chloride, e.g. by passing chlorine through its solution in water:

$$2FeCl_2(aq)+Cl_2(g) \rightarrow 2FeCl_3(aq)$$

(d) Iron(II) sulphide, FeS

Iron(II) sulphide is usually prepared by the direct combination of iron and sulphur:

$$Fe(s) + S(s) \rightarrow FeS(s)$$

In a purer form, it may be obtained by passing hydrogen sulphide through a solution of an iron(II) salt:

$$FeSO_4(aq) + H_2S(g) \rightarrow FeS(s) + H_2SO_4(aq)$$

Iron(II) sulphide is a black solid used in preparing hydrogen sulphide (see section 19.6.1).

(e) Iron(III) chloride, $FeCl_3$

Anhydrous iron(III) chloride is prepared by heating iron in a stream of dry chlorine:

$$2Fe(s) + 3Cl_2(g) \rightarrow 2FeCl_3(s)$$

Iron(III) chloride readily vaporizes and therefore the apparatus shown in figure 24.12 is suitable. If every precaution is taken to exclude moisture, iron(III) chloride vapour condenses in the flask to form blackish crystals. These may be preserved by removing the flask after the preparation and immediately closing it with a well-fitting rubber bung. This is necessary as iron(III) chloride crystals are very deliquescent.

(N.B. Do not confuse the preparations of iron(II) and iron(III) chlorides. Hydrogen chloride is used for the iron(II) compound, whilst chlorine, being an oxidizing agent, gives the iron(III) compound.)

When iron(III) chloride crystals are dissolved in water they give a deep yellow solution from which yellow crystals of hydrated iron(III) chloride may be obtained by evaporation to the point of crystallization.

Iron(III) chloride can be reduced to iron(II) chloride by the use of suitable reducing agents, e.g.

(i) If zinc is added to a solution of iron(III) chloride in hydrochloric acid, it reduces it to iron(II) chloride:

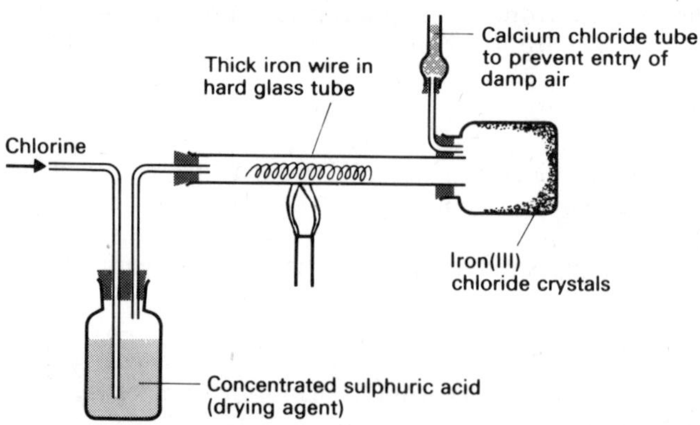

Figure 24.12 Preparation of iron(III) chloride.

$$\begin{bmatrix} Zn + 2FeCl_3 \rightarrow ZnCl_2 + 2FeCl_2 \\ Zn(s) + 2Fe^{3+}(aq) \rightarrow Zn^{2+}(aq) + 2Fe^{2+}(aq) \end{bmatrix}$$

(ii) If hydrogen sulphide is passed through a solution of iron(III) chloride, iron(II) chloride is formed together with a pale yellow deposit of sulphur:

$$\begin{bmatrix} 2FeCl_3 + H_2S \rightarrow 2FeCl_2 + 2HCl + S \\ 2Fe^{3+}(aq) + H_2S(aq) \rightarrow 2Fe^{2+}(aq) + 2H^+(aq) + S(s) \end{bmatrix}$$

(f) Tests for iron(II) and iron(III) salts

(i) Solutions of iron(III) salts are orange or brown in colour. Solutions of iron(II) salts are pale green.

(ii) A solution of an alkali added to a solution of an iron salt precipitates an insoluble hydroxide.

With an iron(II) salt, the precipitate is normally a dirty green in colour.

$$\begin{bmatrix} FeSO_4 + 2NaOH \rightarrow Fe(OH)_2\downarrow + Na_2SO_4 \\ Fe^{2+}(aq) + 2OH^-(aq) \rightarrow \underset{\text{iron(II) hydroxide}}{Fe(OH)_2(s)} \end{bmatrix}$$

At the junction of the liquid with the air, this precipitate becomes brownish due to a partial oxidation to iron(III) hydroxide.

With an iron(III) salt the precipitate is a brick-red colour:

$$\begin{bmatrix} FeCl_3 + 3NaOH \rightarrow Fe(OH)_3\downarrow + 3NaCl \\ Fe^{3+}(aq) + 3OH^-(aq) \rightarrow \underset{\text{iron(III) hydroxide}}{Fe(OH)_3(s)} \end{bmatrix}$$

24.3 Copper

Copper, silver and gold are known as the coinage metals. They are all dense metals of high malleability and ductility, and are excellent conductors of heat and electricity. In addition to having these desirable physical properties, copper, silver and gold are unreactive metals, and therefore of great use when resistance to corrosion is required.

All three metals have been known since prehistoric times. Figure 24.13 is of a Sumerian goat. Like so many metallic objects found on archaeological sites, the goat is made of gold.

Silver and copper are the other metals favoured by ancient craftsmen. This is to be expected as they occur in the uncombined state (gold always occurs thus), or in ores that are readily reduced to the metal. Although they are more resistant to corrosion than many other metals, copper objects discovered on archaeological sites are heavily corroded, but those of silver and especially gold have suffered much less.

24.3.1 Occurrence and extraction of copper

The most important ores of copper are: copper pyrites, $CuFeS_2$; cuprite, Cu_2O; and various basic carbonates, e.g. copper(II) carbonate dihydroxide, (malachite)

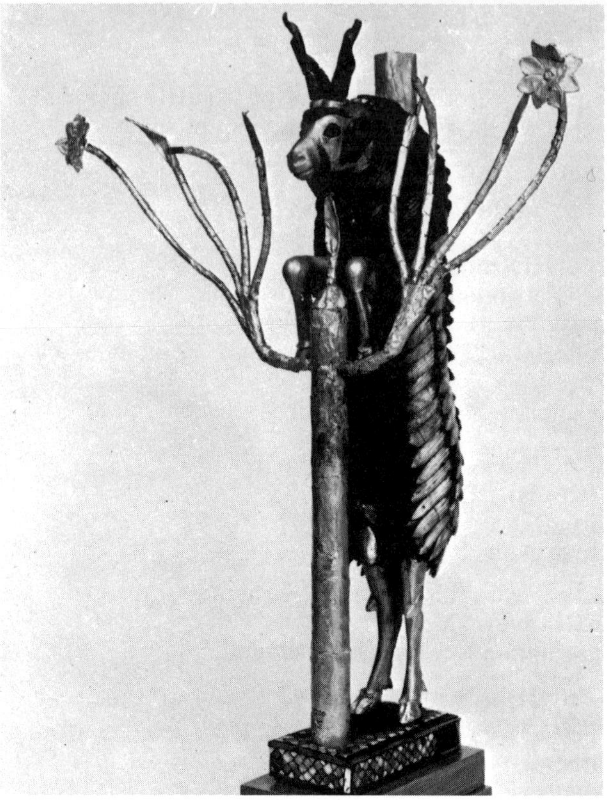

Figure 24.13 Goat and tree from the Royal Cemetery Ur of the Chaldees, c. 2600 B.C. The goat is made of gold. (Photo: by courtesy of the Trustees of the British Museum).

$CuCO_3.Cu(OH)_2$. Limited quantities of native copper are found in many districts.

To extract copper from cuprite or malachite, the ore is strongly heated with coke. Then cuprite is reduced directly to the metal, while the malachite decomposes to give the oxide, which is subsequently reduced:

$$Cu_2O(s) + C(s) \rightarrow 2Cu(l) + CO(g)$$
$$Cu_2CO_3(OH)_2(s) \rightarrow 2CuO(s) + CO_2(g) + H_2O(g)$$
$$CuO(s) + C(s) \rightarrow Cu(l) + CO(g)$$

Most copper is, however, extracted from copper pyrites. This is a more complex process, which consists essentially of two stages:

1. The pyrites is heated in a limited supply of air, when the iron is converted into iron(III) oxide:

$$4CuFeS_2(s) + 9O_2(g) \rightarrow 2Cu_2S(s) + 2Fe_2O_3(s) + 6SO_2(g)$$

Also, part of the copper(I) sulphide is converted to the oxide:

$$2Cu_2S(s) + 3O_2(g) \rightarrow 2Cu_2O(s) + 2SO_2(g)$$

2. On further heating, the remaining copper(I) sulphide reduces the copper(I)

oxide to copper:

$$Cu_2S(s) + 2Cu_2O(s) \rightarrow 6Cu(l) + SO_2(g)$$

24.3.2 Purification of copper

The electrical conductivity of copper is considerably reduced by traces of impurity and, as copper is widely used for electrical purposes, its refining is very important.

Most of the refining is electrolytic. A block of impure copper is made the anode, and a thin sheet of pure copper the cathode, of an electrolytic cell. Both are immersed in a solution of copper(II) sulphate containing sulphuric acid. Therefore the ions present in the solution are: H^+, Cu^{2+}, OH^- and SO_4^{2-}.

On passage of an electric current, H^+ and Cu^{2+} ions move to the cathode, and the Cu^{2+} ions are preferentially discharged:

$$Cu^{2+} + 2e \rightarrow Cu$$

Pure metallic copper is therefore deposited on the cathode (figure 24.14).

OH^- and SO_4^{2-} ions migrate to the anode, but neither type of ion is discharged. Instead the anode gains its electrons by the copper of the anode "dissolving" as copper ions, leaving electrons behind on the anode.

$$Cu \rightarrow Cu^{2+} + 2e$$

This means that for each copper ion discharged at the cathode, one copper ion is formed by the anode, i.e. the cathode of pure copper grows at the expense of the impure copper anode, and the total concentration of copper(II) sulphate in the solution remains unchanged.

Some of the impurities in the copper anode also dissolve. Others, including silver and gold, do not dissolve, but form an "anode slime" at the bottom of the cell.

Figure 24.14 Pure copper cathodes from an electrolytic refining bath. (Photo: International Nickel Co. of Canada Ltd.).

24.3.3 Properties of copper

Copper is a reddish metal of high malleability and ductility. It has a high melting point (1100 °C) and boiling point (2300 °C), and is an exceptionally good conductor of heat and electricity.

Copper is one of the less reactive metals. It does not burn, but when heated in air it becomes coated with black copper(II) oxide, CuO:

$$2Cu(s) + O_2(g) \rightarrow 2CuO(s)$$

Dry air has no effect on copper at ordinary temperatures, but if exposed to damp air for a long time it becomes coated with a green basic carbonate, $Cu_2CO_3(OH)_2$.

24.3.4 Action of acids on copper

Copper is below hydrogen in the activity series, i.e. it forms ions less readily than hydrogen. Thus, if copper is placed in an acid (i.e. in a solution containing hydrogen ions), no hydrogen is liberated.

Copper will react only with those acids which are also oxidizing agents:

(*a*) Hot concentrated sulphuric acid oxidizes copper to copper(II) sulphate, and is itself reduced to sulphur dioxide:

$$Cu(s) + 2H_2SO_4(l) \rightarrow CuSO_4(s) + 2H_2O(l) + SO_2(g)$$

(*b*) Moderately concentrated nitric acid (50 per cent) oxidizes copper to copper(II) nitrate and is itself reduced to nitrogen oxide:

$$\left[\begin{array}{l} 3Cu + 8HNO_3 \rightarrow 3Cu(NO_3)_2 + 4H_2O + 2NO \\ 3Cu(s) + 8H^+(aq) + 2NO_3^-(aq) \rightarrow 3Cu^{2+}(aq) + 4H_2O(l) + 2NO(g) \end{array}\right.$$

This reaction is used in the laboratory preparation of nitrogen oxide (see section 16.4.2(*a*)).

(*c*) Concentrated nitric acid also oxidizes copper to copper(II) nitrate and is itself reduced to nitrogen dioxide:

$$\left[\begin{array}{l} Cu + 4HNO_3 \rightarrow Cu(NO_3)_2 + 2H_2O + 2NO_2 \\ Cu(s) + 4H^+(aq) + 2NO_3^-(aq) \rightarrow Cu^{2+}(aq) + 2H_2O(l) + 2NO_2(g) \end{array}\right.$$

This reaction may be used for preparing nitrogen dioxide (see section 16.5.3(*b*)).

24.3.5 Uses of copper

A large proportion of all the copper produced is used in the electrical industry, e.g. for cables, machinery, etc. It is also used for making cooking utensils, steam-engine boilers and for other purposes where good conduction of heat is essential. Much is used for making copper pipes, hot water tanks, etc.

Copper forms part of many important alloys. These include brass (Cu, Zn); bronze (Cu, Sn); phosphor bronze (Cu, Sn, P); Monel metal (Cu, Ni); and German silver (Cu, Ni, Zn). Modern "silver" coinage is made from a cupronickel alloy containing 75 per cent of copper and 25 per cent nickel. Twelve-sided threepenny bits were made from brass containing 79 per cent copper, 20 per cent zinc and 1 per cent nickel.

24.3.6 Compounds of copper

Copper forms two series of compounds:

(*a*) *copper(II)* (cupric) compounds, in which copper has a valency of *two*.
(*b*) *copper(I)* (cuprous) compounds, in which copper has a valency of *one*.

Copper(II) compounds are by far the most important.

(a) Oxides of copper

(i) Copper(II) oxide, CuO
Copper(II) oxide is formed by heating copper in air. The action is incomplete, and in the laboratory it is better prepared by heating copper(II) carbonate, copper(II) nitrate or copper(II) hydroxide to constant mass:

$$CuCO_3(s) \rightarrow CuO(s) + CO_2(g)$$
$$2Cu(NO_3)_2(s) \rightarrow 2CuO(s) + 4NO_2(g) + O_2(g)$$
$$Cu(OH)_2(s) \rightarrow CuO(s) + H_2O(g)$$

Copper(II) oxide is a black, hygroscopic powder. *It is a basic oxide*, and, although insoluble in water, combines readily with acids to form salts:

$$\left[\begin{array}{l} CuO + H_2SO_4 \rightarrow CuSO_4 + H_2O \\ CuO(s) + 2H^+(aq) \rightarrow Cu^{2+}(aq) + H_2O(l) \end{array}\right.$$

Like other oxides of metals low in the activity series, copper(II) oxide is readily reduced, e.g. by heating in a stream of hydrogen, carbon monoxide or ammonia, or by heating it with carbon:

$$CuO(s) + H_2(g) \rightarrow Cu(s) + H_2O(g)$$
$$CuO(s) + CO(g) \rightarrow Cu(s) + CO_2(g)$$
$$3CuO(s) + 2NH_3(g) \rightarrow 3Cu(s) + N_2(g) + 3H_2O(g)$$
$$CuO(s) + C(s) \rightarrow Cu(s) + CO(g)$$

(ii) Copper(I) oxide, Cu_2O
Copper(I) oxide is usually prepared by reducing a copper(II) compound. *Fehling's solution* is made first by mixing a solution of copper(II) sulphate with a solution containing sodium hydroxide and potassium sodium 2,3-dihydroxy-butanedioate (tartrate). (Tartrate ions stop copper(II) hydroxide precipitating.) This solution is boiled with glucose which is a reducing agent. A red precipitate of copper(I) oxide is formed.

This reaction is used as a test for aldehydes and reducing sugars.

(b) Copper(II) hydroxide, $Cu(OH)_2$
Copper(II) hydroxide is insoluble in water, and is therefore obtained when a solution of an alkali is added to a solution of a copper(II) salt, e.g. by adding sodium hydroxide solution to a copper(II) sulphate solution:

$$\left[\begin{array}{l} CuSO_4 + 2NaOH \rightarrow Cu(OH)_2\downarrow + Na_2SO_4 \\ Cu^{2+}(aq) + 2OH^-(aq) \rightarrow \underset{\text{pale blue precipitate}}{Cu(OH)_2(s)} \end{array}\right.$$

(Ammonium hydroxide is an unsuitable alkali to use, as copper(II) hydroxide combines with it to form a soluble compound (see section *g*).)

Copper(II) hydroxide is unstable and decomposes to give a hydrated oxide even when warmed with water, e.g. if the mixture obtained by mixing the solutions mentioned above is warmed, the blue precipitate rapidly blackens as it changes to the oxide:

$$Cu(OH)_2(s) \rightarrow CuO(s) + H_2O(l)$$

Copper(II) hydroxide is basic and readily combines with acids to form salts, e.g.:

$$\left[\begin{array}{l} Cu(OH)_2 + 2HNO_3 \rightarrow Cu(NO_3)_2 + 2H_2O \\ Cu(OH)_2(s) + 2H^+(aq) \rightarrow Cu^{2+}(aq) + 2H_2O(l) \end{array}\right.$$

(c) Copper(II) sulphate-5-water, blue vitriol, $CuSO_4.5H_2O$

Crystals of this compound are prepared in the laboratory by the action of sulphuric acid on copper(II) oxide, copper(II) carbonate or copper(II) hydroxide. If copper is the starting material it should be dissolved as copper(II) nitrate and hence converted to one of these compounds:

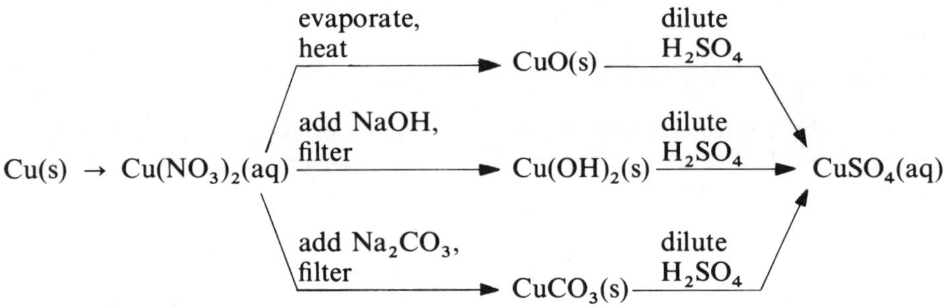

Industrially, much copper(II) sulphate is made by blowing air through a mixture of copper scrap and hot dilute sulphuric acid:

$$2Cu(s) + 2H_2SO_4(aq) + O_2(g) \rightarrow 2CuSO_4(aq) + 2H_2O(l)$$

Copper(II) sulphate crystals have a fine blue colour. When gently heated, they lose their water of crystallization to give a white, anhydrous powder:

$$\underset{\text{Blue}}{CuSO_4.5H_2O(s)} \rightarrow \underset{\text{white}}{CuSO_4(s)} + 5H_2O(g)$$

If a little water is added to the anhydrous salt, the pentahydrate is reformed:

$$\underset{\text{white}}{CuSO_4(s)} + 5H_2O(l) \rightarrow \underset{\text{blue}}{CuSO_4.5H_2O(s)}$$

This reaction is often used to detect the presence of water. Solutions of copper (II) sulphate are also blue in colour. This is a colour characteristic of copper(II) ions in solution.

On very strong heating, anhydrous copper(II) sulphate decomposes to give white fumes of sulphur(VI) oxide and a residue of black copper(II) oxide:

$CuSO_4(s) \rightarrow CuO(s) + SO_3(g)$

Copper(II) sulphate is very poisonous to lower organisms. It is sometimes added in very small quantities to drinking-water to kill algae. A mixture of calcium hydroxide and copper(II) sulphate is used for spraying fruit trees and potatoes to protect them from various pests.

(d) Chlorides of copper

(i) Copper(II) chloride, $CuCl_2$
This unimportant salt is prepared by dissolving copper(II) hydroxide, oxide or carbonate in dilute hydrochloric acid:

e.g. $CuCO_3(g) + 2HCl(aq) \rightarrow CuCl_2(aq) + H_2O(l) + CO_2(g)$

(ii) Copper(I) chloride
Like copper(I) oxide, copper(I) chloride is prepared by reduction.

Experiment. Preparation of copper(I) chloride

Make a solution of copper(II) chloride by mixing 5 cm^3 concentrated hydrochloric acid, 15 cm^3 water and 5 g copper(II) chloride crystals. Boil this solution with 5 g copper for 5–10 minutes. Pour the resulting mixture into a large volume of cold, boiled water. A white precipitate of copper(I) chloride, previously held in solution with the concentrated acid, is then obtained:

$CuCl_2(aq) + Cu(s) \rightarrow 2CuCl(s)$

Alternatively, vary the staring materials and heat together concentrated hydrochloric acid, copper(II) oxide and copper.

In air copper(I) chloride oxidizes to a copper(II) compound.

(e) Copper(II) nitrate, $Cu(NO_3)_2.3H_2O$

This salt, which is very deliquescent, may be prepared by the normal methods for soluble salts. It is of little importance.

(f) Copper(II) carbonate

Pure copper(II) carbonate is not known. If solutions of copper(II) salts and carbonates are mixed, basic carbonates are formed, e.g.

$$\left[\begin{array}{l} 2Na_2CO_3 + H_2O + 2CuSO_4 \rightarrow 2Na_2SO_4 + CuCO_3.Cu(OH)_2 + CO_2 \\ 2CO_3^{2-}(aq) + H_2O(l) + 2Cu^{2+}(aq) \rightarrow Cu_2CO_3(OH)_2(s) + CO_2(g) \end{array}\right.$$

(g) Tests for copper compounds

(i) The flame test, when applied to copper compounds, gives a blue-green coloration. This is illustrated on the back cover of this book.

(ii) Solutions of copper(II) salts are blue in colour. This colour is imparted by copper(II) ions. If ammonium hydroxide solution is gradually added to a solution containing copper(II) ions, first a pale blue precipitate of copper(II) hydroxide is obtained which then dissolves to give an intense blue solution. This is caused by copper(II) ions from the precipitate combining with ammonia from the ammonium hydroxide to form tetraammine copper(II) ions:

$$Cu^{2+} + 4NH_3 \rightarrow Cu(NH_3)_4{}^{2+}$$

Questions on Chapter 24

1. (*a*) Draw a labelled diagram for the preparation and collection of a sample of anhydrous iron(III) chloride given a supply of dry chlorine. Give an equation for the reaction. (No details of the preparation of chlorine are required.)
(*b*) Describe what you would see if a sample of anhydrous iron(III) chloride were heated in a dry test-tube and give the name for the phenomenon described.
(*c*) Name the reagents and give an equation for the preparation of anhydrous iron(II) chloride using the same apparatus as that shown in the answer to part (*a*).
(*d*) Explain how each of the following conversions can be made. In each case, name the reagent or reagents needed, give the conditions for the reaction and state what you would see.
(i) From a solution containing Fe^{3+} ions to one containing Fe^{2+} ions.
(ii) From a solution containing Fe^{2+} ions to one containing Fe^{3+} ions. (*J.M.B.*)

2. Name and give the formulae of *two* common naturally occurring compounds of each of the metals magnesium, zinc and iron.
Give *one* important use of each of these metals, and state the particular property of the metal illustrated by each use.
Give a *brief* account of the reactions occurring in a blast furnace when haematite, Fe_2O_3, is being smelted. (*O.*)

3. (*a*) Design and describe a quantitative experiment to find out whether a new alloy is oxidized by the atmosphere.
What results would you expect to observe if the alloy were (i) easily oxidized, (ii) rust resistant?
(*b*) If the alloy were attacked, what further experiments would you set up to discover which parts of the air cause the reaction?
(*c*) How is air prevented from attacking iron on
(i) the blades of a lawn mower,
(ii) a dustbin,
(iii) cutlery? (*J.M.B.*)

4. Describe carefully the chemistry of the process by which pig-iron is obtained from iron(III) oxide (ferric oxide) ore. (A diagram of the apparatus and technical details are not required.)
Briefly describe *one* way in which pig-iron may be converted into steel.
Give *two* factors which might be important when considering a site for a new iron foundry. (*J.M.B.*)

5. How would you prepare a pure sample of iron(II) sulphate (ferrous sulphate) crystals in the laboratory, starting from iron filings?
A solution contains a mixture of iron(II) and iron(III) chlorides. State what you would see, and explain the reactions that occur when:
(*a*) chlorine is passed into a portion of the solution until there is no further change;
(*b*) another portion of the solution is acidified with hydrochloric acid, then zinc is added and the mixture allowed to stand;
(*c*) excess sodium hydroxide solution is added to the products from (*a*) and (*b*). (*A.E.B.*)

6. Give an account of experiments you would carry out to demonstrate the conditions under which iron rusts.
Explain why zinc protects iron from corrosion and suggest why copper-plating would not have the same effect.

Why does aluminium corrode much less rapidly than iron when exposed to the same atmospheric conditions?

Give *three* important uses for aluminium, each illustrating a different physical property of the metal. (*A.E.B.*)

7. Describe *two* laboratory methods (excluding electrolysis), with an outline of experimental details, which you could use to prepare a sample of copper from copper(II) sulphate as the starting material.

Give a brief account of the refining of copper by electrolysis.

State *two* uses of copper and say what properties of the metal account for these uses. (*D.*)

8. Draw a fully-labelled diagram to illustrate the electrolytic method for the production of pure copper from a block of impure copper. Give ionic equations for the reactions at the electrodes. Explain what happens to any traces of (*a*) iron, (*b*) silver present in the impure copper.

How and under what conditions, does copper react with nitric acid? (*J.M.B.*)

9. How would you prepare from copper (*a*) copper(II) sulphate crystals ($CuSO_4.5H_2O$), (*b*) copper(II) oxide, (*c*) copper(II) sulphide?

Starting with 5 g of copper, what would be the mass of copper(II) sulphate crystals formed in the reaction (*a*)?

Describe the reactions that occur when a solution of sodium hydroxide is added to a solution of copper(II) sulphate (i) when both solutions are cold, (ii) when both solutions are hot. (*B.*)

10. A blue crystalline substance was gently heated in a hard-glass tube. Moisture was seen to condense on the side of the tube, and the substance lost its crystalline shape and blue colour. When heated more strongly, the residue darkened in colour, and dense white fumes were evolved.

A solution of the original substance was made in distilled water, and successive portions were treated as follows: (*a*) caustic soda solution produced a light blue gelatinous precipitate that turned black when the mixture was heated; (*b*) sodium carbonate produced a greenish blue precipitate; (*c*) ammonia solution (0·88) first produced a light blue precipitate that dissolved in an excess of the reagent giving a deep blue coloured solution; (*d*) barium chloride solution produced a white precipitate that proved to be insoluble in an excess of hydrochloric acid.

Give the chemical name and formula of the original blue crystalline substance, and account for all the changes described, giving equations where possible. (*L.*)

25

Organic Chemistry I—Small Molecules

25.1 Introduction

As the name suggests, organic chemistry was originally concerned with chemicals obtained from living organisms. All such chemicals are compounds of carbon. It was once thought that *organic compounds*, i.e. those concerned with living matter, and *inorganic* compounds, i.e. those concerned with non-living matter, were quite distinct. But in 1828 the German chemist, Friedrich Wöhler, prepared the organic compound urea (NH_2CONH_2) from the inorganic compound ammonium cyanate (NH_4OCN). Since then it has been recognized that there is no real difference between organic and inorganic compounds. Therefore *all compounds of carbon*, whether they occur in living matter or not, could be included in *organic chemistry*. However some simple compounds of carbon such as the oxides and carbonates are discussed as inorganic compounds (see chapter 18).

Atoms of carbon show a greater ability to combine with one another, either by forming long chains of atoms or rings of atoms, than do atoms of any other element. Because of this there are many millions of different organic compounds. This ability of carbon to form big molecules has enabled nature to form living matter, and it is most likely that carbon is the basis of all living matter wherever it occurs in the Universe.

25.2 The structures of organic compounds

Carbon has a valency of four. Its simplest compound with hydrogen is methane (CH_4). In a methane molecule each carbon atom is joined to four hydrogen atoms. These hydrogen atoms are arranged at the four corners of a regular tetrahedron with the carbon atom in the middle. Probably the best way to visualise this is with the use of atomic models of the sort found in schools (figure 25.1).

There are various ways in which this three-dimensional structure can be written on a sheet of paper (which is only two-dimensional). A perspective drawing of the molecule with the tetrahedron outlined is the most accurate way (figure 25.2(a)). Another way to show the tetrahedron is to draw as lines the two C–H bonds that lie in the plane of the paper, and to draw as arrow-heads the two C–H bonds that lie above and below the plane of the paper (figure 25.2(b)). Often the shape of the molecule is not important and figure 25.2(c) is a good way of writing the formula–it in no way however shows the three-dimensional nature

Figure 25.1 Model of the methane molecule. (Photo: A. Rendell.)

of the molecule. The formulae in figure 25.2 are *structural formulae*; they show how atoms are arranged within the molecule.

Another simple compound of carbon and hydrogen is ethane, C_2H_6. Figure 25.3 is a photograph of a model of the ethane molecule, and figure 25.4 shows ways of representing its structural formula.

In methane and ethane each carbon atom is joined to four other atoms by four *single bonds*. Compounds such as these are discussed in section 25.3.

A second group of organic compounds is quite different from methane and ethane. They have in them carbon atoms that are joined to three other atoms; ethene (figure 25.5) is a molecule of this type. Here, as each carbon atom must form four bonds, the two carbon atoms are joined together by a *double bond* (figure 25.13(a)). Such compounds are discussed in section 25.4.

There is a third group of compounds that contains carbon atoms that are joined to only two other atoms; ethyne (figure 25.6) is a molecule of this type.

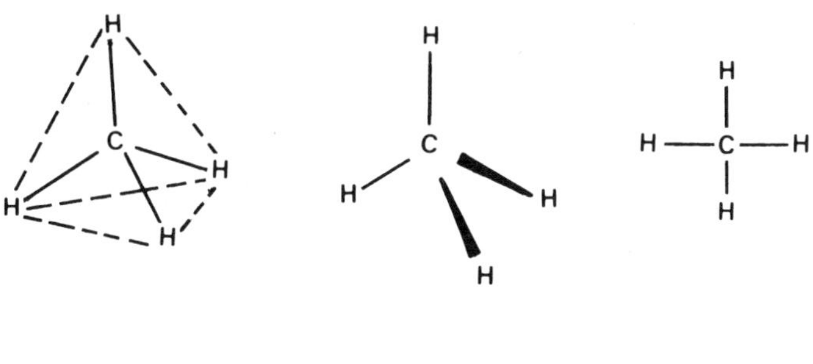

(a) (b) (c)

Figure 25.2 Methane: (a) Three dimensional structural formula; (b) Representation of three dimensional structural formula; (c) Two dimensional representation.

Figure 25.3 Model of the ethane molecule. (Photo: A. Rendell.)

Here, as each carbon atom must form four bonds, the two carbon atoms are joined together by a *triple bond* (H – C ≡ C – H). Such compounds are discussed in section 25.5.

By far the best way to understand the arrangement of atoms in molecules is to build molecules using atomic model kits, or polystyrene balls and match sticks. Most schools have suitable model building materials.

More detailed discussion of the shapes of molecules and the arrangement of electrons in them is given in section 6.2.5.

25.3 Alkanes

Organic compounds that contain only carbon and hydrogen are known as *hydrocarbons*. The hydrocarbons in which all the carbon atoms are joined together in chains by single bonds are said to be *saturated*. They are called *alkanes*

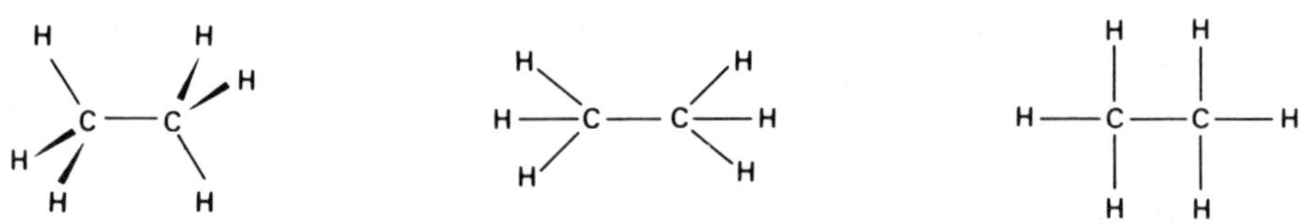

(a) (b) (c)

Figure 25.4 Ethane: (a) Representation of three dimensional structural formula; (b) Two dimensional representation; (c) Two dimensional representation.

Figure 25.5 Model of the ethene molecule. (Photo: A. Rendell.)

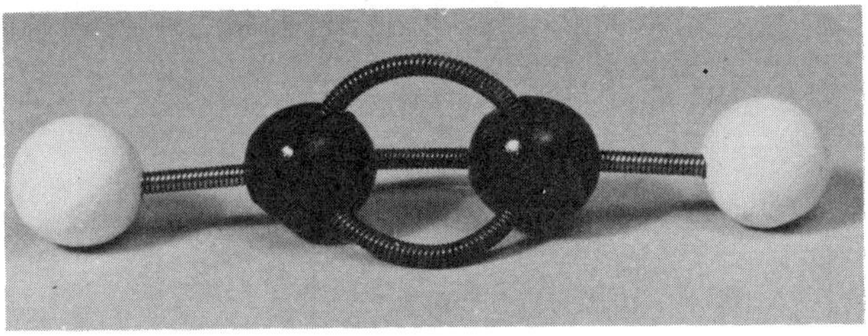

Figure 25.6 Model of the ethyne molecule. (Photo: A. Rendell.)

(previously called the paraffins). The first six alkanes are given in figure 25.7. It should be noted that they all correspond to the general formula C_nH_{2n+2}. For methane n is 1, for ethane n is 2 and so on. A group of compounds such as this to which a general molecular formula can be applied is called an *homologous series*. In an homologous series the physical constants such as boiling points, freezing points and specific gravity change gradually as the molecules get larger, that is as the molecular mass increases.

You will see that the names of *alkanes* all end in *-ane*; i.e., methane, ethane, propane, butane etc. After these first four members of the series the names of the alkanes begin with the Greek prefix for the number of carbon atoms in the chain i.e., pentane (5), hexane (6), heptane (7), octane (8), etc.

The basic name for any alkane or alkane derivative is that obtained by selecting the longest chain of carbon atoms. If there are 3 atoms in the longest carbon chain the molecule will be a propane derivative, if 4 carbon atoms it will be a butane derivative, and so on.

25.3.1 Isomerism

For the alkanes from butane onwards, it is possible to have *different compounds with the same molecular formula but with different structural formulae*. This phenomenon is known as *isomerism*. Consider butane, C_4H_{10}. This has two isomers (figure 25.8(*a*) and (*b*)). The first of these compounds (figure 25.8(*a*)) has a chain of 4 carbon atoms and is called *butane*. In the second compound (figure 25.8(*b*)) the longest carbon chain has only 3 atoms, so it is a derivative of *propane*. Attached to the central atom of this chain of 3 atoms is a *methyl* (CH_3—)

Name	Number of C atoms in chain	Molecular formula	Structural formula	Boiling point (°C)	Freezing point (°C)	Specific gravity (of liquid)
Methane	1	CH_4	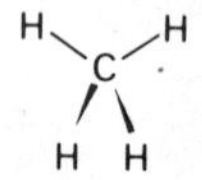	−162	−183	0.42
Ethane	2	C_2H_6	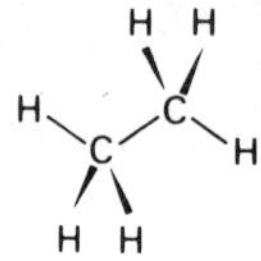	−89	−173	0.55
Propane	3	C_3H_8	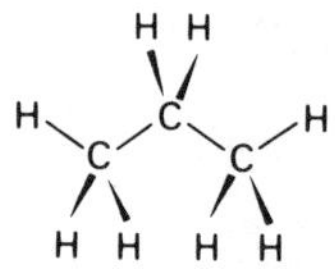	−42	−187	0.58
Butane	4	C_4H_{10}	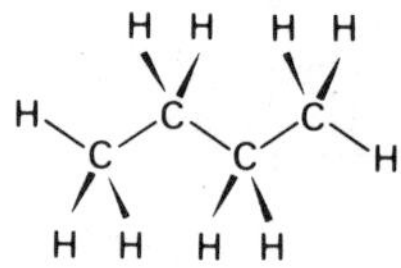	−1	−135	0.58
Pentane	5	C_5H_{12}	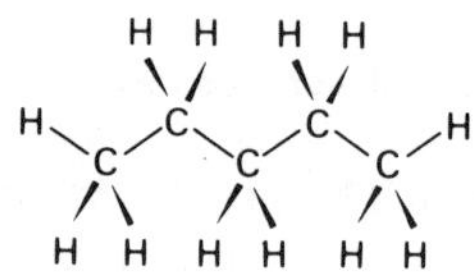	36	−130	0.63
Hexane	6	C_6H_{14}	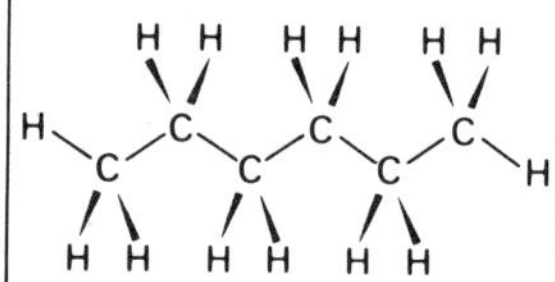	69	−94	0.66

Figure 25.7 The alkanes.

group. If we number the 3 carbon atoms of the chain as 1, 2 and 3, the methyl group is on carbon atom number 2; hence the compound in figure 25.8(*b*) is 2-methylpropane.

For pentane there are three isomers, illustrated in figure 25.9. These are pentane (figure 25.9(*a*)), 2-methylbutane (figure 25.9(*b*)) and 2,2-dimethylpropane (figure 25.9(*c*)). The number of isomers increases rapidly. If there is a total of 6 carbon atoms, i.e., hexane, there are 5 isomers, if there are 10 carbon atoms the number of isomers has grown to 75, and for the molecules with 30 carbon atoms the number of isomers is 4 111 846 793.

25.3.2 The source of alkanes

The only important source of alkanes is crude oil. Oil also gives many other hydrocarbons. Since the internal combustion engine was invented, at the end of

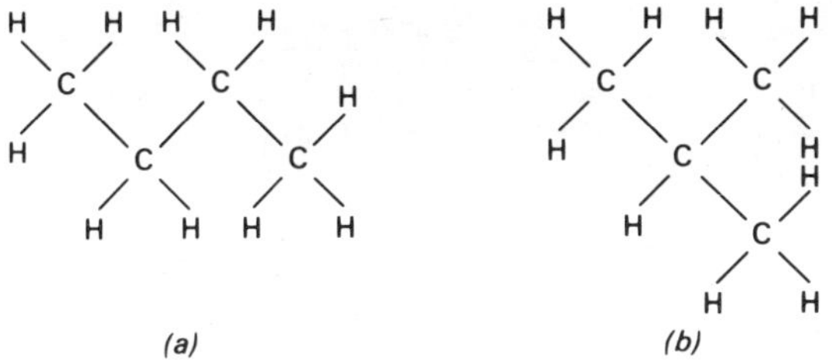

Figure 25.8 Isomerism (C_4H_{10}) (*a*) butane, (*b*) 2-methylpropane.

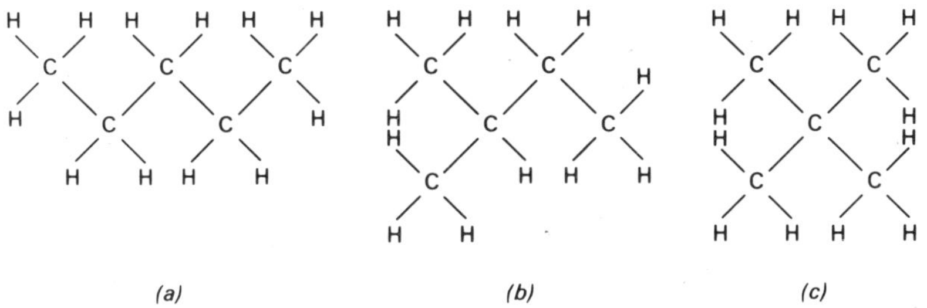

Figure 25.9 Isomerism (C_5H_{12}) (*a*) pentane, (*b*) 2-methylbutane, (*c*) 2,2-dimethylpropane.

the last century, oil has had an increasing importance as a source of petrol. During the last 30 years there have been enormous developments in the use of oil and it is now the primary source of an extremely extensive range of organic chemicals.

Oil was formed by the decomposition of animals and plants that lived in the sea many millions of years ago. These animals and plants had by the process of photosynthesis made sugars and other organic compounds that they required to live (see section 18.6.2). During the course of time high pressures and bacterial action converted these compounds into oil. This oil collected together with the low boiling organic gases and sea water and much of it escaped from the earth's surface, but a little was trapped in amongst impervious rocks (figure 25.10) and these deposits are the oil and gas deposits on which we now depend so much (see figures 28.5 and 28.6).

Methane, the simplest of the alkanes is a gas (figure 25.7) and almost always occurs with liquid oil deposits. To a lesser extent the other volatile alkanes (ethane, propane and butane) are mixed with it. The mixture is known as natural gas. We are now extracting this natural gas from the oil deposits under the North Sea, and it is the gas being used in most homes in Britain. The liquid oil is fractionally distilled using fractionating towers (see section 1.8 and figure 1.13). The oil is distilled into several fractions, each one boiling over a different range (figure 25.11).

In an oil refinery it is possible to convert the larger molecules into smaller ones by breaking carbon–carbon and carbon–hydrogen bonds by the process of *cracking*. Cracking may be achieved just by heating the higher-boiling mixtures under higher pressure or by using a catalyst as well, e.g.,

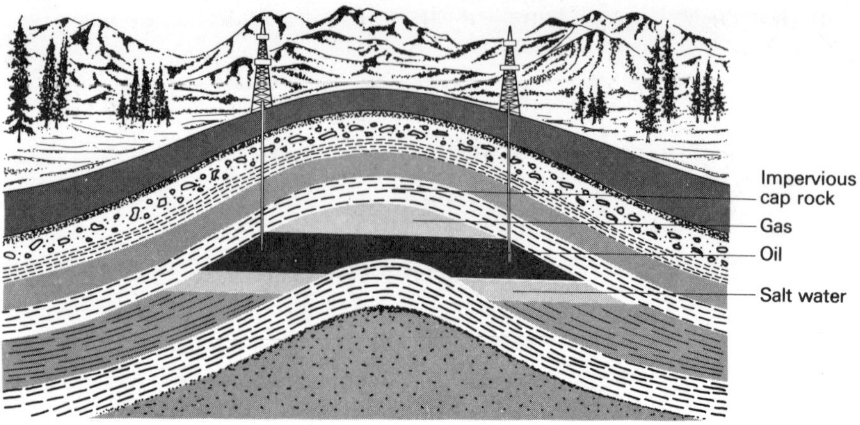

Figure 25.10 Oil, gas and water accumulated in the dome of impervious cap rock.

	Boiling range	Hydrocarbons present	Uses
Gas	→ 25°C	CH_4 – C_4H_{10}	Gaseous fuels; making carbon black
Naphtha	30–100°C	C_5H_{12} – C_7H_{16}	Raw material in chemical industry
Petrol	25–200°C	C_4H_{10} – $C_{12}H_{26}$	Petrols; solvents
Kerosene	100–200°C	C_8H_{18} – $C_{12}H_{26}$	Fuel for jet aircraft; oil fired central heating; cracking
Gas oil/Diesel oil	200–400°C	$C_{15}H_{32}$ – $C_{25}H_{52}$	Diesel fuel; oil fired heating; cracking
Lubricating oil	400–450°C	$C_{20}H_{42}$ –	Lubricating oils and greases; petroleum jelly; waxes
Bitumen		High-molecular-weight hydrocarbons	Asphalt

Figure 25.11 Fractions obtained from crude oil.

$$C_{10}H_{22} \rightarrow C_7H_{16} + C_3H_6$$
$$C_{15}H_{32} \rightarrow C_9H_{20} + C_3H_6 + C_2H_4 + C + H_2.$$

Some of the products of these cracking reactions are not alkanes, but *alkenes* and these are discussed later (section 25.4).

Experiment. Cracking of oil.

Set up the apparatus shown in figure 25.12, and heat the crude oil (medicinal paraffin is excellent) and the porous pot. A gas will be collected in the gas-jar. This gas is a mixture of alkenes, largely ethene (see section 25.4), and will burn with a yellow flame and decolorize bromine water as expected.

There are many books that give more detailed accounts about the processing of crude oil and about the many chemicals that are now obtained from it. Also

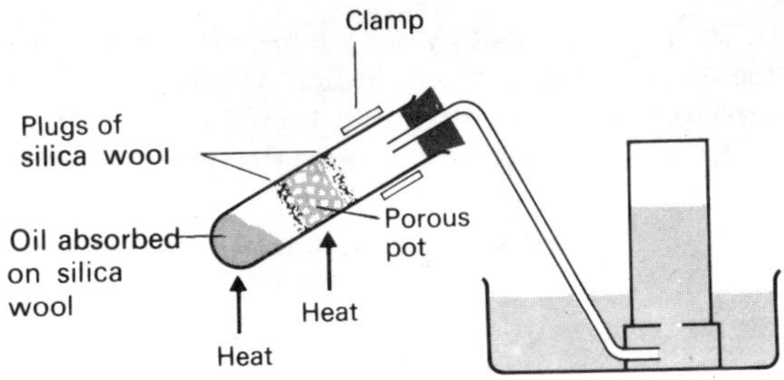

Figure 25.12 Cracking of oil.

the major oil companies give free books and wall-charts to schools. It is likely that you have some in your class-room or laboratory.

25.3.3 The chemistry of alkanes

The alkanes, as their old name paraffins indicated, are not very reactive compounds. Only two reactions are worthy of mention here.

(*a*) They all burn to give carbon dioxide and water: e.g.,

$$CH_4(g) + 2O_2(g) \rightarrow CO_2(g) + 2H_2O(g).$$

A considerable amount of heat is evolved during this reaction ($\Delta H = -802$ kJ mol^{-1}), and this is why the alkanes are widely used as fuels.

(*b*) If an alkane is heated with chlorine, and particularly if ultraviolet light from the sun or an ultraviolet lamp is shone onto a mixture of an alkane and chlorine, then a reaction occurs and the chlorine replaces the hydrogen atoms in the alkane one at a time. For methane, for example, a mixture of four products is formed:

$$CH_4 + Cl_2 \rightarrow CH_3Cl + HCl,$$
chloromethane

$$CH_3Cl + Cl_2 \rightarrow CH_2Cl_2 + HCl,$$
dichloromethane

$$CH_2Cl_2 + Cl_2 \rightarrow CHCl_3 + HCl,$$
trichloromethane (chloroform)

$$CHCl_3 + Cl_2 \rightarrow CCl_4 + HCl.$$
tetrachloromethane (carbon tetrachloride)

These reactions are examples of *substitutions*, i.e., the replacement of hydrogen atoms by other atoms or groups of atoms.

Experiment. The substitution of bromine in hexane

A 2 cm^3 sample of hexane is placed in a boiling-tube and about 0·5 cm^3 of bromine dissolved in tetrachloromethane is added and the solution shaken.

After some time the red bromine colour disappears and hydrogen bromide gas is evolved. This can be detected by holding a moist indicator paper in the vapour in the tube. If an ultra-violet lamp is available then exposing the mixture to ultra-violet light will cause the bromine colour to disappear much more rapidly. Otherwise do the reaction in duplicate and put one tube in sunlight and the other tube in a dark cupboard, and notice how much faster the bromine colour disappears in the tube in the light. When the reaction has taken place substitution will have occurred to give a mixture of bromohexanes.

25.3.4 Alkane derivatives

If a hydrogen atom is removed from an alkane the group remaining is named after the alkane by removing the *-ane* and substituting *-yl*, an alkane becomes an alkyl group. E.g., CH_4 gives CH_3—, the methyl group and C_2H_6 gives C_2H_5—, the ethyl group. In general C_nH_{2n+2} becomes C_nH_{2n+1}—, the alkyl group. Important alkyl derivatives are (*a*) the alkyl halides, $C_nH_{2n+1}X$, where X is a halogen atom, e.g., CH_3I, iodomethane (methyl iodide), (*b*) alcohols, $C_nH_{2n+1}OH$, e.g., C_2H_5OH, ethanol, and (*c*) amines, $C_nH_{2n+1}NH_2$, e.g., $C_3H_7NH_2$, propylamine.

25.4 Alkenes

Alkenes are hydrocarbons that contain carbon–carbon double bonds. They are capable of combining with more hydrogen and because of this they are said to be *unsaturated*. The simplest member of this homologous series, the alkenes, is ethene (ethylene) (figures 25.5 and 25.13(*a*)). The next two members are propene (propylene) and butene (butylene) (figures 25.13(*b*), (*c*) and (*d*) (note that for butene the position of the double bond has been indicated by a – 1 – or – 2 – in the name)). It is evident that alkenes are named in a similar way to alkanes, the *-ene* ending replacing the *-ane* of the saturated hydrocarbons. The general formula of the alkenes is C_nH_{2n}, and the smallest possible value of n is 2 since there must be a C $=$ C group in the molecule.

25.4.1 The sources of alkenes

Both ethene and propene are manufactured by the thermal cracking of alkanes discussed in section 25.3.2.

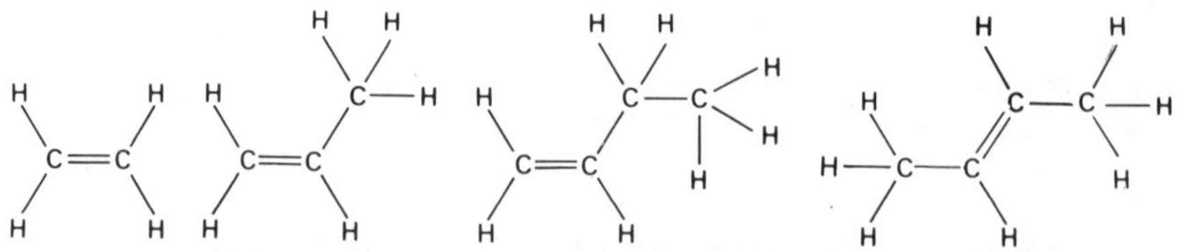

(*a*) (*b*) (*c*) (*d*)

Figure 25.13 (*a*) ethene, (*b*) propene, (*c*) but-1-ene, (*d*) but-2-ene.

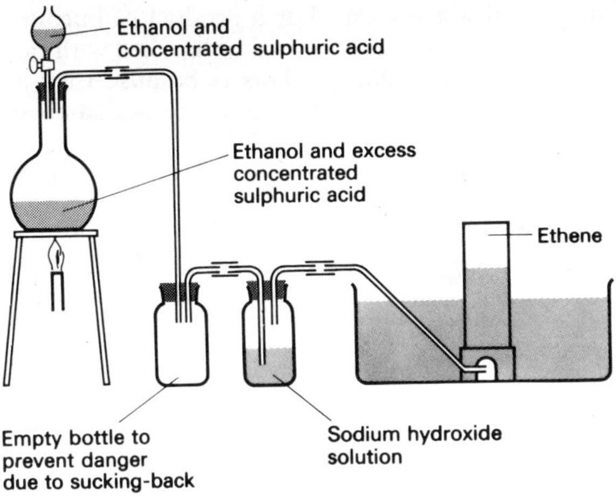

Figure 25.14 Preparation of ethene.

They can also be prepared by the thermal cracking of alcohols. Ethene is prepared from ethanol:

$$C_2H_5OH \rightarrow C_2H_4 + H_2O$$

Experiment. The preparation of ethene

Use the apparatus of figure 25.12 with ethanol instead of the oil and aluminium oxide instead of the porous pot. The gas collected in the gas jar is ethene.

A continuous stream of ethene can be produced by passing ethanol vapour over heated aluminium oxide in a glass tube.

An alternative laboratory preparation is the dehydration of alcohols by concentrated sulphuric acid. Ethanol is heated to 180 °C with excess acid. Sulphur dioxide gas is removed from the ethene by passing the gas stream through sodium hydroxide solution and the ethene is collected over water (figure 25.14).

25.4.2 Properties of alkenes

(a) Physical properties

The simplest alkenes are insoluble gases, and like the alkanes burn to form carbon dioxide and water, and liberate large quantities of heat. Ethene for example reacts:

$$C_2H_4(g) + 3O_2(g) \rightarrow 2CO_2(g) + 2H_2O(g) \qquad (\Delta H = -1322 \text{ kJ mol}^{-1}).$$

When burnt in the laboratory the burning ethene cannot get sufficient oxygen from the air quickly enough for this reaction to take place completely, and it burns with a smoky flame (that is, carbon is formed as a product of burning). N.B., ethene-oxygen mixtures can be explosive and should be handled with care.

The alkenes are more reactive than the alkanes. This is because they are *unsaturated* and reactions take place in which other atoms are added to the carbon atoms that are joined together by the double bond. In *addition reactions* the product will be a saturated compound as the double bond has been destroyed.

(b) Hydrogenation

If a mixture of ethene and hydrogen is passed over a heated catalyst of finely divided nickel then an addition occurs:

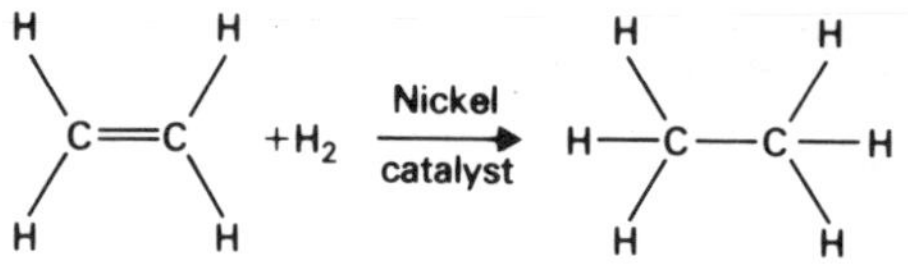

Other alkenes can be hydrogenated in the same way. This reaction is particularly important in the hydrogenation of animal and vegetable oils. Some of these oils, which are esters of propane-1,2,3-triol are unsaturated and may be unpalatable and also unsuitable for soap-making. If they are hydrogenated using a nickel catalyst they are converted to solid fats used for margarine, lard substitute and for making soap. Figure 25.15 shows hydrogenation vessels in which 15 tonne quantities of edible oils are hardened by hydrogenation for use in the manufacture of margarine and cooking fats.

Figure 25.15 Hydrogenation vessels for edible oils at Bromborough on Merseyside. (Photo: Van de Berghs and Jurgens Ltd.).

(c) Halogenation

If some bromine vapour is poured into a jar full of ethene, the brown colour of the bromine soon disappears. In this addition colourless 1,2-dibromoethane is formed:

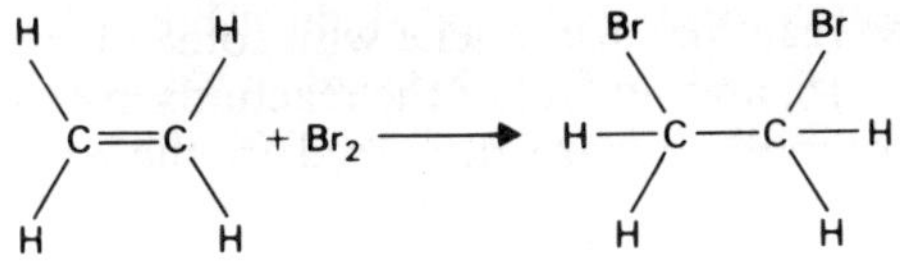

This reaction is often used as a test for unsaturated hydrocarbons.

1,2-dichloroethane is formed by the same reaction if chlorine is used instead of bromine.

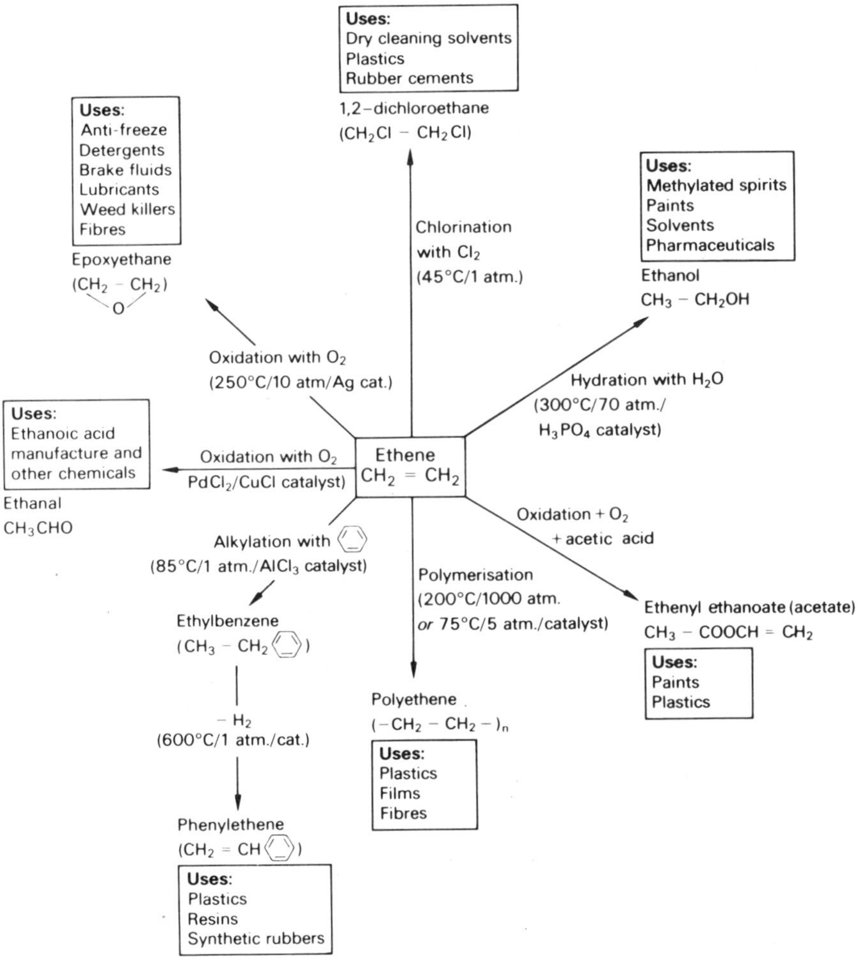

Figure 25.16 Reactions of ethene and uses of products.

(d) Addition of other reagents

Ethene and propene are two of the most important materials in the chemical industry, because to them can be added a great variety of reagents to give many very useful chemicals. A number of these reactions for ethene with some of the uses of the products are given in figure 25.16, and several of the reactions in this figure are of immense industrial importance and are discussed in the next chapter.

25.5 Alkynes

Alkynes are hydrocarbons that contain carbon–carbon triple bonds. The simplest member of this homologous series, and the only one that we need to consider, is *ethyne* (acetylene), $H-C \equiv C-H$ (figure 25.6).

Some ethyne is manufactured by cracking methane. In the laboratory it can be conveniently made by adding water to calcium dicarbide using the apparatus shown in figure 25.17, when the reaction:

$$CaC_2 + 2H_2O \rightarrow Ca(OH)_2 + C_2H_2,$$

takes place.

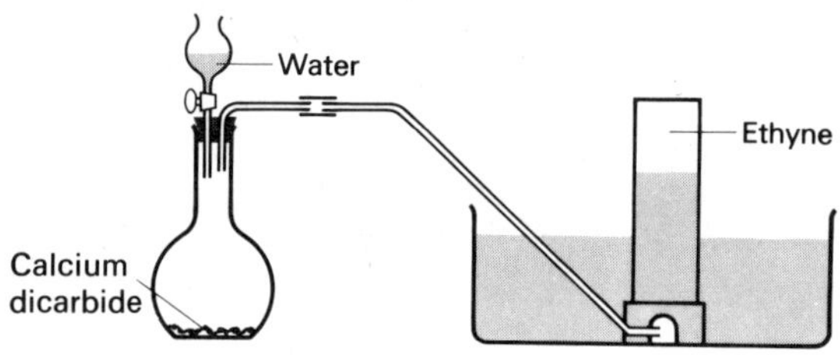

Figure 25.17 Preparation of ethyne.

Ethyne will burn in air with the evolution of a considerable amount of heat. If there is sufficient oxygen the reaction occurs:

$$C_2H_2(g) + 2\tfrac{1}{2}O_2(g) \rightarrow 2CO_2(g) + H_2O(g) \qquad (\Delta H = -1257 \text{ kJ mol}^{-1}).$$

The very hot flame from a mixture of ethyne and oxygen is used for welding and cutting metals. Like ethene, when ethyne is burnt in the laboratory the above reaction is incomplete and the flame is smoky. Mixtures of ethyne and air or oxygen can be dangerously explosive; great care must be taken when using ethyne in the laboratory.

Like ethene, *ethyne* undergoes many addition reactions. Thus if ethyne and hydrogen are passed over a heated nickel catalyst, ethane is formed.

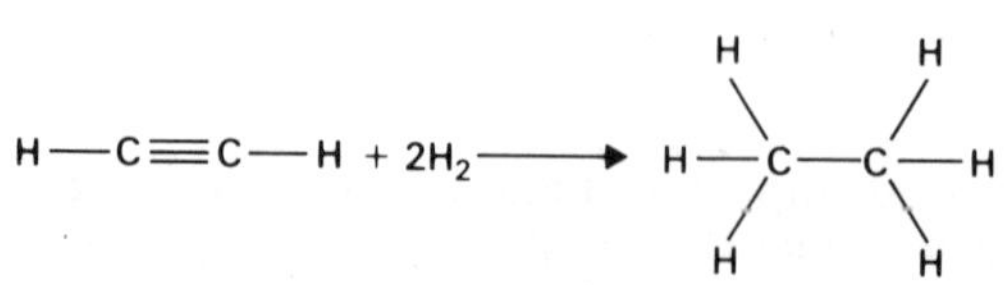

Particularly important are the addition of hydrogen chloride and of ethanoic acid to ethyne. The products, chloroethene and ethenyl ethanoate respectively, are both much used for making plastics (see next chapter).

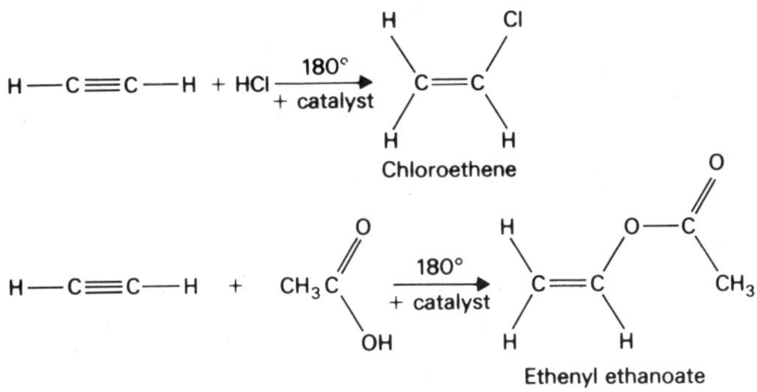

25.6 Alcohols

The alcohols are another homologous series in which a hydrogen atom of an alkane is replaced by a hydroxyl (– OH) group. Figure 25.18 gives the names and structural formulae of some simple alcohols. The alcohols are named by replacing the final *-e* of the parent hydrocarbon with *-ol*. The position of the hydroxyl group is shown by the number of the carbon atom to which the hydroxyl group is attached (see figure 25.18(*c*) and (*d*)). For methanol and ethanol no number is needed. For propanol and higher alcohols *isomerism* is possible.

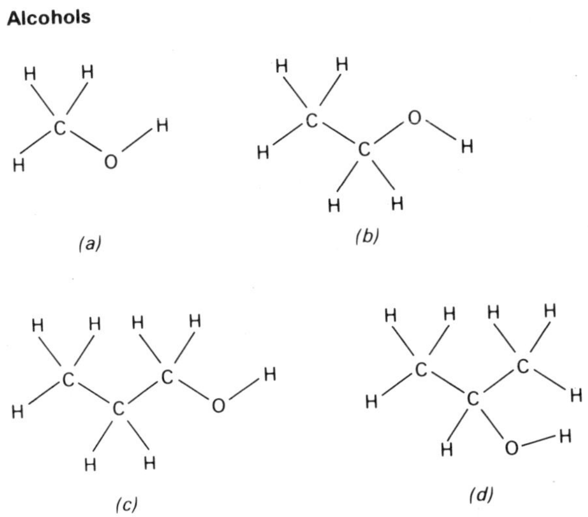

Figure 25.18 (*a*) methanol, (*b*) ethanol, (*c*) propan-1-ol, (*d*) propan-2-ol.

25.6.1 Methanol (figure 25.18(*a*))

Methanol (CH_3OH) is the simplest of the alcohols. It is manufactured by compressing to 300 atmospheres a mixture of carbon monoxide and hydrogen. The mixture is passed over a zinc oxide catalyst at 300 °C:

$$CO + 2H_2 \rightarrow CH_3OH.$$

The carbon monoxide and hydrogen can be obtained from methane and steam (see section 14.3.2) or by the water gas reaction from steam and coke (see section 14.3.1).

Much methanol is oxidized to methanal (formaldehyde), which is used in making plastics such as Bakelite (see next chapter). Some is also added to ethanol in making methylated spirits. It is also being used as a motor fuel (see section 17.6.2(*c*)), a use that will undoubtedly increase since it is a liquid fuel not made from oil, but from gas or coal.

25.6.2 Ethanol (figure 25.18(*b*))

(a) Introduction

Ethanol (C_2H_5OH) is by far the most important of the alcohols. It is the alcohol in beer, wines and spirits such as whisky. The amount of ethanol in drinks is measured in "degrees proof". This is a strange practical unit, first used many years ago when there was no easy method of determining the alcohol content of a liquid such as brandy or whisky, and the test was to pour some onto gun-powder and see if the gun-powder could be burnt. If the gun-powder refused to burn because the amount of water in the liquid was too great, the spirit was said to be "under-proof". The spirit that just allowed the gun-powder to burn was said to be 100° proof spirit. This contains about 60 % ethanol by volume (it contains in fact 57·10 % ethanol by volume). In America a different definition is used and 100° proof spirit contains 50 % ethanol by volume.

In addition to its use in alcoholic drinks, ethanol is an important solvent. It is also used in making many other organic chemicals.

Methylated spirits is ethanol that has been made undrinkable by adding about 5 % of methanol. Several different types of methylated spirits are available, according to the use to which it is put. *Industrial methylated spirit* is colourless and contains just ethanol and methanol. The sort that will be familiar to you is purple because a dye (methyl violet) has been added as well as some pyridine and possibly other evil tasting organic compounds. These are all added to discourage people from drinking it.

(b) Manufacture of ethanol

The two most important methods for the manufacture of ethanol are (i) the fermentation of sugar or starch and (ii) the hydration of ethene.

(i) The fermentation of sugar or starch

This is the process by which ethanol is formed in alcoholic drinks. Much ethanol is still obtained by the fermentation of carbohydrates, particularly starch ($(C_6H_{10}O_5)_n$), and sucrose ($C_{12}H_{22}O_{11}$). The sucrose that is often used is that present in the uncrystallizable residue obtained during sugar refining known as molasses. If molasses are used then yeast is added. This contains an organic catalyst or *enzyme* known as invertase that causes the sucrose to be hydrolyzed into two fermentable isomers—glucose and fructose:

$$\underset{}{C_{12}H_{22}O_{11}} + H_2O \rightarrow \underset{\text{glucose}}{C_6H_{12}O_6} + \underset{\text{fructose}}{C_6H_{12}O_6}.$$

The yeast also supplies another enzyme, zymase, that catalyzes the decomposition of the glucose and fructose into ethanol and carbon dioxide:

$$C_6H_{12}O_6 \rightarrow 2C_2H_5OH + 2CO_2$$

If starch is used as the reactant it must first be extracted from the raw material such as wheat, potatoes, barley or rice by using superheated steam. Malt (i.e., barley that has been partially germinated and then killed by heat) is added to the aqueous extract. This contains the *enzyme* diastase which helps the starch to be hydrolyzed to the sugar maltose:

$$2(C_6H_{10}O_5)_n + nH_2O \rightarrow nC_{12}H_{22}O_{11}$$

Yeast is now added and the enzyme maltase in yeast converts the maltose to glucose:

$$\underset{\text{maltose}}{C_{12}H_{22}O_{11}} + H_2O \rightarrow \underset{\text{glucose}}{2C_6H_{12}O_6}$$

and the enzyme zymase converts the glucose to ethanol and carbon dioxide. When the ethanol content of the liquor reaches about 12 % by volume it poisons the yeast and fermentation stops. The solution can be concentrated by fractional distillation (see section 1.8) until it is mainly ethanol (about 95 % ethanol).

This is the method of manufacture of ethanol preferred by industrial users who want the ethanol for adding to foodstuffs. In fact the ethanol produced this way is less pure than that made by the hydration of ethene, since the process of fermentation produces traces of many other organic molecules. But this is the old traditional method and the foodstuffs industry prefers this *biochemically* produced ethanol.

In the laboratory, ethanol can be obtained by adding yeast to a concentrated glucose solution and fractionally distilling the dilute alcoholic solution obtained.

Experiment. The preparation of ethanol

Put a 10 % solution of glucose in a 250 cm^3 round flat-bottomed flask. Add a little ammonium phosphate, malt extract (two yeast foods) and some yeast. Put a cotton-wool plug in the neck of the flask and leave the solution to ferment in a warm room for about a week. Filter the solution and distil the liquid in a distillation apparatus as figure 1.12. Collect the liquid that boils up to 80 °C. This will be mainly ethanol.

A home experiment. Home brewed beer

Buy a home-brewed beer kit to make 2 gallons of beer (lager beer is recommended). Get one containing a tin of malt extract with added hops, gelatin and beer yeast. Boil about 2 pints of water, add the malt, gelatin and 12 oz of sugar and continue to boil gently for about 5 minutes. Add the mixture to cold water in a clean plastic bucket and make up to 16 pints with cold water. Add the yeast, cover with a cloth and leave in a warm room for about a week when fermentation should have finished. This will be obvious as bubbles of carbon dioxide will not be being produced and the yeast residues will have sunk to the bottom of the bucket. If in doubt whether or not the fermentation has finished leave a few days longer.

Put 1 level teaspoonful of sugar into each of 8 clean lemonade or cider bottles and add the beer, taking care to disturb the yeast residue as little as possible. Fill the bottles leaving about 3 cm at the top. Ensure that a little yeast residue goes into each bottle, but throw most of it away. Seal the bottles and leave for at least 7 days. The small amount of additional sugar will have fermented and the carbon dioxide will dissolve in the beer making it fizzy. (CAUTION *Never add more than 1 level teaspoonful of sugar to 2 pints of beer or the bottle may explode.*) The gelatin will prevent particles of colloidal yeast remaining suspended in solution, and the beer when ready will be clear and sparkling.

Drink in moderation.

*(ii) The hydration of ethene (*not *hydrogenation)*

This reaction, given in figure 25.16, is carried out on a large scale. Most ethanol is made this way, and is then used by the chemical industry for making plastics and other chemicals. Ethene and steam at 70 atmospheres pressure are passed over a catalyst of phosphoric acid absorbed on silica heated to 300 °C. The water adds across the double bond of the ethene to give ethanol:

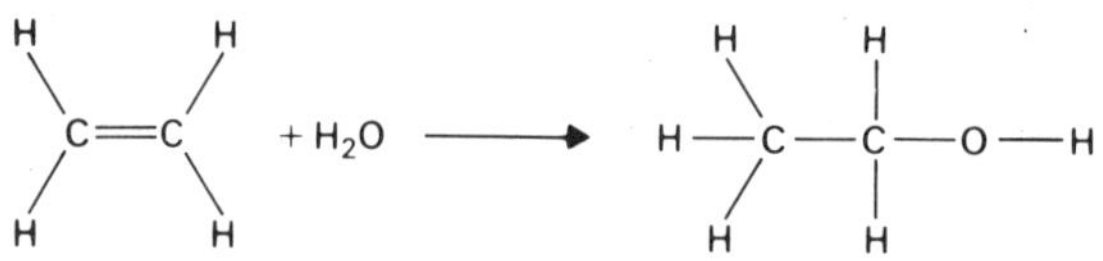

An alternative, and less used method, uses sulphuric acid as the catalyst. Here the compound ethyl hydrogensulphate is first formed, and this on heating with water gives ethanol:

$$C_2H_4 + H_2SO_4 \rightarrow C_2H_5.HSO_4$$
$$C_2H_2.HSO_4 + H_2O \rightarrow C_2H_5OH + H_2SO_4.$$

25.6.3 Properties of alcohols

(a) Physical properties

As with all homologous series, the physical properties of the alcohols show a steady change on passing along the series, e.g., methanol boils at 64 °C, ethanol at 78 °C, propanol at 98 °C and butan-1-ol at 118 °C. These simple alcohols are all water soluble, but the higher alcohols are solids, and, because they have a long alkyl chain, they are insoluble in water.

(b) Reaction with the alkali metals

With sodium and potassium the alcohols react similarly to water but much less readily. The metal salt of the alcohol is produced and hydrogen gas is evolved:

$$2C_2H_5OH + 2Na \rightarrow \underset{\text{sodium ethoxide}}{2C_2H_5ONa} + H_2$$

This reaction can be easily demonstrated by dropping small pieces of sodium into an evaporating dish containing the alcohol. Bubbles of hydrogen are evolved and the metal disappears. Evaporation of the solution gives the solid alkoxide.

(c) Oxidation

(i) The simple alcohols burn readily to give carbon dioxide and water, e.g.:

$$C_2H_5OH + 3O_2 \rightarrow 2CO_2 + 3H_2O.$$

(ii) Ethanol can be oxidized by chemical means to give, firstly, ethanal ($CH_3.CHO$) and then ethanoic acid (CH_3COOH).

If ethanol is warmed with an oxidizing agent such as acidified potassium dichromate(VI) or acidified potassium manganate(VII) solution then reactions such as:

$$C_2H_5OH \xrightarrow{\text{oxidation}} CH_3CHO,$$

and

$$CH_3CHO \xrightarrow{\text{oxidation}} CH_3COOH,$$

take place. The alcohol is oxidized first to ethanal and with more oxidizing agent to ethanoic acid.

These oxidations are readily demonstrated.

Experiment. Oxidation of ethanol

To give ethanal (acetaldehyde)

To about 1 cm^3 of ethanol in a boiling tube add 5 cm^3 of a saturated aqueous solution of potassium dichromate(VI) and 0·5 cm^3 of concentrated sulphuric acid. Warm the tube gently and cautiously smell the characteristic odour of ethanal.

To give ethanoic acid (acetic acid)

Add 40 cm^3 of concentrated sulphuric acid to 60 cm^3 of water (care when mixing). Dissolve 75 g of sodium dichromate(VI) in the diluted acid and put the mixture in the round-bottomed flask of the apparatus given in figure 25.19. Heat the flask on a water or steam bath and slowly add ethanol from a dropping funnel. As the oxidizing agent is in excess the ethanol is oxidized to ethanoic acid.

After all the ethanol has been added and the mixture heated under reflux for a further 10 minutes, allow it to cool and distil in the apparatus as shown in figure 1.8. Collect the distillate above 105 °C; it is an aqueous solution of ethanoic acid.

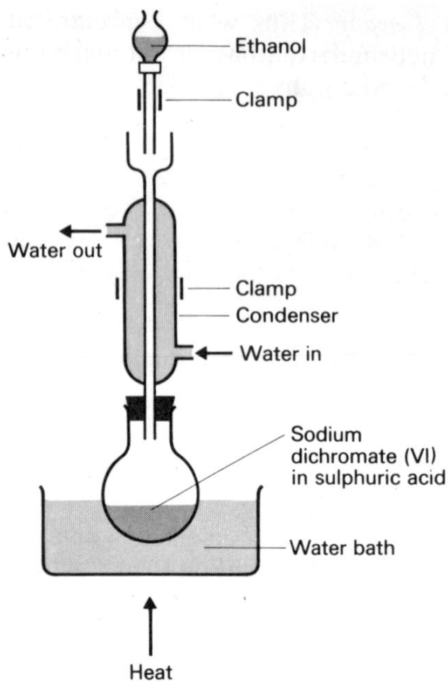

Figure 25.19 Oxidation of ethanol to ethanoic acid.

During the oxidation of the ethanol the orange sodium dichromate(VI) solution turns green as chromium(III) sulphate is formed. This reaction is used in the "breathalyser" test. Exhaled alcoholic vapour turns dichromate(VI) crystals green.

(d) Reaction with phosphorous chlorides

The hydroxyl group of alcohols is replaced by chlorine by reaction with either phosphorous trichloride or phosphorus pentachloride:

$$3C_2H_5OH + PCl_3 \rightarrow \underset{\text{chloroethane}}{3C_2H_5Cl} + \underset{\text{phosphonic acid}}{H_3PO_3}$$

$$C_2H_5OH + PCl_5 \rightarrow C_2H_5Cl + \underset{\text{phosphorus trichloride oxide}}{POCl_3} + HCl$$

These reactions occur with other compounds containing an $-OH$ group. The reaction with phosphorus pentachloride gives the easily identifiable fumes of hydrogen chloride, and is used as a test for the $-OH$ group. Anhydrous compounds must be used because hydrogen chloride is also evolved with water. To illustrate the test add a little solid phosphorus pentachloride to some alcohol in a test-tube. Steamy fumes are formed.

(e) Esterification

One of the most interesting reactions of alcohols is with acids, both organic and inorganic, to form esters. *An ester is a compound formed when all or part of the hydrogen of an acid is replaced by an alkyl group*. Thus when concentrated sulphuric acid is added to ethanol, ethyl hydrogen sulphate is formed by a reversible reaction:

$$C_2H_5OH + H_2SO_4 \rightleftharpoons C_2H_5.HSO_4 + H_2O$$

With glacial ethanoic (acetic) acid, ethyl ethanoate (ethyl acetate) is formed.

$$C_2H_5OH + CH_3C\overset{\displaystyle /\!\!/ O}{-}OH \rightleftharpoons \underset{\text{ethyl ethanoate}}{C_2H_5O.C\overset{\displaystyle /\!\!/ O}{-}CH_3} + H_2O$$

The reaction is catalyzed by hydrogen ions and is, therefore, usually carried out in the presence of concentrated sulphuric acid. This also reacts with the water formed and prevents the reverse reaction.

Experiment. Preparation of ethyl ethanoate

Add a few drops of concentrated sulphuric acid to 2–3 cm^3 of ethanol mixed with 2–3 cm^3 of concentrated ethanoic acid (known as glacial ethanoic acid—because it freezes at above 0 °C to give an icy (glacier-like) solid). Warm the tube gently when the pleasant fruity smell of ethyl ethanoate will be noticed. Pouring the mixture into a little water enables the smell to be identified more readily.

Ethyl ethanoate may also be prepared on a larger scale as follows. Add about 10 cm^3 of concentrated sulphuric acid to an equal volume of ethanol in a distillation flask (*Add with care*—heat is evolved), together with a few pieces of porous pot to ensure smooth boiling later. Connect a condenser and a dropping funnel to the flask as shown in figure 25.20 and place in the dropping funnel 20 cm^3 of ethanol with an equal volume of ethanoic acid. Heat the oil bath to about 150 °C and slowly add the ethanol-acid mixture. Ethyl ethanoate distils over. It will contain ethanol and ethanoic acid as impurities

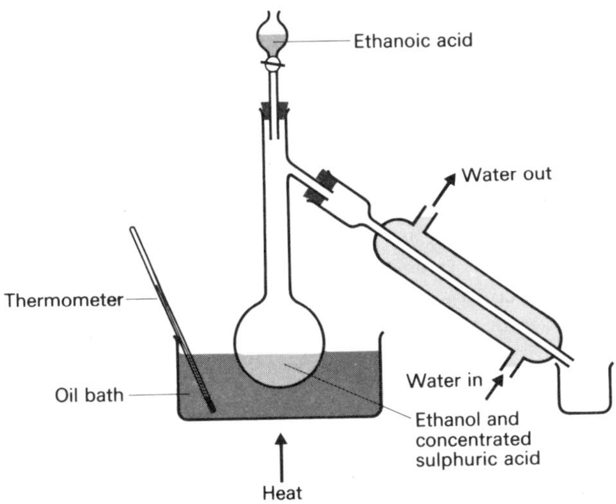

Figure 25.20 Preparation of ethyl ethanoate.

and these can be removed by shaking the distillate with sodium carbonate solution to remove the ethanoic acid and calcium chloride solution to remove the ethanol. The ethyl ethanoate can then be dried over fused calcium chloride and obtained pure by distillation.

Esters are widely distributed in nature. The low molecular weight esters are responsible for the pleasant smell of many fruits and vegetables. Pineapples, for example, contain a little ethyl ethanoate and several other esters. Animal and vegetable fats are esters of propane-1,2,3-triol with high molecular mass acids (see section 21.2.7c(v)).

(f) Dehydration

Water can be removed from ethanol to give ethene:

$$C_2H_5OH \rightarrow C_2H_4 + H_2O$$

There are two methods that can be used. Sulphuric acid can be used as the dehydrating agent, or the ethanol can be dehydrated by heat. Both methods have been given in section 25.4.1 and figures 25.14 and 25.12, under the preparation of ethene.

(g) Physiological effects of ethanol

Ethanol is one of the oldest drugs known. It has no value as a stimulant, its action being that of a sedative. Before the discovery of dinitrogen oxide it was the main anaesthetic in use. In *small* quantities alcohol gives pleasure. It overcomes shyness and subdues fear. However one's judgement is affected by consuming even the smallest amounts – this means, for example, that one should *never* drive whilst under the influence of ethanol.

Extra amounts of alcohol make one *drunk*. A large amount of drink causes acute alcoholic poisoning and the regular heavy drinker suffers from chronic alcoholic poisoning. Many illnesses are caused, wholly or partly, by alcohol, e.g. undernourishment, liver disease, impotence, gout, fatness, delirium tremens, peripheral neuritis and insanity. Some people who drink heavily over a period of years find themselves unable to stop drinking. In the United Kingdom alone there are possibly 750 000 such people, called *alcoholics*, condemned to miserable lives.

Although many of us are able to drink in moderation, alcohol must be regarded as a poison. (Young readers who believe that alcohol is harmless should read "Drinking and Alcoholism" by Dr. Robert Kemp. This is a booklet published by the British Medical Association.)

25.7 Carboxylic acids

This homologous series has the general formula $C_nH_{2n+1}{\cdot}COOH$. The simplest acids are, therefore, HCOOH—called methanoic (formic) acid, CH_3COOH—called ethanoic (acetic) acid, and CH_3CH_2COOH—called propanoic acid.

25.7.1 Manufacture of ethanoic (acetic) acid

Ethanoic acid is manufactured on a large scale either

(*a*) by the oxidation of naphtha (figures 25.21 and 25.11) or of other hydrocarbons such as butane:

Figure 25.21 General view of the ethanoic acid plant at Hull, North Humberside. (Photo: B.P. Chemicals Ltd.).

Hydrocarbon + Oxygen $\rightarrow$ Ethanoic acid, or

(*b*) by the reaction between methanol and carbon monoxide:

$$CH_3OH + CO \xrightarrow{\text{rhodium catalyst}} CH_3COOH$$

A laboratory preparation of ethanoic acid is given in section 25.6.3.

25.7.2 Vinegar

Vinegar is made from diluted ethanol by bacterial action. The alcohol is sprinkled over beech shavings treated with bacteria such as *bacterium aceti*. The ethanol is then oxidized directly to give a 4–5 % by mass solution of ethanoic acid (approx. 0·75M) which is sold as vinegar.

$$C_2H_5OH + O_2 \rightarrow CH_3COOH + H_2O$$

25.7.3 Properties of acids

The simplest acids are liquids and mix with water in all proportions. As with the alcohols, those with high molecular weights and containing long hydrocarbon chains are much less soluble. For example, hexadecanoic acid,

$C_{15}H_{31}.CO.OH$, and octadecanoic acid, $C_{17}H_{35}.CO.OH$ are waxy solids used in making candles.

The carboxylic acids are weak acids. Thus ethanoic acid is slightly ionized when it is dissolved in water to give hydrogen ions and ethanoate ions:

$$CH_3COOH \rightleftharpoons CH_3COO^- + H^+.$$

Notice that it is the hydrogen atom of the hydroxyl group that is acidic and becomes a hydrogen ion.

Ethanoic acid has the typical properties of acids.

(*a*) It has a sour taste and turns blue litmus red.
(*b*) It gives hydrogen with the more electropositive metals, e.g.,

$$2CH_3COOH + Mg \rightarrow \underset{\text{magnesium ethanoate}}{(CH_3COO)_2Mg} + H_2$$

(*c*) It reacts with bases to give salts and water, e.g.,

$$CH_3COOH + NaOH \rightarrow \underset{\text{sodium ethanoate}}{CH_3COONa} + H_2O$$

(*d*) It reacts with carbonates to give salts, water and carbon dioxide, e.g.,

$$2CH_3COOH + K_2CO_3 \rightarrow \underset{\text{potassium ethanoate}}{2CH_3COOK} + H_2O + CO_2$$

(*e*) As we saw in section 25.6.3, with alcohols ethanoic acid gives esters.

25.7.4 Uses of ethanoic acid

Most ethanoic acid is used for making cellulose ethanoate, an important synthetic fibre (see section 26.3.1 (*d*)). It is also used for making transparent wrapping materials, in pharmaceuticals (chemical products that are sold in chemists shops and dispensed by doctors), and in the food industry.

25.8 Amines

The amines are compounds in which one or more hydrogen atoms of an ammonia molecule are replaced by alkyl groups, e.g.:

CH_3NH_2—methylamine, $C_2H_5NH_2$—ethylamine, $(CH_3)_2NH$—dimethylamine, and $(CH_3)_3N$—trimethylamine.

They have similar basic properties to ammonia:

(*a*) The simple amines are gases that dissolve in water to give weakly alkaline solutions:

$$\underset{\text{methylamine}}{CH_3NH_2} + H_2O \rightleftharpoons \underset{\text{methylammonium hydroxide}}{CH_3NH_3OH} \rightleftharpoons CH_3NH_3^+ + OH^-$$

(compare with $NH_3 + H_2O \rightleftharpoons NH_4OH \rightleftharpoons NH_4^+ + OH^-$)

(*b*) The amines react with acids to form salts:

$$C_2H_5NH_2 + HCl \rightarrow C_2H_5NH_3Cl$$
ethylammonium chloride

(compare with $NH_3 + HCl \rightarrow NH_4Cl$).

Questions on Chapter 25

1. (*a*) What name is given to the series of saturated hydrocarbons of general formula C_nH_{2n+2}? Name *two* members of this series and give their molecular formulae.
(*b*) What do you understand by the terms *saturated* hydrocarbon and *unsaturated* hydrocarbon? Illustrate your answer by quoting one suitable chemical reaction for each type.
(*c*) Methane (natural gas) is now used as a domestic fuel.
(i) Calculate the volume of air (20% oxygen) required to burn completely 10 dm^3 of methane. (All gases assumed to be at the same temperature and pressure.)
(ii) Explain briefly why it is dangerous to use a methane-burning appliance in a badly ventilated room.
(*d*) Describe how ethene is converted into *one* product of commercial importance. (*A.E.B.*)

2. Define (*a*) empirical formula, (*b*) molecular formula, (*c*) structural formula, (*d*) isomerism.
Write the full structural formulae of the isomers of pentane (C_5H_{12}).
Explain carefully, with two examples, what you mean by the term "homologous series" How do the physical properties of the members of such a series vary with their position in the series?

3. By drawing their structural formulae show the differences in structure between ethane and ethene (ethylene). Explain how the bond between a carbon atom and a hydrogen atom in these compounds is formed.
By naming the reagents, stating the conditions, and writing an equation for the reaction, describe how ethene could be converted into (*a*) ethane, (*b*) chloro-ethane (ethyl chloride), (*c*) 1,2 dibromoethane (ethylene dibromide), (*d*) ethanol (ethyl alcohol). (*J.M.B.*)

4. Describe how you would obtain a sample of almost pure ethanol from a sugar.
Outline the steps by which ethanol is manufactured from petroleum oil. Name all the chemical processes involved.
Describe *one* chemical and *one* physical test to distinguish between pure water and pure ethanol. State the result of each test on each of these substances. (*J.M.B.*)

5. (*a*) Describe, with the aid of a diagram, a laboratory preparation of ethene. Describe and explain the reaction of this gas with bromine water.
(*b*) Outline the manufacture of ethanol from ethene.

6. Outline one method of manufacturing ethanoic (acetic) acid on a large scale. Write down its empirical formula, molar mass, and structural formula. Give two physical properties of this compound. Give the name and formula, in each case, of the product formed when ethanoic acid reacts with (i) calcium oxide, (ii) ethanol. State two significant differences between these two products. (*O. and C.*)

7. State *one* chemical and *one* physical property of ethane.
Suppose that, in the molecule of ethane, (*a*) one H is replaced by OH, (*b*) one H is replaced by NH_2, (*c*) two H atoms are removed. State, in each case, the name, one chemical property and one physical property of the new compound formed. If two hydrogen atoms in the molecule of ethane are replaced by chlorine atoms, it is possible to obtain two different compounds as products. Explain this. (*O. and C.*)

8. What is the maximum mass of alcohol which could be obtained from 1 kilogram of potatoes containing 16·2% of starch?

9. (*a*) Which homologous series of organic compounds can be represented by the following general formulae:

Series *A*	C_nH_{2n+2}	Series *C*	$C_nH_{2n+1}COOH$
Series *B*	C_nH_{2n}	Series *D*	$C_nH_{2n+1}OH$?

(*b*) Give the name and structural formula of *one* compound in *each* series.
(*c*) Describe reactions by which
(i) a *named* compound of series *B* can be converted to a compound of series *A*;
(ii) a *named* compound of series *D* can be converted to a compound of series *B*.
(*d*) Name an important natural source of compounds of series *A* and give *two* industrial uses of such compounds. (*C*.)

10. Outline how *five* of the following substances can be obtained;
(*a*) ethene (ethylene) from petroleum;
(*b*) ethanol (ethyl alcohol) from ethene;
(*c*) ethene from ethanol;
(*d*) polyethene (polyethylene) from ethene;
(*e*) ethanoic acid (acetic acid) from ethanol;
(*f*) ethyl ethanoate (ethyl acetate) from acetic acid;
(*g*) soap from a fat. (*A.E.B.*)

11. Name the process used to obtain an aqueous solution of ethanol from a sugar, for example, glucose, $C_6H_{12}O_6$. Describe how you would carry out this preparation.

State, without further description, the method you would use to obtain a sample of reasonably pure ethanol from the aqueous solution and indicate the physical bases of this separation.

Write down the *structural* formula of ethanol.

Explain the following, stating the chemical reactions which take place:
(*a*) ethanol burns in air with a pale blue flame,
(*b*) when sodium is added to pure ethanol a colourless gas is evolved,
(*c*) when a mixture of equal volumes of ethanol and acetic acid (ethanoic acid) is warmed with a few drops of concentrated sulphuric acid, a vapour with a pleasant smell is obtained. (*A.E.B.*)

12. 0·282 g of a liquid *A* was found to contain 0·036 g of carbon, 0·006 g of hydrogen and the rest bromine.

0·470 g of *A* occupied 112 cm^3 when vaporized at 273 °C and 1 atmosphere pressure. (Atomic masses: H = 1, C = 12, Br = 80. The molar volume of a gas at s.t.p. is 22·4 dm^3.)
(*a*) Calculate the number of gram-atoms of carbon, hydrogen and bromine present in 0·282 g of *A*.
(*b*) Using the results from (*a*), state the empirical formula of *A*.
(*c*) Calculate the volume that 0·470 g of the vapour would have occupied if measured at s.t.p.
(*d*) Calculate the mass of *A* which would occupy 22·4 dm^3 at s.t.p. and hence state the molecular mass of *A*.
(*e*) What is the molecular formula of *A*.
(*f*) Liquid *A* is one of two isomeric compounds. Name the reagents from which one of these compounds can be made in the laboratory, and write the structural formula for this compound and the equation for the reaction. (*J.M.B.*)

13. Under suitable conditions it is possible to obtain a hydrocarbon, *X*, directly from ethanol (ethyl alcohol). *X* reacts with bromine in the cold to give a liquid *Y*. The relative molecular mass of *Y* is 188 and it contains 12·8% carbon, 2·13% hydrogen, and 85·1% bromine by weight.
(*a*) Calculate (i) the empirical formula, (ii) the molecular formula of liquid *Y*. Name this liquid.

(*b*) Name the hydrocarbon, *X*, and give its *structural* formula.
(*c*) What type of reaction occurs between bromine and the hydrocarbon *X*? Give the equation.
(*d*) Give *one* large-scale use for *X*.
(*e*) State briefly how *X* can be obtained from ethanol.
(*f*) Give *two* other reactions of ethanol. (*A.E.B.*)

14. A mixture of 400 cm^3 of a gas, *A*, and 400 cm^3 of hydrogen reacted completely in the presence of a catalyst to give 400 cm^3 of a new gas, *B*. When 400 cm^3 of this new gas, *B*, were mixed with 400 cm^3 of hydrogen in the presence of a different catalyst, 400 cm^3 of ethane were produced. (All volumes measured under the same conditions of temperature and pressure.)

A liquid, *D*, was produced when *B* and bromine reacted together. Addition of another catalyst to *B* gave a solid, *E*, which did not react with bromine and had a molecular mass of about 20 000.

Explain these results, identify *A*, *B*, *D*, *E* and give equations for the reactions.

How might polyvinyl chloride, i.e. poly (chloroethene), be obtained from *A*? (*J.M.B.*)

26

Organic Chemistry II—Large Molecules

26.1 Introduction

In the last chapter we looked at some simple compounds of carbon. They were all small molecules with molecular masses less than 300. (The smallest molecule was methane (mol. mass = 16) and the largest was octadecanoic acid (mol. mass = 284).) It is possible to form long chains of carbon atoms, both with and without other atoms in the chain, and organic compounds containing such long chains have molecular masses of thousands or even millions.

It is these large molecules, known as *macromolecules* (big molecules), that will be discussed here. They fall into two groups. The first group contains the large molecules that occur in nature, the *natural macromolecules*. These are the substances that make up living plants and animals. The second group contains large molecules that do not occur naturally; these are the *synthetic* or *man-made macromolecules*. Nearly all molecules in this second group have been made for the first time during the last 50 years, and are often called *plastics*. The name "plastic" means a mouldable substance, and it was given to such materials because when they are used they are frequently liquefied, often by heat, and moulded into shape.

Many of these *macromolecules*, both *natural* and *synthetic* have been formed by the process of *polymerization;* they are *polymers*.

A *polymer* is a large molecule made by the process of *polymerization* from many thousand smaller molecules of substances known as *monomers*. In addition to the *polymer* molecule a second product may be formed. This is often water. Hence polymerization may be of the type:

monomer → *polymer*,

or it may be of the type:

monomer → *polymer* + *other product*.

Polymers formed by reactions of the first type are known as addition polymers, because the monomers have simply added together. Polymers formed by reactions of the second type are known as *condensation* polymers because when they are formed some other substance is produced at the same time. Frequently a condensation polymer is formed from two different monomers by the reaction:

monomer(1) + *monomer(2)* → *polymer* + *other product*.

26.2 Synthetic polymers

26.2.1 Addition polymers

(a) Poly(ethene)

Poly(ethene) (commonly called polythene) is an example of an addition polymer.

If ethene is mixed with a trace of oxygen, compressed to a very high pressure (1000 atmospheres) and heated to 100–300 °C the reaction: $nC_2H_4 \rightarrow (C_2H_4)_n$ ($n \sim 1000$), takes place and white "waxy" solid is formed. The poly(ethene) has been formed by an ethene molecule adding to the double bond of another ethene molecule, this process being repeated about a thousand times. The reaction (figure 26.1) gives long chains of CH_2 units. Two of these are the "ethene" unit of the polymer.

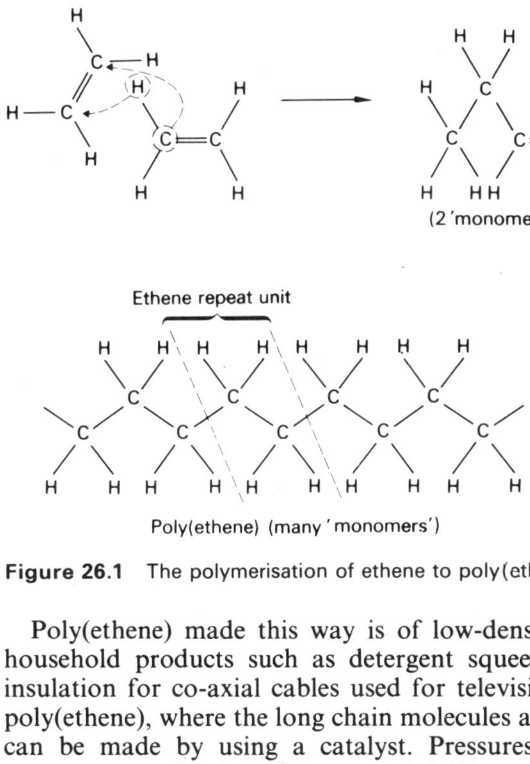

Figure 26.1 The polymerisation of ethene to poly(ethene).

Poly(ethene) made this way is of low-density and is used to make many household products such as detergent squeeze bottles, "plastic" bags, and insulation for co-axial cables used for television aerial leads. A high density poly(ethene), where the long chain molecules are packed more tightly together, can be made by using a catalyst. Pressures of about 5 atmospheres and temperatures of only 75 °C are needed. The reaction is exactly the same (figure 26.1). This high density poly(ethene) is more rigid than the low density material and is used for making, amongst other things, washing-up bowls, dustbins, and domestic bleach bottles.

(b) Poly(propene) or polypropylene

If propene is used instead of ethene in the catalyzed addition reaction then the related polymer poly(propene) is formed (figure 26.2). This is a harder, stiffer plastic than poly(ethene), with a higher melting point. It is used for such articles as pudding basins, some childrens' toys, and "Weedol" containers.

Name	Common Name	Monomer	Polymer repeat unit
Poly (ethene)	Polythene	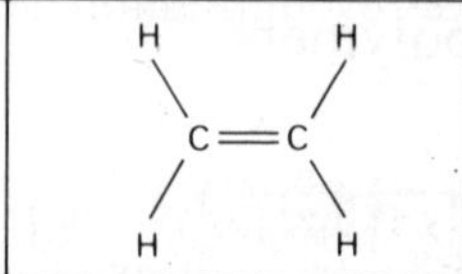	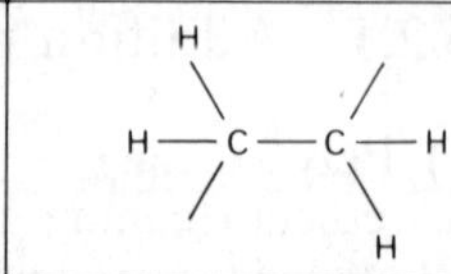
Poly (propene)	Polypropylene	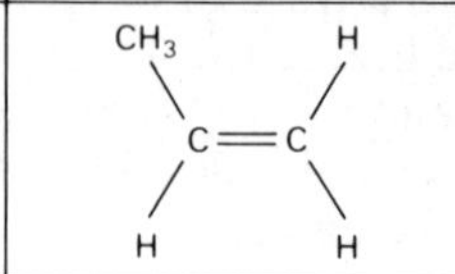	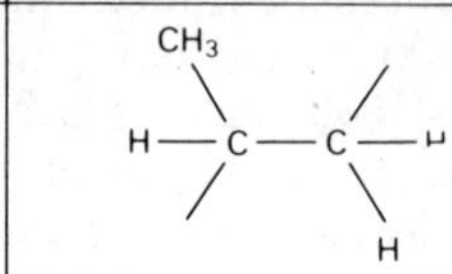
Poly (chloroethene)	Polyvinylchloride (PVC)	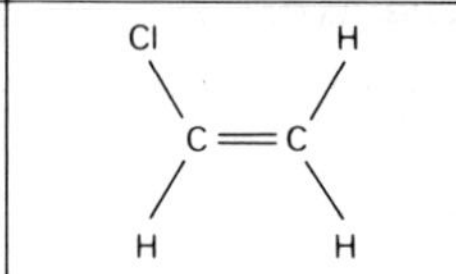	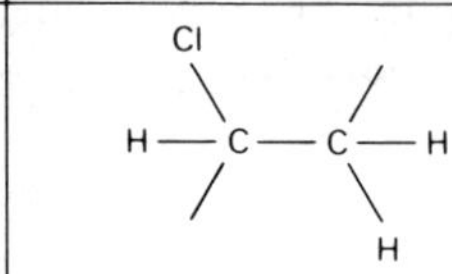
Poly (ethenylethanoate)	Polyvinylacetate (PVA)	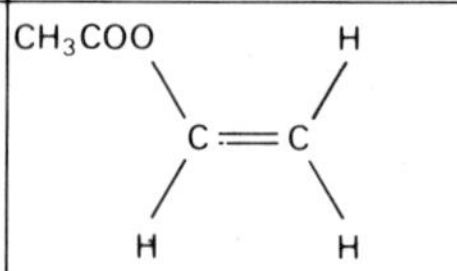	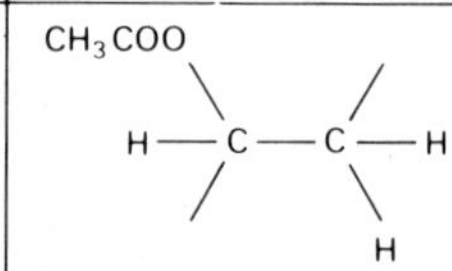
Poly (phenylethene)	Polystyrene	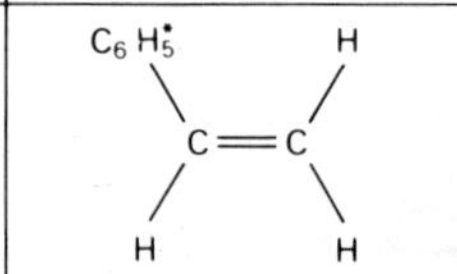	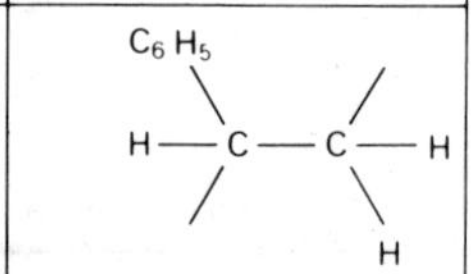
Poly (methyl 2-methyl-propenoate)	Perspex	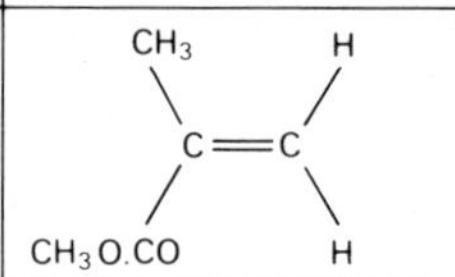	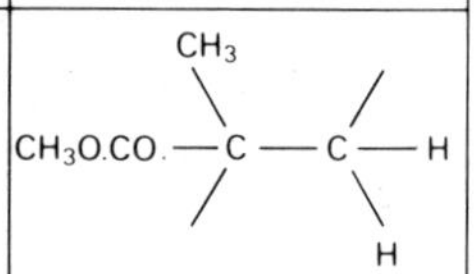
Poly (tetrafluoroethene)	Polytetrafluoroethylene (PTFE)	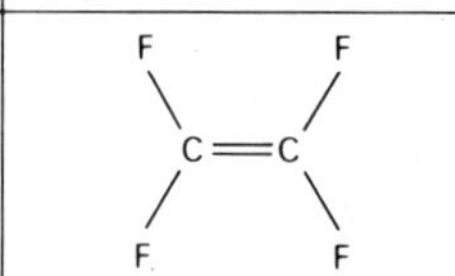	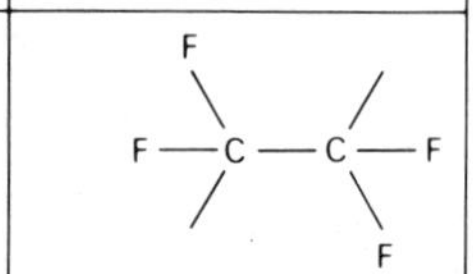
Poly (propenonitrile)	Polyacrylonitrile (acrylic fibre)	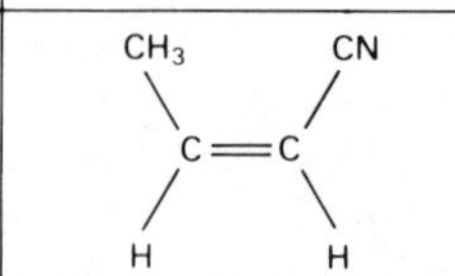	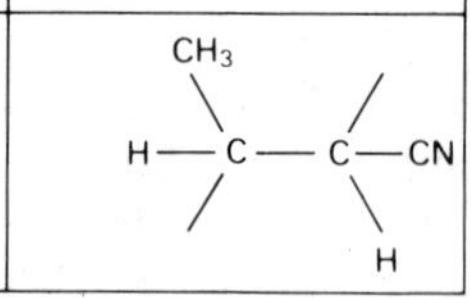

The phenyl group. This is derived from benzene

Figure 26.2 Plastics based on ethene.

(c) Poly(chloroethene) or polyvinyl chloride (PVC)

If chloroethene is heated then it polymerizes to give the polymer poly(chloroethene) (figure 26.2). This plastic may be stiff and hard or may have a plasticizer added, when it is flexible and soft. The stiff variety is widely used for making such things as soft drink bottles, house guttering, gramophone records, and chocolate box liners. The flexible variety is used for, among other things, housepipes, wire and cable insulation, see-through umbrellas and "clingy" wrapping film.

(d) Poly(ethenyl ethanoate) or polyvinyl acetate
If ethenyl ethanoate is polymerized it forms the addition polymer poly(ethenyl ethanoate) (figure 26.2). The molecular mass of the product can be controlled over a considerable range. Poly(ethenyl ethanoate) forms an emulsion with water (see section 1.17), and this emulsion is the basis for emulsion paints.

(e) Poly(phenylethene) or polystyrene
If phenylethene is heated it polymerizes to give the transparent glassy material, poly(phenylethene) (figure 26.2). This is used to make many food containers (e.g., Yoghurt cartons), disposable cups, model construction kits and many other things.

Poly(phenylethene) can be made to appear physically very different by blowing air into it, when expanded poly(phenylethene) (expanded polystyrene) is formed. This is white, opaque and floats on water. It is used for making such things as ceiling tiles, surf-boards and packing for delicate articles such as cameras.

Experiment. Preparation of poly(phenylethene)

Add about 0·1 g of an organic peroxide (e.g., dodecanoyl peroxide) to 10 cm^3 of phenylethene in a test-tube, and put the bottom part of the tube into a boiling water bath. After about half an hour the liquid monomer will have polymerized to colourless, glassy poly(phenylethene) (polystyrene). Break the tube, with care, to get the sample.

(f) Poly(methyl 2-methylpropenoate) or polymethylmethacrylate (perspex)
If the disubstituted ethene, methyl 2-methylpropenoate, is heated it polymerizes to give a hard transparent material with a glossy surface, poly(methyl 2-methylpropenoate) (figure 26.2). This plastic is used to make such things as car rear light units, motor-cycle windshields, "perspex" window panes, and baths.

Experiment. Preparation of poly(methyl 2-methylpropenoate) or perspex

Add about 0·1 g of an organic peroxide (e.g., dodecanoyl peroxide) to 10 cm^3 of methyl 2-methylpropenoate in a test-tube immersed in a water bath at 60–70 °C. After some time (about one hour) the liquid monomer polymerizes to solid polymer. It is necessary to break the tube to obtain the polymer.

Experiment. Depolymerization of Perspex.

If strongly heated the polymer returns to the monomer. Heat some pieces of perspex in a test-tube fitted with a tube that leads into a second test-tube (figure 26.3). Liquid monomer condenses in the second tube.

This experiment can also be carried out using polystyrene when the monomer, phenylethene, is formed.

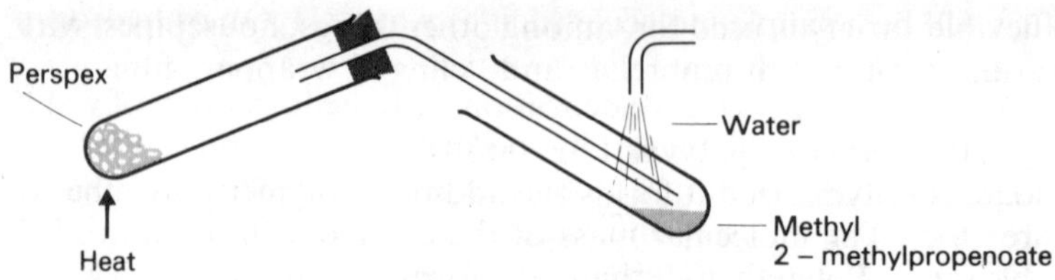

Figure 26.3 Depolymerisation of perspex.

> *Note:* Organic vapours are harmful, and, like all other reactions on organic compounds, these depolymerizations should be carried out in a fume cupboard.

(g) Poly(tetrafluoroethene)

The compound ethene is C_2H_4. There is a compound, tetrafluoroethene, in which all the hydrogen atoms of the ethene have been replaced by fluorine atoms. If this is polymerized a smooth, slippery, soft polymer poly(tetrafluoroethene) is formed. This is used in situations where a non-stick or low friction surface is wanted. It is the non-stick coating on cooking utensils, is used by plumbers as a pipe sealing tape, and as the bearing material for "superfast" toy cars.

(h) Poly(propenonitrile) or polyacrylonitrile

If the ethene derivative with both a methyl and a cyano substituent, propenonitrile, is polymerised a long chain polymer, poly(propenonitrile) is formed (figure 26.2). This polymer can be spun into long fibres and these can be made curly like sheep's wool. The acrylic fibres are, therefore, used as wool substitutes. They are known by trade-marks such as Orlon, Acrilan and Courtelle.

All the above 8 polymers are ethene derivatives and are formed by the addition of the ethene or substituted-ethene to itself.

(i) Polyurethane

One other addition polymer that is very well known is *polyurethane*. This is formed by the addition of a compound known as a diisocyanate to a diol. These add together to give the polyurethane polymer:

— diisocyanate + diol + diisocyanate + diol— →
— diisocyanate-diol-diisocyanate-diol—.

When this polymer is being made a little water is added to the mixture. This reacts with some of the diisocyanate and forms carbon dioxide gas. This blows up the polymer as it is formed to a sponge-like material. The process is just like making a loaf of bread. Polyurethane sponges are to be found in most bathrooms, as well as being used for chair seats and for putting inside the body of motor-cars to stop them from rattling! A rigid polyurethane can also be made. This is injected into house walls to help prevent heat losses.

26.2.2 Condensation polymers

These are formed by the reaction:

monomer(1) + monomer(2) → polymer + other product.

The other product will be a simple molecule such as water (H_2O), or hydrogen chloride (HCl).

We shall see in section 26.3.2 that a condensation with the formation of water is very important in making the naturally occurring macromolecules, the proteins. We saw such a condensation in section 25.6.3 when an alcohol and an acid react together they give an ester and water:

$CH_3COOH +$	HOC_2H_5	$\rightarrow$	$CH_3CO.OC_2H_5$	$+H_2O$
ethanoic acid	ethanol		ethyl ethanoate	water

Two most important synthetic fibres, *nylon* and *terylene*, are made by such a *condensation*.

(a) Nylon

This is a synthetic fibre that, like *proteins*, contains the $-\overset{\overset{\displaystyle O}{\|}}{C}-\overset{\overset{\displaystyle H}{|}}{N}-$ group (see section 26.3.2). The materials for the manufacture of the most common form of nylon are hexanedioic acid (adipic acid) and 1,6-diaminohexane (figure 26.4). When they are heated together a long-chain structure is formed by the acid ($-COOH$) groups reacting with the amino ($-NH_2$) groups (figures 26.4 and 26.5); water being the other reaction product. Other nylons have starting materials similar to the two above.

The use of nylon as a fibre for making clothes, ropes, carpets, etc., is well known; it is also used for making, amongst other things, curtain rail fittings, cupboard hinges, and gear wheels.

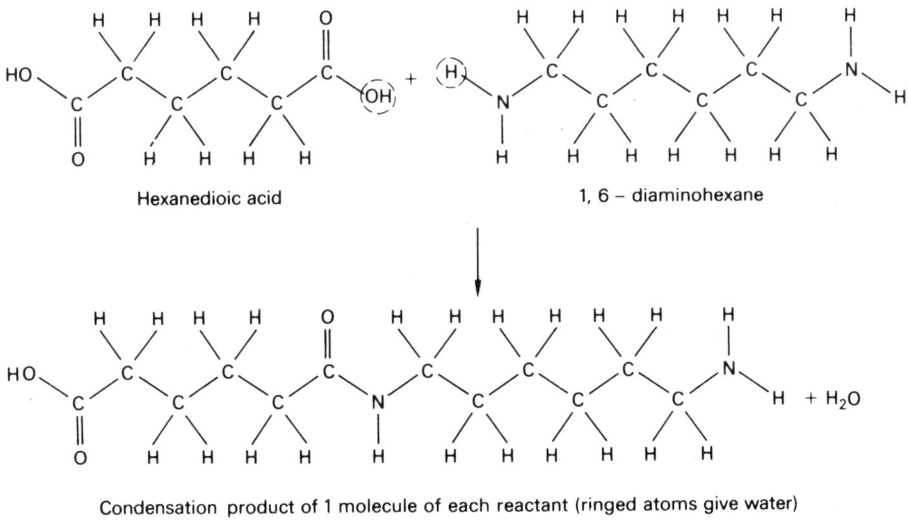

Figure 26.4 The condensation of nylon monomers.

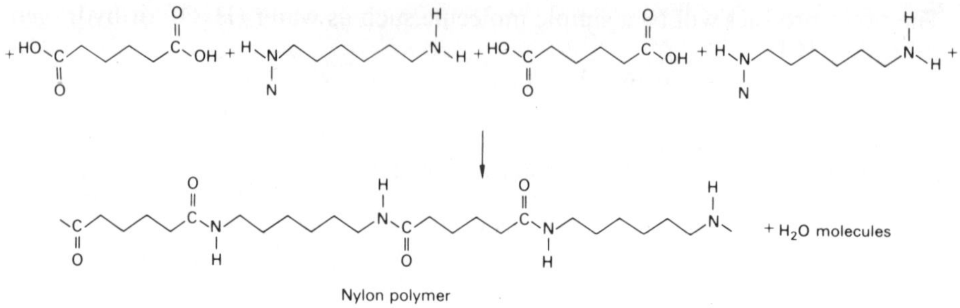

Figure 26.5 Nylon polymer.

Experiment. Preparation of Nylon

To make nylon by a condensation reaction with the formation of water requires high temperatures. A type of nylon can be easily prepared at room temperature by a similar condensation that gives hydrogen chloride as the other product.

Prepare a 2% (by volume) solution of decanedioyl dichloride (sebacoyl chloride) in tetrachloromethane and put 40 cm^3 into a 100 cm^3 beaker. Carefully add 25 cm^3 of an aqueous 10% (by weight) solution of 1,6 diaminohexane when the aqueous solution will float on the heavier tetrachloromethane. Polymerization will occur at the interface between the two solutions and this thin film of polymer can be collected together into a thread with a pair of tweezers. As the polymer layer is removed from the interface more of the solutions come into contact and further polymerization occurs. Hence a long thread of polymer is produced. With care a thread many metres long can be drawn out of the solution (see figure 26.6). (*Warning*. The reagents are toxic and the nylon thread is soaked in them.)

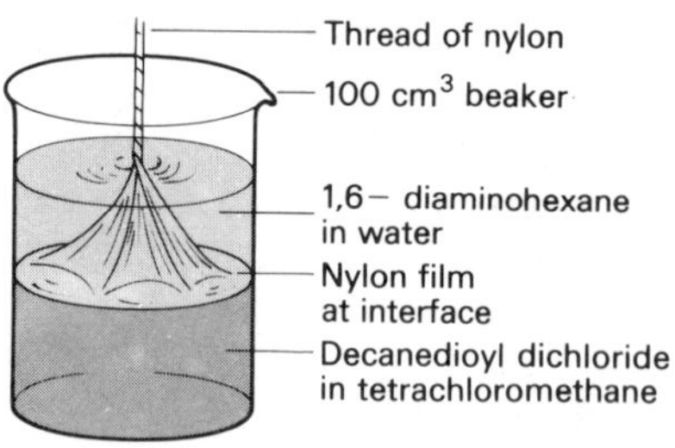

Figure 26.6 Preparation of nylon.

(b) Terylene

This fibre, also known as *Dacron* or *polyester fibre*, is made by the condensation of the acid, benzene-1,4-dicarboxylic acid (terephthalic acid) or one of its derivatives with ethane-1,2-diol (ethylene glycol). Again a long chain structure is formed (figure 26.7).

Terylene is much used as a fibre for making clothes. It can be woven or knitted. Woven terylene is used, for example, for the sails of boats, where a strong rot-

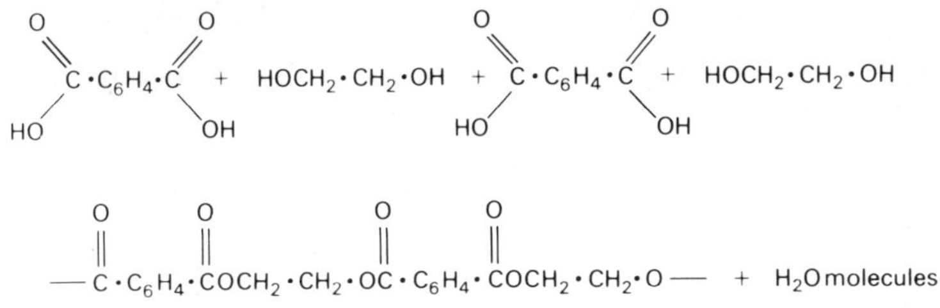

Terylene polymer

Figure 26.7 Terylene polymer.

proof material is required. Woven terylene (Crimplene) is much used for mens' and womens' clothes. Terylene can also be made into a film, and is then used for making recording tape and transparent cooking film.

26.2.3 Identification of plastics

Plastics are now widely used and an interesting group project is to make a collection of different plastics. It is easy to find plastic objects. This experiment is to identify some of them. Common uses of the plastics have been given in the text and this should be used initially to help collect samples.

Experiment. Identification of plastics

Collect together as many plastic articles as you can and identify them by carrying out some simple tests.

Many of the samples that will be used will be bottles and other containers. It is *essential* that these be thoroughly washed and dried before carrying out any tests, not only because the contents may be harmful, but because residues on the plastics will affect their behaviour when burnt.

Tests to be carried out
5 tests are to be carried out and these are labelled (*A*) to (*E*).

(A) Look at the sample
Is it transparent, translucent or opaque?

Opaque "fillers" are often added to plastics. A transparent sample can only be made from one of the materials listed as transparent. However, an opaque material may be a transparent one with a filler.

(B) Feel the sample
Try to bend it, to scratch it with your thumb nail, and feel the surface. Some plastics are flexible, others stiff. Of course, the thickness is important when deciding whether the plastic is flexible or stiff. If your nail scratches the plastic it can be classed as soft. Some plastics have a characteristic "feel"; low density poly(ethene) has a characteristic waxy feel.

(C) Cut the sample
Use a sharp knife. Some cut easily and cleanly, others cut with difficulty and may splinter and have jagged edges. Fillers can sometimes make plastics more difficult to cut.

(D) Drop a small piece of the sample into water
Some will float and some will sink, depending on their density. Ensure that there are no air bubbles on the sample by putting a little detergent in the water and immersing the sample and rubbing its surface under water. Many plastics are produced as foams, full of small air bubbles. These will all float. Check that your sample is not a foam; use a magnifying glass if possible.

(E) Try to burn a small piece of the sample
Use only a *small* piece, hold it in metal tongs, carry out the test over a non-flammable surface, do not smell the sample while it is burning and smell cautiously. *Some plastics burn rapidly and drip molten material, and the fumes given off may be unpleasant and irritating.*

Observe the size and colour of the flame, the amount of smoke, whether molten drips are formed and whether they continue burning, and the odour when the flame has been extinguished. Some plastics are self-extinguishing, that is the flame goes out when the external flame is removed.

Tests such as these cannot always give a positive identification, because fillers and other additives can change the properties of a plastic and sometimes mixtures of plastic materials are used.

Group I. Thermoplastic materials
Thermoplastic materials are materials that can be repeatedly softened by heat, and will harden on cooling.

1. Poly(ethene) or polyethylene (Low and high density)
(*A*) Translucent (transparent when thin); (*B*) Low density is fairly flexible, soft with a "waxy" feel; (*C*) Easily and smoothly cut; (*D*) Floats; (*E*) Not self-extinguishing; molten droplets; blue flame with yellow tip; little smoke; smell of burning candle when flame is extinguished.

2. Poly(propene) or polypropylene
(*A*) Translucent; (*B*) Stiff and hard; (*C*) Easily cut, fairly smooth edges; (*D*) Floats; (*E*) Not self-extinguishing; molten droplets; flame mainly yellow with a trace of clear blue at the bottom; smell of burning candle when flame is extinguished but not identical with that from polyethylene.

3. Poly(chloroethene) or polyvinyl chloride (Without and with plasticizer)
(*A*) Transparent; (*B*) Without plasticizer: stiff, hard; With plasticizer: flexible, soft; (*C*) Fairly or very easy to cut, smooth edges; (*D*) Sinks; (*E*) Burns with difficulty; self-extinguishing; yellow flame, blue-green at bottom edges; unpleasant, irritating odour of hydrochloric acid. The flexible variety may have a strong aromatic smell of the plasticizer.

4. Poly(phenylethene) or polystyrene
(*A*) Transparent (however, fillers and pigments are often added); (*B*) Stiff,

hard, characteristic metallic ring if tapped with a pencil; (*C*) Fairly difficult to cut; (*D*) Sinks; (*E*) Burns readily, not self-extinguishing; drips continue to burn; orange-yellow flame with black, sooty smoke; odour in small amounts said to resemble marigolds (some modified polystyrenes give rubbery odours).

5. Expanded polystyrene
(*A*) Opaque, usually white; (*B*) Can be crumbled and does not recover; (*C*) Crumbles on cutting; (*D*) Very buoyant on water; (*E*) As polystyrene above.

6. Poly(methyl 2-methylpropenoate) or Perspex
(*A*) Can be very clear with a good surface gloss, but will be opaque if fillers or pigments have been added; (*B*) Stiff, hard; (*C*) Splinters on cutting; (*D*) Sinks; (*E*) Burns readily; not self-extinguishing; drips continue to burn, yellow flame with clear edges, not unpleasant, fruity odour.

7. Poly (tetrafluoroethene)
(*A*) Opaque, usually white; (*B*) Smooth and slippery to the touch; fairly soft; (*C*) Easily cut, smooth edges; (*D*) Sinks; (*E*) Will not ignite.

8. Polyurethane foam
(*A*) Opaque; (*B*) Usually soft and flexible, but can be rigid; (*C*) Usually does not crumble on cutting; (*D*) Floats; (*E*) Not self-extinguishing; fairly clear yellow flame; characteristic irritating smell.

9. Nylon
(*A*) Opaque; (*B*) Stiff, hard; (*C*) Fairly easily cut, smooth edges; (*D*) Sinks; (*E*) Difficulty to ignite; not self-extingushing; melts to a free-flowing liquid that drips carrying the flame with it; blue flame with a yellow tip; characteristic odour of burning hair.

10. Terylene
(*A*) Transparent (often contains coloured fillers); (*B*) As a film, stiffer than many other plastic films, strong; (*C*) Fairly easily cut, smooth edges; (*D*) Sinks; (*E*) Burns readily: not self-extinguishing; tendency to drip; yellow smoky flame; odour said to resemble that of burnt raspberry jam.

11. Cellulose ethanoate or cellulose acetate
(*A*) Transparent; (*B*) May be flexible (if a plasticizer is added), or fairly hard; (*C*) Easily cut, smooth edges; (*D*) Sinks; (*E*) Burns readily; may or may not be self-extinguishing, dark yellow flame with some smoke, odour of ethanoic acid (vinegar-like).

12. Cellulose nitrate
(Cellulose nitrate is used for making table tennis balls. It is not discussed elsewhere in this book.)

(*A*) Transparent (a white filler is added); (*B*) Flexible, fairly hard; (*C*) Easily cut, smooth edges; (*D*) Sinks; (*E*) Extremely flammable and very fast burning; yellow flame with some smoke; smell of mothballs from the camphor used as plasticizer.

13. Propenonitrile–buta-1,3-diene copolymer
(This plastic is used for making the casing of many domestic appliances such as food mixers, knobs for radio sets, etc. It is not discussed elsewhere in this book.)

(*A*) Opaque; (*B*) Fairly stiff, fairly hard; (*C*) Fairly easily cut, smooth edges; (*D*) Sinks; (*E*) Burns readily; normally not self-extinguishing, orange-yellow flame with black smoke; odours similar to polystyrene mixed with a bitter odour due to propenonitrile and a rubbery odour due to buta-1,3-diene.

Group II. Thermosetting materials
There is a second group of plastics known as *thermosetting*. When these are made, by mixing the monomers and heating (for some of them) chemical reactions occur to give the hard, solid polymer. Once this has happened the plastic cannot be softened again.

Tests for the four most common thermosetting plastics are given here, as samples of them will undoubtedly occur when a collection of plastics is made.

14. Urea-methanal (Urea-formaldehyde) and Melamine-methanal (2,4,6-triamino-1,3,5,-triazine-formaldehyde)
(Urea-methanal is used for light-coloured domestic electric fittings such as switch covers, bulb holders and plugs. Melamine is used for tableware such as "Melaware" and "Gaydon".)

(*A*) Opaque; usually light in colour; (*B*) stiff, hard; (*C*) Flakes on cutting; (*D*) Sinks; (*E*) Burns with difficulty, self-extinguishing; swells, cracks and turns white at edges of burnt portion, pale yellow flame with light blue-green edges, sharp odour of methanal and also, sometimes, a fish-like smell.

15. Phenol-methanal or phenol-formaldehyde
(Used for dark-coloured electric light fittings, domestic iron and cooking pan handles.)

(*A*) Opaque, dark coloured; (*B*) Stiff, hard; (*C*) Flakes on cutting; (*D*) Sinks; (*E*) Burns with difficulty; self-extinguishing; odour of phenol (carbolic acid).

16. Polyester resin
(Used with glass fibre for boat and car body manufacture and repairs.)

(*A*) Normally the glass fibre reinforcement can be seen; (*B*) Stiff, hard; (*C*) Difficult to cut, brittle; (*D*) Sinks; (*E*) Burns readily; often self-extinguishing, smoky flame, fruity odour.

The details given in this experiment are taken from the booklet "Identification of plastics for schools", by I.C.I. plastics division. Teachers who want more information about plastics are advised to contact The Academic Liaison Officer, I.C.I.Plastics Division, P.O.Box 6, Bessemer Road, Welwyn Garden City, Herts.

26.2.4 Fabrication of plastics

The *synthetic polymers*, that we have just been describing are commonly called *plastics*. The name means mouldable substance. Plastics are made into recognisable articles (i.e. fabricated) by heating in a mould under pressure. The methods used depend on whether the plastic is in Group I or Group II of the experiment above, on whether it is *thermoplastic*, that is a plastic which can be repeatedly softened by heat, and will harden on cooling, or on whether it is *thermosetting*, that is a plastic made by mixing and heating monomers together, and once formed cannot be softened again.

The commonest types of moulding processes are (a) injection moulding, (b) blow moulding, (c) vacuum moulding, (d) extrusion, (e) compression and (f) spinning.

(a) Injection moulding
The thermoplastic is melted and forced under pressure through a hole into the unheated mould. The plastic cools in the mould. Injection moulded articles often have a piece of plastic from the injection hole attached to them.

(b) Blow moulding
A length of softened thermoplastic tube is put into the hollow split mould. The mould closes and seals off one end of the tube. Compressed air then blows the plastic to the shape of the mould. Blow-moulding is used to make bottles and other containers with a small opening. The seam mark where the two halves of the mould fit together is often visible on the finished article.

(c) Vacuum moulding
Heated thermoplastic sheet is sucked into a mould by a vacuum. Complicated shapes can be formed in a plastic sheet by this technique. For example, the lining of chocolate boxes, with a shape to fit each chocolate, is made in this way.

(d) Extrusion
The thermoplastic material is fed into a heater and is continually pushed forward by a rotating screw. The molten plastic is forced through a hole, the shape of which determines the shape of the finished article. Plastic piping and plastic bags are made in this way.

(e) Compression
Thermosetting plastic articles are made by compressing the monomer mixture in a heated mould, when the polymerisation occurs and the plastic sets in the shape of the mould.

(f) Spinning
All fibres are made in this way. The fibre plastics are thermoplastics and are extruded through plates containing many fine holes (called *spinnerets*) as sets of long continuous fibres.

Experiment. Fabrication of plastics

See if you can decide which method has been used to fabricate each of the plastic articles that you have collected together for the previous experiment.

26.3 Natural polymers

Polymers are essential constituents of plants and animals. Two, in particular are very important, these are *carbohydrates* and *proteins*.

26.3.1 Carbohydrates

Carbohydrates are formed in plants by *photosynthesis* (see section 18.6.2) and are then eaten by animals. Thus they provide an indirect way by which animals use the sun's energy. The name *carbohydrate* has been given to this group of compounds because they all have the general formula $C_x(H_2O)_y$, e.g., glucose, $C_6H_{12}O_6$, and sucrose, $C_{12}H_{22}O_{11}$.

There are two main groups of carbohydrates, *sugars* and *non-sugars.*

(a) Sugars

These are all sweet, soluble solids. The simplest sugars are all isomers with the formula $C_6H_{12}O_6$ and include glucose (figure 26.8(*a*)) and fructose. All common carbohydrates are based on a unit of 6 carbon atoms, and therefore these simple sugars that contain only one unit are called *mono*-saccharides.

Other sugars have the formula, $C_{12}H_{22}O_{11}$, e.g., cane sugar (sucrose) (figure 26.8(*b*)) and maltose, and as they are built up of two units of 6 carbon atoms they are called *di*-saccharides.

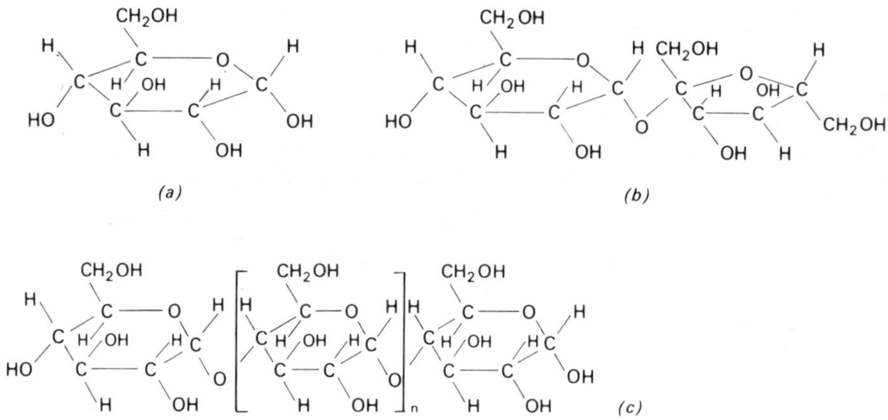

Figure 26.8 (*a*) glucose, (*b*) sucrose (cane-sugar), (*c*) starch.

(b) Non-sugars

These are insoluble, tasteless polymeric compounds. They include starch (figure 26.8(*c*)) and cellulose and have the general formula $(C_6H_{10}O_5)_n$, where *n* is a large number. These carbohydrates are called *poly*saccharides; they are basically sugar molecules that have condensed together:

$$nC_6H_{12}O_6 \rightarrow (C_6H_{10}O_5)_n + nH_2O$$

(c) Breakdown of starch

In section 25.6.2 the breakdown of starch into maltose, glucose, and finally ethanol is discussed. The first two stages result from hydrolysis, i.e., the reaction with water, catalyzed by enzymes.

Starch is present in many common foods. For example, 80% by weight in rice, 40% by weight in bread and 20% by weight in potatoes. When you eat such carbohydrates the enzymes in your body break down the starch to di- and mono-saccharides. The hydrolysis of starch to maltose begins in your mouth, catalyzed by the enzyme *amylase* that is in saliva. When the starch reaches your stomach the dilute hydrochloric acid that is present there catalyzes the hydrolysis to glucose.

Experiment. The hydrolysis of starch

Prepare a solution of starch by boiling 1 g of soluble starch in 100 cm^3 of water.

To 5 cm^3 of the starch solution in a test-tube add a few drops of saliva, and place the tube in a water-bath at between 35–40 °C.

To 5 cm^3 of the starch solution in a second test-tube add a few drops of dilute hydrochloric acid and place this tube in the water-bath.

After a few minutes add a spot of each solution separately to a solution of iodine in potassium iodide. Starch produces a dark-blue colour with iodine. If this is not produced then hydrolysis is complete.

The products of the hydrolysis can be identified by paper-chromatography using the method described in section 1.10. The solvent to be used is a mixture of 1 part water, 1 part ethanoic acid and 3 parts propan-2-ol, by volume. Spots of glucose and maltose solutions should also be put on the chromatography paper. The chromatogram should be left until the solvent front has reached the top of the paper. It is then dried and developed by spraying in a fume cupboard with a solution of 1 % phenylamine (aniline), 1 % diphenylamine and 10 % phosphoric(V) acid in propanone (acetone). After spraying, the chromatogram should be warmed when coloured spots show the position of the sugars.

If the spots are not visible the solution used for preparing the chromatogram was too dilute. Prepare another chromatogram, repeating the spotting several times, and drying the paper between applications.

The chromatograms should show that with saliva as the catalyst the starch has been hydrolyzed to maltose, and with hydrochloric acid to glucose.

(d) Cellulose and cellulose ethanoate (cellulose acetate)

Cellulose is a larger polymer than starch of the same general formula $(C_6H_{10}O_5)_n$. The fibrous parts of plants are made of the long cellulose molecules that collect together in bundles. Cotton and linen are examples of natural plant fibres. Wood is made up of cellulose and a related carbohydrate called lignin.

Cellulose is the main constituent of paper. The cellulose fibres from wood chips are mixed with china clay, aluminium sulphate and other materials and a dilute solution is poured onto a moving gauze-belt. The water drains away and the strip of cellulose and other materials is rolled and heated by steam to give paper. Figure 26.9 is a view inside a paper mill.

We cannot digest cellulose; that which we do eat passes straight through our bodies. But grass-eating animals such as cows can digest it because they have in their stomachs the necessary bacteria. We therefore let the cow convert the cellulose to beef-steaks that we can digest.

Cellulose will dissolve in sodium hydroxide solution and this dissolved cellulose will react with carbon disulphide to give a viscous liquid. If this liquid is injected as a thin stream into dilute acid the cellulose is reformed as long fibres.

Figure 26.9 A paper-making machine showing the gauze belt in the foreground, the press-section where some of the water is removed from the paper, and finally the hooded drying section. (Photo: Reed Paper and Board (U.K.) Ltd.).

These are known as *viscose rayon*. If a thin film is injected into the acid a transparent film known as *cellophane* is formed.

Cellulose will also react with ethanoic acid to form cellulose ethanoate (cellulose acetate). This can be dissolved in propanone and injected as a thin stream into heated chambers when the cellulose ethanoate fibre is formed. This is called *acetate rayon* and resembles natural silk more closely than any other artificial fibre. Cellulose ethanoate polymer is also used as the base of photographic film, and for the handles of workshop tools.

26.3.2 Proteins

Proteins are polymers of chains of amino-acids and they occur in all living tissues. Some simple amino-acids are illustrated in figure 26.10; you will see that they are compounds of carbon, hydrogen, oxygen and nitrogen containing an acid (—COOH) group and an amino (—NH_2) group. Amino-acids can also contain sulphur or phosphorus. These amino-acids condense together in the same way that the monomers form the nylon polymer (figure 26.5). Each

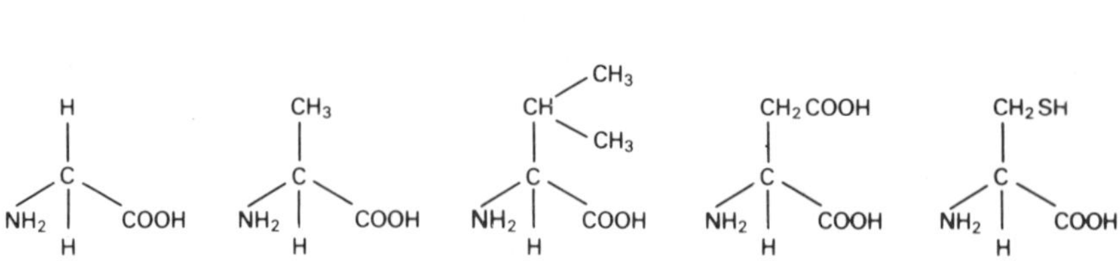

Glycine (aminoethanoic acid) — Alanine (2-aminopropanoic acid) — Valine (2-amino-3-methyl butanoic acid) — Aspartic acid (aminobutanedioic acid) — Cysteine (2-amino-3-thio-propanoic acid)

Figure 26.10 Some simple amino-acids.

—COOH group reacts with an—NH_2 group to give a molecule of water and the two molecules link together through a $-\overset{\mathrm{H}}{\overset{|}{\mathrm{N}}}-\overset{\mathrm{O}}{\overset{\|}{\mathrm{C}}}$ linkage:

$$-\mathrm{N}{<}^{\mathrm{H}}_{\mathrm{H}} + {}^{\mathrm{O}}_{\mathrm{HO}}{>}\mathrm{C}- \;\rightarrow\; -\overset{\mathrm{H}}{\overset{|}{\mathrm{N}}}-\overset{\mathrm{O}}{\overset{\|}{\mathrm{C}}}- + \mathrm{H_2O}.$$

This is called a *peptide* linkage, and a protein is a *poly*-peptide.

When a poly-saccharide was formed (section 26.3.1) the *mono*-saccharides that joined together were all identical. When a protein is formed, many different amino-acids join together (up to 25) and in a particular protein the amino acids that are present are in a specific sequence. The number of possible proteins is enormous, and this great variety enables each living organism to have its own individuality.

Proteins have molecular masses from about 10 000 to several million. Their structure is determined by breaking them up, using enzymes or acids, into the amino-acids from which they were formed. This is illustrated in figure 26.11, where, for simplicity, parts of the protein not involved in the hydrolysis have been represented by non-chemical symbols (circle, square, triangle, etc.). You can readily see from the figure that the amino-acids only differ by the group attached to the carbon atom between the—COOH and—NH_2 groups.

It will be clear that, even if only 25 amino-acids have been isolated from proteins, the number of different possible arrangements of these amino-acids in substances whose molecular masses may be several million is very large, i.e., there are very many different proteins. Some long protein chains do not have a

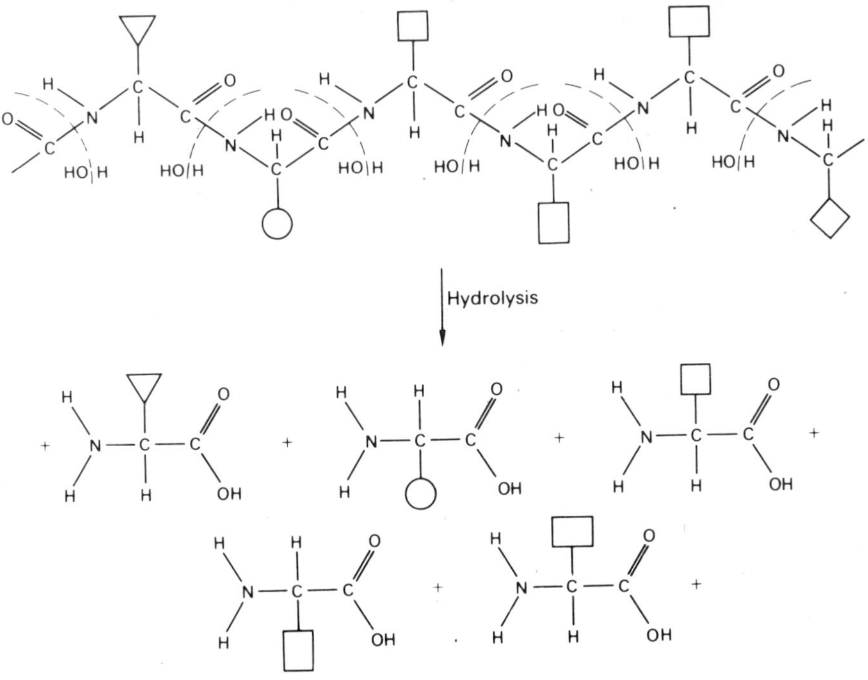

Figure 26.11 Hydrolysis of a protein to amino-acids.

zig-zag structure as in figure 26.11, but curl round and round like a spiral staircase. This spiral structure is an important feature of them and responsible for some of the important properties for which proteins are responsible in living organisms, such as the organisms' ability to reproduce and grow.

26.4 Rubbers

Another important group of polymers are *rubbers*, both *natural* and *synthetic*.

26.4.1 Natural rubber

This is the sap, or *latex*, of the rubber tree, (figure 26.12). It is a hydrocarbon polymer built up from the compound, methylbuta-1,3-diene (isoprene) (figure 26.13(*a*)). Rubber latex is soft and sticky and quite unsuitable for most purposes. When heated to 140 °C a well-mixed mixture of natural rubber and sulphur combine to give a much stronger and better wearing product. This process, known as *vulcanization* (figure 26.13(*b*)) results in sulphur atoms holding the long rubber chains together. The properties of rubber can be modified still further by addition of other substances such as carbon black (see section 17.4.4).

Figure 26.12 Latex from a rubber tree being collected, on a plantation in Malaysia. (Photo: Paul Popper Ltd.)

26.4.2 Synthetic rubber

When rubber supplies became scarce during the World War 2, chemists invented synthetic rubbers. These have certain properties that are superior to natural rubber, so even after natural rubber supplies increased again the production of synthetic rubbers also increased.

The first synthetic rubber made was *neoprene*. This is poly(2-chlorobuta-1,3-diene) (figure 26.13(c)). The monomer, 2-chlorobuta-1,3-diene (chloroprene), is made from ethyne (figure 26.13(*c*)). It is used for making hoses for chemicals, petrol and oil as it is highly resistant to chemical action.

Most synthetic rubber made is a ploymer containing *buta-1,3-diene*. The polymer produced from a mixture of buta-1,3-diene and *phenylethene* (styrene) (figure 26.13(*d*)) gives a rubber with good wear resistance and is used for car tyres.

A rubber produced from 2-*methylpropene* (isobutylene) with 1–2% methylbuta-1,3-diene (isoprene) gives a polymer that after vulcanization is less than a fortieth as permeable by gases as natural rubber. This *butyl rubber* (figure 26.13(*e*)) is much used for inner tubes for bicycle and motor car tyres. If you have a bicycle with one butyl inner tube and the other inner tube of some other rubber you have probably discovered for yourself this superior property of butyl rubber.

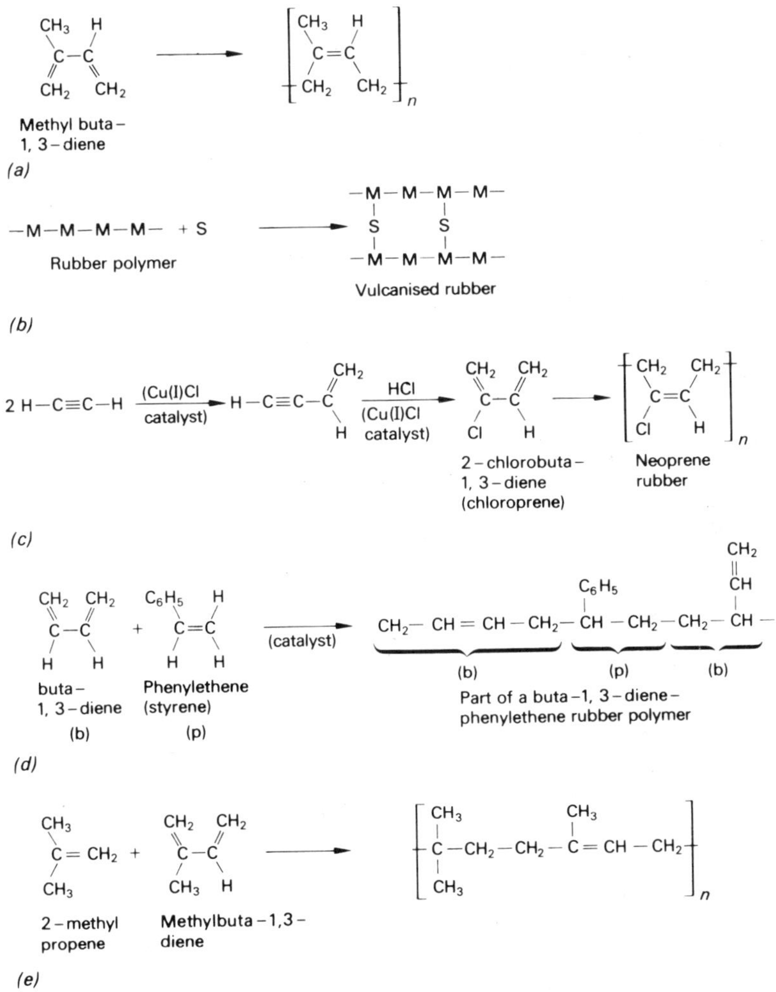

Figure 26.13 (*a*) Natural rubber, (*b*) Vulcanisation, (*c*) Neoprene rubber, (*d*) Butadiene-styrene rubber, (*e*) Butyl rubber.

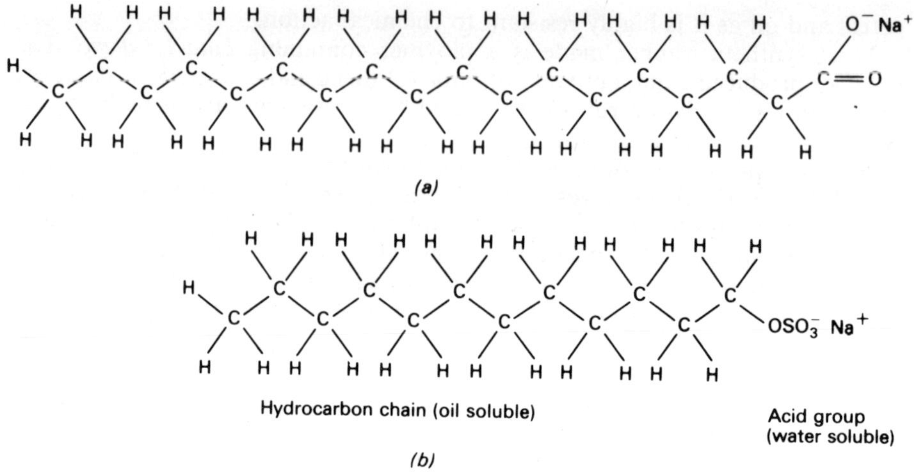

Figure 26.14 (*a*) Sodium stearate (a soap), (*b*) "Teepol" (a synthetic detergent).

26.5 Synthetic detergents

Soap and most modern synthetic detergents (figure 26.14) consist of long molecules. One end of the molecule is made up of a group of atoms that tend to be soluble in water. The rest of the molecule is a long hydrocarbon chain that causes insolubility in water but solubility in oils.

In soap the water-soluble group is the sodium salt of the acid (—COOH) group, while in many synthetic detergents it is the sodium sulphonate group ($-SO_3^-Na^+$) (figure 26.14).

These detergents, soap included, act in several ways. By lowering the surface tension of water, they enable materials to be thoroughly wetted. They also emulsify oil and grease particles by the hydrocarbon chains dissolving in the grease leaving the water-soluble groups on the outside. These particles are then attracted to the water and can be washed away.

Modern detergents have the advantage over soap that they do not react with the salts present in hard water and are therefore less wasteful.

Questions on Chapter 26

1. Write down a formula to show how a section of a protein molecule is built up and explain what happens to it on hydrolysis.

A certain protein has a molecular mass of 1 500 000. What is the mass of one molecule of this protein? (One mole contains $6{\cdot}02 \times 10^{23}$ molecules)

2. Write short notes on each of the following: polymerization, proteins, nylon, fermentation.

3. Write an essay on "Petroleum as a source of organic chemicals".

4. (*a*) Write the structural formula of ethene.

(*b*) Describe the reaction which takes place between bromine and ethene or a compound of similar structure.

(*c*) Explain the formation of either poly(ethene) or poly(phenylethene). (Details of conditions are not required). Give a large scale use of each of these substances.

(*d*) Outline one process by which ethene is obtained from natural sources. (*C.*)

5. Poly (phenylethene) decomposes on heating to yield the monomer $C_6H_5CH{:}CH_2$.

(*a*) What volume of gas, at 273 °C and 1 atmosphere pressure, would be expected for each gram of poly(phenylethene) decomposed?

(*b*) Phenylethene boils at about 150 °C. Describe a method for distilling, indicating how you would check the boiling point.

(*c*) What simple chemical test could be made to confirm that phenylethene is an unsaturated compound?

(The volume of one mole of gas at s.t.p. is 22·4 dm³. C = 12; H = 1) (*L.*)

6. What are isomers? Write the structural formulae and the names of the isomers of C_4H_{10}.

"Poly(ethene) is a polymer". Carefully explain the meaning of this statement.

Describe in outline the steps by which poly(ethene) is obtained from petroleum oil.

What is the chief chemical characteristic of poly(ethene) which makes it so useful?

Why is the heavy organic chemical industry, which makes chemicals from petroleum, centred near the coast? (*J.M.B.*)

27

Identification of Common Gases and Ions

27.1 Introduction

Chemists often wish to find out what is present in a substance or a mixture of substances. They do this by *analysis*.

Finding out the *amounts* of substances present is called *quantitative analysis*. One type of quantitative analysis is *volumetric analysis* (see chapter 12).

Finding out *which* substances are present is called *qualitative analysis*. There are many tests which can be used to identify a substance. For example you can usually tell if a substance is a metal because you know what most metals look like. If you see a pale yellow-green gas or a violet vapour you will probably know immediately what they are. Many solids too, like copper(II) salts, have characteristic colours.

The smell of a substance is sometimes sufficient for identification. Ammonia and hydrogen sulphide have characteristic odours–once smelt, never forgotten! Also the organic acid, ethanoic acid, the important constituent of vinegar, has an odour that most of us recognize readily. Many organic compounds have such characteristic odours. All smells are due to chemical compounds, often very small amounts, that are present in the air.

The taste of a compound can also be used to identify it, but we certainly do not advice you to taste an unknown substance as it might be a poison! Even the feel of a substance may help you to identify it. Polyethene, for example, has a "waxy" feel. Also sodium hydroxide solution feels "soapy" because it dissolves the surface layers of your fingers. Like tasting we do not think that you should feel unknown chemicals.

The research chemist and the industrial chemist use many sophisticated methods to identify substances and you may meet some of these in the Sixth form.

In this chapter we are concerned only with the tests used for identifying simple compounds and ions. Many of the tests are described elsewhere in the book; they are, for convenience, brought together in this chapter.

27.2 Tests for gases

Common gases can be identified by carrying out the tests given in table 27.1. Some of these tests give positive results for more than one gas and, where possible, you should always carry out two tests. Often the smell of the gas is sufficient to be almost certain of its identity but, of course, you need to be familiar with the smell.

You will probably be able to perform some of the following tests in your school

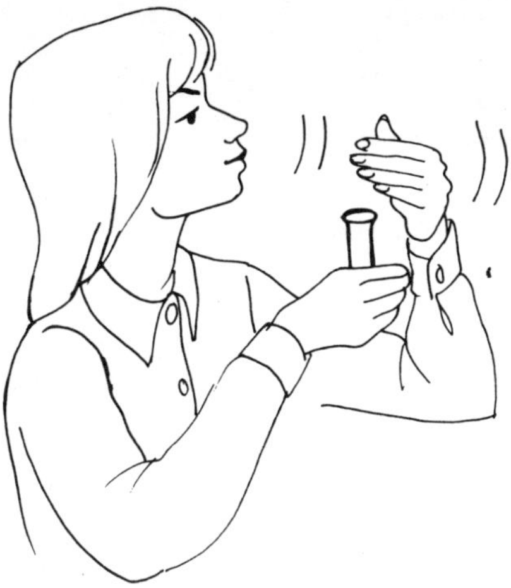

Figure 27.1 How to smell a gas. (Drawing: Mary Randell).

laboratory. You should always smell gases with care as many of them are very poisonous. This does not mean that the small amount needed for you to identify its odour will do you any harm, but you must always breathe in as little as possible. Do this by taking shallow breaths. Do not hold your nose right over the tube or other vessel containing the gas but fan some of the gas towards you with your hand (see figure 27.1).
Note: Those with chest complaints must be especially careful.

27.3 Tests for common chemical compounds

27.3.1 Appearance

Sometimes the appearance of a substance gives a clue to its identity, particularly if it is a coloured compound.

Most compounds consist of colourless crystals. These often appear white when they exist as a powder. A few are coloured. The most common coloured compounds are given in table 27.2.

27.3.2 The flame test

Some metal ions can be identified by the colour of light they give out when heated. Beautiful coloured flames can be obtained, the colour depending on which metal ion is present.

Add a few drops of concentrated hydrochloric acid to a little of the compound being tested. Dip the end of a piece of nichrome wire in the mixture and then place the wire in a non-luminous Bunsen flame. The metal ions that can be detected in this way are given in table 27.3 and the remarkable photographs by Aubrey Rendell on the back cover of this book show flames coloured by these metals.

TABLE 27.1 TESTS FOR GASES

Gas	Test	Result of test
Hydrogen	Collect in an inverted test-tube, apply lighted taper	Burns with a sharp pop
Oxygen	Place glowing wooden splint into gas	Splint relights
Carbon dioxide	Pass gas into calcium hydroxide solution (lime water)	Lime water turns "milky" due to fine precipitate of calcium carbonate
Chlorine	(a) Place moist blue litmus paper in the gas (b) Smell gas – TAKE CARE	Litmus turns red and is then bleached Characteristic odour – VERY POISONOUS
Hydrogen chloride	(a) Dip a glass rod into concentrated ammonium hydroxide solution and place in the gas (b) Place moist blue litmus paper in the gas	Dense white smoke of ammonium chloride formed Paper turns red and is then bleached
Ammonia	(a) Dip a glass rod into concentrated hydrochloric acid and place in the gas (b) Place moist red litmus paper in the gas (c) Smell gas – TAKE CARE	Dense white smoke of ammonium chloride formed Litmus turns blue Characteristic odour – VERY POISONOUS
Sulphur dioxide	(a) Place a filter paper soaked in potassium manganate(VII) solution in the gas (b) Place a filter paper soaked in potassium dichromate(VI) solution in the gas (c) Smell gas – TAKE CARE	Colour changes from purple to colourless Colour changes from orange to green Characteristic odour of burning sulphur – VERY POISONOUS
Hydrogen sulphide	(a) Place a filter paper soaked in lead(II) nitrate or ethanoate solution in the gas (b) Place filter paper soaked in potassium manganate(VII) solution in the gas (c) Place filter paper soaked in potassium dichromate(VI) solution in the gas (d) Smell gas – TAKE CARE	Dark brown or black or silvery stain of lead(II) sulphide formed Colour changes from purple to colourless Colour changes from orange to green Characteristic odour of bad eggs – VERY POISONOUS
Nitrogen dioxide	(a) Colour (b) Smell gas – TAKE CARE	Brown gas Characteristic odour – VERY POISONOUS
Nitrogen	No positive response to any simple test	

TABLE 27.2 COLOURED COMPOUNDS

Colour	Compound
Blue	Copper(II) salts (hydrated)
Pale green	Iron(II) salts (hydrated)
Yellow	Iron(III) salts (hydrated)
Yellow	Lead(II) oxide
Dark brown	Lead(IV) oxide
Red	Dilead(II) lead(IV) oxide (red lead)
Yellow when hot, white when cold	Zinc oxide
Black	Copper(II) oxide, manganese(IV) oxide
Purple	Potassium manganate(VII)
Orange	Potassium dichromate(VI)

TABLE 27.3 COLOURS OF FLAMES

Colour	Metal ion
Lilac	Potassium
Crimson	Lithium
Light green	Barium
Brick red	Calcium
Red	Strontium
Blue-green	Copper
Intense yellow	Sodium*

* often present as an impurity in other substances.

27.3.3 Action of heat

Heating a substance and carefully observing what happens can help to identify certain ions. A small quantity of the substance is heated in a small test-tube called an ignition tube. A cork fitted with a delivery tube is ready should a test for carbon dioxide prove necessary (figure 27.2). The possible observations and what they signify are given in table 27.4.

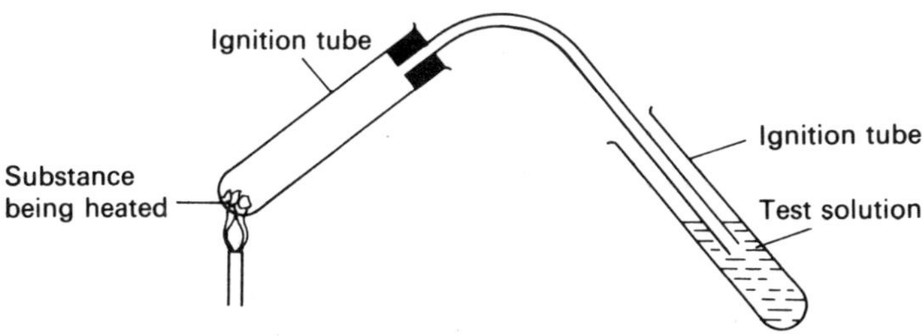

Figure 27.2 Heating a solid substance.

TABLE 27.4 ACTION OF HEAT ON A SUBSTANCE

Observation	Conclusion
Residue is yellow when hot and white when cold	Probably zinc oxide formed by heating a zinc salt
Blue solid gives a black residue	Copper(II) salt on heating gives copper(II) oxide
White solid gives a yellow residue	Possibly a lead(II) salt on heating giving lead(II) oxide
White solid that sublimes	Ammonium salt
Solid melts; oxygen evolved	Sodium or potassium nitrate
Oxygen and brown nitrogen dioxide evolved	Most nitrates
Carbon dioxide evolved	Carbonate or hydrogencarbonate
Ammonia evolved	Ammonium salt
Sulphur dioxide evolved	Some sulphites and a few sulphates

27.3.4 Solubility

Shake a small quantity of the substance (about the size of a pea) with half a test-tube full of distilled water. The following solubility rules are worth remembering:

(*a*) All nitrates are soluble.
(*b*) All the common chlorides are soluble except those of silver and lead.
(*c*) All the common sulphates are soluble except those of barium and lead. Calcium sulphate is only slightly soluble.
(*d*) All the common carbonates are insoluble except those of sodium, potassium and ammonium.
(*e*) All common salts of sodium, potassium and ammonium are soluble.
(*f*) Lead(II) nitrate and lead(II) ethanoate are the only common soluble lead salts.

27.3.5 Action of dilute hydrochloric acid

Add a little dilute hydrochloric acid to a little of the substance in a test-tube. If no effervescence occurs in the cold, warm the tube carefully. Test any gases evolved as described in table 27.1. The significance of any observations is given in table 27.5.

TABLE 27.5 ACTION OF DILUTE HYDROCHLORIC ACID ON A SUBSTANCE

Observation	Conclusion
Vigorous effervescence; carbon dioxide evolved	Carbonate or hydrogencarbonate
Sulphur dioxide evolved	Sulphite or hydrogen sulphite
Hydrogen sulphide evolved	Sulphide

27.3.6 Tests for cations (positively charged ions)

The following tests will identify these ions: ammonium (NH_4^+), calcium (Ca^{2+}), magnesium (Mg^{2+}), aluminium (Al^{3+}), iron(II) (Fe^{2+}), iron(III) (Fe^{3+}), zinc

(Zn^{2+}), copper (Cu^{2+}) and lead (Pb^{2+}). Other common cations can be identified by their flame tests.

The first stage is to obtain a solution of the substance by trying, in order, the following solvents:

(i) *Distilled water* Use a small amount of the substance (about the size of a pea) and dissolve in half a test-tube full of water.

(ii) *Dilute hydrochloric acid* Use the same amount of substance and dissolve in as little dilute hydrochloric acid as possible. Dilute with distilled water to half a test-tube full of solution.

(iii) *Dilute nitric acid* Proceed as with (ii).

Divide the solution into four portions.

(a) To one portion of the solution add dilute hydrochloric acid. The possible result is given in table 27.6.

TABLE 27.6 ACTION OF DILUTE HYDROCHLORIC ACID ON A SOLUTION OF A SUBSTANCE

Observation	Conclusion
White precipitate	Lead(II) chloride from a lead(II) salt. To confirm filter off the lead(II) chloride. It will dissolve in hot water and crystallize out on cooling

(*Note.* Silver salts also give a white precipitate (see section 27.3.4). The precipitate is soluble in dilute ammonium hydroxide solution. However you are unlikely to be given a silver compound to analyse.)

(b) To a second portion of the solution add dilute hydrochloric acid and pass hydrogen sulphide into the mixture obtained (or add a solution of hydrogen sulphide in water). Possible results are given in table 27.7.

(c) To a third portion of the solution add sodium hydroxide solution drop by

TABLE 27.7 ACTION OF HYDROGEN SULPHIDE ON AN ACIDIC SOLUTION OF A SUBSTANCE

Observation	Conclusion
Black precipitate	Copper(II) or lead(II) sulphide from a copper(II) or lead(II) salt
Very faint cloudiness	Deposit of sulphur formed by oxidation of the hydrogen sulphide e.g. by iron(III) ions or nitric acid

drop with shaking until the solution is alkaline (test by removing a little of the solution on a glass rod and placing on red litmus paper). Note the colour of any precipitate. Then add an excess of the sodium hydroxide solution and note whether or not the precipitate dissolves on shaking.

(d) With the fourth portion of the original solution repeat this test using ammonium hydroxide solution.

Observations from these tests and what they signify are given in table 27.8.

TABLE 27.8 ACTION OF SODIUM HYDROXIDE AND AMMONIUM HYDROXIDE ON A SOLUTION OF A SUBSTANCE

Addition of a few drops of sodium hydroxide or ammonium hydroxide solution	Excess of sodium hydroxide solution	Excess of ammonium hydroxide	Cation present in solution
Red-brown precipitate of iron(III) hydroxide	Precipitate remains	Precipitate remains	Iron(III)
Green-white precipitate of iron(II) hydroxide; oxidizes to red-brown in contact with air	Precipitate remains	Precipitate remains	Iron(II)
White precipitate of magnesium or calcium hydroxide	Precipitate remains	Precipitate remains	Magnesium or Calcium
White precipitate of lead or aluminium hydroxide	Precipitate dissolves	Precipitate remains	Lead or Aluminium
White precipitate of zinc hydroxide	Precipitate dissolves	Precipitate dissolves	Zinc
Pale blue precipitate of copper(II) hydroxide	Precipitate remains	Precipitate dissolves to give a deep blue solution	Copper(II)
No precipitate	Ammonia gas on heating		Ammonium

27.3.7 Tests for anions

(a) Chloride, bromide and iodide

(i) Dissolve a small quantity of the substance (about the size of a pea) in half a test-tube of distilled water. Add silver nitrate solution. If a precipitate is formed one of these halides is present.

Divide the suspended precipitate into two parts. To one part add amonium hydroxide solution. Leave the other part in sunlight for at least ten minutes.

The observations and what they signify are given in table 27.9.

TABLE 27.9 SILVER HALIDE PRECIPITATES

Colour of precipitate	Solubility in ammonium hydroxide solution	Effect of sunlight	Conclusion
White	Soluble	Goes violet	Chloride
Pale yellow	Appears insoluble*	Darkens	Bromide
Yellow	Insoluble	No change	Iodide

* Silver bromide does dissolve to some extent. If a little of the precipitate is filtered off, it will dissolve in *concentrated* ammonium hydroxide solution.

(ii) Bromide and iodide can also be identified by adding a little chlorine water to a solution in water of a bromide or iodide. Then bromine or iodine is liberated colouring the solution orange-red or brown. If a little tetrachloromethane is then shaken with the solution, most of the bromine or iodine leaves the water to dissolve in the tetrachloromethane, colouring it brown if bromine is present and violet if iodine is present.

(b) Sulphate

To a solution of the substance in dilute hydrochloric acid add barium chloride or barium nitrate solution. A white precipitate of barium sulphate is formed if a sulphate is present.

If the substance is dissolved in water, sulphides, sulphites and carbonates also give a white precipitate. These white precipitates dissolve on the addition of acid.

(c) Nitrate

(i) Make a solution of the substance alkaline with excess sodium hydroxide solution and heat. If ammonia is evolved this indicates that an ammonium salt is present. In these circumstances the mixture must be heated until no more ammonia is evolved before proceeding.

Then add a small amount of Devarda's alloy (a metal powder containing copper, aluminium and zinc) and heat. If a nitrate is present ammonia gas is evolved because of the reduction of the nitrate ions.

(ii) The 'brown ring test' is also used to identify the nitrate ion (see section 16.6.6).

(d) Carbonate, hydrogencarbonate, sulphite and sulphide

These are identified by the gases evolved when a sample of the substance is treated with dilute hydrochloric acid (table 27.5)

27.4 Recording your results

Always write down the result of a test at the time that you do it. Do not try to remember what happened until the end of a series of tests; it is likely that you will remember things incorrectly. Also if you record your observations when you are doing the experiment it is possible to repeat tests that appear to give conflicting evidence. If you do not think about your experiments until after you have left the laboratory it is not possible to repeat tests or carry out further tests that you think are desirable.

Set out your results clearly. One of the best ways is to use three columns. In the first column you can give the "test", in the second column the "observation" and in the third column the "conclusion" that you draw from the test. At the end of a series of tests give a "final conclusion" stating either the identity of the substance or why you cannot identify it from your tests.

28

Chemistry and Society

28.1 The growth of technology

This chapter is concerned with the impact of chemistry on society. In this book there are a number of topics that everyone has heard of, even if they don't really understand what the words mean. Plastics, nuclear fission, petroleum, iron and steel, and fertilizers are just a few of the topics that you study as a chemist.

People who have heard of these things have done so because, like you and me, they are part of a *technological society*. What this means is that we are now a society that depends on science and scientists.

28.2 The growth of the population

The population of the world is growing at an enormous rate (figure 28.1). It took several million years (until 1850) for the world population to reach 1 billion (1 000 000 000). This number had doubled by 1930, (only a further 80 years), and by 1976, (only 46 years later), doubled again. Even today the population of the world has reached such a level that it is putting a noticeable strain on the natural resources of the planet–on our air, our water, our mineral resources, our energy resources, our food resources. Let us examine each of these in turn.

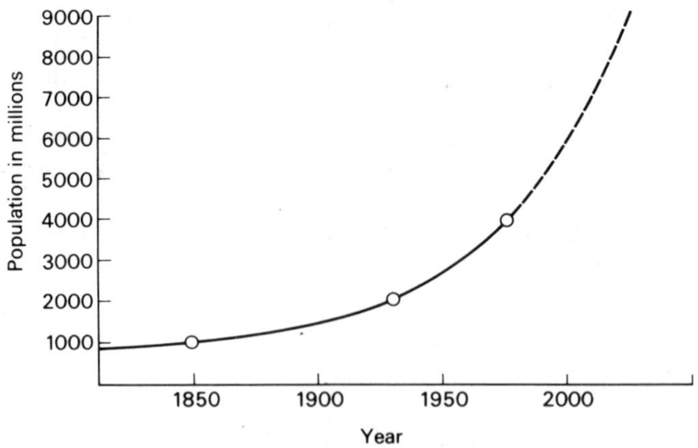

Figure 28.1 The growth of the population of the earth.

28.3 Air

28.3.1 Introduction

Air has been discussed in Chapter 13 and we saw there that it is a mixture of nitrogen and oxygen with small amounts of the noble gases, carbon dioxide and water vapour. Now the increasing world population needs and desires warm homes and manufactured goods and also the ability to move readily over the earth's surface. This means that gases and dusts have begun to contaminate the atmosphere. *Air pollution*, its nature, causes and effects are discussed in this section.

If air pollutants are produced on a small scale they are so diluted in the atmosphere that they become harmless. Today's problems are caused by the large growth in the population, its concentration into towns and its increasing use of energy.

28.3.2 Sources of air pollution

The materials in the air that are not there naturally i.e. the pollutants, are produced largely by power and heat generating plants, by motor vehicles and by industrial processes.

(a) Power and heat generation

It was when *coal* started to be widely used for heating homes and industry that *pollution from burning* became serious. Most coal contains sulphur, and when it is burnt *sulphur dioxide* and *sulphur(VI) oxide* are formed:

$$S + O_2 \rightarrow SO_2$$

$$2SO_2 + O_2 \rightarrow 2SO_3$$

For several hundred years these have been the *major air pollutants*. Although it is possible to remove the sulphur from the *fuel-oil* used in oil-burning power stations, it is uneconomic to do so, and these also contribute to the sulphur oxides in the air.

Soot, i.e. small solid particles, also gets into the atmosphere when burning occurs.

Nature also puts sulphur oxides into the atmosphere from volcanoes (see figure 28.7) and by biological decay. These natural emissions are equal to those formed from coal and oil (about 120 million tonnes each year) but they are spread out over the land surface of the earth. In contrast the sulphur oxides in the air from burning are highly concentrated near to their source.

(b) Motor vehicles

The consumption of petroleum and gaseous fuels has continued to increase during the past sixty years and because of this new pollutants have been introduced into the air. Fumes emitted from a car exhaust will contain not only carbon dioxide and water vapour–the products of the complete burning of petrol–but also carbon monoxide, soot, unburnt petrol, nitrogen oxides and lead compounds.

(i) Carbon monoxide and soot

These are produced by the incomplete combustion of petrol vapour in the car engine. If octane (one of the hydrocarbons in petrol) is burnt in an ample supply of air, carbon dioxide and water vapour are formed:

$$2C_8H_{18}+25\,O_2 \rightarrow 16CO_2+18H_2O$$

If the air supply is restricted, as in a car engine, the combustion is incomplete i.e. some carbon monoxide and soot (i.e. carbon particles) are formed instead of carbon dioxide.

(ii) Nitrogen oxides
These are produced in the car engine from the nitrogen and oxygen in air. At the high temperatures in the engine, some nitrogen and oxygen will combine to give nitrogen oxide:

$$N_2+O_2 \rightleftharpoons 2NO$$

When this cools down as it leaves the exhaust it combines with more oxygen to give nitrogen dioxide:

$$2NO+O_2 \rightleftharpoons 2NO_2$$

(iii) Lead compounds
These are present because lead is deliberately added to petrol to increase its *octane-number*. The octane-number (named after the hydrocarbon octane which has been given an octane number of 100) is a measure of the *anti-knock* properties of the petrol. It is the number often displayed on petrol pumps. "Star-rating" is also a measure of octane number (e.g. 4-star petrol has an octane number between 97 and 100). If the octane-number of the petrol used by a particular engine is too low then spontaneous *pre-ignition* or *knocking* occurs. This means that the petrol-air mixture in the engine ignites too soon i.e. before the spark-plug fires. Knocking can cause serious damage to an engine and to prevent it, petrol with a higher octane-number must be used. The explosion of a petrol-air mixture is a complicated process but it has been found that knocking can be stopped if small quantities of lead compounds are added to the petrol. That is, by adding these compounds the octane-number of the petrol is increased and higher quality petrol is produced for a very small cost.

The lead compounds that are added to the petrol (organic lead compounds e.g. tetraethyl lead(IV)) are converted by combustion in the car engine to lead(IV) oxide and lead. These would deposit in the engine but if dichloro- or dibromoethane is also added to the petrol, lead(II) chloride or lead(II) bromide is formed. These are volatile enough to be emitted with the exhaust gases and form solid particles on cooling. Several grams may be produced from each gallon of petrol.

Unburnt petrol will also be present in the fumes from car exhausts.

(c) Industrial processes

Many chemical manufacturing plants and other factories emit into the air varying amounts of hundreds of different substances. These depend on the type of factory. Certain paper-making processes, for example, use sulphides, and hydrogen sulphide gas is produced. Volatile metal salts are produced during mineral smelting. We expect there is a factory near you which pollutes the atmosphere. Problems connected with industrial processes are discussed further in section 28.9.

TABLE 28.1 ESTIMATED AMOUNTS OF AIR-POLLUTANTS AND THEIR SOURCES (U.S.A). (IN MILLIONS OF TONNES PER YEAR)

Source	Dust particles	Sulphur oxides	Nitrogen oxides	Carbon monoxide	Hydrocarbons
Power stations and houses	8	22	9	2	0·5
Cars, trains, ships, etc.	1	1	7	58	15
Refuse disposal	1	0·1	0·5	7	1·5
Industry	7	7	0·2	9	8
Total	17	30	17	76	25

28.3.3 The extent of air pollution

Air pollutants can be divided into *particulate matter* (i.e. solids and fine clouds of liquids such as those in fogs) and *gases*. Table 28.1 is an estimate of the amounts of 5 major pollutants and their sources. This table (which is for America, but the data will be similar for other industrialised countries) shows that sulphur oxides from power stations and domestic fires and carbon monoxide and unburnt fuel from motor-cars are the most serious air-pollutants, although the oxides of nitrogen, because they are toxic, also present a serious problem.

The table does not include the vast amount of *carbon dioxide*, above the natural concentration, that is also put into the atmosphere by burning, nor does it include the many other pollutants that are emitted by factories and homes, or the lead compounds emitted by motor-car exhausts.

28.3.4 The effect and control of air-pollution

Plants are more sensitive to air-pollution than are animals. One of the first signs of severe pollution is poor growth by plants, general lack of vigour and even death. Here we concentrate on the affects of pollutants in air on animals. Our respiratory system is not designed to receive air contaminated with dust, sulphur dioxide and other gases. Many of these gases are soluble in water and dissolve in the body. Many respiratory diseases are in part due to air pollution.

(a) Sulphur oxides

(i) Effect

Sulphur dioxide dissolves in the water in the atmosphere to form sulphurous acid and sulphur(VI) oxide to form sulphuric acid. *Acid-fogs* were a major cause of respiratory diseases particularly in the nineteenth century when more coal was burned in domestic fires than today. As recently as 1952 a build-up of smoke, fog and sulphur oxides (commonly called *smog*) in London caused the death of 4000 people more than expected statistically.

Sulphurous and sulphuric acids also damage buildings particularly those made of limestone or marble (i.e. calcium carbonate). Sulphuric acid forms a sticky layer on the surface of buildings to which soot in the atmosphere readily adheres. It also reacts with calcium carbonate to form calcium sulphate and carbon dioxide (section 18.4.2 (b)). Calcium sulphate has a greater volume than calcium carbonate and expansion causes a peeling away of the surface layer of the building. Also calcium sulphate dissolves in water and is slowly removed by rain. The beautiful decorations on the outside of many cathedrals and churches are often made of limestone as it is an ideal rock for carving. You will have seen how many of these have, over the course of time, been destroyed in this way.

(ii) Control

Pollution by sulphur oxides can only be controlled by removing sulphur from the fuels or the sulphur oxides from the waste gases. Since the smogs mentioned above, Acts of Parliament have required *smokeless fuels* i.e. fuels from which the sulphur and other volatile impurities have been removed, to be used on domestic fires in many areas.

Crude oil, like coal, contains sulphur. The middle boiling fractions (figure 25.11) can contain up to 1 %. It is easier to remove the sulphur from liquid oil than from solid coal and this is done with the light grade oils used by car-engines and oil-burning appliances, and consequently sulphur dioxide pollution does not occur with these. The sulphur is removed by passing the oil with hydrogen at a high temperature and pressure over a catalyst of cobalt and molybdenum oxides. Then the sulphur, which is present as organic sulphur compounds, combines with the hydrogen to give hydrogen sulphide. The hydrogen sulphide is then burnt in a limited supply of air to form sulphur:

$$2H_2S + O_2 \rightarrow 2H_2O + 2S$$

The sulphur is used for manufacturing sulphuric acid.

It is, however, uneconomic to remove the sulphur from either the coal or the heavy grade oil used in power-stations. In power-stations the fine particles of ash in the waste gases are removed using electrostatic precipitors (section 1.5). The sulphur oxides are sometimes removed during combustion or from the waste gases, for example by mixing limestone with the fuel or by passing the gases through a suspension of limestone in water when the reaction:

$$CaCO_3 + SO_3 \rightarrow CaSO_4 + CO_2$$

occurs. At most power stations the vast amounts of sulphur dioxide cannot be removed economically, and are emitted into the air. High chimneys are used in the hope that the gases will move into the high atmosphere, mix and become so diluted that they are harmless. It has, however, been suggested that the sulphur dioxide from British power-stations is responsible for the increase in recent years in the sulphur dioxide in the air over Scandinavia.

(b) Carbon monoxide and soot

(i) Effect

Carbon monoxide is a major pollutant (see table 28.1). It poisons by starving the body of oxygen (see section 18.5.2 (b)) but it is not stored in the body. Animals which have been exposed to carbon monoxide quickly recover when its source is removed and there is no evidence that carbon monoxide in the atmosphere causes a significant amount of poisoning.

Soot particles are a major contributor to respiratory problems, particularly in high concentration, in fogs.

(ii) Control

As carbon monoxide and soot is formed in car-engines, much effort is now being put into designing car-engines so that the amount of carbon monoxide produced is as little as possible. Even with a well-designed engine this amount can vary depending on the ratio of petrol to air, and one of the purposes of "tuning" a car engine (adjusting the carburettor) is to get the carbon monoxide concentration in the exhaust gases to the smallest possible value. Many countries now have a law which requires that the carbon monoxide in car exhaust gases be below a "maximum permitted concentration".

The amount of carbon monoxide and smoke emission will depend on the conditions. You will all have seen the black smoke from heavy lorries as their engines struggle to pull them up a hill.

No carbon monoxide, and hydrocarbons, would be emitted by motor car exhausts if they could be completely oxidized to carbon dioxide and water:

$$2CO + O_2 \rightarrow 2CO_2$$

$$2C_8H_{18} + 25O_2 \rightarrow 16CO_2 + 18H_2O$$

Two methods are now being used to do this:

1. In the *exhaust thermal reactor* air is injected into the hot exhaust gases and complete oxidation of the mixture occurs as it passes through the reactor.
2. In the *catalytic converter* air is also injected into the hot exhaust gases but the reaction is catalysed by platinum or a metal oxide. This enables a lower temperature to be used.

(c) Nitrogen oxides

(i) Effect

Nitrogen dioxide reacts with water and oxygen to give nitric acid. This adds to the acidic pollution caused by the oxo-acids of sulphur.

However, this is not the most serious form of pollution caused by nitrogen dioxide. It can, under the right conditions, react with hydrocarbons and other impurities in the air to give a complex mixture known as *photochemical smog*. One substance in this smog is trioxygen (O_3) which penetrates to the lungs and interferes with the passage of oxygen to the blood.

(ii) Control

Pollution by nitrogen oxides can only be reduced *either* by decreasing the amount of air in the air-petrol mixture used by the car engine *or* by reducing the combustion temperature. In either case this would increase pollution by carbon monoxide, smoke and unburnt fuel. So there is no easy solution to this problem.

(d) Lead compounds

(i) Effect

Even in small amounts lead has harmful effects. It is a *cumulative poison* i.e. our bodies are not able to dispose of it. Therefore, the amount of lead in us increases throughout our lifetime.

(ii) Control
Pollution by lead can be stopped by not adding lead compounds to petrol. Then the petrol would have a lower octane number unless a hydrocarbon such as benzene (both expensive and carcinogenic) were added.

Some cars, e.g. those driven by the authors, use low octane petrol (two-star) but these operate at lower temperatures and their exhaust fumes contain more carbon monoxide and unburnt hydrocarbons. So this is another vicious circle!

(e) Unburnt petrol

(i) Effect
The hydrocarbons that make up petrol are, as we have seen, partly responsible for photochemical smog.

(ii) Control
The amount of unburnt fuel can be minimised by "tuning" the car-engine. Not only is it desirable not to have hydrocarbons as pollutants, but the motorist does not want to waste his expensive fuel by allowing it to pass unused through the engine of his car.

The reduction in the amount of unburnt hydrocarbons in the car exhaust gases can only be achieved by complete combustion to carbon dioxide and water. Ways in which this can be done are discussed in (b) above.

(f) Carbon Dioxide
In addition to the pollutants just discussed a further serious potential danger is becoming apparent. We are putting more and more carbon dioxide into the atmosphere, by burning carbon containing fuels. Carbon dioxide occurs naturally in the atmosphere as the product of respiration. The amount of carbon dioxide in the atmosphere is kept approximately constant as photosynthesis by plants takes in carbon dioxide (section 18.6.2) and, as it dissolves in water, it is in equilibrium with the seas. However, the quantities of carbon dioxide now being produced are too vast for the plants to convert back into oxygen and the carbon dioxide content of the atmosphere is slowly increasing. Carbon dioxide prevents heat escaping from the earth and if its concentration gets too high, the average temperature of the earth will rise. This would mean that eventually the antarctic ice would melt, and the level of the seas would rise until eventually they covered most of the land.

28.4 Water

28.4.1 Introduction

Water is one of the most abundant substances on earth, and there should be enough to meet all conceivable needs. Nevertheless, water shortages are becoming increasingly common. One reason is that water is not distributed uniformly over the earth. Some areas have too much and others, like the desert regions, have too little. In areas where there are ample water supplies the poor quality of much of the water, because of pollution, means that even in these areas water shortages do occur.

Water occurs on earth as a liquid in the sea, rivers, lakes and streams, as a vapour in the atmosphere and as a solid in the polar ice. The composition of the water in these different locations and the chemical and physical properties of water have been discussed in chapter 11. In this section we shall examine *water pollution*, the common pollutants, and what effect they have on the water in which they are dissolved.

28.4.2 The water cycle

The distribution of water on the earth is shown in table 28.2. Most, by far, is in the seas. Very little of the total is in lakes, rivers and streams and even less in the atmosphere. Except for the large amounts locked up in the polar ice-caps, the water is continually moving from one place to another. Groundwater goes into the streams and rivers. These run into the seas. Evaporation of the seas gives the water vapour in the atmosphere. This forms the clouds and is eventually precipitated as rain on the land and the *water cycle* starts again.

TABLE 28.2 THE DISTRIBUTION OF WATER ON EARTH

Sea	97·6 %
Polar ice	1·8 %
In soil (ground-water)	0.55 %
Lakes, rivers and streams	0·015 %
Atmosphere	0.001 %

Man uses water in his home, for industry and for agriculture. After it has been used it is returned to the rivers and sea. It often returns in a much worse condition than when extracted. Domestic and industrial waste water is usually treated in a sewage works before being returned to the rivers. The major purpose of such treatment is to remove those substances from the water that use up the dissolved oxygen gas that is essential for fish and other aquatic life.

28.4.3 The major water pollutants and their effects

There are three main types of water pollution:

(i) Chemical pollution where dissolved oxygen has been removed so that the water cannot support aquatic life or where toxic chemicals are dissolved in the water.
(ii) Physical pollution where the water temperature has been increased or oil spilt on the surface of the water.
(iii) Biological pollution where there is a disturbance of the balance of aquatic life in the water or unchecked plant growth.

These types of pollution are connected with one another. For example, if chemicals that use up oxygen are in the water then fish that require oxygen will disappear (figure 28.2). The presence of fish such as salmon or trout in river water is a good measure of how clean the water is. It is good news that fish are now returning to rivers which were recently too polluted; salmon can now be found in the River Thames in London.

Figure 28.2 Dead fish being collected from the Rhine at Düsseldorf. (Photo: Keystone Press.)

The most common pollutants in the United Kingdom are *detergents, phosphates, nitrates, mercury* and *cadmium, heat* and *oil.*

(a) Detergents

A detergent is a substance used for cleaning. Soap (see section 21.2.7 (c)) has been used for centuries and is present in waste water. The bacterial and other treatments at sewage works remove soap from water. Some modern synthetic detergents (section 26.5), which are better than soap for some cleaning purposes, especially in hard water areas, are however *not* removed by the bacteria i.e. they are not *biodegradable* (capable of being decomposed by bacteria or other biological means). They remain in the water and pass into rivers. Their presence is shown by foam forming on the water surface particularly where rivers run over weirs. In recent years biodegradable synthetic detergents have been introduced and the problems of river foam have decreased.

(b) Phosphates

Phosphates are added to detergents mainly to soften the water (section 16.8.7). An important substance used in this way is pentasodium triphosphate, $Na_5P_3O_{10}$. The triphosphate ions or similar ions in compounds like this react with calcium and magnesium ions in the hard water to form soluble and stable complex ions. These ions do not react with soap and therefore the water is softened. Furthermore no deposit of calcium or magnesium carbonate forms from the hydrogencarbonates in temporarily hard water (see section 22.2.4).

Many household washing powders (Daz, Surf, Tide etc.) contain 20–30 % of phosphates. These phosphates are not removed by sewage treatment processes and remain in the water. Phosphates are a plant food and in enclosed areas of water cause plankton and algae to grow much more rapidly. They in turn support larger plant and fish life. Eventually the production of dead plant and animal matter is greater than can be decomposed by bacterial action. The excess sinks. The accumulated organic matter reduces the oxygen content of the water and the mass of floating plants on the water prevents light entering the water. Thus photosynthesis which produces oxygen is also reduced. Aquatic life dies and the water becomes of very poor quality. In severe cases bacteria of a different sort that do not require oxygen decompose nitrogen containing matter to nitrogen, sulphur containing matter to hydrogen sulphide and carbon containing matter to methane (marsh gas).

Detergent mixtures that do not contain phosphates have been made, but the compounds used as substitutes for the phosphates can cause other pollution effects so phosphate containing detergents are still widely used.

Over half the phosphates in water comes from laundry waste. In addition there is phosphorus in animal waste which is converted to phosphates at the sewage works. Also agricultural fertiliser contains phosphates. There is no way in which these phosphates can be removed from the water once they are there.

(c) Nitrates

Nitrates can enter water by oxidation of ammonia in sewage by nitrifying bacteria (see section 16.7), and from fertilizers, containing nitrates, which have been spread on fields. The second source of nitrates has become more significant in recent years as more and more fertilizer has been used to increase the yield of foodcrops (see section 28.7.2(c)). Like phosphates, nitrates are a plant food and cause plankton and algae to grow more rapidly.

An excess of nitrates in drinking water can cause serious illness particularly in young babies. Bacteria in the stomach converts the nitrate ion, NO_3^-, to the nitrite ion, NO_2^-. Nitrite ions can react with blood and prevent it from combining with oxygen. The resulting disease can cause suffocation.

(d) Mercury and cadmium

Pollution by mercury and cadmium is very serious in certain parts of the world. Mercury is used in the manufacture of sodium hydroxide and chlorine from sodium chloride (see section 21.2.7(c)), and although the mercury is recycled, 100% efficiency is not achieved and the waste waters contain mercury salts. Mercury salts are also used in paper-making. In areas where this is major industry, in Sweden, for example, the natural waters, including the sea in many places, are contaminated with so much mercury that it is dangerous to eat fish caught there, as the fish concentrate the mercury in their bodies. Within recent years, in Japan, a number of people have died from eating mercury-contaminated fish. In a similar way cadmium from industrial waste is a dangerous metal to have in water. Sea-animals also concentrate this metal, for example the mussels in the Bristol Channel have such a high cadmium content that they must not be eaten.

(e) Heat

You might not think that heating up water is a form of pollution, but the consequences can be quite serious. Warm water is discharged from most power stations; their need for a large volume of cooling water is the reason why they are always located on a river or a lake. The discharged water from a power-station is usually between 5 and 15 °C hotter than the intake water. The effect of this only modest increase in temperature is to decrease the solubility of oxygen in the water and to increase the rate of plant growth. It also affects the type of fish and other aquatic animals living in the heated water. Most aquatic animals are cold-blooded and are unable to regulate their body temperature; they can only survive in water that is within a narrow temperature range. The major effect, therefore, is to upset the natural balance of plant and various marine species in the water around the power station.

A little of the heated water can be used to beneficial effect. Around several power stations, fish-farms and greenhouses have been built. The heated water is used for breeding eels and other fish that grow better in warm water, and for growing tomatoes. Also some fish prefer warmer water, and the area around the discharge from a power plant is often a good spot for fishing.

(f) Oil

Pollution of the seas by oil is becoming a major problem as more and more oil is transported in larger and larger tankers (figure 28.3), and more oil is being

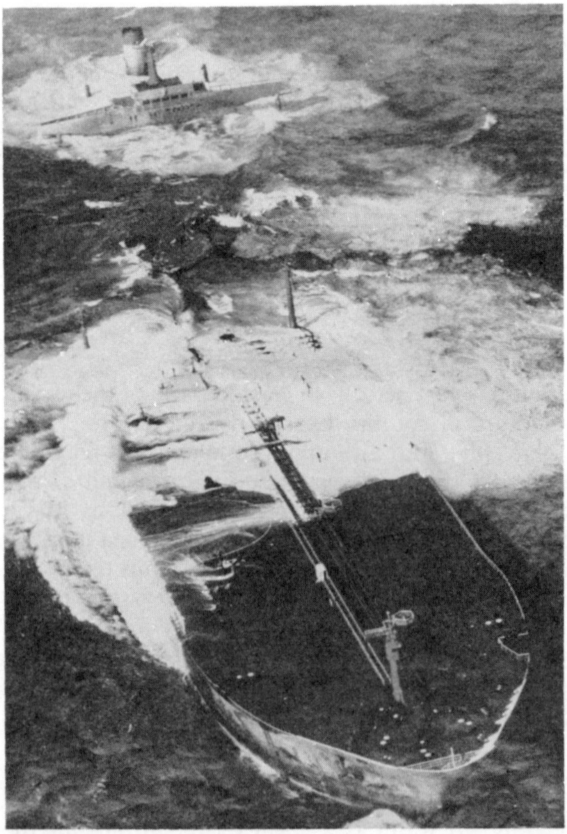

Figure 28.3 The super-tanker 'Amoco Cadiz' aground off the Brittany coast in March 1978, her cargo of 200 000 tons of oil spilling into the sea. (Photo: Magnum photos.)

obtained from wells in the seabed (figure 28.5). Most of these oil spillages happen accidentally, but some oil pollution occurs when ships, illegally, wash and empty their oil tanks at sea. There is no one satisfactory method for cleaning up oil from the sea. Spraying with detergents that break up the oil and allow bacteria to digest it is the method most often used, but the detergents can upset marine life, particularly close to land, and oil spillages that drift onto beaches must often be removed by digging up the oil and burning or burying it.

28.4.4 Drinking water

If you want a drink of water, you do not usually get it from a lake, a river or a stream. It comes out of the kitchen tap. This water has come from a lake, river, stream or well and has been purified to make it fit to drink. It is likely that the water has been filtered through beds of sand to remove suspended solids and to allow the bacteria that live in the sand to remove harmful impurities. It will also have had chlorine added to it. This oxidizes other impurities that might give bad tastes and smells and disinfects the water by killing any bacteria that remain in it.

The water may also be treated in several other ways. It may be passed through activated carbon to remove organic compounds that give the water taste, colour or smell, and it may have oxygen bubbled through it to oxidize other impurities.

The processes that take place in a sewage treatment plant and in a drinking water treatment plant are very similar. The major difference is the quality of the water that enters and leaves the two plants. As the sewage plant will discharge its effluent into a river that may later be the input water to a drinking water treatment plant you can see that some water can pass through this cycle many times before it escapes to the sea and returns to the land again as rain.

28.5 Mineral resources

28.5.1. Introduction

Our natural resources of air, water, minerals, energy and food divide themselves into two groups–*renewable* and *non-renewable*. Food is renewable–the stock of plants and animals that we eat can be replenished. The stock of an energy source such as wood can also be replenished. Hydroelectric power or solar energy do not use up the earth's resources. *But the bulk of our present energy resources, together with our air, water and minerals, cannot be replaced*–they are *not renewable*. Our stock of these was determined either when the earth was formed (i.e. minerals, air and water) or during the early history of the earth (i.e. fuels). Air and water can be used over and over again. Our mineral resources and energy resources such as oil and coal can only be used once–once they are used they are lost for ever. The problems of the use of our energy resources are considered in section 28.6. Here we consider our *mineral resources*. When used they are often spread throughout the world and to recover and reuse them is often impossible or economically not worthwhile. The conservation and reuse of these resources is discussed in section 28.5.4.

28.5.2. The abundance and distribution of the elements

Table 28.3 gives the composition of the earth's crust. From it you can see that just 8 elements make up 99% of it, and the other 81 elements the remaining 1%.

TABLE 28.3 THE COMPOSITION OF THE EARTH'S CRUST

Oxygen	45·6%	Sodium	2·27%
Silicon	27·3%	Potassium	0·84%
Aluminium	8.36%	Titanium	0·63%
Iron	6·22%	Hydrogen	0·15%
Calcium	4·66%	Phosphorus	0·11%
Magnesium	2·76%	Manganese	0·10%

All others < 0·1%

Oxygen and silicon, that make up silica (sand) and silicate minerals, are by far the most abundant elements. Carbon, which is essential to all living matter as the basic element of organic compounds, is very rare (only 0·02%). Most of this, 99·8% of it, exists as calcium carbonate (chalk) (figure 28.4) and as carbon in other rocks. 0·195% of the remaining 0·2% is the carbon in coal and petroleum and only 0·005% of the carbon exists as part of living matter.

Figure 28.4 The chalk cliffs at Flamborough Head, North Humberside. (Photo: R. Baldwin).

Many of the metals on which our civilization is based, well known metals like copper, lead, zinc and tin, are very scarce (see table 28.4). As some of these are so scarce you might wonder why they have become so important. There are three reasons. Firstly, nature has not distributed the elements evenly over the surface of the earth, but has concentrated many of them in deposits known as *ore-bodies*. It is these ores that are mined to supply the metals. Secondly, the common metals were those that were easy to extract from their ores, and thirdly, the common metals have very desirable properties.

TABLE 28.4 SOME METALS IN THE EARTH'S CRUST (% OF TOTAL)

Aluminium	8·36%	Copper	0·007 %
Iron	6·22%	Zinc	0·007 %
Titanium	0·63%	Cobalt	0·003 %
Manganese	0·10%	Lead	0·001 %
Chromium	0·01%	Tin	0·000 2%
Nickel	0·01%	Uranium	0·000 2%

28.5.3 The exploitation of mineral resources

(a) Distribution of mineral resources

It is from our mineral resources that we get all the elements. These include the metals and the non-metals. All of the metals, with the exception of gold and platinum, exist in the earth's crust as compounds combined with non-metals. There are many different compounds but the majority are metal oxides, metal silicates or metal sulphides.

Vast quantities of metals are combined with silicon and oxygen in silicates (section 18.7.3), but they are difficult to extract. Oxides and sulphides are the compounds from which metals are usually extracted, e.g. iron from iron oxide (section 24.1), aluminium from aluminium oxide (section 23.1.2) and zinc from zinc sulphide (section 23.3.2). Some of the non-metals exist in the earth's crust as the free element. These include the gases in the atmosphere, oxygen, nitrogen and the noble gases. Sulphur also exists uncombined with any other element (section 19.2). Chlorine is found combined with sodium, in the vast deposits of rock salt and in sea-water (section 21.1). The sea, which contains many elements (table 11.1), is also the source of other elements, in particular bromine (section 20.2.3) and magnesium (section 22.1.3), and it will undoubtedly become an important source of other elements in years to come. The methods of extraction of the important elements are given throughout this book as each element is discussed.

(b) Economic mineral resources

Whether or not an element is extracted from a particular mineral depends on (i) the ease of its extraction, (ii) the concentration in which it exists and (iii) its value. We shall confine discussion of these three points to the metals.

(i) The ease of extraction of a metal

Metals occur in their ores as *positive ions*. To obtain the free metal they must be *reduced*, that is electrons must be added to them (section 15.8.2). The various methods of reduction are described throughout the book. The method will depend upon the position of the metal in the *activity series*. It can be seen from table 8.2 that the least reactive metals, platinum and gold at the bottom of the table, are the ones that exist uncombined with any other element. The next metal, mercury, can be obtained from its oxide by heat alone. Carbon can be used as the reducing agent for the next group of metals (copper, lead, iron and zinc), but for the most electropositive metals (aluminium, the alkaline earth and alkali metals) it is necessary to use electrolysis to reduce the metal ion to the free metal.

The nature of the other elements with which the metal is combined will have a major influence on the ease of extraction. It is particularly difficult to extract metals from silicates, as they often contain many elements all very strongly bonded together. Although the vast bulk of the earth's crust is silicate, it is as yet not an economic source of the metals.

(ii) The concentration of the metal in its ore

If the metal exists as its oxide, as it usually does, it is obviously there in a fixed concentration. But the pure oxides are rarely found in nature. In an ore the metal oxide will be mixed with other substances, such as sand, clay and rocks. Before the chemical extraction of the metal from its oxide can begin the ore must be mined and concentrated by removing this unwanted material. Clearly, the higher the concentration of the metal oxide in the ore, the easier and cheaper will be the mining and concentration processes.

Sometimes it is economic to use a source of the metal of low concentration. For example, magnesium is extracted from sea-water where its concentration is only 0·13 %, rather than from certain silicates where its concentration is 30 %.

The purity of the ore is another factor that must be considered. If the metal oxide occurs with other metal oxides in the ore it may be that they will all be reduced and the metal obtained will be very impure and require complex and costly refining before it is suitable for use.

(iii) The value

The value of the metal is another factor that determines whether time, effort and money is put into extracting it from the earth's crust. There is firstly the value of the metal because it has desirable properties. Certain metals are very good for particular purposes. Copper, for example, for plumbing, lead in batteries, and silver for photographic films are almost irreplaceable and this makes these metals very valuable.

A second type of value is also used. This is the value of the metal because people desire to have it. Gold is the best example. Gold is not a particularly useful metal although there are certain uses for which gold is the best, jewellery and fillings in teeth for example. It is not these uses that make gold so valuable, it is that people want to have gold. It has had a fascination for man for thousands of years. This is due to its low chemical reactivity and the fact that it can be found uncombined. It is the one possession that cannot be destroyed (you will know that this statement is untrue as no element can be destroyed–what is really meant that it is very difficult to convert gold into its compounds) and has been used as the primary form of money since early civilisations. This *monetary value* of gold means that much time and effort is spent in extracting gold from very low concentration ores in the earth. Having been obtained, it is then immediately buried in the earth again in bank vaults!

(c) Exploitation of a mineral resource

Before a mineral is mined and the metal extracted it must first be decided whether or not the particular ore-body is an *economic resource*. Even before that the ore-body must be found. Geologists are continuously surveying the earth's surface for new mineral resources. Not only the land surfaces but surfaces under the seas are being systematically surveyed. The discovery of oil under the seas around Britain is a dramatic example of the rewards of such surveys.

When an ore-body has been discovered, extensive sampling and testing will be carried out to determine its extent and value. These factors together with many other economic and environmental considerations will be used to arrive at a decision whether to mine the ore (or exploit the resource). Factors to be considered therefore are:

(i) What is the extent of the ore deposit?
(ii) How desirable is the element that will be obtained? Is it a rare and valuable element in short supply or a less valuable one of which supplies are ample?
(iii) What is the location of the ore-deposit in relation to the plant where the ore will be processed?

(iv) What new plant will need to be built and what will it cost?
(v) How near is the processing plant to the users of the metal that will be extracted?
(vi) How near is the ore-body to the transportation links i.e. the road, railways, canals or sea?
(vii) What are the available energy and water resources that will be needed for the mining processes?
(viii) What is the likely impact on the environment in mining the ore?

Many of these decisions are difficult ones to take. Competing factors and interests have to be balanced one against the other before a final decision can be taken. A good example is the long debate that has followed the discovery of extensive deposits of coal in West Yorkshire (the Selby coalfield) and in the Vale of Belvoir.

28.5.4 The conservation of mineral resources

(a) Mineral stocks

As we have seen some scarce metals like copper, zinc, lead, and tin have become widely used. The stocks of many of these metals, that still exist in deposits where the metal is of high enough concentration to make it recoverable (*exploitable metal stocks*), are surprisingly small. Table 28.5 shows the expected life of known deposits, assuming a steady increase in population and demand. Some of these metal reserves will last for only a few more years–we have almost exhausted our supplies of lead, zinc and tin. The metals that exist in huge quantities in the earth's crust, iron and particularly aluminium, are mainly spread over the earth in too low concentrations to be recovered economically at present. The high concentration deposits of aluminium will only last about another 150 years.

TABLE 28.5 ESTIMATED LIFETIMES OF METAL RESERVES*

Iron	550 years	Tin	15 years
Chromium	550 years	Gold	15 years
Aluminium	150 years	Silver	15 years
Manganese	125 years	Zinc	12 years
Nickel	100 years	Lead	10 years
Copper	30 years		

* These figures are for ores in which the concentration of the metal is high enough for it to be extracted by present methods. For some, particularly iron and aluminium, there are abundant supplies of less rich ores: (Note that other books may have data different from those given here: there are no agreed figures available).

(b) Future possibilities

With our exploitable metal stocks being rapidly reduced what can be done? The various possibilities are now discussed.

(i) Alternative sources can be found and used. As we have said earlier there are vast quantities of metals, and non-metals, locked up in the silicates. New extraction methods will undoubtedly be developed. It is very important to realise that we are at present extracting metals from the ores that yield metal most

readily, i.e. with the least energy. *To exploit resources that are now uneconomic will require more energy* (see next section).

The sea and under the sea can be mined more than at present. At present only sodium, magnesium, chlorine and bromine are extracted from the sea. The sea is a potential source of many other elements. Other elements, particularly manganese, exist as ores on the ocean-bed and these will undoubtedly be mined in years to come.

(ii) Alternative materials can be used. We can reduce our dependence on the scarcer metals. Aluminium, a plentiful metal, is now being used to replace copper for many applications. Silver coinage has been almost entirely replaced by copper and nickel. Metals can often, and with added advantages, be replaced by plastics. Many parts of cars are now made of plastic materials that do not corrode and can be moulded into much more complex shapes and textures than the metals which they have replaced.

There is, however, some limitation on when alternative materials can be used. Certain metals are used because of their *specific chemical properties*. There is, as yet, no replacement for lead in car batteries, or silver in photographic film.

(iii) Salvage and recycling can help conserve our scarce resources. There is an increasing interest in reusing materials rather than throwing them away. Undoubtedly there is much that can be done, and some refuse collection organisations separate different types of waste, e.g. glass and plastics, to be reused. Whether or not it is worthwhile collecting and recycling a particular material is an economic decision, and what is uneconomic one year may become economic when that material becomes more expensive. For example, the rapid increase in the price of oil, and consequently of plastics which are made of oil, has meant that it has become increasingly attractive to separate plastic materials from household waste and sell them back to the manufacturers who can recycle them.

Sometimes the cost of collection is more than the saving by recycling. Glass is a good example. Glass is made of silicates (section 18.8), and the raw material is very plentiful. Nevertheless it needs less energy to melt glass to use again, than to make new glass, and it is therefore sometimes worthwhile recycling glass. However the cost of collection and transport of glass bottles from the consumer back to the glass-works may be more than the saving at the factory and this makes recycling uneconomic. It is done in certain areas, because consumers, without payment, return their glass-bottles to central collecting points (bottle-banks).

If recycling is to be carried out, the cooperation of manufacturers is also needed. The metal used in a recycled metal object needs to be reasonably pure for the process to be economic. Metal objects containing two or more metals cannot be easily recycled, as the metals must be separated. There are many millions of beer and lemonade cans thrown away each year. As now made, these are difficult to recycle since different metals are used. Some cans are made of steel, with aluminium ends and contain a soldered joint (containing lead and tin).

Also some metals when used are so finely spread over the earth that recycling is quite impossible. Lead used in petrol is a very good example.

Finally it is worth noting that refuse can be used in other ways than for recycling. Processes now exist for converting the carbon containing material (paper, animal and vegetable matter) into a solid fuel or oil; these processes will no doubt be developed further as oil becomes more expensive (see next section).

28.6 Energy Resources

28.6.1 Current energy resources

During the past few years everyone has become aware of the increasing shortage of sources of energy. In section 9.1 we explained that energy is "that which makes things move". A living world requires things to move, i.e. it must get energy. Much of this energy comes from the sun. The sun makes plants grow, and animals get their energy by eating plants. In addition to the sun the major energy sources at this present stage of the history of the earth are the *fossil fuels*, i.e. coal (section 17.6.3), oil (section 17.6.2 and figure 28.5) and natural gas (section 17.6.1 and figure 28.6). As we shall see the period of time during which energy can be obtained in this way will be quite short. The only long term energy sources are the sun, the earth's heat, nuclear fission and nuclear fusion. Minor energy sources might be the winds, the rivers and the sea.

The primary source of almost all our present energy is the sun. The coal, petroleum and natural gas that we use came from plants and animals as part of the carbon cycle (section 18.6). This required the light of the sun for photosynthesis (see section 18.6.2). The commonest of these fossil fuels is coal (70 %). Petroleum and natural gas comprise another 15 %, the remainder consisting of oil mixed in certain rocks (known as tar sands and shale oil). Stocks

Figure 28.5 Production platform in BP's Forties oilfield in the North Sea. (Photo: British Petroleum.)

Figure 28.6 The "Methane Princess" carrying liquefied natural gas from the Arab states to Great Britain. (Photo: Vickers Ltd.).

of such fuels are quite small. How long they will last depends on how carefully we use them. It seems likely that petroleum will last for 20–80 years and coal for 200–300 years.

It is our duty to conserve such limited stocks by using other energy sources whenever possible. For motor vehicles and aeroplanes there is no known satisfactory substitute for the liquid fuels obtained from crude oil. We should therefore try to use economical vehicles and other methods of transport (e.g. bicycles or trains). The fossil fuels also need to be conserved because they (particularly oil) are the starting materials used by the *petrochemical industry* for making many useful substances such as plastics and synthetic fibres. If we wish to have enough of the many chemicals which can be made from oil, it is not wise to burn it as a fuel. But of course, our resources should not be squandered in any way.

There are of course political reasons as well for reduced dependence on oil. Much world tension is caused because some countries, e.g. those in the Middle East, have more oil than they need for their present purposes, and others do not have enough.

In the previous section, we explained that some essential metals will soon be in short supply if we do not use lower grade ores. Extraction from low grade ores will require more energy. In fact despite all our efforts, for one reason and another, the world's need for energy will increase. We must therefore exploit *alternative energy sources*. These are discussed in the next section.

28.6.2 Alternative energy sources

(a) The sun

Until recently the direct use of the sun, i.e. solar heating, has been little and limited to such uses as heating water and generating electricity in satellites and other spacecraft. Vast installations are needed for solar energy to replace fossil fuels to a noticeable extent. Suitable sites would be desert areas which receive most sunshine and are little used. The earth's atmosphere absorbs much sunlight and one ambitious scheme is to have a very large energy converter circulating as a satellite around the earth outside the atmosphere. Its electrical energy generated by the sun would be 'beamed' to collectors on the earth.

It is the sun's energy which helps to convert carbon dioxide and water into carbohydrates by photosynthesis (section 18.6.2). Plants are grown with the deliberate intention of burning them to regain the energy or of converting them to other substances which can be burnt subsequently. Some fast growing water plants can be converted to methane and methanol can be obtained from wood.

(b) The wind, rivers and sea

Using the wind and rivers is another way of using the energy of the sun. The sun causes the winds and the sun transfers water from the sea back into the rivers. Winds are too erratic to be used as a major energy source, i.e. windmills will not solve our energy problem! Hydroelectric power, i.e. power obtained from rivers, is important in certain areas, for example in Scandinavia, but there are few untapped sources.

Potentially the sea is much more important. If the up and down motion of the waves can be converted into a similar motion of a piece of machinery, this could be used to generate electricity. A number of machines to do this have been developed and are being tested in several countries.

Another possibility is to use the fact that the temperature of the sea at its surface is higher than in the depths to operate a *heat engine*. A low boiling liquid such as ammonia circulates in a closed system. It is vaporized by the warmer water and passes through a turbine to generate electricity. Then the cold water liquefies it again and the cycle is repeated.

The energy of the tides, which is using the energy of the moon's motion, is used to generate electricity in the world's only tidal-electric plant in Brittany in France.

(c) The earth's heat

There is a lot of heat in the centre of the earth. Some of this is spectacularly released at the surface by volcanoes (figure 28.7). The use of some of the earth's heat, known as geothermal energy, is important in certain parts of the world, notably Iceland, New Zealand and America, where geysers and hot springs occur.

(d) Nuclear energy

It is very clear that when the fossil fuels become exhausted none of the alternative supplies discussed above can provide all the energy that the world will need. The obvious long term source of energy is that in the atom, obtained through *nuclear fission* or *nuclear fusion*. These have been discussed in detail earlier (section 6.3.5). *Nuclear fission* is now being used in nuclear power stations to generate electricity in many parts of the world (figure 28.8). In some of these the nuclear fuel plutonium is produced from the uranium-238. These *fission reactors* do, however, have major drawbacks. The most serious is the production of large quantities of very dangerous radioactive waste materials that will remain dangerous for hundreds of years. How and where this waste material is to be stored are major problems to be solved if the number of nuclear-fission power plants is to be greatly increased.

The most hopeful energy source for the future is *nuclear fusion*. At present

Figure 28.7 The volcano "Surtur" (named after an old heathen Norse God) erupting and sending a large cloud of ash, rocks, smoke and flames into the air. In 1963 "Surtur" rose out of the floor of the Atlantic Ocean off the coast of Iceland to form the new volcanic island, Surtsey. (Photo: Peter Kitson, Camera Press Ltd.).

Figure 28.8 The fast nuclear reactor at Dounreay, Caithness, Scotland. (Photo: U.K.A.E.A.).

nuclear fusion occurs in atomic bombs and in suns (see section 6.3.5 (b)). Enormous technical problems are involved in getting nuclear fusion to occur in the laboratory and controlling it when it does occur. It is unlikely that it can be developed for very many years.

28.7 Food resources

28.7.1 Introduction

The energy source for living machines like you is food. The food is slowly "burnt" in your body forming carbon dioxide and water and releasing heat which provides the energy for your body to move. This process is called *respiration*. It is discussed further in section 15.5.

For the first two million years on earth man lived by eating the plants and wild animals. By providing food in this way the maximum population that the earth could support would be about 10 million; this is fewer people than live in London today. About 10 000 years ago, man began to grow his own food, and to keep animals to provide food. With this change to becoming a farmer the capacity of the earth to provide food increased enormously, and with the growth of the population the amount of cultivated land has steadily increased. Today about 10% of the land is ploughed, watered and fertilized by man, and this land provides the food for about 4 000 000 000 people (figure 28.1). In addition to the food from the land we get a little food by fishing in the sea and rivers.

28.7.2 Modern agriculture

Modern agriculture depends on four branches of technology: mechanisation, irrigation, fertilization and the chemical control of weeds and insects. The last two depend on chemistry.

(a) Mechanisation

In early days the farmer did all the work with his own hands. The energy expended in farming his crops was that of the food which he ate. He then started using animals, particularly the horse and the ox, to help in tilling the land, planting and harvesting the crops. The energy input into the farming had increased–that due to the food eaten by the animals was now being used. Over large areas of the world this is still the state of farming today; it is done by man and animals. In the "developed" world an enormous change took place about 200 years ago with the discovery of the coal-burning steam-engine followed by the petrol-engine. By using coal and oil the energy input into farming was increased enormously. The effects were clearly seen. The number of workers employed on the farm decreased to a small fraction of what it was; the amount of food produced for each farm-worker increased enormously. Man was using our energy resources to produce food; the farm had become *mechanised.*

(b) Irrigation

Large areas of the earth would be very fertile, except that there is insufficient water to support plant growth. There are desert areas which once every few years, when it rains, become a mass of green as the plant seeds spring into growth. With his increasing need for new agricultural land man has moved into these desert regions bringing water with him by *irrigation*. Rivers have been diverted, dams built and lakes formed, and canals have been cut to take water from one place to another. In a similar way large areas of swamp-land have been drained to make agriculture possible, and land has been reclaimed from the sea.

(c) Fertilization

The productive capacity of the land has been increased enormously by fertilizers. To begin with *natural fertilizer* i.e. animal and vegetable waste, was used, but so much is needed that *chemical fertilizers* (or *artificial fertilizers*) are now used in large quantities (see section 16.7(d)).

In addition to water, plants require many elements for healthy growth. Only three of these, *nitrogen, phosphorus* and *potassium* are needed in more than trace amounts. All three are found naturally in the soil:

Nitrogen compounds in dead animals and plants and in animal excreta are converted by decay to ammonium salts. These are oxidised by bacteria to nitrates. Also nitrifying bacteria convert atmospheric nitrogen directly to nitrates. It is the nitrate ion that is used by plants. Its production as part of the nitrogen cycle is discussed in section 16.7.

Phosphorus exists in soil as organic phosphorus compounds, and largely as calcium phosphates.

Potassium occurs as the potassium ion, although most of the potassium ions in soil are firmly held as part of the compounds making up clays (silicates) and cannot be used by plants.

These three elements, *nitrogen, phosphorus* and *potassium*, are called the *essential elements*. Nitrogen is required for the green parts of plants (i.e. the leaves and the stem), phosphorus for healthy root growth and potassium for flower and fruit formation. They are the three elements provided by a chemical fertilizer. Often chemical fertilizers also contain sulphur, which is an essential trace element. Much sulphur is obtained by plants from the sulphur dioxide in the air.

Many different substances are used in chemical fertilizers. The most important nitrogen containing compounds are ammonium nitrate and ammonium sulphate. The phosphorus compounds used are calcium superphosphate and bone ash (ash formed when animal bones are burnt). Potassium sulphate, potassium nitrate and potassium chloride are the sources of most of the potassium in fertilizers. The three elements are usually found together in a *"balanced" fertilizer*, i.e. a mixture of different substances containing nitrogen, phosphorus and potassium. Different amounts of each of the three elements are put into the different fertilizers depending on the crop. For example for a lawn, green growth is needed and a lawn fertilizer will be high in nitrogen, whereas for tomatoes fruit growth is most important and a tomato fertilizer will be high in potassium. The composition of a fertilizer is often given as the *"NPK value"*. This gives the amount of nitrogen (N), phosphorus (P) and potassium (K) in the mixture. Unfortunately the figure given to indicate the amount of phosphorus present is the percentage of phosphorus(V) oxide, P_2O_5 (which is not present!) and the figure for potassium is the percentage of potassium oxide, K_2O (which does not exist!). Hence a 4-8-6 fertilizer contains 4 % by mass of nitrogen, 8 % by mass of phosphorus(V) oxide and 6 % by mass of potassium oxide. (This is because many salts used to be thought of as compounds of oxides e.g. calcium phosphate as $3CaO.P_2O_5$ and potassium sulphate as $K_2O.SO_3$. The true percentages of the elements can easily be calculated. Thus 6 % of potassium oxide

indicates $\frac{(2 \times 39)}{(2 \times 39) + 16} \times 6\%$ of potassium.)

Both the phosphorus and potassium in fertilizer are made from *non-renewable mineral resources* (section 28.5), and it could be that it will not be possible to meet the demand for more fertilizer to grow more food as the world population grows. Phosphorus in particular is being continually moved from the land (where it is available in concentrated deposits) to the sea (where it is much diluted and therefore not available). The need to recover phosphorus from water and recycle it is obvious. Also chemical fertilizers should not be used in excess. Too much is not only wasteful of our natural resources but excess fertilizer can dissolve in the ground-water and run off, contaminating rivers, lakes and drinking water (see section 28.4.3(b)).

In addition to using fertilizers the farmer will also put lime (calcium hydroxide) or limestone (calcium carbonate) on his land. This is done to control the pH of the soil. Some soils are acidic, particularly those with a high peat content. Certain plants, e.g. heathers, rhododendrons and azaleas, grow best in such acid soils but a neutral or slightly alkaline soil is best for growing vegetables. The basic compounds calcium hydroxide or calcium carbonate are therefore used to neutralize any acidity in the soil. Care has to be taken in not using too much calcium hydroxide or the soil will become too alkaline which is also undesirable.

(d) Chemical control of weeds and insects

Many different chemicals are used to prevent the growth of undesirable weeds amongst food-crops, and to kill the great variety of insects that destroy these crops. These chemicals are often complicated and have not been dealt with in this book, but you will have heard of DDT (*d*ichloro *d*iphenyl *tri*chloroethane) and probably at home you use chemical methods to kill the weeds in your lawn and the insects in the garden. The chemicals used can be dangerous, however, if misused. DDT was widely used twenty-five years ago. Only afterwards was it realised that it was not destroyed by nature but accumulated in the earth, in plants and then in animals, until it eventually killed them. In some countries, for example America, the use of DDT is now forbidden.

It is very difficult to know whether any particular chemical should be used. DDT, for example, has killed probably millions of birds, but it has also greatly reduced the terrible disease of malaria by killing the malaria-carrying insects. More recently the use of certain chemicals that control weed growth in forests, at roadsides and in gardens has been questioned, as during its manufacture a related chemical, dioxin, is produced, and dioxin, even in minute quantities, is a deadly poison. On balance the gain of using chemicals on farms outweighs the disadvantages. Without insecticides the loss of food would be enormous. For example, it has been estimated that without chemically controlling insects a quarter of the corn and vegetables grown would be lost, and even now a large proportion of the world's population is short of food.

28.8 Population control

In the previous sections we have looked in some detail at some of the strains that the increasing population is putting on the air, the water, and the mineral, energy and food resources of our planet. As these problems follow directly from the rapidly expanding population, the best solution would be to stop this expansion. Nature will continue to do this on a small scale. Millions of people have died in famines and floods over past centuries, but the population has continued to expand despite these local disasters. This is largely because of our success. We have over the years increased the average length of life by developments in medicine, fed people better by expanding food resources, and enabled them to buy more food by earning more money. We have kept people warmer by expanding our energy consumption, and educated them better so that they are more able to lead a healthy life. But these successes have brought with them the problems that we have been discussing. Should we turn back the clock 200 years? Almost everyone would say "no". Indeed, many countries that have not developed as much as others, are striving hard themselves to become technological societies.

Much drastic action is needed to prevent the massive population explosion that is now threatening the earth. In many countries the rate of birth of babies has been voluntarily reduced. A combination of education of future parents, like you, and chemical methods of controlling the size of families, has meant that now many people often choose to have smaller families. But in many countries there is suspicion and opposition to restricting the size of families. The population explosion, looked at as a global problem, is regarded by most experts with growing pessimism; there is little that can be done. Only massive changes in the cultural, religious and economic beliefs of some countries, combined with massive education and propaganda in others, will prevent the impending disasters.

28.9 The Chemical industry

28.9.1 Introduction

Our advanced society depends upon the chemical industry. Its products are used by all other industries. Yet although we are so dependent on the chemical industry its image is often a poor one–it is associated with noise and dirt and smells. This is hardly fair!

We have seen that the major problems of decreasing resources and pollution of the air, water and the land are caused by an increasing population demanding a higher standard of living. The chemical industry, like all other industry, responds to the demands of the consumer–it does not make what it cannot sell. It can be argued that only the skill of scientists, including chemists, can prevent the threatened shortages.

Chemical factories, like other factories, are not very attractive. Care must be taken in siting them and of course they cause pollution. They use massive quantities of energy to carry out the chemical reactions required to produce metals, plastics, detergents and the thousands of other chemicals we want. In the process they will release heat and waste materials into the environment. But, as we have seen, these *local pollutants* are insignificant compared with the pollution we all cause in keeping warm and in moving about.

Some of the factors to be considered in siting a chemical factory and some of the more important industrial chemical processes and products are considered in the rest of the chapter.

28.9.2 The location of a chemical plant

The choice of a site for a chemical factory depends upon a number of factors. Some of these are economic i.e. the choice will depend on the comparative costs of choosing different locations. These costs are directly related to what the plant will produce, what raw materials are needed, where the markets are for its products and the source of its energy supplies. Other factors will be environmental. These are related to the uses of the land around the chemical plant.

Questions like these should be answered in choosing a site:

(a) Are the raw materials readily available?
If the raw material is an ore, it might be possible to build the plant very close to where the ore is mined. Then the cost of transporting much waste material with an ore will be kept to a minimum.

Often the product of one part of the chemical industry is the raw material for another chemical process, e.g. sulphuric acid and sodium hydroxide are used in many chemical processes. It is therefore often found that industries requiring chemicals as raw materials are located close to the industries producing them.

(b) If raw materials are not available locally can they easily be transported to the site?
If the raw material is to be imported e.g. oil, is the site near a port? Is the site well placed for road, rail or canal transport?

(c) Are energy supplies available?
The type of energy supply will depend on the particular process–it may be coal, oil or electricity. Some processes require much energy–the extraction of aluminium from its ore being a good example. Aluminium extraction plants are located close to supplies of (relatively) cheap electricity, often far from the source of the aluminium ore.

(d) Is there a good supply of water of the required quality?
Most chemical plants need copious supplies of water. Some require very pure water (e.g. the nuclear energy industry), others require a lower grade of water largely for cooling.

(e) Is there a ready supply of labour?
Considering the quantity and value of the equipment that the chemical industry uses, and of the products that it produces, the chemical industry uses relatively little labour compared with many other industries (e.g. the building and food-processing industries). Also the few very highly-skilled workers that are needed to supervise the operation of a chemical plant are often willing to move home to where the industry is located. Consequently the supply of labour is not usually an important factor when deciding on the location of a chemical plant.

(f) Is it easy to get the products to the consumer?
If the product is not to be used locally is the site well placed for road, rail or sea transport?

Usually it is cheaper to transport the finished product to where it is to be used, than to transport the raw-materials. Consequently the existence of transport links is more important than locating the industry where the products will be used. Also many products are used all over the country or the world.

(g) What will be the impact of the factory on the environment?
There are many aspects to this question. Will the factory cause much pollution to air or water? Will the plant be very noticeable or will it blend in with its surroundings? What are the other uses of the area–is it an industrial area, a local beauty spot or sea-side resort? If the other factors allow this, the environmental factors may require the plant to be sited in amongst other industry, away from the more attractive parts of the countryside.

28.9.3 Some industrial processes

In this section we look at some of the important industrial processes and their location in Britain.

(a) The iron and steel industry
This has been discussed in detail in section 24.2.

The raw materials for making iron and supplying the energy are iron ore, coke, limestone and oxygen. The industry established itself where there were iron ore deposits. Usually there were coal deposits nearby from which coke was made. In the very early days of iron making, charcoal, made from wood, was used instead of coke; such iron works, e.g. in Kent, exist now as archaeological sites. Many of the original sources of iron ore are now exhausted or the quality of the ore (i.e. an iron content of about 30%) is not high enough for modern iron-making processes (for example at Millom in Cumbria, Corby in Northamptonshire, Scunthorpe in Humberside). Increasingly, high-grade iron ore with an iron content of 60–65% is being imported into Britain from all over the world using large freighters (figure 28.9). This has meant a continuing shift of the industry from the areas of our native ore deposits to places near the large coastal ore-terminals on Teeside, Humberside, Clydeside and in South Wales (figure 28.10).

(b) The aluminium industry
Aluminium is extracted from molten aluminium oxide by electrolysis (section 23.1.2). Bauxite (impure aluminium oxide) is the raw material and is imported from countries such as Guyana and Jamaica. Electrolysis requires electricity, and aluminium plants are located near to cheap supplies. The Highlands of Scotland are a source of hydro-electric power and major plants are at Invergorden and Fort William (figure 28.10) both of which are also near the sea with good access for the import of bauxite.

(c) The salt-based industries (sodium carbonate, sodium hydroxide and chlorine)
Salt (sodium chloride) is the raw material, together with limestone, coal and ammonia for the manufacture of sodium carbonate by the *ammonia-soda (Solvay)* process. It is also the raw material for the manufacture of chlorine and sodium hydroxide (section 21.2.7(*c*)). In Britain most of the salt used is rock salt. Salt fields are found in various parts of the country, particularly under Cheshire, Durham, Staffordshire and in Somerset. These industries grew up on the Cheshire saltfield (figure 21.1). This location was also convenient for supplies of limestone and coal. It is also very close to the major market, the Lancashire textile industry, which uses the chlorine to bleach cotton and the sodium hydroxide to make soap.

Figure 28.9 Unloading iron-ore at Port Talbot, South Wales. (Photo: British Steel.)

Other industries that needed these products (see figure 21.6) have since grown up in the same area, being located close to the supply of their raw materials. In this way has arisen the major industrial complex on Merseyside.

(d) The Haber process and the manufacture of nitric acid

Ammonia, largely for fertilizer manufacture, is made by the Haber process (section 16.3.2) by the direct combination of nitrogen and hydrogen. The nitrogen is obtained by the fractional distillation of liquid air and the hydrogen from water gas or, more recently, from natural gas (see section 14.3). Being near to a coal field (for water gas manufacture) was a major economic factor and the industry was therefore established on Tees-side (figure 28.10). This was also conveniently near the sea for the export of fertilizer. Now that natural gas is used for making hydrogen, Tees-side is still a good location near to North Sea gas.

Nitric acid is manufactured from ammonia (section 16.6) and the plant was naturally built close to those making ammonia.

(e) The manufacture of sulphuric acid

Sulphuric acid is made by the oxidation of sulphur dioxide by the contact process (section 19.7.3). The sulphur dioxide was almost all made by burning sulphur which had to be imported. As one of the major uses of the acid was to make the fertilizer ammonium sulphate by reacting it with ammonia, an obvious location was Tees-side (figure 28.10). This was a fortunate choice since much sulphuric acid has also been made from anhydrite (calcium sulphate), large quantities of which have been found underneath the Tees-side plant.

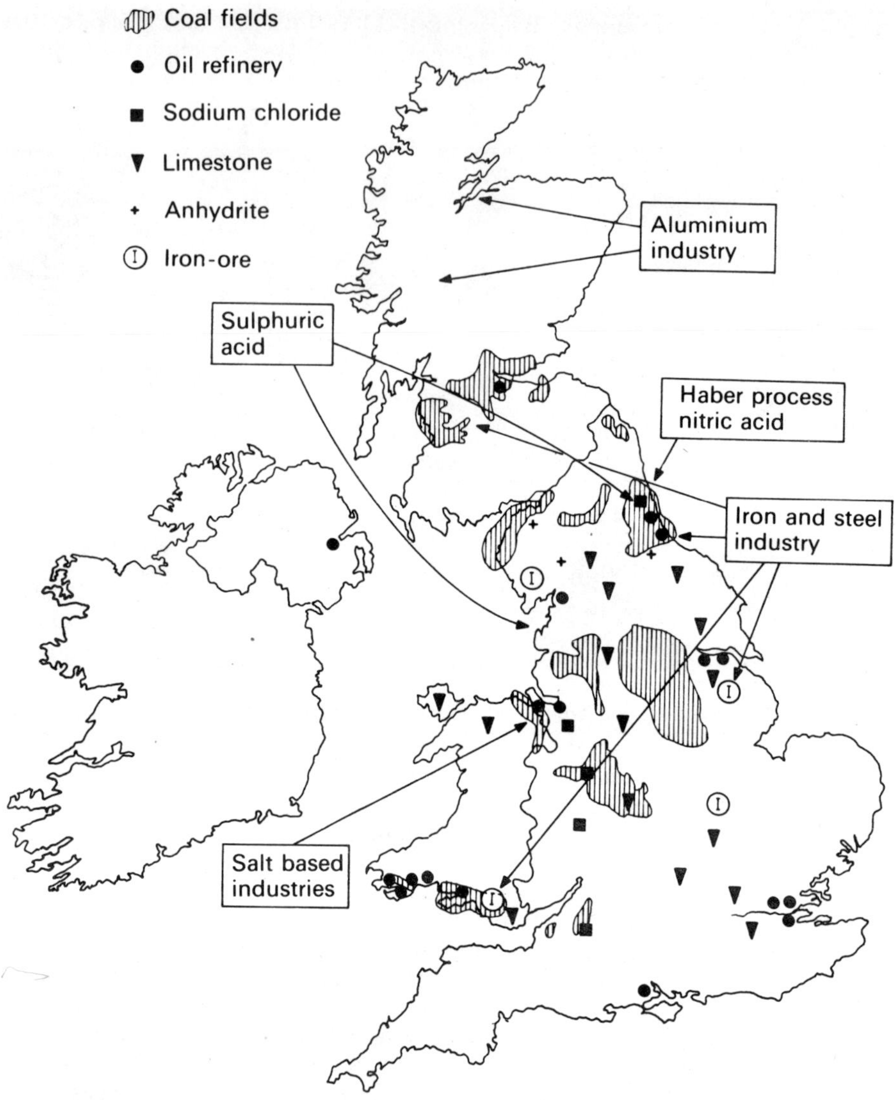

Figure 28.10 Mineral resources and location of chemical industries in the United Kingdom.

Sulphuric acid is used by many other industries and now there are several sulphuric acid plants on sites throughout the country.

(f) The petrochemical industry

This is a very varied industry making an enormous range of organic compounds, from distillates of crude oil (section 25.3.2) to plastics and synthetic fibres (section 26.2). The raw material is crude oil and the market for the products is very diverse. It is an industry which because it processes an energy source can usually make its own energy. Balancing these factors, the best place for the industry is near the ports where the raw material, oil, can be brought to the plant in tankers. The industry is situated around major river estuaries.

28.9.4 Some industrial products

There are many products of the chemical industry that have not been discussed in detail in this book. This chapter is completed by mention of some of these.

(a) Plastics and synthetic fibres (see chapter 26)
Without plastics and synthetic fibres the world as you know it would be a very difficult place. It is not *just* as substitutes that they are of value. Often they have novel properties, properties not found in natural materials. For example, synthetic fibres are stronger and less affected by water and oil than any natural fibre. You might like to imagine your home without polyethene, nylon, terylene, polyvinylchloride, perspex and other plastics.

One snag with waste plastic materials is that they do not corrode. Scrap metal articles slowly corrode away as the metal is oxidized by the air and rain. Scrap made of vegetable or animal matter (wood, bone etc) also slowly disappears as water and bacteria decompose it; it is said to be *bio-degradable*. However, plastics are not attacked by oxygen, water or bacteria; they are *not bio-degradable*. This is obvious when you walk along the sea-shore; you are sure to find plastic objects washed up by the tide.

(b) Medicines and drugs
Probably the biggest impact that the chemical industry has had on man is that it has enabled him to have a longer life. Improved hygiene and better food are the biggest factors in this country but medicine and drugs have been of major help in combating disease. Chemical anaesthetics and analgesics relieve pain and make modern surgery possible. These are the products of chemistry.

(c) Means of communication
We need to communicate–both physically and intellectually.

Physical communication is the ability to travel and to move materials about. The chemical industry has helped build the machinery for travelling, and chemical energy, as we have seen, provides the power for the cars, trains, aeroplanes and ships.

Equally important is intellectual communication, the ability to communicate in speech and writing with each other. The manufacture of paper (section 26.3.1(*d*)) has played a critical role here–paper has a very important place in our civilisation. Most information is at present communicated in writing on paper; we can record our discoveries and inventions for others to read and in this way man has passed on his knowledge from generation to generation. Before paper was invented man had to pass on his experience by word of mouth or had to carve it on stones or clay. Imagine your task in carrying this book around if this were still so!

In recent years the chemical industry has helped in the development of, what is likely to be the most significant invention of this century–the micro-chip. Micro-chips are made of pieces of silicon with a thin film of silicon dioxide on them (section 18.7). Arsenic, boron, and other elements are then reacted with the

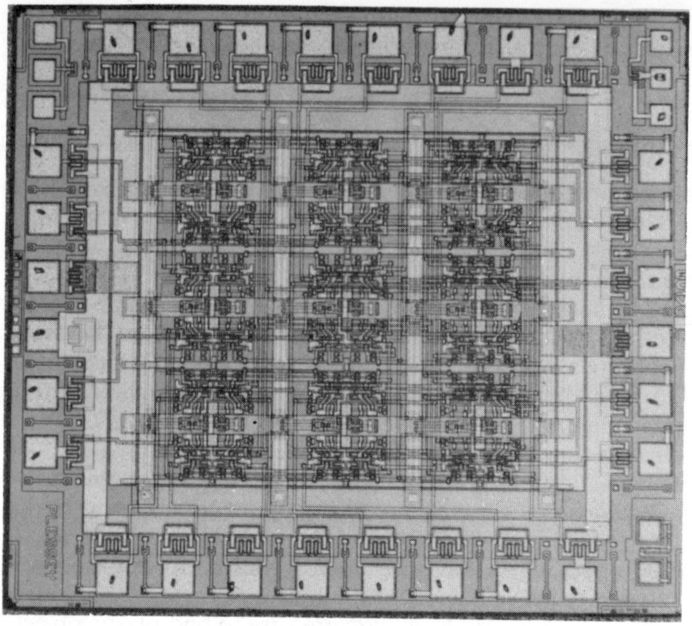

Figure 28.11 A silicon micro-chip. This chip is 2.5 mm square and contains approximately 200 transistors. (Photo: Plessey Research.)

silicon to form many thousands of electronic circuits (figure 28.11). Finally a layer of aluminium metal is put onto the chip to make the contacts. In this way a computer can be built up on a piece of silicon no bigger than 5 mm square. With the micro-computer available, new means of communication will be developed. Even books may one day become obsolete.

Perhaps by using their micro-computer to communicate with a library computer your children will be able to read a future edition of this book on their television screen. A doubtful blessing!

Questions on Chapter 28

1. Which one of the following atmospheric pollutants (i–iv) arises from (a) the use of an 'anti-knock' additive in motor-fuel? (b) incomplete combustion of petrol in a motor car engine? (c) reactions of the constituents of air at high temperatures? (d) coal-burning power stations?

(i) carbon monoxide
(ii) lead(II) bromide
(iii) nitrogen dioxide
(iv) sulphur dioxide

State briefly why each of these substances is regarded as a pollutant.

2. The enthalpies (heats) evolved for the combustion of hydrocarbon gases are:

$$CH_4 + 2O_2 \rightarrow CO_2 + 2H_2O \quad \Delta H = -890\ \text{kJ mol}^{-1}$$
$$C_3H_8 + 5O_2 \rightarrow 3CO_2 + 4H_2O \quad \Delta H = -2220\ \text{kJ mol}^{-1}$$
$$2C_4H_{10} + 13O_2 \rightarrow 8CO_2 + 10H_2O \quad \Delta H = -5740\ \text{kJ mol}^{-1}$$

(One mole of gas at room temperature and pressure occupies 22·4 l)

(i) Calculate the heat liberated by the complete combustion of 1 g of methane and 1 g of butane.
(ii) Calculate the heat liberated by the complete combustion of 1 litre of methane measured at s.t.p..
(iii) Why are these gases often described as "clean" fuels?
(iv) The proportion of carbon dioxide in the air is generally less than 0·05 % even though large quantities enter the atmosphere from the burning of carbon fuels. Why is this so?
(v) Why is ethene unsuitable as a fuel?

3. During the exploitation of a natural resource it is necessary to (a) search and survey for the resource, (b) decide whether or not to exploit the resource, (c) build a mine and extraction plant, (d) purify the resource, (e) use the resource.
(i) Which stage of these 5 could be expensive and yet yield no commercial return?
(ii) If the ore is contaminated with sulphur how could process (d) affect the environment?
(iii) If the resource is a metal oxide what chemical process is necessary to produce the metal?
(iv) If it is a metal high in the activity series (e.g. sodium or potassium) how would process (iii) be achieved?
(v) Give 2 factors that would have to be considered in deciding where to build the plant for a process such as (iii).
(vi) What, besides the cost of the raw material, is the other chief running cost of an extraction process.
(vii) Why will recycling become more important in the future?

4. (a) What are the names and percentages by volume of the two major elemental gases in the air?
(b) Green plants cannot use one of these gases directly. Which one?
(c) What group of plants are able to use this gas with the help of bacteria?
(d) Artificial fertilizers are used to aid plant growth. Give the name and formula of one chemical compound suitable for such use which contains the element named in (b), and calculate the percentage mass of that element in the compound.
(e) What detrimental effect upon the environment has become apparent with the use of such fertilizers?

5. What is a "fossil fuel"? Name three which are in use today.
We are continually exhorted to "save fuel". Explain the reasons why it is particularly important that the substances you name should be used sparingly as fuels.

6. Zinc is extracted from zinc blende at the port of Avonmouth.
(a) Suggest reasons why this site was chosen.
(b) What pollution problems (i) in the atmosphere, (ii) in the Bristol Channel, do you think have had to be overcome?
(c) What other allied industries would you expect to exist near Avonmouth?

7. You are required in a school debate to argue the case for an increased use of rail transport as opposed to road transport. What environmental and conservation arguments would you be able to use?

8. What are the processes involved in the manufacture of aluminium from bauxite? Three possible sites for an aluminium plant are shown on the map. List the advantages and disadvantages of each site and say which one you would choose.

9. (a) Ammonia is used in making artificial fertilisers. Give the name of a reagent which could be reacted with ammonia to produce a fertiliser. Give two other compounds which might be mixed with this product in order to produce a general fertiliser.
(b) A typical artificial fertiliser might have the following analysis: 12 % N, 8 % P_2O_5, 10 % K_2O.
What does this mean and why is this sort of information given on packets of fertiliser?

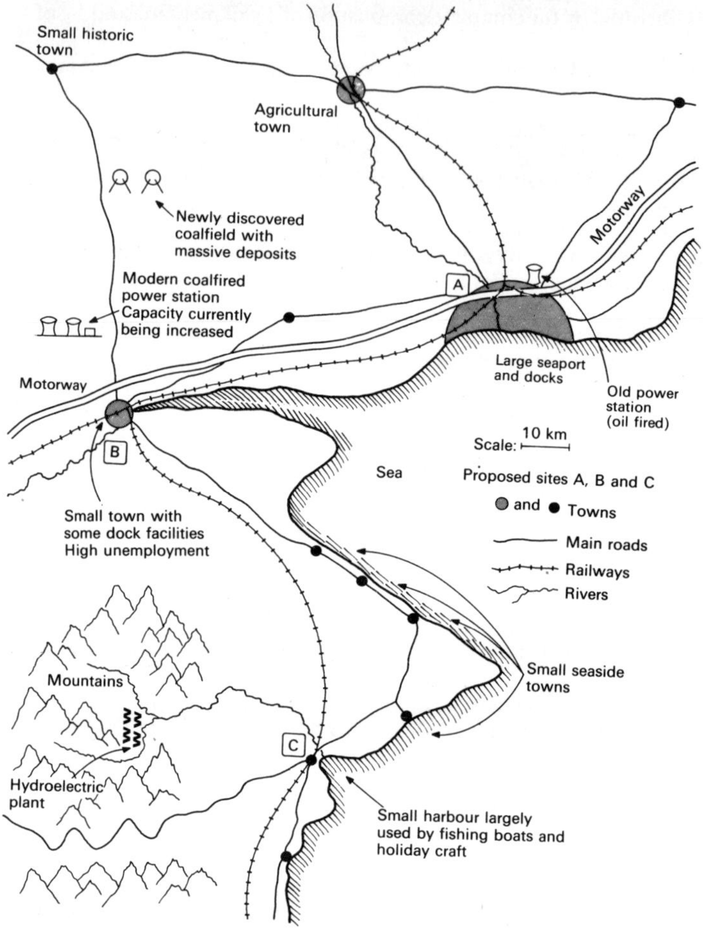

(c) Describe briefly how the large scale use of artificial fertilisers has had a detrimental effect on some aspects of the environment.
(d) How is ammonia manufactured?
10. Write an essay on "Energy Resources".
11. Explain the factors that decide whether a particular ore-body is an "economic mineral resource".
What considerations are necessary in deciding whether or not to exploit this resource?
12. Discuss the various factors involved in deciding where to locate a chemical plant. In your answer consider in particular the siting of (a) a blast furnace for making iron, (b) an aluminium plant, (c) a petroleum refinery.
13. What is the water-cycle? State the main pollutants of water and say what their effects are.
14. Briefly describe the distribution of the elements in the earths crust.
Why are certain quite rare elements in common use?
What can be done to conserve our stocks of these useful rare elements?

TABLE OF RELATIVE ATOMIC MASSES, ATOMIC NUMBERS AND SYMBOLS

Based on the Atomic Mass of $^{12}C = 12$

The values for atomic masses given in the Table apply to elements as they exist in nature.

Alphabetical Order in English

Element	Symbol	Atomic Number	Atomic Mass
Actinium	Ac	89	(227)
Aluminium	Al	13	26·9815
Americium	Am	95	(243)
Antimony	Sb	51	121·75
Argon	Ar	18	39·948
Arsenic	As	33	74·9216
Astatine	At	85	(210)
Barium	Ba	56	137·34
Berkelium	Bk	97	(247)
Beryllium	Be	4	9·0122
Bismuth	Bi	83	208·981
Boron	B	5	10·81
Bromine	Br	35	79·904
Cadmium	Cd	48	112·40
Caesium	Cs	55	132·906
Calcium	Ca	20	40·08
Californium	Cf	98	(251)
Carbon	C	6	12·011
Cerium	Ce	58	140·12
Chlorine	Cl	17	35·453
Chromium	Cr	24	51·996
Cobalt	Co	27	58·9332
Copper	Cu	29	63·546
Curium	Cm	96	(247)
Dysprosium	Dy	66	162·50
Einsteinium	Es	99	(254)
Erbium	Er	68	167·26
Europium	Eu	63	151·96
Fermium	Fm	100	(253)
Fluorine	F	9	18·9984
Francium	Fr	87	(223)
Gadolinium	Gd	64	157·25
Gallium	Ga	31	69·72
Germanium	Ge	32	72·59
Gold	Au	79	196·967
Hafnium	Hf	72	178·49
Helium	He	2	4·0026
Holmium	Ho	67	164·930
Hydrogen	H	1	1·007 97
Indium	In	49	114·82
Iodine	I	53	126·9045
Iridium	Ir	77	192·2
Iron	Fe	26	55·847
Krypton	Kr	36	83·80
Lanthanum	La	57	138·91
Lawrencium	Lr	103	(257)
Lead	Pb	82	207·19
Lithium	Li	3	6·941
Lutetium	Lu	71	174·97
Magnesium	Mg	12	24·305
Manganese	Mn	25	54·9380
Mendelevium	Md	101	(256)
Mercury	Hg	80	200·59
Molybdenum	Mo	42	95·94
Neodymium	Nd	60	144·24
Neon	Ne	10	20·179
Neptunium	Np	93	237·04
Nickel	Ni	28	58·71
Niobium	Nb	41	92·906
Nitrogen	N	7	14·0067
Nobelium	No	102	(254)
Osmium	Os	76	190·2
Oxygen	O	8	15·9994
Palladium	Pd	46	106·4
Phosphorus	P	15	30·9738
Platinum	Pt	78	195·09
Plutonium	Pu	94	(242)
Polonium	Po	84	(210)
Potassium	K	19	39·102
Praseodymium	Pr	59	140·908
Promethium	Pm	61	(147)
Protactinium	Pa	91	231·036
Radium	Ra	88	226·025
Radon	Rn	86	(222)
Rhenium	Re	75	186·2
Rhodium	Rh	45	102·905
Rubidium	Rb	37	85·47
Ruthenium	Ru	44	101·07
Samarium	Sm	62	150·35
Scandium	Sc	21	44·956
Selenium	Se	34	78·96
Silicon	Si	14	28·086
Silver	Ag	47	107·868
Sodium	Na	11	22·9898
Strontium	Sr	38	87·62
Sulphur	S	16	32·064
Tantalum	Ta	73	180·949
Technetium	Tc	43	98·9062
Tellurium	Te	52	127·60
Terbium	Tb	65	158·925
Thallium	Tl	81	204·37
Thorium	Th	90	232·038
Thulium	Tm	69	168·934
Tin	Sn	50	118·69
Titanium	Ti	22	47·90
Tungsten	W	74	183·85
Uranium	U	92	238·03
Vanadium	V	23	50·941
Xenon	Xe	54	131·30
Ytterbium	Yb	70	173·04
Yttrium	Y	39	88·906
Zinc	Zn	30	65·37
Zirconium	Zr	40	91·22

	0	1	2	3	4	5	6	7	8	9	Differences								
											1	2	3	4	5	6	7	8	9
10	0000	0043	0086	0128	0170	0212	0253	0294	0334	0374	4	8	12	17	21	25	29	33	37
11	0414	0453	0492	0531	0569	0607	0645	0682	0719	0755	4	8	11	15	19	23	26	30	34
12	0792	0828	0864	0899	0934	0969	1004	1038	1072	1106	3	7	10	14	17	21	24	28	31
13	1139	1173	1206	1239	1271	1303	1335	1367	1399	1430	3	6	10	13	16	19	23	26	29
14	1461	1492	1523	1553	1584	1614	1644	1673	1703	1732	3	6	9	12	15	18	21	24	27
15	1761	1790	1818	1847	1875	1903	1931	1959	1987	2014	3	6	8	11	14	17	20	22	25
16	2041	2068	2095	2122	2148	2175	2201	2227	2253	2279	3	5	8	11	13	16	18	21	24
17	2304	2330	2355	2380	2405	2430	2455	2480	2504	2529	2	5	7	10	12	15	17	20	22
18	2553	2577	2601	2625	2648	2672	2695	2718	2742	2765	2	5	7	9	12	14	16	19	21
19	2788	2810	2833	2856	2878	2900	2923	2945	2967	2989	2	4	7	9	11	13	16	18	20
20	3010	3032	3054	3075	3096	3118	3139	3160	3181	3201	2	4	6	8	11	13	15	17	19
21	3222	3243	3263	3284	3304	3324	3345	3365	3385	3404	2	4	6	8	10	12	14	16	18
22	3424	3444	3464	3483	3502	3522	3541	3560	3579	3598	2	4	6	8	10	12	14	15	17
23	3617	3636	3655	3674	3692	3711	3729	3747	3766	3784	2	4	6	7	9	11	13	15	17
24	3802	3820	3838	3856	3874	3892	3909	3927	3945	3962	2	4	5	7	9	11	12	14	16
25	3979	3997	4014	4031	4048	4065	4082	4099	4116	4133	2	3	5	7	9	10	12	14	15
26	4150	4166	4183	4200	4216	4232	4249	4265	4281	4298	2	3	5	7	8	10	11	13	15
27	4314	4330	4346	4362	4378	4393	4409	4425	4440	4456	2	3	5	6	8	9	11	13	14
28	4472	4487	4502	4518	4533	4548	4564	4579	4594	4609	2	3	5	6	8	9	11	12	14
29	4624	4639	4654	4669	4683	4698	4713	4728	4742	4757	1	3	4	6	7	9	10	12	13
30	4771	4786	4800	4814	4829	4843	4857	4871	4886	4900	1	3	4	6	7	9	10	11	13
31	4914	4928	4942	4955	4969	4983	4997	5011	5024	5038	1	3	4	6	7	8	10	11	12
32	5051	5065	5079	5092	5105	5119	5132	5145	5159	5172	1	3	4	5	7	8	9	11	12
33	5185	5198	5211	5224	5237	5250	5263	5276	5289	5302	1	3	4	5	6	8	9	10	12
34	5315	5328	5340	5353	5366	5378	5391	5403	5416	5428	1	3	4	5	6	8	9	10	11
35	5441	5453	5465	5478	5490	5502	5514	5527	5539	5551	1	2	4	5	6	7	9	10	11
36	5563	5575	5587	5599	5611	5623	5635	5647	5658	5670	1	2	4	5	6	7	8	10	11
37	5682	5694	5705	5717	5729	5740	5752	5763	5775	5786	1	2	3	5	6	7	8	9	10
38	5798	5809	5821	5832	5843	5855	5866	5877	5888	5899	1	2	3	5	6	7	8	9	10
39	5911	5922	5933	5944	5955	5966	5977	5988	5999	6010	1	2	3	4	5	7	8	9	10
40	6021	6031	6042	6053	6064	6075	6085	6096	6107	6117	1	2	3	4	5	6	8	9	10
41	6128	6138	6149	6160	6170	6180	6191	6201	6212	6222	1	2	3	4	5	6	7	8	9
42	6232	6243	6253	6263	6274	6284	6294	6304	6314	6325	1	2	3	4	5	6	7	8	9
43	6335	6345	6355	6365	6375	6385	6395	6405	6415	6425	1	2	3	4	5	6	7	8	9
44	6435	6444	6454	6464	6474	6484	6493	6503	6513	6522	1	2	3	4	5	6	7	8	9
45	6532	6542	6551	6561	6571	6580	6590	6599	6609	6618	1	2	3	4	5	6	7	8	9
46	6628	6637	6646	6656	6665	6675	6684	6693	6702	6712	1	2	3	4	5	6	7	7	8
47	6721	6730	6739	6749	6758	6767	6776	6785	6794	6803	1	2	3	4	5	5	6	7	8
48	6812	6821	6830	6839	6848	6857	6866	6875	6884	6893	1	2	3	4	4	5	6	7	8
49	6902	6911	6920	6928	6937	6946	6955	6964	6972	6981	1	2	3	4	4	5	6	7	8
50	6990	6998	7007	7016	7024	7033	7042	7050	7059	7067	1	2	3	3	4	5	6	7	8
51	7076	7084	7093	7101	7110	7118	7126	7135	7143	7152	1	2	3	3	4	5	6	7	8
52	7160	7168	7177	7185	7193	7202	7210	7218	7226	7235	1	2	2	3	4	5	6	7	7
53	7243	7251	7259	7267	7275	7284	7292	7300	7308	7316	1	2	2	3	4	5	6	6	7
54	7324	7332	7340	7348	7356	7364	7372	7380	7388	7396	1	2	2	3	4	5	6	6	7
	0	1	2	3	4	5	6	7	8	9	1	2	3	4	5	6	7	8	9

	0	1	2	3	4	5	6	7	8	9	Differences 1	2	3	4	5	6	7	8	9
55	7404	7412	7419	7427	7435	7443	7451	7459	7466	7474	1	2	2	3	4	5	5	6	7
56	7482	7490	7497	7505	7513	7520	7528	7536	7543	7551	1	2	2	3	4	5	5	6	7
57	7559	7566	7574	7582	7589	7597	7604	7612	7619	7627	1	2	2	3	4	5	5	6	7
58	7634	7642	7649	7657	7664	7672	7679	7686	7694	7701	1	1	2	3	4	4	5	6	7
59	7709	7716	7723	7731	7738	7745	7752	7760	7767	7774	1	1	2	3	4	4	5	6	7
60	7782	7789	7796	7803	7810	7818	7825	7832	7839	7846	1	1	2	3	4	4	5	6	6
61	7853	7860	7868	7875	7882	7889	7896	7903	7910	7917	1	1	2	3	4	4	5	6	6
62	7924	7931	7938	7945	7952	7959	7966	7973	7980	7987	1	1	2	3	3	4	5	6	6
63	7993	8000	8007	8014	8021	8028	8035	8041	8048	8055	1	1	2	3	3	4	5	5	6
64	8062	8069	8075	8082	8089	8096	8102	8109	8116	8122	1	1	2	3	3	4	5	5	6
65	8129	8136	8142	8149	8156	8162	8169	8176	8182	8189	1	1	2	3	3	4	5	5	6
66	8195	8202	8209	8215	8222	8228	8235	8241	8248	8254	1	1	2	3	3	4	5	5	6
67	8261	8267	8274	8280	8287	8293	8299	8306	8312	8319	1	1	2	3	3	4	5	5	6
68	8325	8331	8338	8344	8351	8357	8363	8370	8376	8382	1	1	2	3	3	4	4	5	6
69	8388	8395	8401	8407	8414	8420	8426	8432	8439	8445	1	1	2	2	3	4	4	5	6
70	8451	8457	8463	8470	8476	8482	8488	8494	8500	8506	1	1	2	2	3	4	4	5	6
71	8513	8519	8525	8531	8537	8543	8549	8555	8561	8567	1	1	2	2	3	4	4	5	5
72	8573	8579	8585	8591	8597	8603	8609	8615	8621	8627	1	1	2	2	3	4	4	5	5
73	8633	8639	8645	8651	8657	8663	8669	8675	8681	8686	1	1	2	2	3	4	4	5	5
74	8692	8698	8704	8710	8716	8722	8727	8733	8739	8745	1	1	2	2	3	4	4	5	5
75	8751	8756	8762	8768	8774	8779	8785	8791	8797	8802	1	1	2	2	3	3	4	5	5
76	8808	8814	8820	8825	8831	8837	8842	8848	8854	8859	1	1	2	2	3	3	4	5	5
77	8865	8871	8876	8882	8887	8893	8899	8904	8910	8915	1	1	2	2	3	3	4	4	5
78	8921	8927	8932	8938	8943	8949	8954	8960	8965	8971	1	1	2	2	3	3	4	4	5
79	8976	8982	8987	8993	8998	9004	9009	9015	9020	9025	1	1	2	2	3	3	4	4	5
80	9031	9036	9042	9047	9053	9058	9063	9069	9074	9079	1	1	2	2	3	3	4	4	5
81	9085	9090	9096	9101	9106	9112	9117	9122	9128	9133	1	1	2	2	3	3	4	4	5
82	9138	9143	9149	9154	9159	9165	9170	9175	9180	9186	1	1	2	2	3	3	4	4	5
83	9191	9196	9201	9206	9212	9217	9222	9227	9232	9238	1	1	2	2	3	3	4	4	5
84	9243	9248	9253	9258	9263	9269	9274	9279	9284	9289	1	1	2	2	3	3	4	4	5
85	9294	9299	9304	9309	9315	9320	9325	9330	9335	9340	1	1	2	2	3	3	4	4	5
86	9345	9350	9355	9360	9365	9370	9375	9380	9385	9390	1	1	2	2	3	3	4	4	5
87	9395	9400	9405	9410	9415	9420	9425	9430	9435	9440	0	1	1	2	2	3	3	4	4
88	9445	9450	9455	9460	9465	9469	9474	9479	9484	9489	0	1	1	2	2	3	3	4	4
89	9494	9499	9504	9509	9513	9518	9523	9528	9533	9538	0	1	1	2	2	3	3	4	4
90	9542	9547	9552	9557	9562	9566	9571	9576	9581	9586	0	1	1	2	2	3	3	4	4
91	9590	9595	9600	9605	9609	9614	9619	9624	9628	9633	0	1	1	2	2	3	3	4	4
92	9638	9643	9647	9652	9657	9661	9666	9671	9675	9680	0	1	1	2	2	3	3	4	4
93	9685	9689	9694	9699	9703	9708	9713	9717	9722	9727	0	1	1	2	2	3	3	4	4
94	9731	9736	9741	9745	9750	9754	9759	9763	9768	9773	0	1	1	2	2	3	3	4	4
95	9777	9782	9786	9791	9795	9800	9805	9809	9814	9818	0	1	1	2	2	3	3	4	4
96	9823	9827	9832	9836	9841	9845	9850	9854	9859	9863	0	1	1	2	2	3	3	4	4
97	9868	9872	9877	9881	9886	9890	9894	9899	9903	9908	0	1	1	2	2	3	3	4	4
98	9912	9917	9921	9926	9930	9934	9939	9943	9948	9952	0	1	1	2	2	3	3	4	4
99	9956	9961	9965	9969	9974	9978	9983	9987	9991	9996	0	1	1	2	2	3	3	3	4
	0	1	2	3	4	5	6	7	8	9	1	2	3	4	5	6	7	8	9

LIST OF APPROXIMATE RELATIVE ATOMIC MASSES

(*For use in solving the problems*).

Element	
Aluminium	27
Barium	137
Bromine	80
Calcium	40
Carbon	12
Chlorine	35·5
Copper	63·5
Hydrogen	1
Iodine	127
Iron	56
Lead	207

Element	
Magnesium	24
Mercury	201
Nitrogen	14
Oxygen	16
Phosphorus	31
Potassium	39
Silver	108
Sodium	23
Sulphur	32
Tin	119
Zinc	65

THE PERIODIC TABLE

KEY

Relative atomic mass →	63 55	
Atomic number →	29Cu	← Symbol
	Copper	← Name

Period	Group 1	2											3	4	5	6	7	0
								1.008 1H Hydrogen										4.003 2He Helium
2	6.939 3Li Lithium	9.012 4Be Beryllium											10.811 5B Boron	12.01 6C Carbon	14.00 7N Nitrogen	16.00 8O Oxygen	19.00 9F Fluorine	20.18 10Ne Neon
3	22.99 11Na Sodium	24.31 12Mg Magnesium	←					Transition metals				→	26.98 13Al Aluminium	28.09 14Si Silicon	30.97 15P Phosphorus	32.06 16S Sulphur	35.45 17Cl Chlorine	39.95 18Ar Argon
4	39.10 19K Potassium	40.08 20Ca Calcium	44.96 21Sc Scandium	47.90 22Ti Titanium	50.94 23V Vanadium	52.00 24Cr Chromium	54.94 25Mn Manganese	55.85 26Fe Iron	58.93 27Co Cobalt	58.71 28Ni Nickel	63.55 29Cu Copper	65.37 30Zn Zinc	69.72 31Ga Gallium	72.59 32Ge Germanium	74.92 33As Arsenic	78.96 34Se Selenium	79.90 35Br Bromine	83.80 36Kr Krypton
5	85.47 37Rb Rubidium	87.62 38Sr Strontium	88.91 39Y Yttrium	91.22 40Zr Zirconium	92.91 41Nb Niobium	95.94 42Mo Molybde-num	98 91 43Tc Technetium	101.1 44Ru Ruthenium	102.9 45Rh Rhodium	106.4 46Pd Palladium	107.9 47Ag Silver	112.4 48Cd Cadmium	114.8 49In Indium	118.7 50Sn Tin	121.8 51Sb Antimony	127.6 52Te Tellurium	126.9 53I Iodine	131.3 54Xe Xenon
6	132.9 55Cs Caesium	137.3 56Ba Barium	(see below) 57-71	178.5 72Hf Hafnium	180.9 73Ta Tantalum	183.9 74W Tungsten	186.2 75Re Rhenium	190.2 76Os Osmium	192.2 77Ir Iridium	195.1 78Pt Platinum	197.0 79Au Gold	200.6 80Hg Mercury	204.4 81Ti Thallium	207.2 82Pb Lead	209.0 83Bi Bismuth	(210) 84Po Polonium	(210) 85At Astatine	(222) 86Rn Radon
7	(223) 87Fr Francium	226.0 88Ra Radium	(see below) 89-103															

Lanthanides →	138.9 57La Lanthanum	140.1 58Ce Cerium	140.9 59Pr Praseody-mium	144.2 60Nd Neodymium	(147) 61Pm Promethium	150.4 62Sm Samarium	152.0 63Eu Europium	157.3 64Gd Gadolinium	158.9 65Tb Terbium	162.5 66Dy Dysprosium	164.9 67Ho Holmium	167.3 68Er Erbium	168.9 69Tm Thulium	173.0 70Yb Ytterbium	175.0 71Lu Lutetium
Actinides →	(227) 89Ac Actinium	232.0 90Th Thorium	231.0 91Pa Protactinium	238.0 92U Uranium	237.0 93Np Neptunium	(242) 94Pu Plutonium	(243) 95Am Americium	(247) 96Cm Curium	(247) 97Bk Berkelium	(251) 98Cf Californium	(254) 99Es Einsteinium	(253) 100Fm Fermium	(256) 101Md Mendele-vium	(254) 102No Nobelium	(257) 103Lw Lawrencium

Answers to questions

Chapter 3

1. (*a*) 24 g mol^{-1}, (*b*) 80 g mol^{-1}, (*c*) 56 g mol^{-1} (*d*) 103 g mol^{-1}, (*e*) 200 g mol^{-1}, (*f*) 102 g mol^{-1}
2. (*a*) $6{\cdot}02 \times 10^{23}$, (*b*) $6{\cdot}02 \times 10^{23}$, (*c*) $6{\cdot}02 \times 10^{23}$
3. (*a*) $6{\cdot}02 \times 10^{23}$, (*b*) $6{\cdot}02 \times 10^{23}$, (*c*) $6{\cdot}02 \times 10^{23}$
4. (*a*) $3{\cdot}01 \times 10^{23}$, (*b*) $12{\cdot}04 \times 10^{23}$, (*c*) $6{\cdot}02 \times 10^{21}$
5. (*a*) 1, (*b*) 1, (*c*) 2, (*d*) 3, (*e*) 0·25, (*f*) 0·5
6. (*a*) calcium, (*b*) magnesium, (*c*) the same number, (*d*) nitrogen, (*e*) iodine, (*f*) nitrogen.

Chapter 4

1. 56; 95; 144; 44; 100; 100; 100; 174; 107; 152; 187·5
2. (*a*) Mg, 60 % O, 40 %
   (*b*) H, 2·04 %; S, 32·65 %; O, 65·3 %
   (*c*) Ca, 36·0 %; Cl, 64·0 %
   (*d*) Na, 27·4 %; H, 1·19 %; C, 14·3 %; O, 57·1 %
   (*e*) Cu, 33·9 %; N, 14·9 %; O, 51·2 %
   (*f*) K, 31·8 %; Cl, 29·0 %; O, 39·2 %
   (*g*) N, 35·0 %; H, 5·0 %; O, 60·0 %
3. (*a*) 80 g; (*b*) 6·4 g; (*c*) 14·4 g; (*d*) 8·1 g
4. (*a*) 45·3 %; (*b*) 36·3 %; (*c*) 55·9 %
5. (*a*) K_2CO_3; (*b*) PbN_2O_6; (*c*) $CuCl_2$; (*d*) NaCl; (*e*) HN_3; (*f*) H_2N; (*g*) $FeSO_4{\cdot}7H_2O$; (*h*) $K_4FeC_6N_6$; (*i*) $N_2CaCl_2H_6$; (*j*) $MgSO_4{\cdot}7H_2O$
6. (*i*) 3×10^{22}; (ii) $1{\cdot}2 \times 10^{23}$; (iii) $1{\cdot}5 \times 10^{23}$
7. (*a*) 4·79 g; (*b*) 66·5 %
8. (*a*) 18·5 g; (*b*) 4·69 g
9. (*a*) (i) 0·88 g; (ii) 1·30 g; (*b*) 28·6 g
10. (*a*) $N_2H_4O_3$; (*b*) $M.7H_2O$; (*c*) 8 %
11. 2
12. (i) 8·40 g; (ii) 5·13 g
13. 13·9; 50 cm^3
14. $6{\cdot}02 \times 10^{18}$

Chapter 5

1. A : B = 3 : 1
3. (i) Sr_2O_3; P_2O_5 (iv) 44·8 litres

Chapter 6

6. (ii) 0·8 g

Chapter 7

1. (*a*) 409·5 cm^3; (*b*) 682·5 cm^3; (*c*) 205 cm^3; (*d*) 205 cm^3; (*e*) 317 cm^3; (*f*) 260 cm^3
3. (*a*) NH_2; (*b*) N_2H_4
4. (*a*) 71·2; (*b*) 28; (*c*) 58
5. 42·5 cm^3
6. (i) 13·25 g; (ii) 2·8 dm^3
7. (i) 11·15 g; (ii) 560 cm^3
8. (i) 8 g; (ii) 2·8 litres
9. (*a*) NO; (*b*) (i) 28; (ii) CO
10. (i) 30 cm^3; (ii) 20 cm^3
11. 5 cm^3 and 10 cm^3; 15 cm^3 and 10 cm^3
12. 2·91 g; 2·09 g; 1070 cm^3
13. (i) 400 cm^3 oxygen; (ii) 200 cm^3 ammonia
14. C_2H_4

15. 22; 35·5; 8·5; 15; 2·11 dm^3
16. 350; 4; 208
17. 2·72 dm^3
18. 16·0
19. 16·0
20. (i) 32
21. 80; 9
23. (*a*) 2·49 dm^3; (*b*) 4·98 dm^3

Chapter 8
5. (*a*) 2
7. 0·54 g
9. (*c*) (i) 965; (ii) 0·01 g; (iii) 0·3175 g
12. (*a*) 2; 1
14. (i) 4·74 g; (ii) 1·671 H_2; 0·841 O_2

Chapter 9
1. (*a*) 10·45 kJ; (*b*) 50·16 kJ; (*c*) 20·9 kJ; (*d*) 20·9 kJ
2. (*a*) 6·02 kJ; (*b*) 12·0 kJ; (*c*) 11·01 kJ; (*d*) 67·6 kJ
3. (*a*) 22 kJ; (*b*) 45·5 kJ; (*c*) 103·0 kJ; (*d*) 575 kJ
4. −391 kJ
5. −1250 kJ
6. +25·08 kJ
7. 3·4 °C
9. −57·3 kJ
12. 56·8 °C
13. −460 kJ

Chapter 10
1. 100 g
2. 10·8 cm^3/minute
5. 4 g
9. (*d*) (i) 795 cm^3; (ii) 1000 cm^3
12. (*b*) (i) 2·24 litres

Chapter 11
4. 6
6. 5

Chapter 12
1. (*a*) (i) 25 cm^3; (ii) 200 cm^3; (iii) 90 cm^3
(*b*) (i) 5 cm^3; (ii) 30 cm^3; (iii) 90 cm^3
2. (*a*) (i) 0·4M; (ii) 0·12M; (iii) 0·08M
(*b*) (i) 0·3M; (ii) 0·96M; (iii) 0·125M
3. (i) 9·8; (ii) 3·27; (iii) 8·17
4. (i) 4·24; (ii) 4·24; (iii) 4·24
5. (i) 2·1 g; (ii) 2·1 g; (iii) 2·1 g; (iv) 4·2 g
6. 2
7. (*a*) 24; (*b*) 7
8. 0·094M
9. 86·7%
10. (*b*) 0·128M; 8·064 g dm^{-3}
11. 40
12. (*c*) 20 cm^3; 224 cm^3
13. (i) 0·2M; (ii) 0·216M; (iii) 100; (iv) 39

Chapter 13
1. $33\frac{1}{3}$%

Chapter 14

1. (*a*) 345 cm^3; (*b*) 345 cm^3; (*c*) 345 cm^3; (*d*) 487 cm^3

Chapter 15

2. (iv) 1120 cm^3

Chapter 16

2. (i) 0·5 g; (ii) 0·5 dm^3 or 1·0 dm^3
4. 15 cm^3
13. (*f*) 6·29 g; (*g*) 1·25 g

Chapter 17

5.504 cm^3; 0·45 g

Chapter 18

2. (i) 20 cm^3
3. 628 cm^3
4. (*a*) 45 cm^3; (*b*) 15 cm^3
8. 20 cm^3 carbon dioxide; 10 cm^3 oxygen

Chapter 19

8.(i) 32 g; (ii) 112 dm^3

Chapter 20

7. (*b*) 160; 2

Chapter 22

5. 3 : 2
13. 670 g; 660 g

Chapter 23

1. 1·8 g

Chapter 24

9. 19·6 g

Chapter 25

1. 100 dm^3
8. 0·092 kg
12. (*b*) CH_2Br
    (*d*) 188 g
    (*e*) $C_2H_4Br_2$

Index